SEMICONDUCTOR OPTOELECTRONICS
Physics and Technology

Electronics and VLSI Circuits

Also available from McGraw-Hill

Schaum's Outline Series in Electronics & Engineering

Most outlines include basic theory, definitions and hundreds of example problems solved in step-by-step detail, and supplementary problems with answers.

Related titles on the current list include:

Analog & Digital Communications
Basic Circuit Analysis
Basic Electrical Engineering
Basic Electricity
Basic Mathematics for Electricity & Electronics
Digital Principles
Electric Circuits
Electric Machines & Electromechanics
Electric Power Systems

Electromagnetics
Electronic Circuits
Electronic Communication
Electronic Devices & Circuits
Electronic Technology
Feedback & Control Systems
Introduction to Digital Systems
Microprocessor Fundamentals

Schaum's Solved Problems Books

Each title in this series is a complete and expert source of solved problems with solutions worked out in step-by-step detail.

Related titles on the current list include:

3000 Solved Problems in Calculus
2500 Solved Problems in Differential Equations
3000 Solved Problems in Electric Circuits
2000 Solved Problems in Electromagnetics

2000 Solved Problems in Electronics
3000 Solved Problems in Linear Algebra
2000 Solved Problems in Numerical Analysis
3000 Solved Problems in Physics

Available at most college bookstores, or for a complete list of titles and prices, write to:

Schaum Division
McGraw-Hill, Inc.
1221 Avenue of the Americas
New York, NY 10020

SEMICONDUCTOR OPTOELECTRONICS
Physics and Technology

Jasprit Singh
University of Michigan

McGraw-Hill, Inc.
New York St. Louis San Francisco Auckland Bogotá Caracas
Lisbon London Madrid Mexico City Milan Montreal New Delhi
San Juan Singapore Sydney Tokyo Toronto

The editor was George T. Hoffman;
the production supervisor was Richard A. Ausburn.
The design manager was Joseph A. Piliero.
The cover was designed by Teresa Singh.
All illustrations were done by Teresa Singh.
R. R. Donnelley & Sons Company was printer and binder.

SEMICONDUCTOR OPTOELECTRONICS
Physics and Technology

 This book is printed on recycled, acid-free
paper containing 10% postconsumer waste.

1 2 3 4 5 6 7 8 9 0 DOH DOH 9 0 9 8 7 6 5 4

ISBN 0-07-057637-8

Library of Congress Catalog Card Number: 94-78280

About the Author

Jasprit Singh received his Ph.D. in solid state physics from the University of Chicago. He has carried out research in solid state electronics at the University of Southern California, Wright Patterson Air Force Base, and the University of Michigan, Ann Arbor, where he is currently a professor in the Department of Electrical Engineering and Computer Science. His research interests cover the area of semiconductor materials and their devices for information processing. He is also the author of "Physics of Semiconductors and Their Heterostructures," McGraw-Hill (1993); and "Semiconductor Devices, An Introduction," McGraw-Hill (1994).

To
Nirala and Nihal

Contents

3 DOPING AND CARRIER TRANSPORT   113

4 OPTICAL PROPERTIES OF SEMICONDUCTORS — 170

5 EXCITONIC EFFECTS AND MODULATION OF OPTICAL PROPERTIES 234

6 SEMICONDUCTOR JUNCTION THEORY 286

7 OPTOELECTRONIC DETECTORS 336

8 NOISE AND THE PHOTORECEIVER 399

9 THE LIGHT EMITTING DIODE

13 OPTICAL COMMUNICATION SYSTEMS: DEVICE NEEDS 645

E OPTICAL WAVES IN WAVEGUIDES AND CRYSTALS 705

Preface

In 1988 the trans-Atlantic undersea cable TAT-8 was placed between North America and Europe using semiconductor optoelectronic components in a most demanding and harsh environment. The unsuccessful lawsuit of the satellite communication companies to block TAT-8 signified that the powerful electronic devices had met their match in communication applications. Today optoelectronic networks promise 500 TV stations (to the utter disgust of some and utter delight of others) where you can interactively demand movies, boxing matches, and order a variety of goodies. Indeed, when Wall street takeover/merger specialists use the words "optical networks" as freely as "aggressive growth mutual funds," the message is clear—optoelectronics is here!

With the maturing of semiconductor optoelectronic technology, the number of Universities offering optoelectronic courses and the breadth and depth of these courses is rapidly increasing. However, in this transient mode, it is not yet established what sort of knowledge base is necessary to train students of this field. A typical electrical engineer takes about 20 courses directly related to his or her field to get a Bachelors and about 10 more to get a Masters. Optoelectronics has not yet reached the stature where Universities can devote a similar effort for optoelectronics students. Of course, some of the traditional electronics courses are important for the student of optoelectronics as well. To develop a thorough understanding of optoelectronic devices, a user must have the following understanding:

Semiconductor Technology: It is important for the student to understand the state of technology and how devices are fabricated. It is also important to appreciate the challanges faced by the difficult technologies of optoelectronic integrated circuits (OEICs) or quantum wire lasers.

Physics of Semiconductors: The physics behind concepts such as effective mass, mobility, absorption coefficient, bandstructure, etc., should be understood. Also, the importance of semiconductor alloys and heterostructures should be appreciated. The interactions of photons and electrons, and concepts such as spontaneous and stimulated emission, excitonic and electro-optic effects are to be understood.

Semiconductor Optoelectronic Devices: The physical interactions between photons and semiconductors have to be exploited to design and optimize a variety of information processing devices.

System Needs and New Device Challanges: Since optoelectronic devices are not very mature at this stage (compared to electronic devices), it is important for the student to know what improvements can be made and the resulting payoffs. For this it is important to understand system needs.

It is admittedly difficult for any textbook to address all the needs outlined above in any real depth. As a result, textbooks tend to focus on one or two areas without providing the student a global picture. In this textbook, I have touched upon all the needs while providing an in depth coverage of two key topics—device physics and device design. The reader will find that the areas of technology and systems is covered in enough detail to provide a much needed appreciation of the challenges faced here. The areas of semiconductor physics, electron-photon interactions, and optoelectronic devices are covered in great depth.

This book is written primarily as a textbook for one or more optoelectronic courses. However, where appropriate, I have provided discussions on the state of the art issues. By offering a balanced discussion and about 150 worked examples, I hope that this textbook will not only serve the coursework needs for students, but would also be a long term resource for their future careers.

This manuscript was typed by Ms. Izena Goulding, to whom I am extremely grateful. The figures, cover design, and the typesetting of this book were done by Teresa Singh, my wife. She also provided the support without which this book would not be possible. I am also indebted to my students, past and present: Dr. John Hinckley, Prof. Songcheol Hong, Dr. Mark Jaffe, Dr. John Loehr, Prof. Yeeloy Lam, and Mr. Igor Vurgaftman. I am extremely grateful to George Hoffman, my editor, for providing me valuable input from an outstanding group of referees. The referees were generous in providing positive criticism which I believe greatly benefited the book. I would like to thank Professor Joe Campbell of the University of Texas at Austin, Professor James Coleman of the University of Illinois at Urbana-Champaign, Professor Karl Hess of the University of Illinois at Urbana-Champaign, Professor Marek Osinski of the University of New Mexico, and Dr. Daniel Renner of the Ortel Corporation. I am also grateful to Professor Pallab Bhattacharya of the University of Michigan for valuable discussions.

This book shares some sections with the other two books, "Physics of Semiconductors and Their Heterostructures" (1993) and "Semiconductor Devices: An Introduction" (1994) which I have written, published by McGraw-Hill.

An Instructor's Manual is available to professors wishing to use this text. This manual has solutions to the end of chapter problems. In addition, a computer disc is available to address the following class of problems: i) electron and hole levels in quantum wells (4 band k.p model for holes); ii) optical confinement factor for a waveguide; iii) laser gain in quantum well lasers; iv) effect of strain on quantum well bandstructure and laser performance.

Please write, on your department stationary, to McGraw-Hill for a copy of this manual.

Jasprit Singh

INTRODUCTION

I.1 THE INFORMATION AGE

"Knowledge is power," says an ad campaign for a new business which exploits computer hookups and the latest in medical technologies to provide a health care package. Information and its distilled form—knowledge—has become a survival tool for all of us. It is hard to imagine a world without computers, satellites, undersea fiber networks, televisions, fax machines, laser printers, and a myriad of other information processing tools. And, of course, just a few decades ago none of these gadgets were available. What has made us so dependent on information and its rapid processing? Whether we like it or not, most of the livelihoods of workers in industrially developed countries are intimately tied to the ability to access and process information—preferably before our competition does it!

Regardless of what the information is, we want to process it faster, with less inaccuracies, at a lower cost, with a system which consumes less space, etc.. In fact, the driving forces for new technologies can range from the need to diagnose tumors in the brain to the diagnosis of a poisonous gas in a factory. It is estimated that at present (mid-nineties), over ten trillion bits of information are generated each day! Most of these bits can be attributed to the television industry, which should not be surprising to the reader.

The modern age of information processing has been ushered in by the electronic devices, particularly the mass produced high density semiconductor devices. These devices process information at blinding speeds, crunching numbers at speeds of millions of instructions per second. Electronic devices are deeply entrenched in any information processing system, and for good reason. Can optoelectronic devices, which exploit light and electrons, make inroads into the domain of electronic devices? Over the last decade, optoelectronic devices have started to make an impact on the information processing scene. This impact has been felt most in the area of information communication and information storage and retrieval. However, so far the impact has been less in the area of "intelligent" information processing. To understand the challenges facing optoelectronics and the potential payoffs, we examine the demands placed upon devices that are to be successful in information processing.

I.2   DEMANDS OF THE INFORMATION AGE

As noted in the previous section, we live in an age where acquiring, manipulating, and transmitting information is of utmost importance. In Fig. I.1 we show some of the important functions that need to be carried out in order to survive in the information age. Regardless of what medium is used to produce the devices—water, electrons, photons, beads, etc.,—the devices must be able to provide at least some of these functions. Let us briefly examine these functions and see what sort of requirements they put on devices.

Information Reception/Detection

This is, of course, one of the most important functions in an information system. For example, in our own case we receive information about the world we live in by our eyes, nose, skin, ears, and tongue. Our five senses allow us to obtain important information about our surroundings and this information is conveyed to our main processing unit— our brain. Devices which hope to serve as sensors/detectors must have a well-defined response to an external input. They must convert the input information into a form which can be used for further processing.

Information Enhancement/Amplification

Very often the information that is received is of either very poor quality, or is too "weak" to be directly useful. In such cases, the information must be amplified or enhanced. We often ask people to repeat themselves louder, since we cannot hear them well. Many hearing impaired people need hearing aids which can amplify the sound coming in. Thus, amplifiers are an essential part of an information processing system. To be able to amplify information, the device must have the very important characteristic of *gain*, i.e., a small change in input should result in a large change in output. In times bygone, a message was often conveyed by beating a code on drums. However, every half a mile or so, one had to have a new drummer who would hear the faint drums and send forth a more amplified code to the next drummer. In this particular example, it is interesting to note that the response of the second drummer to the input is *non-linear*. Thus, if for some reason he hears a much fainter drumbeat (due to the wind direction), he still beats out a signal of the same high strength.

Gain and non-linear response to information are highly desirable properties of devices. A non-linear response can distinguish between two closely spaced pieces of information, as shown in Fig. I.2. A linear response between input and output would not be able to distinguish the inputs as well. Moreover, if the inputs were noisy, as shown by the shaded area, the non-linear response would still maintain a very high degree of separation.

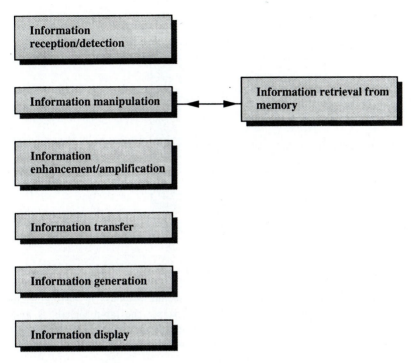

Figure I.1: Necessary functions to survive in the "information age."

Information Manipulation

This is, of course, the most important aspect of information processing. When some information comes in, much of it may be redundant or in a form that is not appropriate. Invariably, it has to be manipulated, which may mean carrying out processes like addition, subtraction, division, multiplication, comparison with previous information, extracting a "signature" of the information, etc.

Memory

Memory is obviously essential in an information processing system. We are all frustrated when, at some crucial moment, our memory fails us. The process of learning, comparing, selecting, and reusing information all require memory. The memory device should be able to store information by, perhaps, changing the state of the device, and then one should be able to retrieve the information (i.e., be able to WRITE/READ). This page that you are reading is a form of a memory—perhaps the most influential kind of memory in the history of mankind. The walls of the caves in which our ancestors lived thousands of years ago were a kind of memory.

Memory is an area where semiconductors have been most challenged. Even in high technology applications, semiconductor memories are not the only game in town. Optical memories based on a plastic disc (the compact disc), magnetic tape memories,

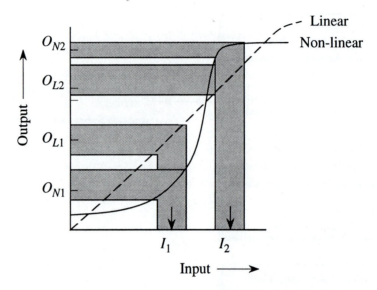

Figure I.2: The advantages of a non-linear input-output response in distinguishing closely-spaced inputs. A linear response system cannot separate the input signals I_1 and I_2 (with response O_{L1} and O_{L2}) as well as the non-linear response (with response O_{N1} and O_{N2}. Also, the non-linear device has a better noise immunity, as shown by the shaded areas.

magnetic bubble memories, etc., all form important spokes of the memory technology, along with semiconductor memories.

Information Transfer

An important function in an information processing system is to be able to transfer the information or the results obtained after manipulating the information into a storage or memory. For example, when an image is seen by our eye, the information is sent to the brain via the optic nerve in a well-defined sequence so that the brain knows that there is a door to the right, a chair in front, a blackboard to the left, etc. Any scrambling in this transfer process can lead to serious disabilities. Such transfer devices are most useful when two or three dimensional information must pass through a one dimensional path. A most popular case where this happens is in the case of a video camera. In a video camera, a two dimensional view is recorded and sent sequentially via charge coupled devices (CCDs) to a memory. Since a precise sequence is maintained in this transfer, the scene can later be easily reconstructed if the sequence is known.

Information Generation

Another important step in information technology is the generation of information. We generate information by a variety of means, ranging from our speech, to hand actions, facial expressions, writing, etc. Of course, in each case the real information is generated by the firing of neurons in the brain.

Information can be generated in semiconductor technology by the semiconductor laser or by a microwave device. By coupling semiconductor technology with other technologies, information can be generated in the form of sound waves as well.

Information Display
The saying, "A picture is worth a thousand words," seems to be coming more and more valid as the amount of information becomes greater and greater. Often in our daily life, a single facial expression conveys more information than any speech or writing could. Displaying information is extremely important and has great impact on human experience. Consider the enormous sum of money spent by companies on advertisements. Displays need not just be pictures—they can be words conveying information as well. Display technology is one of the fastest growing technologies in recent years. Nations and companies vie fiercely to obtain an edge in display technology. New display technologies, such as high density television (HDTV), flat panel displays, programmable transparencies, etc., hold keys to the economic success of many companies. Graphic workstations have already transformed the lives of designers of houses, automobiles, and microelectronic chips. Semiconductor technology has coupled extremely well with liquid crystal technology to produce displays. Also for active displays and light sources, semiconductor devices such as LEDs and laser diodes serve an important need.

I.3 DEMANDS ON ACTIVE DEVICES

Based on our previous discussions of the information processing needs, the demands on active devices are summarized in Fig. I.3. In order for a device to succeed, it must display a physical phenomenon that leads to one or more of the properties mentioned in Fig I.3. Let us briefly discuss all of these requirements, along with other requirements coming from fabrication and market forces, in this section.

Non-Linear Response
Whether the task to be accomplished by a device is image enhancement or simple addition of two numbers, the key property a device must possess is the ability to distinguish between two "closely spaced" pieces of information. We have already discussed, in the previous section via Fig. I.2, the advantage of a non-linear response over a linear response.

Another consequence of a nonlinear response is manifested when two inputs are simultaneously inputted into the device. Consider the two inputs to have a form

$$I_{in} = I_1 \cos \omega_1 t + I_2 \cos \omega_2 t$$

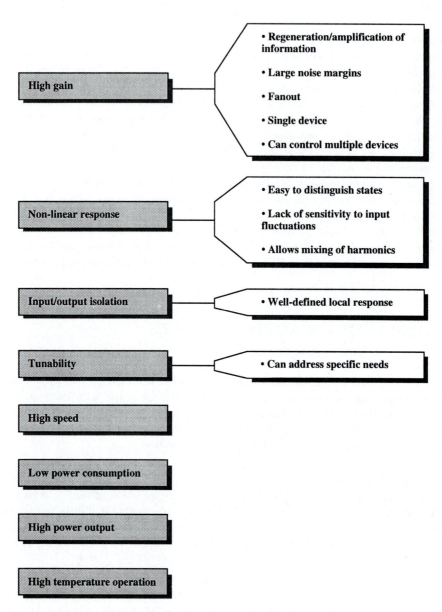

Figure I.3: Requirements of a useful device.

and let us assume a simple physical response of the form

$$I_o = \alpha I + \beta I^2$$

where α is the linear response coefficient and β represents a non-linear coefficient. Of course, the real output function may be more complicated. The output of the device to the input is now

$$I_o = \alpha \left[I_1 \cos \omega_1 t + I_2 \cos \omega_2 t\right] + \beta \left[I_1^2 \cos^2 \omega_1 t + I_2^2 \cos^2 \omega_2 t + 2I_1 I_2 \cos \omega_1 t \cos \omega_2 t\right]$$

Noting that

$$\cos(a)\cos(b) = \frac{1}{2}\left[\cos(a+b) + \cos(a-b)\right]$$

we have

$$
\begin{aligned}
I_o &= \alpha \left[I_1 \cos \omega_1 t + I_2 \cos \omega_2 t\right] + \beta \left[\frac{I_1^2}{2}(1 + \cos 2\omega_1 t)\right. \\
&+ \left. \frac{I_2^2}{2}(1 + 2\cos(\omega_2 t) + I_1 I_2 \left\{\cos(\omega_1 + \omega_2) + \cos(\omega_1 - \omega_2)t\right\}\right]
\end{aligned}
$$

The output now has signals of frequencies not only ω_1 and ω_2, but also $2\omega_1$, $2\omega_2$ as well as most importantly $(\omega_1 + \omega_2)$ and $(\omega_1 - \omega_2)$. Thus the non linearity is able to "mix" the two input signals and provide sum and difference signals. This is extremely useful for communication where a signal to be transmitted (say at audio frequencies) can be "carried" on a high frequency and then decoded.

Gain in the Device

If a physical phenomenon is to lead to a viable device, an important requirement is the presence of gain in the output-input relation. Thus, a small change in the input should produce a large change in the output. In most applications of devices, a "regeneration" of information is required, i.e., the output of previous step is used to generate another output in successive steps. The ability to introduce gain is one of the most powerful ingredients of most electronic devices, as we shall briefly discuss. Lack of gain has also been one of the pitfalls of many promising exotic device ideas.

The ability to provide gain (or amplify the incoming signal) is not only important for signal regeneration, it is also very useful, along with nonlinear effects, in providing large noise tolerances, especially in digital technologies. One of the great strengths of digital technologies comes from this "noise immunity" of the processed signal at each step of the processing. Thus, in most processing systems, an analog signal is converted to a digital signal (by A/D converters) in spite of the fact that a slight loss of accuracy is suffered in this conversion. However, after this initial loss, the digital signal can then be processed by many complex steps and its integrity is maintained by the presence of high gain devices.

Another advantage of a high gain device is that a single output can drive a number of other devices (i.e., have a large "fanout"). This is an obvious plus, since complex circuitry can be designed based on a large fanout.

Input-Output Isolation

We have all had an experience when, during a conversation, a listener suddenly starts speaking while we are still speaking, causing great confusion. The human body is designed in such a way that it is difficult to shut off our ears and so it is usually not possible to attain input-output isolation during conversation. For devices, the isolation between the input signal and the output signal is essential for most applications. This ensures a well defined response of the device and the system for a given set of inputs. One of the key attractions of three terminal devices (that are well designed) is the isolation between the gate (or base in a bipolar device) of an FET and the output at the drain (or the collector). This is often not possible in two terminal devices, causing serious difficulties in technologies that have attempted to use these devices.

Tunable Response

One of the driving forces of the modern technology is tunability of device response. This may involve optical devices that can emit in important frequency windows (e.g., blue or green light for displays; 1.55 μm lasers for low loss transmission in fibers, etc.), or special structures which have unusual input-output characteristics. A tunable response, especially if it is achieved while maintaining the other device requirements, is of great importance for the development of technology.

High Speed Operation

The ability to process information at a faster and faster rate is, of course, the main reason for advances in device technology. Physical phenomenon which are "fast" and can be harnessed to produce viable devices are constantly being researched. However, it must be realized, based on the discussions of this section, that every "femtosecond" phenomenon is not going to lead to a device. Thus, while it may be popular to discuss how a femtosecond laser may be able to transmit all the information in the world's libraries in a second, usually such claims do not survive a closer scrutiny.

The issue of speed is usually a simple one in some technologies, such as microwave technology or optical communication. Here, "faster" materials (e.g., Si $\rightarrow$ GaAs/InGaAs $\rightarrow$) drive the technology for microwave applications. However, in the case of the all purpose computation systems, the speed issue is rather complex and "architecture and layout" play as much, if not more, of a role in the overall speed of a system. Let us not forget that one of the fastest computers (at least for some operations, like recognition, association, etc.)—the human brain—has devices that can only switch in tens of milliseconds!

Low Power Operation

In addition to high speed, an important consideration for a physical phenomenon is the power consumed in the process. For many applications, it is the power-delay product that is of most significance. Low power requirements are not only important, since less power is demanded from the various sources of power, but less dissipation of power also translates into lower costs for heat sinking and being able to introduce a higher density of devices.

Low Noise

A very important property of a device is the intrinsic noise level that is present during device operation. The presence of noise obviously distorts the signal and often conveys a totally wrong set of information. A number of exotic device concepts are based on phenomenon which is intrinsically noisy and therefore not sufficient for a reliable device. The importance of noise is increasing rapidly in technology and as devices get smaller and new physical phenomenon is introduced into device design, the noise problem is likely to grow in importance.

Special Purpose Devices

A very important consideration in device technology is, of course, to find physical phenomena which produce devices capable of functioning in domains where current devices cannot operate. For example, Si devices cannot operate above ~ 150 C because of high intrinsic carrier concentration and related effects. A drive for new materials thus exists to produce devices which can operate at temperatures where strong needs exist. For example, these devices can be used to monitor functions, as in engines of machinery, or in oil wells, etc., where temperatures are high.

The need for higher density optical memory is driving the search for laser materials which can emit at shorter wavelengths.

Other driving forces for special applications include absorption dips in the dispersion curves of transmission of waves through optical fibers. In fact, the need for new materials can appear "suddenly," often driven by a specific application. For example, if the optical communication technology switches from glass fibers to plastic fibers, the laser material of choice may have to switch also.

In addition to processing requirements, one has to contend with system requirements where issues of heat sinking, packaging, integration, etc., are as critical as the device performance. Finally, one comes to the market forces and the inertia that a device technology must face. A system that can be inserted into existing technology with the least amount of perturbation will have a far greater chance of acceptance.

I.4   ELECTRONICS AND INFORMATION SYSTEMS

Electronic devices have dominated the modern information processing systems. Essentially all the demands placed on devices discussed in the previous section can be met by electronic devices. Field effect transistors (FETs) and bipolar junction transistors (BJTs) provide high gain and are extremely fast. They are widely used for microwave devices as amplifiers and oscillators, digital switches, and memories. Two terminal Gunn diodes, tunnel diodes, IMPATTS, etc., are used for signal generations at hundreds of gigahertz.

Electronics does, however, have some vulnerable spots. The electronic circuits are formed by connecting devices to each other using metallic inter-connects. This limits the inter-connectivity of the devices, and system architecture calling for massive inter-connections are very difficult to implement. An optics-based system would not have such problems.

Another dificulty faced by electronics is in the transmission of information over very long distances. For such transmissions (e.g., telecommunications), cables made from metals (e.g., copper) are needed. The system is quite expensive, cannot carry a large number of information channels, and requires repeaters after a kilometer or so because of severe signal decay. This is an area where optoelectronics has made a most significant impact.

Electronics also suffers from external electromagnetic interference (EMI) effects. For example, a surge of current induced due to a lightning bolt, can have a disastrous effect on an electronic system. Once again, an optics-based system would not be affected by EMI.

Another vulnerability of electronics comes from the fact that, electrons being charged particles, suffer a lot of scattering as they move in a material. While this is acceptable for current transistors, future devices based on "interference" effects will be seriously limited by scattering. Photons suffer very little scattering, and interference effects can be exploited for "functional" devices.

From the short discussion of this section, it is clear that while electronics dominates the information systems, in certain areas, other technologies can play a significant role.

I.5 THE PROMISE OF OPTICAL INFORMATION PROCESSING

The notion of using light for information processing has been of great interest to engineers for a long time. Light has many properties that make it very attractive for information processing. Some of these properties shown in Fig. I.4 are:

i) Immunity to electromagnetic interference: Since light particles carry no charge, electromagnetic activities, such as lightning and other potential discharges which can play havoc with electrical signals, have essentially no effect on optical signals;

ii) Non-interference of crossing light signals: Two unrelated light beams can cross one another and emerge with little effect on each other—a property that could be exploited in very high density information processing. In electronic signals, two crossing signals will have serious effects and cause loss of information;

iii) Promise of high parallelism: The benefits of optics, as far as parallelism is concerned, is obvious to us when we see an image and are able to process it in parallel to make real time decisions, like crossing a busy road. Of course, we do not know how exactly the human brain exploits the parallelism, but this is one of the great challenges for computer scientists;

iv) High speed/high bandwidth: Optical pulses have been produced with widths of only a few femtoseconds! In principle, such short pulses could be exploited for a variety of high speed applications;

v) Signal (beam) steering: Optical beams can be steered quite easily by use of lens or holograms. This is difficult or impossible to do for electron beams in a reasonable manner. The beam steering phenomenon can, in principle, allow one to reconfigure interconnections in very short times and thus, generate circuits which can be flexible (or functional) in real time;

vi) Special functions devices: This is a most exciting property of optical devices which has great potential in high speed information processing. An important example in this case is the lens which, when used with a proper object to image relationship, can produce a Fourier transform of the object image. This property is exploited in numerous recognition based systems. Another example is the use of optics in spectrum analyzers which exploits the special diffraction properties of light;

vii) Wave nature of light: Since light suffers little scattering over long distances (compared to electrons), its wave nature can be readily exploited for special purpose devices. In electronics, the wave nature of electrons comes into picture only when the device dimensions are below ~500 Å, since scattering effects smear out the electron's phase over longer distances;

viii) Nonlinear interactions: A number of materials have a strongly nonlinear response to optical intensity and can be exploited for devices;

ix) Ease of coupling with electronics: This is one of the most important features of optics and one that has paid most dividend so far. Optical and electronic interactions can easily be merged in semiconductor devices. This has led to the most important optoelectronic devices vis., the laser, the detector, and the modulator.

It is obvious from the above brief discussion that optics has the potential of playing a major role in the information processing age. Nevertheless, so far it has not fully met this potential. Only in special purpose systems (spectrum analyzers, remote imaging by radar recognition, etc.,) is optics used to carry out processing of information. In most other areas, optics plays a more passive role, serving as carrier of information while leaving all the active processing to electronics. This is not to say that optical logic and processing devices have not been developed. It is just that in the serial oriented computation that we are currently used to, these devices have not favorably competed with existing electronic devices. It may well be that conceptual changes in computing are required; ones which will exploit the inherent parallelism of optics before optical devices become viable in general purpose computation.

So far, optics has played a very important role in a number of areas shown in Fig. I.5. These areas include:

i) Memory: Information is stored digitally on optical discs (compact discs or CDs) as tiny "bumps" which can be read by a solid state laser. This has greatly revolutionized the music industry, as well as the general information storage industry. However, the laser in the CD player still has to be backed up by an electronic chip which does all the signal processing and controls the audio output;

ii) Optical communication: This has been the most important area where optoelectronic devices have made inroads into modern technology. This is also an area which has given impetus to compound semiconductor research and development. Optical fibers are rapidly replacing the traditional copper cables for carrying telephone conversations and television programs;

iii) Local area networks (LANs): This is another area where optical interconnects between local computers, telephones, etc., are making office buildings and factories more efficient and capable of handling high volume information;

iv) Printing and desktop publishing: This area has received a great boost with the availability of laser printers;

v) Guidance and control: Laser guided weapons and unmanned flying crafts have become important components of modern armies;

vi) Photonic switching and interconnects: The use of optical devices in chip-to-chip interconnects is becoming increasingly feasible.

In addition to the above areas where optics has already made inroads into technology, a number of areas where optics is expected to make impact are listed in Fig. I.6 These new areas will require the development of new optical devices based on novel physical phenomenon as well as new breakthroughs in computing concepts.

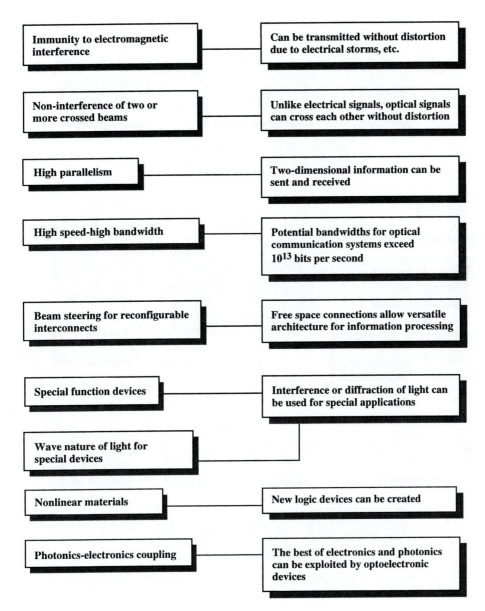

Figure I.4: Special features of light and optical devices which make optics an attractive medium for information processing.

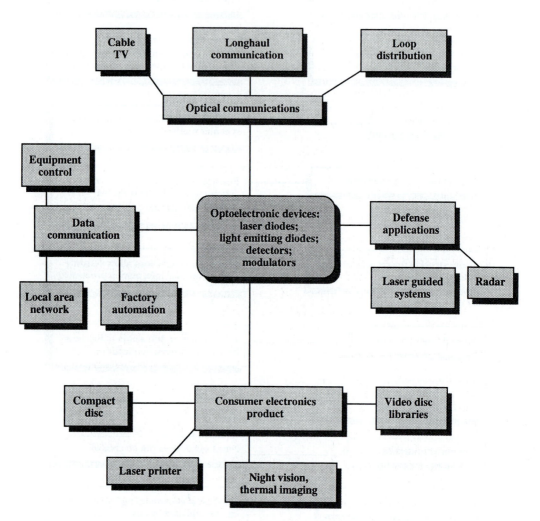

Figure I.5: Some areas where optoelectronics has made impacts in modern technology.

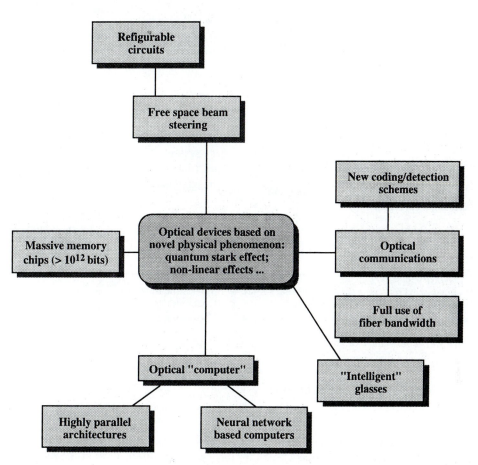

Figure I.6: Areas where optoelectronic systems are expected to make an impact in the future.

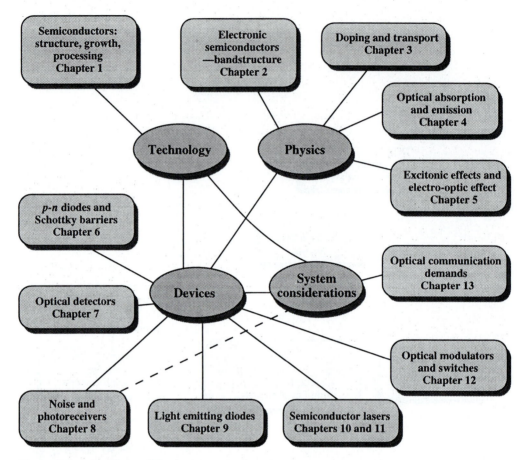

Figure I.7: A flowchart of the topics covered in this book.

I.6   ROLE OF THIS BOOK

In Fig. I.7 we show the topics covered in this book. This book could be used to teach a one semester course focusing on optoelectronic devices to graduate level students. It could also be used to teach a two semester course covering optical processes in semiconductors and optoelectronic devices. A guideline for instructors will be provided later in this section. Some of the important features of this text are listed below.

Solved Examples: The textbook has about 150 worked examples. This feature should be useful not only in a classroom setting, but also for practicing engineers and applied scientists who may want to learn more about modern semiconductor devices.

Difficulty Levels of the Topics: The various topics in this text are placed in three categories and are identified by three symbols discussed below:

$\longrightarrow$: These topics are relatively easy and appropriate for a lower level one semester course on optoelectronics for beginning graduate students.

$\rightsquigarrow$: The sections are somewhat difficult and/or appropriate for students who have had some introduction to semiconductor optoelectronic devices. In an introductory class, the instructor may decide to skip these sections entirely or just summarize the results without going through the mathematical derivations. These topics can be covered in a second course on optoelectronics.

$\mathcal{R}$: The sections are presented for review or informative reading. The instructors may choose to assign these sections as reading assignments.

Units: The book uses SI units throughout. Many worked out examples provide the units at each step. The student may notice the use of centimeters at some places and meters at others. Also, the energy unit is Joules in some places and electron volt at others. These are to conform to standard practices and should not cause the student any difficulty.

I.7 GUIDELINES FOR INSTRUCTORS

In the Figs. I.8–I.10 a guideline is provided for a one semester course (about 35 one hour lectures) focusing on optoelectronic devices. For such a course it has to be assumed that the student has familiarity with basic semiconductor concepts such as doping, conduction and valence bands, p-n diodes etc. These topics are covered in this text but can be given as reading assignments to the students. Detailed derivations of photon-electron interactions would not be covered in such a course.

A two semester sequence on "optical phenomena in semiconductors and their use in optoelectronic devices" can cover essentially all of this text-book.

The book can also be used for special courses on topics, such as "photon-semiconductor interactions and semiconductor lasers" or "optical processes in semiconductors."

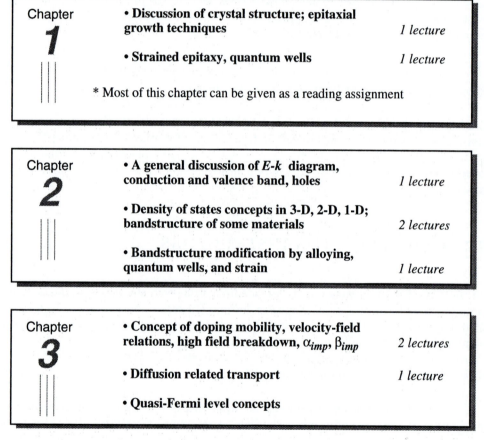

Figure I.8: Suggested topics for a 1 semester course oriented towards basic optoelectronic devices. A 2 semester course could cover all the topics of this text.

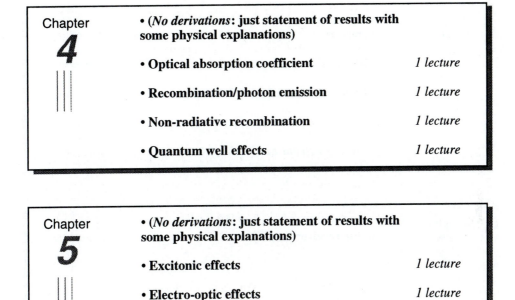

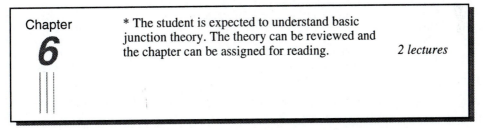

Figure I.9: Suggested topics for a 1 semester course oriented towards basic optoelectronic devices (con't.).

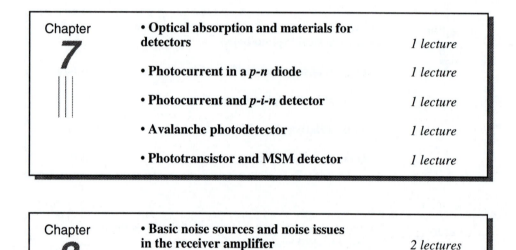

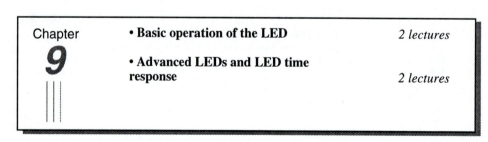

Figure I.10: Suggested topics for a 1 semester course oriented towards basic optoelectronic devices (con't.).

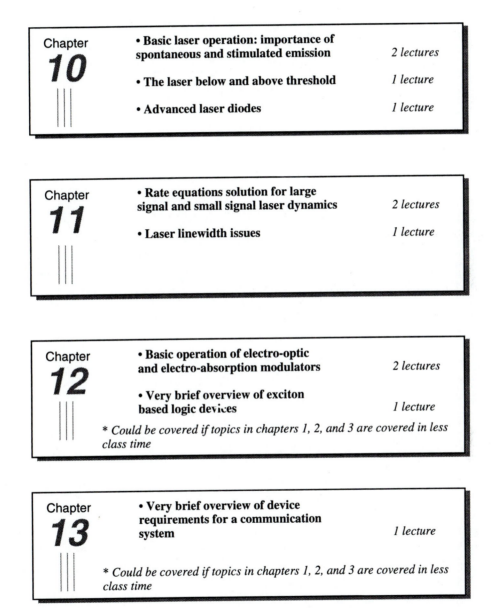

Chapter **10**
- **Basic laser operation: importance of spontaneous and stimulated emission** — *2 lectures*
- **The laser below and above threshold** — *1 lecture*
- **Advanced laser diodes** — *1 lecture*

Chapter **11**
- **Rate equations solution for large signal and small signal laser dynamics** — *2 lectures*
- **Laser linewidth issues** — *1 lecture*

Chapter **12**
- **Basic operation of electro-optic and electro-absorption modulators** — *2 lectures*
- **Very brief overview of exciton based logic devices** — *1 lecture*

* Could be covered if topics in chapters 1, 2, and 3 are covered in less class time

Chapter **13**
- **Very brief overview of device requirements for a communication system** — *1 lecture*

* Could be covered if topics in chapters 1, 2, and 3 are covered in less class time

Figure I.11: Suggested topics for a 1 semester course oriented towards basic optoelectronic devices (con't.).

CHAPTER
1

SEMICONDUCTORS: CRYSTAL STRUCTURE AND TECHNOLOGY ISSUES

1.1   INTRODUCTION

This textbook deals with the optoelectronic devices which are designed to provide the key components of the information age. The devices we will address are all based on semiconductors. Semiconductors are currently the basis of most electronic devices used in information processing and it makes sense to use the same materials for optoelectronic devices. In this chapter we will discuss the physical characteristics of semiconductors. We will also discuss how semiconductors are manufactured and give an overview of the various techniques used to fabricate devices. We will also discuss some of the challenges that remain to be overcome in the device fabrication arena, particularly in regard to optoelectronic devices.

1.2   THE COMPLEXITY OF SOLID STATE ELECTRONICS

$\mathcal{R}$ In semiconductor devices, whether electronic or optoelectronic, we are interested in the behavior of a very large number of negatively charged electrons moving through positively charged fixed ions. In general, it is difficult to solve a problem where there are a large number of interacting particles.

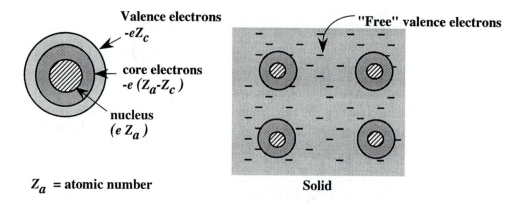

Electrons in a solid

Figure 1.1: A conceptual picture of an atom showing the nucleus with charge eZ_a, the core electrons and the valence electrons. In a solid the valence electrons are "free" and are capable of charge conduction. The electron concentration in solids is quite high.

To appreciate the enormity of the problem let us examine the number density of electrons involved. An element contains 6.022×10^{23} atoms per mole (the Avogadro's number). If ρ is the density of the material, the number of moles per unit volume are ρ/A, where A is the atomic mass. We now assume that the number of electrons that are free to conduct current is Z_c where Z_c are the number of electrons in the outermost shell (i.e., the valence of the element) of the atom as shown in Fig. 1.1. The electron density for the conduction electrons is now

$$n = 6.022 \times 10^{23} \frac{Z_c \rho}{A} \tag{1.1}$$

For most materials, this number is $\sim 10^{23}$ cm^{-3}! There is an enormously large density of free conduction electrons in the material. The spacing between electrons at such densities is 1–2 Åfor most materials!

How does one go about solving such an enormously complex problem? This problem is, indeed, insolvable if there were no simplifying feature in the system. Fortunately, there is a very important simplifying feature in semiconductors that we will be dealing with. This feature is the symmetry and order present in the structure of the semiconductors. The electrons are not moving through a random distribution of ions, but through a well defined periodically arranged distribution. The semiconductors we are interested in have crystalline structures which, as we shall see, endows them with enormous order and periodicity. *This periodicity allows us to reduce the seemingly impossible task of a large number of electrons in semiconductors to a very manageable problem.* The first step in understanding semiconductors is to understand the underlying periodicity of crystals.

1.3 PERIODICITY OF A CRYSTAL

$\mathcal{R}$ Crystals are made up of identical building blocks, the block being an atom or a group of atoms. While in "natural" crystals the crystalline symmetry is fixed by nature, new advances in crystal growth techniques are allowing scientists to produce artificial crystals with modified crystalline structure. These advances depend upon being able to place atomic layers with exact precision and control during growth, leading to "superlattices". The underlying periodicity of crystals is the key which controls the properties of the electrons inside the material. Thus by altering crystalline structure artificially, one is able to alter electronic properties.

To understand and define the crystal structure, two important concepts are introduced. The *lattice* represents a set of points in space which form a periodic structure. Each point sees an exact similar environment. The lattice is by itself a mathematical abstraction. A building block of atoms called the *basis* is then attached to each lattice point yielding the crystal structure.

An important property of a lattice is the ability to define three vectors $\mathbf{a}_1$, $\mathbf{a}_2$, $\mathbf{a}_3$, such that any lattice point $\mathbf{R}'$ can be obtained from any other lattice point $\mathbf{R}$ by a translation

$$\mathbf{R}' = \mathbf{R} + m_1\mathbf{a}_1 + m_2\mathbf{a}_2 + m_3\mathbf{a}_3 \qquad (1.2)$$

where m_1, m_2, m_3 are integers. Such a lattice is called Bravais lattice. The entire lattice can be generated by choosing all possible combinations of the integers m_1, m_2, m_3 . The crystalline structure is now produced by attaching the basis to each of these lattice points.

$$\boxed{lattice + basis = crystal\ structure} \qquad (1.3)$$

The translation vectors $\mathbf{a}_1$, $\mathbf{a}_2$, and $\mathbf{a}_3$ are called primitive if the volume of the cell formed by them is the smallest possible. There is no unique way to choose the primitive vectors. One choice is to pick

$\mathbf{a}_1$ to be the shortest period of the lattice
$\mathbf{a}_2$ to be the shortest period not parallel to $\mathbf{a}_1$
$\mathbf{a}_3$ to be the shortest period not coplanar with $\mathbf{a}_1$ and $\mathbf{a}_2$

It is possible to define more than one set of primitive vectors for a given lattice, and often the choice depends upon convenience. The volume cell enclosed by the primitive vectors is called the *primitive unit cell*.

Because of the periodicity of a lattice, it is useful to define the symmetry of the structure. The symmetry is defined via a set of point group operations which involve

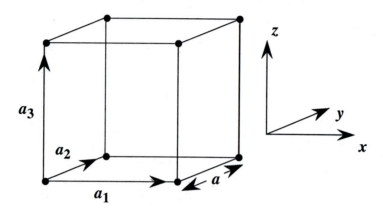

Figure 1.2: A simple cubic lattice showing the primitive vectors. The crystal is produced by repeating the cubic cell through space.

a set of operations applied around a point. The operations involve rotation, reflection and inversion. The symmetry plays a very important role in the electronic properties of the crystals. For example, the inversion symmetry is extremely important and many physical properties of semiconductors are tied to the absence of this symmetry. As will be clear later, in the diamond structure (Si, Ge, C, etc.), inversion symmetry is present, while in the Zinc Blende structure (GaAs, AlAs, InAs, etc.), it is absent. Because of this lack of inversion symmetry, these semiconductors are piezoelectric, i.e., when they are strained an electric potential is developed across the opposite faces of the crystal. In crystals with inversion symmetry, where the two faces are identical, this is not possible. The lack of inversion symmetry leads to electro-optic effects that can be exploited to design optical switches. This will be discussed in Chapters 5 and 12.

1.4   BASIC LATTICE TYPES

$\mathcal{R}$ The various kinds of lattice structures possible in nature are described by the symmetry group that describes their properties. Rotation is one of the important symmetry groups. Lattices can be found which have a rotation symmetry of 2π, $\frac{2\pi}{2}$, $\frac{2\pi}{3}$, $\frac{2\pi}{4}$, $\frac{2\pi}{6}$. The rotation symmetries are denoted by 1, 2, 3, 4, and 6. No other rotation axes exist; e.g., $\frac{2\pi}{5}$ or $\frac{2\pi}{7}$ are not allowed because such a structure could not fill up an infinite space.

There are 14 types of lattices in 3D. These lattice classes are defined by the relationships between the primitive vectors a_1, a_2, and a_3, and the angles α, β, and γ between them. The general lattice is triclinic ($\alpha \neq \beta \neq \gamma, a_1 \neq a_2 \neq a_3$) and there are 13 special lattices. Table 1.1 provides the basic properties of these three dimensional lattices. We will focus on the cubic lattice which is the structure taken by all semiconductors.

System	Number of lattices	Restrictions on conventional cell axes and singles
Triclinic	1	$a_1 \neq a_2 \neq a_3$ $\alpha \neq \beta \neq \gamma$
Monoclinic	2	$a_1 \neq a_2 \neq a_3$ $\alpha = \gamma = 90° \neq \beta$
Orthorhombic	4	$a_1 \neq a_2 \neq a_3$ $\alpha = \beta = \gamma = 90°$
Tetragonal	2	$a_1 = a_2 \neq a_3$ $\alpha = \beta = \gamma = 90°$
Cubic	3	$a_1 = a_2 = a_3$ $\alpha = \beta = \gamma = 90°$
Trigonal	1	$a_1 = a_2 = a_3$ $\alpha = \beta = \gamma < 120°, \neq 90°$
Hexagonal	1	$a_1 = a_2 \neq a_3$ $\alpha = \beta = 90°$ $\gamma = 120°$

Table 1.1: The 14 Bravais lattices in 3-dimensional systems and their properties.

There are 3 kinds of cubic lattices: simple cubic, body centered cubic, and face centered cubic.

Simple cubic: The simple cubic lattice shown in Fig. 1.2 is generated by the primitive vectors

$$a\mathbf{x}, a\mathbf{y}, a\mathbf{z} \tag{1.4}$$

where the $\mathbf{x}$, $\mathbf{y}$, $\mathbf{z}$ are unit vectors.

Body centered cubic: The bcc lattice shown in Fig. 1.3 can be generated from the simple cubic structure by placing a lattice point at the center of the cube. If $\hat{\mathbf{x}}, \hat{\mathbf{y}}$, and $\hat{\mathbf{z}}$ are three orthogonal unit vectors, then a set of primitive vectors for the body-centered cubic lattice could be

$$a_1 = a\hat{\mathbf{x}}, a_2 = a\hat{\mathbf{y}}, a_3 = \frac{a}{2}(\hat{\mathbf{x}} + \hat{\mathbf{y}} + \hat{\mathbf{z}}). \tag{1.5}$$

A more symmetric set for the bcc lattice is

$$a_1 = \frac{a}{2}(\hat{\mathbf{y}} + \hat{\mathbf{z}} - \hat{\mathbf{x}}), a_2 = \frac{a}{2}(\hat{\mathbf{z}} + \hat{\mathbf{x}} - \hat{\mathbf{y}}), a_3 = \frac{a}{2}(\hat{\mathbf{x}} + \hat{\mathbf{y}} - \hat{\mathbf{z}}) \tag{1.6}$$

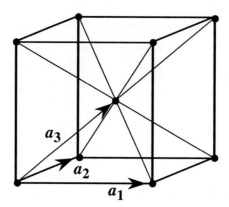

Figure 1.3: The body centered cubic lattice along with a choice of primitive vectors.

Face Centered Cubic: Another equally important lattice for semiconductors is the *face-centered cubic* (fcc) Bravais lattice. To construct the face-centered cubic Bravais lattice add to the simple cubic lattice an additional point in the center of each square face (Fig. 1.4).

A symmetric set of primitive vectors for the face-centered cubic lattice (see Fig. 1.4) is

$$a_1 = \frac{a}{2}(\hat{y} + \hat{z}), a_2 = \frac{a}{2}(\hat{z} + \hat{x}), a_3 = \frac{a}{2}(\hat{x} + \hat{y}) \qquad (1.7)$$

The face-centered cubic and body-centered cubic Bravais lattices are of great importance, since an enormous variety of solids crystallize in these forms with an atom (or ion) at each lattice site. Essentially all semiconductors of interest for electronics and optoelectronics have fcc structure.

1.4.1   The Diamond and Zinc Blende Structures

Most semiconductors of interest for electronics and optoelectronics have an underlying fcc lattice. However, they have two atoms per basis. The coordinates of the two basis atoms are

$$(000) \ and \ (\frac{a}{4}, \frac{a}{4}, \frac{a}{4}) \qquad (1.8)$$

Since each atom lies on its own fcc lattice, such a two atom basis structure may be thought of as two inter-penetrating fcc lattices, one displaced from the other by a translation along a body diagonal direction ($\frac{a}{4} \frac{a}{4} \frac{a}{4}$).

Figure 1.5 gives details of this important structure. If the two atoms of the

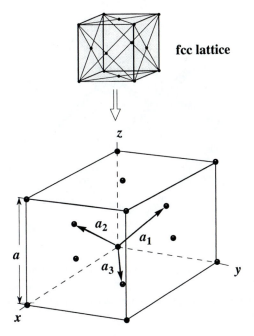

Figure 1.4: Primitive basis vectors for the face centered cubic lattice.

basis are identical, the structure is called diamond. Semiconductors such as Si, Ge, C, etc., fall in this category. If the two atoms are different, the structure is called the Zinc Blende structure. Semiconductors such as GaAs, AlAs, CdS, etc., fall in this category. Semiconductors with diamond structure are often called elemental semiconductors, while the Zinc Blende semiconductors are called compound semiconductors. The compound semiconductors are also denoted by the position of the atoms in the periodic chart, e.g., GaAs, AlAs, InP are called III-V (three-five) semiconductors while CdS, HgTe, CdTe, etc., are called II-VI (two-six) semiconductors.

Some semiconductors crystallize in the wurtzite structure which is discussed in problem 10 of this chapter. The structure is shown in Fig. 1.26. The technologies of these semiconductors is not yet well developed, although they have important potential for high power electronic devices and short wavelength optical devices.

1.4.2   Notation to Denote Planes and Points in a Lattice: Miller Indices

A simple scheme is used to describe lattice planes, directions and points. For a plane, we use the following procedure:

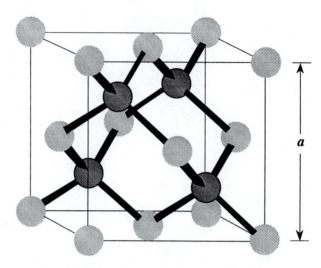

Figure 1.5: The zinc blende crystal structure. The structure consists of the interpenetrating fcc lattices, one displaced from the other by a distance $(\frac{a}{4}\frac{a}{4}\frac{a}{4})$ along the body diagonal. The underlying Bravais lattice is fcc with a two atom basis. The positions of the two atoms is (000) and $(\frac{a}{4}\frac{a}{4}\frac{a}{4})$.

(1) Define the x, y, z axes (primitive vectors).

(2) Take the intercepts of the plane along the axes in units of lattice constants.

(3) Take the reciprocal of the intercepts and reduce them to the smallest integers.

The notation (hkl) denotes a family of parallel planes.

The notation (hkl) denotes a family of equivalent planes.

To denote directions, we use the smallest set of integers having the same ratio as the direction cosines of the direction.

In a cubic system the Miller indices of a plane are the same as the direction perpendicular to the plane. The notation [ ] is for a set of parallel directions; < > is for a set of equivalent direction. Fig. 1.6 shows some examples of the use of the Miller indices to define planes.

EXAMPLE 1.1 The lattice constant of silicon is 5.43 Å. Calculate the number of silicon atoms in a cubic centimeter. Also calculate the number density of Ga atoms in GaAs which has a lattice constant of 5.65 Å.

Silicon has a diamond structure which is made up of the fcc lattice with two atoms on each lattice point. The fcc unit cube has a volume a^3. The cube has eight lattice sites at the

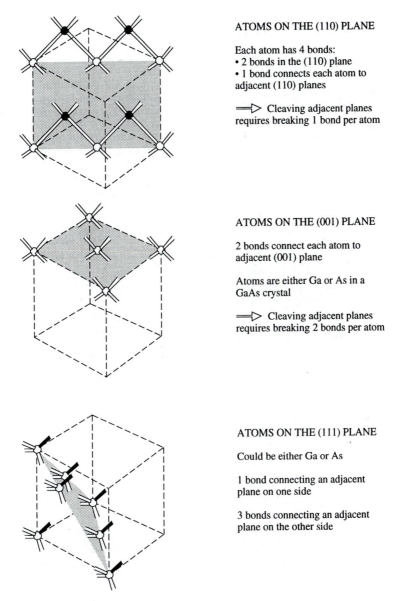

ATOMS ON THE (110) PLANE

Each atom has 4 bonds:
• 2 bonds in the (110) plane
• 1 bond connects each atom to
adjacent (110) planes

⟹ Cleaving adjacent planes
requires breaking 1 bond per atom

ATOMS ON THE (001) PLANE

2 bonds connect each atom to
adjacent (001) plane

Atoms are either Ga or As in a
GaAs crystal

⟹ Cleaving adjacent planes
requires breaking 2 bonds per atom

ATOMS ON THE (111) PLANE

Could be either Ga or As

1 bond connecting an adjacent
plane on one side

3 bonds connecting an adjacent
plane on the other side

Figure 1.6: Some important planes in the cubic system along with their Miller indices. This figure also shows how many bonds connect adjacent planes. This number determines how easy or difficult it is to cleave the crystal along these planes.

cube edges. However, each of these points is shared with eight other cubes. In addition, there are six lattice points on the cube face centers. Each of these points is shared by two adjacent cubes. Thus the number of lattice points per cube of volume a^3 are

$$N(a^3) = \frac{8}{8} + \frac{6}{2} = 4$$

In silicon there are two silicon atoms per lattice point. The number density is, therefore,

$$N_{Si} = \frac{4 \times 2}{a^3} = \frac{4 \times 2}{(5.43 \times 10^{-8})^3} = 4.997 \times 10^{22} \text{ atoms/cm}^3$$

In GaAs, there is one Ga atom and one As atom per lattice point. The Ga atom density is, therefore,

$$N_{Ga} = \frac{4}{a^3} = \frac{4}{(5.65 \times 10^{-8})^3} = 2.22 \times 10^{22} \text{ atoms/cm}^3$$

There are an equal number of As atoms.

EXAMPLE 1.2 In semiconductor technology, a Si device on a VLSI chip represents one of the smallest devices while a GaAs laser represents one of the larger devices. Consider a Si device with dimensions $(5 \times 2 \times 1)$ μm^3 and a GaAs semiconductor laser with dimensions $(200 \times 10 \times 5)$ μm^3. Calculate the number of atoms in each device.

From Example 1.1 the number of Si atoms in the Si transistor are

$$N_{Si} = (5 \times 10^{22} \text{ atoms/cm}^3)(10 \times 10^{-12} \text{ cm}^3) = 5 \times 10^{11} \text{ atoms}$$

The number of Ga atoms in the GaAs laser are

$$N_{Ga} = (2.22 \times 10^{22})(10^4 \times 10^{-12}) = 2.22 \times 10^{14} \text{ atoms}$$

An equal number of As atoms are also present in the laser.

EXAMPLE 1.3 Calculate the surface density of Ga atoms on a Ga terminated (001) GaAs surface.

In the (001) surfaces, the top atoms are either Ga or As leading to the terminology Ga terminated (or Ga stabilized) and As terminated (or As stabilized), respectively. A square of area a^2 has four atoms on the edges of the square and one atom at the center of the square. The atoms on the square edges are shared by a total of four squares. The total number of atoms per square is

$$N(a^2) = \frac{4}{4} + 1 = 2$$

The surface density is then

$$N_{Ga} = \frac{2}{a^2} = \frac{2}{(5.65 \times 10^{-8})^2} = 6.26 \times 10^{14} \text{ cm}^{-2}$$

EXAMPLE 1.4 Calculate the height of a GaAs monolayer in the (001) direction.

In the case of GaAs, a monolayer is defined as the combination of a Ga and As atomic layer. The monolayer distance in the (001) direction is simply

$$A_{m\ell} = \frac{a}{2} = \frac{5.65}{2} = 2.825 \text{ Å}$$

1.5 ARTIFICIAL STRUCTURES: SUPERLATTICES AND QUANTUM WELLS

$\mathcal{R}$ So far in this chapter we have discussed crystal structures that are present in natural semiconductors. These structures are the lowest free energy configuration of the solid state of the atoms. Since the electronic and optical properties of the semiconductors is completely determined by the crystal structure, scientists have been intrigued with the idea of fabricating artificial structures or superlattices. Since the mid-70's, these ideas have been gaining ground, inspired by the pioneering work of Esaki and Tsu at IBM. The key to growing artificial structures with tailorable crystal structure and hence tailorable optical and electronic properties has been the progress in hetero-epitaxy. Heteroepitaxial crystal growth techniques such as molecular beam epitaxy (MBE) and metal-organic chemical vapor deposition (MOCVD) have made a tremendous impact on semiconductor physics and technology. From very high speeds, low noise electronic devices used for satellite communications to low threshold lasers for communication, semiconductor devices are being made by these techniques. Although, so far, only compound semiconductors have benefitted from these growth techniques, it appears that silicon technology is on the threshold of using hetero-epitaxy for faster devices, by combining Si with Si-Ge alloys.

MBE or MOCVD are techniques which allow monolayer ($\sim$3 Å) control in the chemical composition of the growing crystal. These techniques will be examined in brief in Section 1.10. Nearly every semiconductor extending from zero bandgap (α-Sn,HgCdTe) to large bandgap materials such as ZnSe,CdS, etc., has been grown by epitaxial techniques such as MBE and MOCVD. In MBE, atoms or molecules of the species to be grown impinge upon the substrate in high vacuum. In MOCVD, the impinging species are complex molecules containing the atoms which are to form the crystal and a dissociative chemisorption reaction occurs at the surface.

Since the heteroepitaxial techniques allow one to grow heterostructures with atomic control, one can change the periodicity of the crystal in the growth direction. This leads to the concept of superlattices where two (or more) semiconductors A and B are grown alternately with thicknesses d_A and d_B respectively. The periodicity of

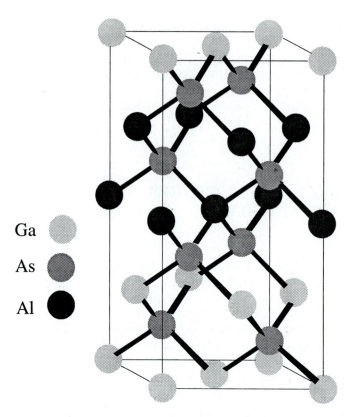

Ga

As

Al

Figure 1.7: Arrangement of atoms in a $(GaAs)_2(AlAs)_2$ superlattice grown along (001) direction.

the lattice in the growth direction is then $d_A + d_B$. A $(GaAs)_2$ $(AlAs)_2$ superlattice is illustrated in Fig. 1.7. It is a great testimony to the precision of the new growth techniques that values of d_A and d_B as low as monolayer have been grown. The presence of these superlattices is revealed in X-ray diffraction (to be discussed later), as well as in the new electronic and optical properties.

Superlattices that have been grown can be placed in three general categories: i) lattice matched; ii) strained, and iii) strained with intermediate substrate. In the lattice matched superlattices (e.g., GaAs/AlAs; HgTe/CdTe), there is a very good match between the two components forming the superlattice. Thus, there is no significant distortion in the individual unit cells making up the superlattice. These superlattices are easiest to grow and the highest quality superlattices fall in this category. The strained superlattices can be classified as those where the components A and B have different lattice constants and are grown on a substrate B. In this case the thickness of the coherent superlattice is limited by the competition between the strain energy in the A region and the dislocation formation energy. For the coherent or pseudomorphic (i.e., the substrate crystallinity is maintained) structure, the region A is under strain which distorts the cubic symmetry. These distortions play a very important role in changing semiconductor properties. We will discuss important issues in strained epitaxy in Section 1.12.

It is important to point out that the most widely used heterostructures are not superlattices but quantum wells, in which a single layer of one semiconductor is sandwiched between two layers of a larger bandgap material. Such structures allow one to exploit special quantum effects that have become very useful in electronic and optoelectronic devices as will be discussed in Chapters 4 and 5.

1.6   SURFACES: IDEAL VERSUS REAL

$\mathcal{R}$ So far we have only considered infinite crystals with no boundaries. In reality, of course, the semiconductors are finite and have surfaces which play a very important role in their properties. Obviously, as a semiconductor structure is being grown, growth occurs at its surface. Also, when contacts are placed on semiconductors to make connection to the outside world, the nature of the surface is extremely important. In fact, in both of these cases mentioned above, apparently minor changes in the surface structure can drastically alter the physical properties.

Naively, it may appear that as long as the semiconductor is clean (no impurities), the surface structure should simply be defined by the bulk crystal structure. One simply needs to disconnect the bonds along a surface to form the surface structure. Such a surface would be called the ideal surface and almost never occurs in nature! Upon a little reflection, it is not surprising that this is the case. The bulk crystal structure is decided by the internal chemical energy of the atoms forming the crystal with a certain number of nearest neighbors, second nearest neighbors, etc. At the surface, the number of neighbors is suddenly altered. Thus the spatial geometries which were providing the lowest energy configuration in the bulk may not provide the lowest energy configuration at the surface. Thus, there is a readjustment or "reconstruction" of the surface bonds towards an energy minimizing configuration.

An example of such a reconstruction is shown for the GaAs surface in Fig. 1.8. The figure (a) shows an ideal (001) surface where the topmost atoms form a square lattice. The surface atoms have two nearest neighbor bonds (Ga-As) with the layer below, four second neighbor bonds (e.g., Ga-Ga or As-As) with the next lower layer, and four second neighbor bonds within the same layer. In a "real" surface, the arrangement of atoms is far more complex. We could denote the ideal surface by the symbol C(1×1), representing the fact that the surface periodicity is one unit by one unit along the square lattice along [110] and [$\bar{1}$10]. The reconstructed surfaces that occur in nature are generally classified as C(2×8) or C(2×4) etc., representing the increased periodicity

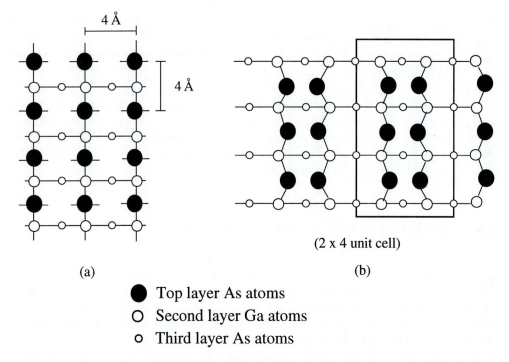

(a)   (b)

● Top layer As atoms
○ Second layer Ga atoms
o Third layer As atoms

Figure 1.8: The structure (a) of the unreconstructed GaAs (001) arsenic-rich surface. The missing dimer model (b) for the GaAs (001) (2×4) surface. The As dimers are missing to create a 4 unit periodicity along one direction and a two unit periodicity along the perpendicular direction.

along the $[\bar{1}10]$ and $[110]$ respectively. The C(2×4) case is shown schematically in Fig. 1.8b, for an arsenic stabilized surface (i.e., the top monolayer is As). The As atoms on the surface form dimers (along $[\bar{1}10]$ on the surface to strengthen their bonds. In addition, rows of missing dimers cause a longer range ordering as shown to increase the periodicity along the $[110]$ direction to cause a C(2×4) unit cell. The surface periodicity is directly reflected in the x-ray diffraction pattern.

A similar effect occurs for the (110) surface of GaAs. This surface has both Ga and As atoms (the cations and anions) on the surface. A strong driving force exists to move the surface atoms and minimize the surface energy. Reconstruction effects also occur in silicon surfaces, where depending upon surface conditions a variety of reconstructions are observed. Surface reconstructions are very important since often the quality of the epitaxial crystal growth depends critically on the surface reconstruction.

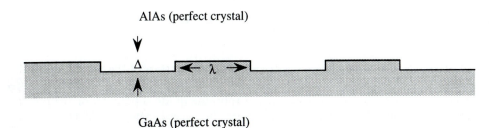

AlAs (perfect crystal)

GaAs (perfect crystal)

Figure 1.9: A schematic picture of the interfaces between materials with similar lattice constants such as GaAs/AlAs. No loss of crystalline lattice and long range order is suffered in such interfaces. The interface is characterized by islands of height Δ and lateral extent λ.

1.7   INTERFACES

$\mathcal{R}$ Like surfaces, interfaces are an integral part of semiconductor devices. We have already discussed the concept of heterostructures and superlattices which involve interfaces between two semiconductors. These interfaces are usually of high quality with essentially no broken bonds, except for dislocations in strained structures (to be discussed later). There is, nevertheless, an *interface roughness* of one or two monolayers which is produced because of either non-ideal growth conditions or imprecise shutter control in the switching of the semiconductor species. The general picture of such a rough interface is as shown in Fig. 1.9 for epitaxially grown interfaces. The crystallinity and periodicity in the underlying lattice is maintained, but the chemical species have some disorder on interfacial planes. Such a disorder is quite important in many electronic and opto-electronic devices.

One of the most important interfaces in electronics is the Si/SiO_2 interface. This interface and its quality is responsible for essentially all of the modern consumer electronic revolution. This interface represents a situation where two materials with very different lattice constants and crystal structures are brought together. However, in spite of these large differences the interface quality is quite good. It appears that the interface has a region of a few monolayers of amorphous or disordered Si/SiO_2 region creating fluctuations in the chemical species (and consequently in potential energy) across the interface. This interface roughness is responsible for reducing mobility of electrons and holes in MOS devices. It can also lead to "trap" states, which can seriously deteriorate device performance if the interface quality is poor.

Finally, we have the interfaces formed between metals and semiconductors. Structurally, these important interfaces are hardest to characterize. These interfaces are usually produced in presence of high temperatures and involve diffusion of metal elements along with complex chemical reactions. The "interfacial region" usually extends over several hundred Angstroms and is a complex non-crystalline region.

1.8   DEFECTS IN SEMICONDUCTORS

$\mathcal{R}$ In the previous section we have discussed the properties of the perfect crystalline structure. In real semiconductors, one invariably has some defects that are introduced due to either thermodynamic considerations (nothing in life is perfect!) or the presence of impurities during the crystal growth process. In general, defects in crystalline semiconductors can be characterized as i) point defects; ii) line defects; iii) planar defects and iv) volume defects. These defects are detrimental to the performance of electronic and optoelectronic devices and are to be avoided as much as possible. We will give a brief overview of the important defects.

Point Defects
A point defect is a highly localized defect that affects the periodicity of the crystal only in one or a few unit cells. There are a variety of point defects, as shown in Fig. 1.10. An important point in defects is the vacancy that is produced when an atom is missing from a lattice point. The vacancy defects are present in any crystal and their concentration is given roughly by the thermodynamics relation

$$\frac{N_{vac}}{N_{Tot}} = exp \left(-\frac{E_{vac}}{k_B T} \right) \tag{1.9}$$

where N_{vac} is the vacancy density, N_{Tot} the total site density in the crystal, E_{vac} the vacancy formation energy and T, the crystal growth temperature. The vacancy formation energy is approximately 2.0 eV for most semiconductors.

An important point defect in compound semiconductors such as GaAs is the anti-site defect in which one of the atoms, say Ga, sits on the arsenic sublattice instead of the Ga sublattice. Such defects (denoted by Ga_{As}) can be a source of reduced device performance.

Other point defects are interstitials in which an atom is sitting in a site that is in between the lattice points as shown in Fig. 1.10, and impurity atoms which involve a wrong chemical species in the lattice. In some cases the defect may involve several sites forming a defect complex.

Line Defects or Dislocations
In contrast to point defects, line defects (called dislocations) involve a large number of atomic sites that can be connected by a line. Dislocations are produced if, for example, an extra half plane of atoms are inserted (or taken out) of the crystal as shown in Fig. 1.11. Such dislocations are called edge dislocations. Dislocations can also be created if there is a slip in the crystal so that part of the crystal bonds are broken and reconnected with atoms after the slip.

Dislocations can be a serious problem, especially in the growth of strained

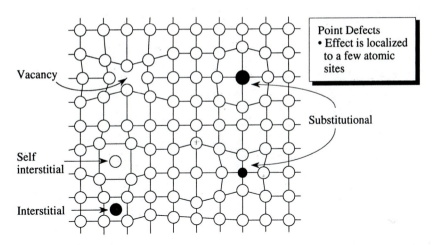

Figure 1.10: A schematic showing some important point defects in a crystal. (Adapted from J.W. Mayer and S.S. Lau, *Electronic Material Science: For Integrated Circuits in Si and GaAs*, MacMillan, New York (1990).)

heterostructures (to be discussed later). In optoelectronic devices, dislocations can ruin the device performance and render the device useless. Thus the control of dislocations is of great importance.

Planar Defects and Volume Defects

Planar defects and volume defects are not important in single crystalline materials, but can be of importance in polycrystalline materials. If, for example, silicon is grown on a glass substrate, it is likely that polycrystalline silicon will be produced. In the polycrystalline material, small regions of Si (∼ a few microns in diameter) are perfectly crystalline, but are next to microcrystallites with different orientations. The interface between these microcrystallites are called grain boundaries. Grain boundaries may be viewed as an array of dislocations.

Volume defects can be produced if the crystal growth process is poor. The crystal may contain regions that are amorphous or may contain voids. In most epitaxial techniques used in modern optoelectronics, these defects are not a problem. However, the developments of new material systems such as diamond (C) or SiC are hampered by such defects.

EXAMPLE 1.5 Consider an equilibrium growth of a semiconductor at a temperature of 1000 K. The vacancy formation energy is 2.0 eV. Calculate the vacancy density produced if the site density for the semiconductor is 2.5×10^{22} cm^{-3}.

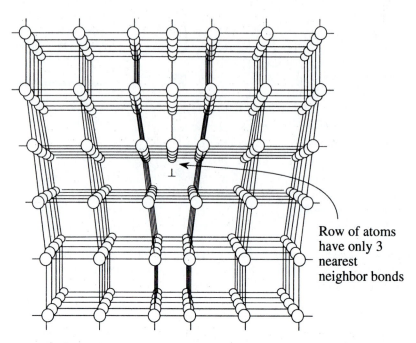

Row of atoms
have only 3
nearest
neighbor bonds

Figure 1.11: A schematic showing the presence of a dislocation. This line defect is produced by adding an extra half plane of atoms. At the edge of the extra plane, the atoms have a missing bond.

The vacancy density is

$$
\begin{aligned}
N_{vac} &= N_{Tot} exp \left(-\frac{E_{vac}}{k_B T} \right) \\
&= (2.5 \times 10^{22} \ cm^{-3}) \ exp \left(-\frac{2.0 \ eV}{0.0867 \ eV} \right) \\
&= 2.37 \times 10^{12} \ cm^{-3}
\end{aligned}
$$

This is an extremely low density and will have little effect on the properties of the semiconductor. The defect density would be in mid 10^{15} cm^{-3} range if the growth temperature was 1500 K. At such values, the defects can significantly affect device performance.

1.9   BULK CRYSTAL GROWTH

$\mathcal{R}$ Bulk crystal growth techniques are used mainly to produce substrates on which devices are eventually fabricated. While for some semiconductors like Si and GaAs (to some extent for InP) the bulk crystal growth techniques are highly matured; for most other semiconductors it is difficult to obtain high quality, large area substrates. The aim of the bulk crystal growth techniques is to produce single crystal boules with as large a diameter as possible and with as few defects as possible. In Si the boule diameters have reached 30 cm with boule lengths approaching 100 cm. Large size substrates ensure low cost device production.

For the growth of boules from which substrates are obtained, one starts out with a purified form of the elements that are to make up the crystal. One important technique that is used is the Czochralski (CZ) technique. In the Czochralski technique shown in Fig. 1.12, the melt of the charge (i.e., the high quality polycrystalline material) is held in a vertical crucible. The top surface of the melt is just barely above the melting temperature. A seed crystal is then lowered into the melt and slowly withdrawn. As the heat from the melt flows up the seed, the melt surface cools and the crystal begins to grow. The seed is rotated about its axis to produce a roughly circular cross-section crystal. The rotation inhibits the natural tendency of the crystal to grow along certain orientations to produce a faceted crystal.

The CZ technique is widely employed for Si, GaAs, and InP and produces long ingots (boules) with very good circular cross-section. For Si up to 100 kg ingots can be obtained. In the case of GaAs and InP the CZ technique has to face problems arising from the very high pressures of As and P at the melting temperature of the compounds. Not only does the chamber have to withstand such pressures, also the As and P leave the melt and condense on the sidewalls. To avoid the second problem one seals the melt by covering it with a molten layer of a second material (e.g., boron oxide) which floats on the surface. The technique is then referred to as liquid encapsulated Czochralski, or the LEC technique.

A second bulk crystal growth technique involves a charge of material loaded in a quartz container. The charge may be composed of either high quality polycrystalline material or carefully measured quantities of elements which make up a compound crystal. The container called a "boat" is heated till the charge melts and wets the seed crystal. The seed is then used to crystallize the melt by slowly lowering the boat temperature starting from the seed end. In the gradient-freeze approach the boat is pushed into a furnace (to melt the charge) and slowly pulled out. In the Bridgeman approach, the boat is kept stationary while the furnace temperature is temporally varied to form the crystal. The approaches are schematically shown in Fig. 1.13.

The easiest approach for the boat technique is to use a horizontal boat. However, the shape of the boule that is produced has a D-shaped form. To produce circular cross-sections vertical configurations have now been developed for GaAs and InP.

In addition to producing high purity bulk crystals, the techniques discussed above are also responsible for producing crystals with specified electrical properties. This may involve high resistivity materials along with *n*- or *p*-type materials. In Si it is difficult to produce high resistivity substrated by bulk crystal growth and resistivities are

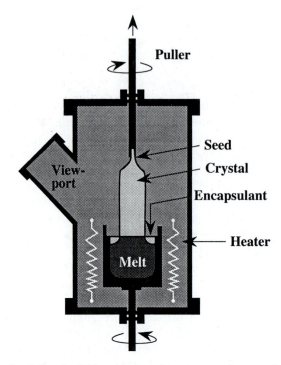

Figure 1.12: Schematic of Czocharlski-style crystal grower used to produce substrate ingots. The approach is widely used for Si, GaAs and InP.

usually $<10^4$ Ω-cm. However, in compound semiconductors carrier trapping impurities such as chromium and iron can be used to produce material with resistivities of $\sim 10^8$ Ω cm. The high resistivity or semi-insulating (SI) substrates are extremely useful in device isolation and for high speed devices. For n- or p-type doping carefully measured dopants are added in the melt.

The availability of high quality substrates is essential to any device technology. Other than the three materials mentioned above (Si, GaAs, InP) the substrate fabrication of semiconductors is still in its infancy. Since epitaxial growth techniques used for devices require close lattice matching between the substrate and the overlayer, non- availability of substrates can seriously hinder the progress of a material technology. This is, for example, one of the reasons of slow progress in large bandgap semiconductor technology necessary for high power-high temperature electronic devices and short wavelength semiconductor lasers.

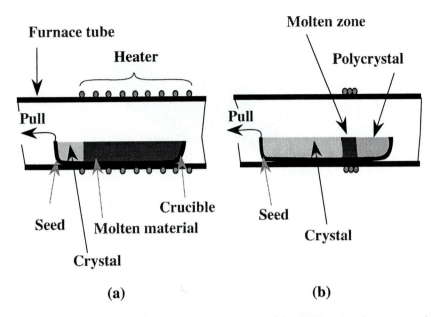

Figure 1.13: Crystal growing from the melt in a crucible: (a) solidification from one end of the melt (horizontal Bridgeman method); (b) melting and solidification in a moving zone.

1.10 EPITAXIAL CRYSTAL GROWTH

$\longrightarrow$ The substrates that result once the bulk grown semiconductor boule is sliced and capped are almost never used directly for devices. Invariably an epitaxial layer or epilayer is grown which may be a few microns in thickness. The work epitaxy comes from the Greek word "epi" (upon) and "taxis" (ordered) meaning the ordered continuation of the substrate crystal. The epitaxial growth techniques have a very slow growth rate (as low as a monolayer per second for some techniques) which allow one to control very accurately the dimensions in the growth direction. In fact, in techniques like molecular beam epitaxy (MBE) and metal organic chemical vapor deposition (MOCVD), one can achieve monolayer ($\sim 3\mathring{A}$) control in the growth direction. This level of control is essential for the variety of heterostructure devices that are being used in optoelectronics. The epitaxial techniques are also very useful for precise doping profiles that can be achieved. In fact, it may be argued that without the advances in epitaxial techniques that have occurred over the last two decades, most of the developments in semiconductor physics would not have occurred. Table 1.2 gives a brief view of the various epitaxial techniques used along with some of the advantages and disadvantages.

Liquid Phase Epitaxy (LPE)
LPE was an epitaxial growth technique of choice until the 70's when it gradually started to be replaced by other techniques. LPE is still used for growth of crystals such as

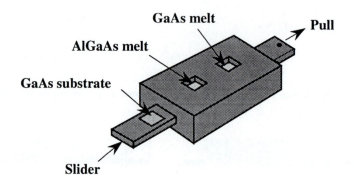

Figure 1.14: A schematic of the LPE growth of AlGaAs and GaAs. The slider moves the substrate, thus positioning itself to achieve contact with the different melts to grow heterostructures.

HgCdTe for long wavelength detectors and AlGaAs for double heterostructure lasers. As shown in Table 1.2, LPE is a close to equilibrium technique in which the substrate is placed in a quartz or a graphite boat and covered by a liquid of the crystal to be grown (see Fig. 1.14). The liquid may also contain dopants that are to be introduced into the crystal. LPE is often used for alloy growth where the growth follows the equilibrium solid-liquid phase diagram. By precise control of the liquid composition and temperature, the alloy composition can be controlled. Because LPE is a very close to equilibrium growth technique, it is difficult to grow alloy systems which are not miscible or even grow heterostructures with atomically abrupt interfaces. Nevertheless heterostructures where interface is graded over 10-20 Å can be grown by LPE by sliding the boat over successive "puddles" of different semiconductors. For many applications such interfaces are adequate and since LPE is a relatively inexpensive growth technique, it is used widely in many commercial applications.

Vapor Phase Epitaxy (VPE)
A large class of epitaxial techniques rely on delivering the components that form the crystal from a gaseous environment. If one has molecular specy in a gaseous form with partial pressure P, the rate at which molecules impinge upon a substrate is given by

$$F = \frac{P}{\sqrt{2\pi m k_B T}} \sim \frac{3.5 \times 10^{22} P(torr)}{\sqrt{m(g)T(K)}} mol./cm^2 - s \qquad (1.10)$$

where m is the molecular weight and T the cell temperature. For most crystals the surface density of atoms is $\sim 7 \times 10^{14}$ cm^{-2}. If the atoms or molecules impinging from the vapor can be deposited on the substrate in an ordered manner, epitaxial crystal growth can take place.

The VPE technique is used mainly for homoepitaxy and does not have the additional apparatus present in techniques such as MOCVD for precise heteroepitaxy. As an example of the technique, consider the VPE of Si. The Si containing reactant silane

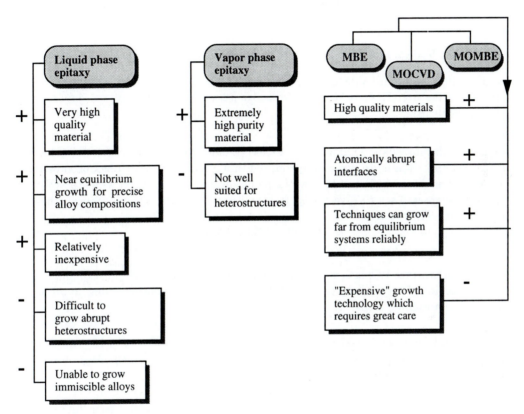

Table 1.2: A schematic of the various epitaxial crystal growth techniques and some of their positive and negative aspects.

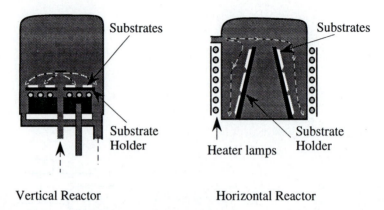

Figure 1.15: Reactors for VPE growth. The substrate temperature must be maintained uniformly over the area. This is achieved better by lamp heating. A pyrometer is used for temperature measurement.

(SiH_4) or dichlorosilane (SiH_2Cl_2) or trichlorosilane ($SiHCl_3$) or silicon tetrachloride ($SiCl_4$) is diluted in hydrogen and introduced into a reactor in which heated substrates are placed as shown in Fig. 1.15. The silane pyrolysis to yield silicon while the chlorine containing gases react to give $SiCl_2$, HCl and various other silicon-hydrogen-chlorine compounds. The reaction

$$2SiCl_2 \rightleftharpoons Si + SiCl_4 \qquad (1.11)$$

then yields Si. Since HCl is also produced in the reaction, conditions must be tailored so that no etching of Si occurs by the HCl. Doping can be carried out by adding appropriate hydrides (phosphine, arsine, etc.,) to the reactants.

An important consideration in VPE is safety related, since hydrogen, which is produced in the deposition, can explode in contact with any oxygen. Also, almost all the reactants are highly toxic.

VPE can be used for other semiconductors as well by choosing different appropriate reactant gases. The reactants used are quite similar to those employed in the MOCVD technique discussed later.

Molecular Beam Epitaxy (MBE)
Molecular beam epitaxy (MBE) is one of the most important epitaxial techniques as far as heterostructure physics and devices are concerned. Almost every semiconductor (other than a few very large bandgap semiconductors) has been grown by this technique. MBE is a high vacuum technique ($\sim 10^{-11}$ torr vacuum when fully pumped down) in which crucibles containing a variety of elemental charges are placed in the growth chamber (Fig. 1.16). The elements contained in the crucibles make up the components of the crystal to be grown as well as the dopants that may be used. When a crucible

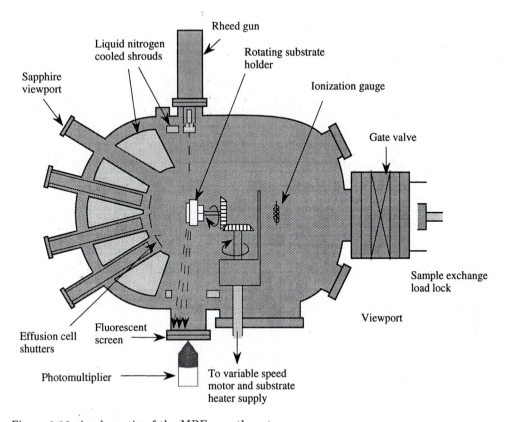

Figure 1.16: A schematic of the MBE growth system.

is heated, atoms or molecules of the charge are evaporated and these travel in straight lines to impinge on a heated substrate.

The growth rate in MBE is ∼1.0 monolayer per second and this slow rate coupled with shutters placed in front of the crucibles allow one to switch the composition of the growing crystal with monolayer control. However, to do so, the growth conditions have to be adjusted so that growth occurs in the monolayer by monolayer mode rather than by 3-dimensional island formation. This requires that the atoms impinging on the substrate have enough kinetics to reach an atomically flat profile. Thus the substrate temperature has to be maintained at a point where it is high enough to provide enough surface migration to the incorporating atoms, but not so high as to cause entropy controlled defects.

Since no chemical reactions occur in MBE, the growth is the simplest of all epitaxial techniques and is quite controllable. However, since the growth involves high vacuum, leaks can be a major problem. The growth chamber walls are usually cooled by liquid N_2 to ensure high vacuum and to prevent atoms/molecules to come off from the chamber walls.

The low background pressure in MBE allows one to use electron beams to monitor the growing crystal. The reflection high energy electron diffraction (RHEED) techniques relies on electron diffraction to monitor both the quality of the growing substrate and the layer by layer growth mode. As each monolayer gets filled up, one can see this reflected in the RHEED intensity by the naked eye!

While MBE is a simple and elegant growth technique (conceptually), it cannot be used conveniently for all semiconductors. For example, phosphides are often not grown by MBE due to the danger in handling elemental phosphorus. Also elements with very low vapor pressures are difficult to use since it is not easy to heat the crucibles beyond ∼1500 K. Silicon epitaxy in MBE, for example, requires e-beam evaporation where an electron beam is used to knock off Si atoms for growth.

MBE is a relatively safe technique and has become the technique of choice for the testing of almost all new ideas on heterostructure physics.

Metal Organic Chemical Vapor Deposition (MOCVD)
Metal organic chemical vapor deposition (MOCVD) is another important growth technique widely used for heteroepitaxy. Like MBE, it is also capable of producing monolayer abrupt interfaces between semiconductors. A typical MOCVD system is shown in Fig. 1.17. Unlike in MBE, the gases that are used in MOCVD are not made of single elements, but are complex molecules which contain elements like Ga or As to form the crystal. Thus the growth depends upon the chemical reactions occurring at the heated substrate surface. For example, in the growth of GaAs one often uses Triethyl Gallium and Arsine and the crystal growth depends upon the following reaction:

$$Ga(CH_3)_3 + AsH_3 \rightleftharpoons GaAs + 3CH_4 \tag{1.12}$$

One advantage of the growth occurring via a chemical reaction is that one can use lateral temperature control to carry out local area growth. Laser assisted local area growth is also possible for some materials and can be used to produce new kinds of device structures. Such local area growth is difficult in MBE.

There are several varieties of MOCVD reactors. In the atmospheric MOCVD the growth chamber is essentially at atmospheric pressure. One needs a large amount of gases for growth in this case, although one does not have the problems associated with vacuum generation. In the low pressure MOCVD the growth chamber pressure is kept low. The growth rate is then slower as in the MBE case.

The use of the MOCVD equipment requires very serious safety precautions. The gases used are highly toxic and a great many safety features have to be incorporated

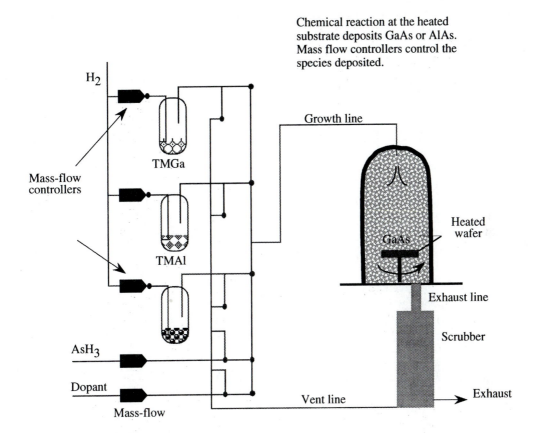

TMGa : Gallium containing organic compound
TMAl : Aluminum containing organic compound
AsH_3 : Arsenic containing compound

Figure 1.17: Schematic diagram of an MOCVD system employing alkyds (trimethyl gallium (TMGa) and trimethyl aluminum (TMAl) and metal hydride (arsine) material sources, with hydrogen as a carrier gas.

to avoid any deadly accidents. Safety and environmental concerns are important issues in almost all semiconductor manufacturing since quite often one has to deal with toxic and hazardous materials.

In addition to MBE and MOCVD one has hybrid epitaxial techniques often called MOMBE (metal organic MBE) which try to combine the best of MBE and MOCVD. In MBE one has to open the chamber to load the charge for the materials to be grown while this is avoided in MOCVD where gas bottles can be easily replaced from outside. Additionally, in MBE one has occasional spitting of material in which small clumps of atoms are evaporated off on to the substrate. This is avoided in MOCVD and MOMBE.

EXAMPLE 1.6 Consider the growth of GaAs by MBE. The Ga partial pressure in the growth chamber is 10^{-5} Torr, and the Ga cell temperature is 900 K. Calculate the flux of Ga atoms on the substrate.

The mass of Ga atoms is 70 g/mole. The flux is (from Eqn. 1.10)

$$F = \frac{3.5 \times 10^{22} \times 10^{-5}}{\sqrt{70 \times 900}} = 5.27 \times 10^{14} \, \text{atoms/cm}^2$$

Note that the surface density of Ga atoms on GaAs is $\sim 6.3 \times 10^{14}$ cm^{-2}. Thus, if all of the Ga atoms were to stick, the growth rate would be $\sim$0.8 monolayer per second. This assumes that there is sufficient arsenic to provide As in the crystal.

1.11   STRAINED HETEROEPITAXY

$\longrightarrow$ We have noted earlier that in an epitaxial process, the overlayer that is grown on the substrate could have a lattice constant that may differ from that of the substrate. Such epitaxy is called strained epitaxy and is one of the important emerging areas of crystal growth studies. The motivation for strained epitaxy is two fold:

i) Incorporation of built-in strain: When a lattice mismatched semiconductor is grown on a substrate and the thickness of the overlayer is very thin (this will be discussed in detail later), the overlayer has a built-in strain. This built-in strain has important effects on the electronic and optoelectronic properties of the material and can be exploited for high performance devices.

ii) Availability of alternate substrate: We have noted that in semiconductor technology, high quality substrates are only available for Si, GaAs and InP (quartz substrates are also available and used for some applications). Most semiconductors are not lattice-matched to these substrates. How can one grow these semiconductors epitaxially? One solution that has emerged is to grow the overlayer on a mismatched substrate. If the conditions are right, a lot of dislocations are generated and eventually the overlayer forms its own substrate. This process (to be discussed later in this section) allows a tremendous flexibility in semiconductor technology. Not only can it, in principle, resolve the substrate availability problem, it also allows the possibility of growing GaAs on Si, CdTe on GaAs, etc. Thus different semiconductor technologies can be integrated on the same wafer.

Strained Heteroepitaxy for Built-in Strain

Consider a case where an overlayer with lattice constant a_L is grown on a substrate with lattice constant a_S. This situation is shown schematically in Fig. 1.18. The strain between the two materials is defined as

$$\epsilon = \frac{a_S - a_L}{a_L} \tag{1.13}$$

Consider a conceptual exercise where you are depositing a monolayer of the overlayer on the substrate. If the lattice constant of the overlayer is maintained to be a_L, it is easy to see that after every $1/\epsilon$ bonds between the overlayer and the substrate, either a bond is missing or an extra bond appears as shown in Fig. 1.18b. In fact, there would be a row of missing or extra bonds since we have a 2-dimensional plane. These defects are the dislocations. The presence of these dislocations costs energy to the system since a number of atoms do not have proper chemical bonding at the interface.

An alternative to the case shown in Fig. 1.18b is the case shown in Fig. 1.18c. Here all the atoms at the interface of the substrate and the overlayer are properly bonded by adjusting the in-plane lattice constant of the overlayer to that of the substrate. This causes the overlayer to be under strain and the system has a certain amount of strain energy. *This strain energy grows as the overlayer thickness increases.* In the strained epitaxy, the choice between the state of the structure shown in Fig. 1.18b and the state shown in Fig. 1.18c is decided by free energy minimization considerations. Theoretical and experimental studies have focussed on these considerations for over six decades, and the importance of these studies has grown since the advent of heteroepitaxy. The general observations can be summarized as follows:

i) If the lattice mismatch ϵ is less than a critical value ϵ_{C1} (~ 0.1), the first monolayer of the overlayer can be grown on a thick substrate without dislocations. If the lattice mismatch lies between ϵ_{C1} and ϵ_{C2} (~ 0.14), the first monolayer of the overlayer has dislocations, although a metastable state can be produced in which there are no dislocations and there is perfect registry between the overlayer and the substrate. If $\epsilon > \epsilon_{C2}$, the first monolayer always has an array of dislocations.

ii) For small lattice mismatch ($\epsilon < 0.1$), the overlayer initially grows in perfect registry with the substrate, as shown in Fig. 1.18c. However, as noted before, the strain energy will grow as the overlayer thickness increases. As a result, it will eventually be

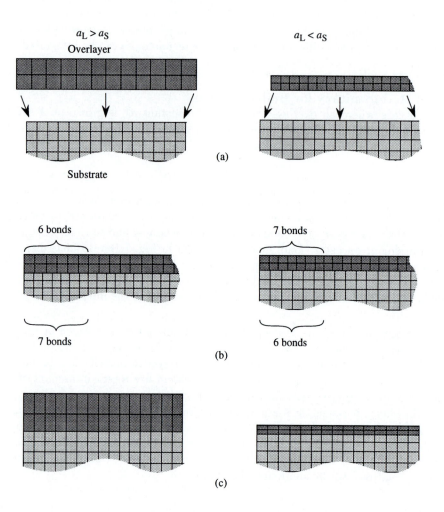

Figure 1.18: (a) The conceptual exercise in which an overlayer with one lattice constant is placed without distortion on a substrate with a different lattice constant. (b) Dislocations are generated at positions where the interface bonding is lost. (c) The case is shown where the overlayer is distorted so that no dislocation is generated.

favorable for the overlayer to generate dislocations. In simplistic theories this occurs at an overlayer thickness called the critical thickness, d_c, which is approximately given by

$$d_c \cong \frac{a_S}{2|\epsilon|} \tag{1.14}$$

In reality, the point in growth where dislocations are generated is not so clear cut and depends upon growth conditions, surface conditions, dislocation kinetics, etc. However, one may use the criteria given by Eqn. 1.14 for loosely characterizing two regions of overlayer thickness for a given lattice mismatch. Below critical thickness, the overlayer grows without dislocations and the film is under strain. Above critical thickness, the film has a dislocation array, and after the dislocation arrays are generated, the overlayer grows without strain with its free lattice constant.

In Fig. 1.19 we summarize how the discussions presented here impact upon semiconductor optoelectronic technology. We will see in the next chapter that as a crystal is strained, its electronic and optical properties can be altered. Thus, by growing an overlayer on a substrate with a small lattice mismatch, one can incorporate strain in a film as long as the thickness is below the critical thickness. One needs a strain value of about 1% so that the critical thickness is a few hundred angstroms (at the most). This usually involves growing a quantum well structure in which the strained region forms the well region and is surrounded by a lattice matched region.

While strained epitaxy below critical thickness is an extremely powerful tool for tailoring the optoelectronic properties of semiconductors, epitaxy beyond the critical thickness is important to provide new effective substrates for new material growth. As shown in Fig. 1.19, for these applications the key issues center around ensuring that the dislocations generated stay near the overlayer-substrate interface and do not propagate into the overlayer as shown in Fig. 1.20. A great deal of work has been done to study this problem. Often thin superlattices in which the individual layers have alternate signs of strain are grown to "trap" or "bend" the dislocations. It is also useful to build the strain up gradually.

EXAMPLE 1.7 Estimate the critical thickness for $In_{0.3}Ga_{0.7}As$ grown on a GaAs substrate.

The lattice constant of an alloy is given by the Vegard's law:

$$
\begin{aligned}
a(In_{0.3}Ga_{0.7}As) &= 0.3a_{InAs} + 0.7a_{GaAs} \\
&= 5.775 \ \text{Å}
\end{aligned}
$$

The strain is

$$\epsilon = \frac{5.653 - 5.775}{5.653} = -0.022$$

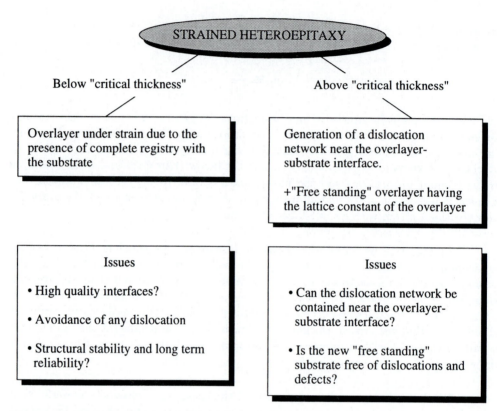

Figure 1.19: Important regimes in strained heteroepitaxy. The issues of importance in below critical thickness epitaxy and above critical thickness epitaxy are identified.

The critical thickness is approximately

$$d_c = \frac{5.653 \ \mathring{A}}{2(0.022)} = 128 \ \mathring{A}$$

This thickness is quite adequate for most devices and can be used to make useful quantum well devices. If, on the other hand, the strain is, say, 5%, the critical thickness is ~ 50 Å, which is too thin for most useful device applications.

1.12 LITHOGRAPHY

$\mathcal{R}$ Modern solid state electronics and optoelectronics owe their great success to device fabrication techniques which can produce extemely complicated devices with high yield. A variety of active and passive devices can be fabricated on the same wafer using

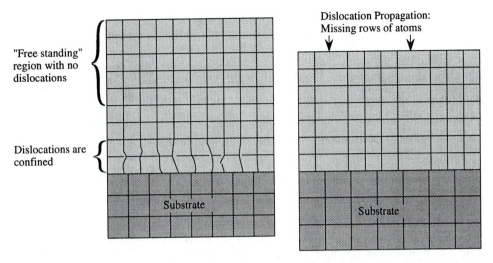

Figure 1.20: Strained epitaxy above critical thickness. On the left hand side is shown a structure in which the dislocations are confined near the overlayer-substrate interface. This is a desirable mode of epitaxy . On the right hand side, the dislocations are penetrating the overlayer, rending it useless for most optoelectronic applications.

lateral patterning techniques based on lithography. Since crystal growth processes do not produce any controlled lateral variations in material properties, lithographic techniques are needed to alter the lateral properties of the wafer. Modern solid state circuits (ICs) contain millions of devices on the same chip. Lithography plays a central role in the success of the solid state technology. Lithographic techniques allow one to fabricate complex and dense circuits. They also allow one to obtain high speed performance by reducing key device dimensions. Indeed, advances in lithographic techniques are one of the most important driving forces for advances in device performance. Optoelectronic devices and chips have not yet reached the state that electronic chips have, in terms of integration. However, most of the lithographic techniques are shared by the two technologies.

The technique of lithography to alter lateral properties of a surface is an ancient one. Many lithograph masterpieces hang around the world's museums. "Batik" paintings and t-shirts also rely on the concepts of lithography. The lithography process involves taking a certain design created in a computer or by an artist and transferring it onto the semiconductor wafer. A number of steps are involved in this process. While new advances in lithographic techniques are introducing continual changes in the individual steps, the following discussion will provide an overview of the current state of the art.

1.12.1   Photoresist Coating

The semiconductor wafer in its virgin form has little irreversible sensitivity to optical
or electron beams. Such a sensitivity is needed if a pattern of device and circuits is to
be transferred to the wafer. To make the wafer (which is usually covered by a thin oxide
film or some other dielectric passivation material) sensitive to an image, a photoresist is
spread on the wafer by a process called spin coating. For the resist to be reliable it must
satisfy three criteria: i) it must have good bonding to the substrate; ii) its thickness
must be uniform; and iii) the thickness should be reliably controlled over different wafer
runs.

Spin-coating has emerged as the most reliable technique for photoresist application. As shown in Fig. 1.21, a small puddle of the resist is applied to the center of the
wafer which is held to a spindle by a vacuum chuck. The spindle is now spun at a rate
of 2000-8000 rpm for 10 to 60 seconds. During the first couple of seconds of spinning,
most of the resist is thrown off and carefully drained away. The remaining resist forms
a thin layer whose thickness is controlled by the spin speed (thickness $\propto \frac{1}{\sqrt{\omega}}$, where ω is
the spin frequency). An edge bead is formed which is several times the thickness of the
film. A variety of details are introduced (e.g. variable spin-speed; applying the puddle
to an already spinning wafer, etc.) to obtain more uniform resists. The thickness of the
resists are usually in the range of 0.7 to 1.0 μm. For very fine lithography thicker films
are deposited.

It must be noted that the resist is not deposited on untreated wafers since the
untreated wafers typically have hydroxyl (OH) groups attached to the surface. These
produce poor adhesion and must be removed by a prebake at 200-250 C for 30-60
minutes. An adhesion promoting primer is then applied before resist application. For
SiO_2 covered wafers, hexamethyl disilazane (HMDS) is used. Once the resist is applied,
it is soft baked at 90 to 100 C to improve adhesion to the oxide.

Once the resist is ready, it is exposed to an optical image through a mask (to
be discussed next) for a certain exposure time as shown in Fig. 1.22. The resist is then
developed by washing it in a solvent which dissolves away the regions of higher solubility.
A resist which becomes more soluble when exposed to illumination is called positive. Its
image is identical to the opaque image on the mask plate. A resist that loses solubility
when illuminated is called negative.

The first materials which proved to have a high sensitivity and etchant resistance for successful microelectronic photoresists were negative resists. These were based
on polyisoprene in the form of cyclized rubber. The resist consists of isoprene molecules
(a hydrocarbon containing double carbon bonds and the basic building block of rubber) and a photo-sensitive N_3 group radical. When light falls on the resist, the radical
causes formation of long chains from the individual isoprene hydrocarbons causing poly-

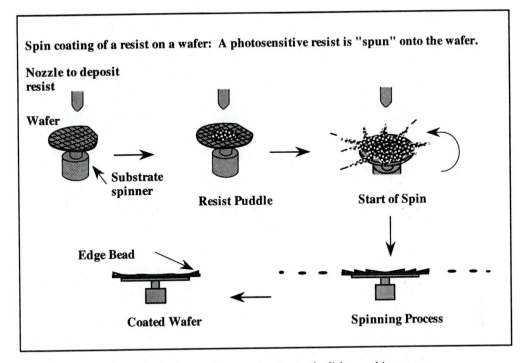

Figure 1.21: The process of spin coating a wafer starts the lithographic process.

merization of the of the isoprene. The polymerized region is highly insoluble while the unexposed part is soluble.

The negative resist has an advantage that its solubility is almost digital. There is a tremendous contrast between exposed and unexposed regions. However, since it is based on chain formation, its resolution is limited to 2-3 μm.

The positive resist is not based on chain formation and can produce much sharper resolution. The resist relies upon molecules where N_2 is bonded to a hydrocarbon with benzene like structure. Exposure to light causes the dissociation of the N_2 group producing a carboxylic acid. This can be dissolved and washed away in an alkaline solution such as dilute sodium hydroxide.

Once a resist is in place, one needs to expose it with an optical image containing the pattern to be transferred to the wafer. This requires the fabrication and use of a mask.

EXAMPLE 1.8 A negative resist is made up of molecules with a mass of 10^5 amu. The molecule is a straight chain of CH_2 units with a C-C spacing of 1.54 Å. The resolution of the

Transference of an image to the resist by using a mask and etching of the exposed regions.

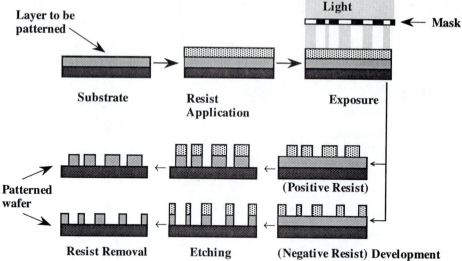

Figure 1.22: The exposure and etching process which allows one to transfer a pattern to the wafer. (After W.S. Ruska, *Microelectronic Processing, An Introduction to the Manufacture of Integrated Circuits*, McGraw-Hill, New York (1988).)

resist is about the length of one molecule. Estimate this resolution. The atomic mass of C is 12 and of H is 1.

Each unit has a mass of 14. Thus, the number of units in the molecule is

$$n = \frac{10^5}{14} = 7143 \text{units}$$

The total length is then

$$
\begin{aligned}
\ell = 7143 \times 1.54 &= 1.1 \times 10^4 \text{ Å} \\
&= 1.1 \ \mu m
\end{aligned}
$$

This resist is adequate for most technologies, but not for sub-micron technology.

1.12.2   Mask Generation and Image Transfer

Like the negative in the photography, the mask allows one to transfer the complicated circuit pattern on to the sensitive resist deposited on the wafer. Once the design of a device is done, a large scale (100 times or even larger) replica is prepared on paper and transferred to rubylith which is a transparent plastic sheet coated with a thin layer of red plastic. The pattern transfer involves cutting out the thin unwanted red plastic. In the early days of microelectronics, the desired pattern was cut by hand and reduction cameras would then reduce the pattern to its desired value. These days, as shown in Fig. 1.23 computer aided design (CAD) software programs allow one to directly generate a design tape which is then analyzed and "written" on to a mask plate. The pattern on the final mask plate can be generated optically or by electron beam writing. The electron beam (e beam) produces much finer features and can be controlled with greater precision.

The mask plate which is later used to repeatedly transfer patterns to different wafers must be transparent (at proper positions) to the radiation used for final pattern transfer. The mask should also have good mechanical and thermal properties. Quartz or borosilicate glasses are often used for ultraviolet light lithography. The regions of the mask that are to be opaque are covered with metallic chrome or iron oxide.

Usually only the basic building block of the device or circuit is produced first on the mask. This single pattern is called a reticle, and a step and repeat camera (or stepper) is used to create the entire pattern from the reticle. The reticle can be used to generate a wafer size mask plate or, in some cases, may be directly used to generate the entire pattern on the wafer by the stepper, as shown in Fig. 1.23.

Once the mask plate is made and the resist is deposited, the next step is to transfer the pattern on the mask to the wafer. For feature sizes greater than 0.25 μm, one uses optical equipment for the pattern transfer. The limit of $\sim$0.25 μm is governed by the wavelength of the light available. Most materials become opaque to light once the wavelength goes below $\sim$1000 Å. If electromagnetic radiation is to be used, one must go down to x-ray lithography with $\lambda \sim$40-80 Å. Another way to decrease feature size is by use of e-beam lithography.

The patterns to be printed in microelectronics can be classified as windows and lines. The windows may be used to allow doping diffusion or to make connections with a conductor through an insulator. The conducting interconnects and resistors are usually formed into line patterns. The imaging process involves transferring an ideal image to the mask and then from the mask to the wafer. Once the image is transferred, the resist is developed and etching is carried out. In all these steps, the ideal pattern in the designers mind is gradually lost. This occurs even if the design considerations are well within the image transferring technology.

1.13   ETCHING

$\mathcal{R}$ Like the chisel and the drill for the carpenter and the sculptor, etchants are an important tool for the microelectronics processing engineer. The etchants allow one to remove material in a selective manner, once the resist has been patterned. The choice and control of the etching process is crucial if the features in the resist film are to become a part of the substrate.

An ideal etching process must be able to remove a layer of semiconductor from the region where there is no resist. The etchant should not attack the resist, nor should it penetrate under the resist causing undercuts. It should also attack only one layer and should be self limiting (e.g., should etch SiO_2 and not Si, etc.). A number of techniques are developed to carry out etching. These are discussed below. Once etching is finished, one must remove the resist material by either other etchants or by the "lift off" technique. We will now briefly review the etching approaches.

1.13.1   Wet Chemical Etching

The simplest and most commonly employed etching technique is the wet chemical etching in which the wafer is simply soaked in a liquid chemical which dissolves away the semiconductor. The etching process involves a chemical reaction in which the elements of the film to be etched react with the etching solution. The reaction products can then be rinsed away. The rate of the etching is proportional to the etch time and is usually isotropic. The isotropic nature of the wet etching is not very suitable for devices where very sharp sidewalls are to be produced.

The etch rate depends upon the concentration of the chemicals used in the etchant and the temperature of the solution. The etchants are usually either acids (Hydrofluoric acid– HF; Nitric acid–HNO_3; Acetic acid–$H_4C_2O_2$; Sulfuric acid–H_2SO_4; Phosphoric acid–H_3PO_4) or alkaline solutions of Ammonium Hydroxide (NH_4OH). The etchants are usually diluted in water according to well established recipes.

In many devices, an important film to be etched is SiO_2. This film is etched with HF solutions and the ease and control of this process is one of the reasons for the success of the Si-technology. Usually the HF is buffered with NH_4F to produce buffered oxide etch (BOE) which is stable and has a very long shelf life.

An important film that is often used for passivation is silicon nitride (Si_3N_4). This film can also be etched by BOE, but the rate is much slower than the SiO_2 rate. Thus, if a Si_3N_4 film is on top of an SiO_2 film and only the Si_3N_4 film is to be etched, one cannot use BOE. In such cases H_3PO_4 is used.

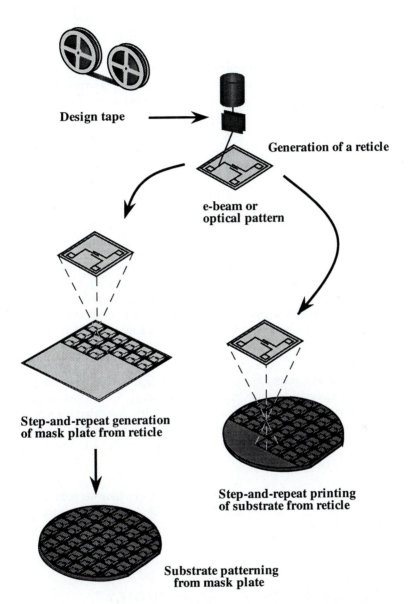

Figure 1.23: The processes used in the generation of a mask for lithography. (After W.S. Ruska, *Microelectronic Processing, An Introduction to the Manufacture of Integrated Circuits,* McGraw-Hill, New York (1988).)

Since Aluminum is commonly used for metal interconnects, one often has to etch it. A mixture of phosphoric acid, nitric acid and acetic acid is usually used if the underlying crystal is Si. However, since this etchant attacks GaAs, one uses hydrochloric acid for etching Al over GaAs substrates.

Silicon and polycrystalline silicon can be etched by HF-HNO$_3$ mixtures while GaAs is etched by using bromine in methanol or hydrogen peroxide mixed with sulfuric acid in water. A number of selective etches have also been developed for the heterostructure technology.

While chemical etching is simple and inexpensive, it is not compatible with submicron technology or technology which demands deep anisotropic etching.

1.13.2   Plasma Etching

Plasma etching resolves one of the main problems with wet chemical etching viz feature size control. It also provides efficient etching for a wide variety of films including those that are difficult to etch by wet chemicals. The plasma is produced by passing an rf electrical discharge through a gas at a low pressure. The rf discharge creates ions and electrons. The ions can be used to interact with the elements in the substrate and cause etching.

The ions being charged particles can be accelerated in the electric field and be made to bombard the substrate with controlled energy. If the ion energy is large, the ions simply sputter off atoms from the surface in a rather unselective manner. However, this provides extremely anisotropic etching with almost no undercutting effects. At low energies the ions can cause chemical reactions at the surface and cause removal of atoms selectively.

In typical plasmas one introduces fluorine or chlorine containing gases. The fluorocarbons (e.g., CF$_4$) or silicon tetrafluoride (SiF$_4$) and silicon tetrachloride (SiCl$_4$) are often used for the plasma. Once the plasma is formed fluorine (or chlorine) ions are produced which cause the etching process.

1.13.3   Reactive Ion Beam Etching (RIBE)

An important tool in microelectronic technology is the ion-implantation which is widely used to dope semiconductors. The implanter accelerates ions to a prechosen energy and shoots them into the semiconductor by controlling the energy. The depth at which the ions are embedded can be controlled. The ion-implanter can also be used for etching by using appropriate ions and it provides focusing of the ion beam. At low energies, the ions can be used to selectively etch very small feature sizes. The advantages are the same as for plasma etching.

1.13.4   Ion Beam Milling

The ion beam milling is another application of the ions in "chiseling" off material from a substrate. The ion milling requires a focussed beam of ions of energy high enough that the ions can physically knock out atoms from the film. The focus is on removing the atoms physically rather than through a chemical reaction. The process is thus highly directional and can produce extremely anisotropically etched structures. Ion milling is particularly advantageous if the etching involves small patterns, very steep walls or materials which are relatively inert and cannot be etched by chemical reactions.

The ion milling is dominated by geometric effects as shown in Fig. 1.24. In Fig. 1.24a, we show a substrate with a resist in the process of being etched by a perpendicular ion-beam. The impact of the beam causes the substrate material to fly off randomly. Some of the material can get redeposited on the etched sides forming "ears." Ions bounding off the edges can cause "trenches" to be formed around the resist pattern. These effects are controlled by impinging the beam at a slight angle and rotating the substrate.

The resolution of ion beam milling is controlled by the ion-beam spot size and can reach 0.1 μm or less.

EXAMPLE 1.9 A 0.2 μm film of AlGaAs is to be etched. The etching rate is 1.0 μm/hour. Calculate the time needed for the etching. If the rate changes by -1%, calculate the number of monolayers of AlGaAs that will remain unetched.

The time needed for the etching is

$$t = \frac{0.2}{1.0} = 0.2 \text{ hour}$$

A -1% error in the etch rate means that the thickness etched is

$$d = 0.2 \times 0.99 = 0.198 \ \mu m$$

The unetched film remaining is

$$\Delta d = 0.2 - 0.198 = 0.002 \ \mu m$$
$$= 20 \ \mathring{A}$$

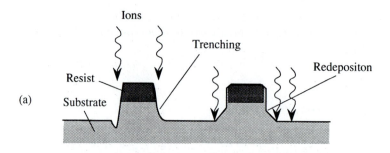

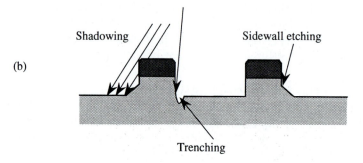

Figure 1.24: The importance of geometric effects in ion beam milling. In (a) the perpendicular incident beam can produce trenching effects as well as redeposition causing sidewall "ears;" (b) if the beam comes at an oblique angle and the substrate is rotated, the trenching and "ear" formation can be balanced.

A monolayer of AlGaAs is 2.83 Å, so that seven monolayers of film will remain. Such thickness errors may appear small, but are unacceptable in many heterostructure devices.

1.14   EPITAXIAL REGROWTH

$\mathcal{R}$ The spectacular growth of semiconductor microelectronics owes a great deal to the concept of the integrated circuit. The ability to fabricate transistors, resistors, inductors and capacitors on the same wafer is critical to the low cost and high reliability we have come to expect from microelectronics. It is natural to expect similar dividents from the concept of the optoelectronic integrated circuit (OEIC). In the OEIC, the optoelectronic device (the laser or detector or modulator) would be integrated on the same wafer with an amplifier or logic gates.

One of the key issues in OEICs involves etching and regrowth. As we will see later, the optoelectronic devices have a structure that is usually not compatible with the structure of an electronic device. The optimum layout then involves growing one of the device structures epitaxially and then masking the region to be used as, say, the optoelectronic device and etching away the epitaxial region. Next a regrowth is done to grow the electronic device with a different structure. The process is shown schematically in Fig. 1.25. While this process looks simple conceptually, there are serious problems associated with etching and regrowth.

A critical issue in the epitaxial growth of a semiconductor layer is the quality of the semiconductor-vacuum interface. This semiconductor surface must be "clean," i.e., there should be no impurity layers (e.g., an oxide layer) on the surface. Even if a fraction of a monolayer of the surface atoms have impurities bonded to them, the quality of the epitaxial layer suffers drastically. The growth may occur to produce microcrystalline regions separated by grain boundaries or may be amorphous in nature. In either case, the special properties arising from the crystalline nature of the material (to be discussed in the next chapter) are then lost.

In Section 1.6 we had discussed the issue of surface reconstruction. In most epitaxial growths, not only is it important to have a clear surface from which to start with, it is also important to maintain a special reconstruction during growth. The reconstructions are a representation of the vapor pressure (e.g., the Ga to As ratio in the growth of GaAs) and are, therefore, critical to high quality growth.

The issue of surface cleanliness and surface reconstruction can be addressed when one is doing a single epitaxial growth. For example, a clean wafer can be loaded into the growth chamber and the remaining impurities on the surface can be removed by heating the substrate. The proper reconstruction (which can be monitored by RHEED) can be ensured by adjusting the substrate temperature and specy overpressure. Now consider the problems associated with etching after the first epitaxial growth has occurred. As the etching starts, foreign atoms or molecules are introduced on the wafer as the semiconductor is etched. The etching process is quite damaging and as it ends, the surface of the etched wafer is quite rough and damaged. In addition, in most growth techniques the wafer has to be physically moved from the high purity growth chamber to the etching system. During this transportation, the surface of the wafer may collect some "dirt." During the etching process this "dirt" may not be etched off and may remain on the wafer. As a result of impurities and surface damage, when the second epitaxial layer is grown after etching, the quality of the layer suffers.

A great deal of processing research in OEICs focusses on improving the etching/regrowth process. So far the OEICs fabricated in various laboratories have performances barely approaching the performance of hybrid circuits. Clearly the problem of etching/regrowth is hampering the progress in OEIC technology.

It may be noted that the etching regrowth technology is also important in creating quantum wires and quantum dots which require lateral patterning of epitaxial

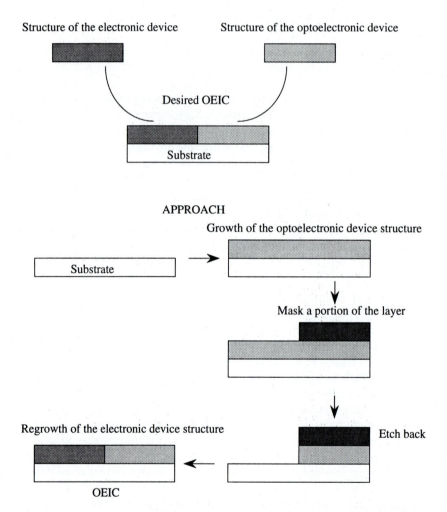

Figure 1.25: The importance of regrowth is clear when one examines the difference in the structure of electronic and optoelectronic devices. Etching and regrowth is essential for fabrication of optoelectronic integrated circuits (OEIC).

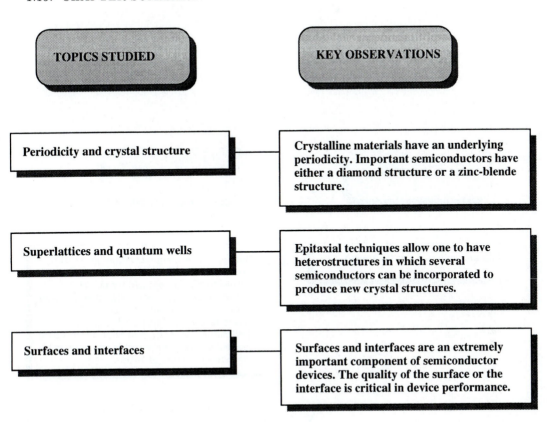

TOPICS STUDIED	KEY OBSERVATIONS
Periodicity and crystal structure	Crystalline materials have an underlying periodicity. Important semiconductors have either a diamond structure or a zinc-blende structure.
Superlattices and quantum wells	Epitaxial techniques allow one to have heterostructures in which several semiconductors can be incorporated to produce new crystal structures.
Surfaces and interfaces	Surfaces and interfaces are an extremely important component of semiconductor devices. The quality of the surface or the interface is critical in device performance.

Table 1.3: Summary table.

layers.

1.15 CHAPTER SUMMARY

In this chapter we have discussed the important structural properties of semiconductors and their heterostructures. The semiconductors we will be dealing with in most optoelectronic devices have a zinc blende or diamond structure. We have discussed the important growth techniques used in producing the semiconductors. We have also identified the important techniques used to fabricate devices. Tables 1.3 to 1.5 give an overview of the issues that have emerged from this chapter.

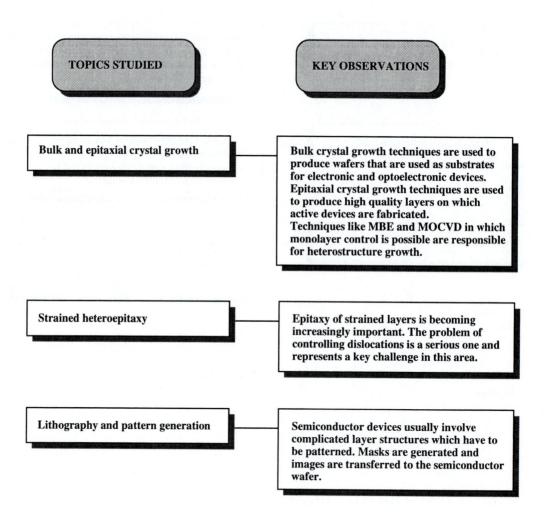

Table 1.4: Summary table.

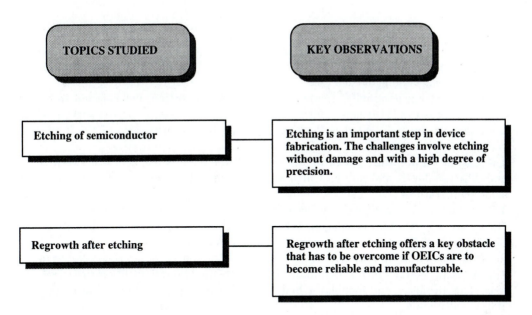

Table 1.5: Summary table.

1.16 PROBLEMS

Section 1.4

1.1 a) Find the angles between the tetrahedral bonds of a diamond lattice.

b) What are the direction cosines of the (111) oriented nearest neighbor bond along the x,y,z axes.

1.2 Consider a semiconductor with the zinc blende structure (such as GaAs).

a) Show that the (100) plane is made up of either cation or anion type atoms.

b) Draw the positions of the atoms on a (110) plane assuming no surface reconstruction.

c) Show that there are two types of (111) surfaces: one where the surface atoms are bonded to three other atoms in the crystal, and another where the surface atoms are bonded to only one. These two types of surfaces area called the A and B surfaces, respectively.

1.3 Suppose that identical solid spheres are placed in space so that their centers lie on the atomic points of a crystal and the spheres on the neighboring sites touch each other. Assuming that the spheres have unit density, show that density of such spheres is the following for the various crystal structures:

$$
\begin{aligned}
fcc &: \quad \sqrt{2}\pi/6 = 0.74 \\
bcc &: \quad \sqrt{3}\pi/8 = 0.68 \\
sc &: \quad \pi/6 = 0.52 \\
diamond &: \quad \sqrt{3}\pi/16 = 0.34
\end{aligned}
$$

1.4 Calculate the number of cells per unit volume in GaAs (a = 5.65 Å). Si has a 4% larger lattice constant. What is the unit cell density for Si? What is the number of atoms per unit volume in each case?

1.5 A Si wafer is nominally oriented along the (001) direction, but is found to be cut 2° off, towards the (110) axis. This off axis cut produces "steps" on the surface which are 2 monolayers high. What is the lateral spacing between the steps of the 2° off-axis wafer?

1.6 Conduct a literature search to find out what the lattice mismatch is between GaAs and AlAs at 300 K and 800 K. Calculate the mismatch between GaAs and Si at the same temperatures.

1.7 In high purity Si crystals, defect densities can be reduced to levels of 10^{13} cm^{-3}. On an average, what is the spacing between defects in such crystals? In heavily doped Si, the dopant density can approach 10^{19} cm^{-3}. What is the spacing between defects for such heavily doped semiconductors?

1.8 A GaAs crystal which is nominally along (001) direction is cut 8° off towards (110) axis. This produces one monolayer high steps. If the step size is to be no more than 100 Å, calculate θ.

1.9 Assume that a Ga-As bond in GaAs has a bond energy of 1.0 eV. Calculate the energy needed to cleave GaAs in the (001) and (110) planes.

1.10 In this chapter, we have considered only the fcc Bravias lattice, since most semiconductors have this underlying structure. However, some semiconductors have the hexagonal close-packed structure. Semiconductors such as BN, AlN, GaN, InN, SiC, etc., crystallize in the structure.

The hcp structure is formed as shown in Fig. 1.26a. A close-packed layer of spheres is formed with centers at points A. A second layer of spheres is placed on top of this with centers at points B. The third layer can be placed on points A (giving rise to the hcp structure), or points C (giving rise to the fcc structure). The hcp has the primitive cell of the hexagonal lattice with two atoms on the basis as shown in Fig. 1.26b. The primitive cell has primitive vectors $a_1 = a_2$ and the c-axis (vector a_3 is parallel to c) normal to the a_1, a_2 plane. One atom is at the origin and the other, at the point

$$r = \frac{2}{3}a_1 + \frac{1}{3}a_2 + \frac{1}{2}a_3$$

Show that the ratio $c/a, (a = a_1, a_2)$ is given by $\sqrt{8/3} = 1.633$. The values of these lattice constants for several semiconductors are given below. These semiconductors are often said to have the Wurtzite structure.

Section 1.9

1.11 A serious problem in the growth of a heterostructure made from two semiconductors is due to the difficulty in finding a temperature at which both semiconductors can grow with high quality. Consider the growth of HgTe and CdTe which is usually grown at ~ 600 K. Assume that the defect formation energy in HgTe is 1.0 eV and in CdTe is 2.0 eV. Calculate the density of defects in the heterostructure with equal HgTe and

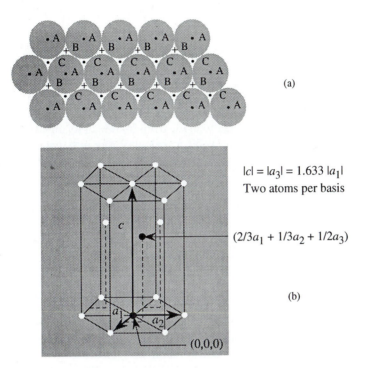

$|c| = |a_3| = 1.633\,|a_1|$

Two atoms per basis

$(2/3a_1 + 1/3a_2 + 1/2a_3)$

(a)

(b)

Figure 1.26: Hexagonal close-packed structure.

CdTe.

1.12 Calculate the defect density in GaAs grown by LPE at 1000 K. The defect formation energy is 2.0 eV.

1.13 Why are entropy considerations unimportant in dislocation generation?

Section 1.11

1.14 Consider the (001) MBE growth of GaAs by MBE. Assuming that the sticking coefficient of Ga is unity, calculate the Ga partial pressure needed if the growth rate has to be 1 μm/hr. The temperature of the Ga cell is 1000 K.

1.15 A surface emitting laser (SEL) structure is produced by epitaxially growing a reflector on either side of an active laser region. the reflector, called a distributed Bragg reflector, consists of a periodic array of GaAs and AlGaAs layers; the periodicity, a, being given by $2a = \lambda$. In an SEL, $\lambda = 8000$ Å and the total SEL thickness is 54 $\lambda + 1.0$ μm. If the flux of cations (Ga, Al) is 5×10^{14} cm^{-2}s^{-1} and all the atoms get incorporated to form the crystal, how long will it take to grow the SEL? Assume that the As atoms are present in sufficient quantity and growth rate is controlled by cation flux only.

1.16 In the growth of GaAs/AlAs structures in a particular MBE system, the background pressure of Ga when the Ga shutter is off is 10^{-7} Torr. If the growth rate of

AlAs is 1 μm/hr, what fraction of Ga atoms are incorporated in the AlAs region? The Ga cell is at 1000 K.

Section 1.12

1.17 Estimate the critical thickness for growth of GaAs on Si.

1.18 A strained quantum well laser has to made from $In_x Ga_{1-x}As$ on a GaAs substrate. If the minimum thickness of the region is 50 Å, calculate the maximum composition of In that can be tolerated. Assume that the lattice constant of the alloy can be linearly interpolated from its components.

1.19 Assume that in a semiconductor alloy, the lattice constant scales as a linear weighted average. Find the composition of the $In_x Ga_{1-x}As$ alloy that lattice matches with an InP substrate.

1.20 Calculate the critical thickness for the growth of AlAs on a GaAs substrate.

Section 1.13

1.21 To produce lateral confinement of electronic states in a semiconductor, the feature sizes of the device have to be $\sim$100 Å. Calculate the energy of photons and electrons that can be used in such a lithography.

1.22 In optoelectronics, the smallest feature sizes of typical lasers are $\sim$5 μm. However, for laser known as distributed feedback (DFB) lasers, the smallest feature size is $\sim$1000 Å. Discuss the kind of lithography systems that can be used for the two kinds of lasers.

1.23 In a particular resist spinner, the initial deposit of resist on a 3 inch GaAs wafer is 0.2 cm high. After a spin at 4000 rpm, the resist thickness is 1 μm. How much resist has been wasted in this spin?

Section 1.14

1.24 Selective etchents are often used in microelectronic technology. Consider an etch that etches GaAs at the rate of 1 μm/hr and $Al_{0.3}Ga_{0.7}As$ at 500 Å/hr. AlGaAs is used as an etch stop in a heterostructure where 0.5 μm of GaAs has to be etched. If the AlGaAs layer is 30 Å, what should be the tolerance in the etching time?

1.17 REFERENCES

* **Crystal Structures**

  - J. M. Buerger, *Introduction to Crystal Geometry*, McGraw-Hill (1971).
  - M. Lax, *Symmetry Principles in Solid State and Molecular Physics*, Wiley (1974). Has a good description of the Brillouin zones of several structures in Appendix E.

- J. F. Nye, *Physical Properties of Crystals*, Oxford (1985).
- F. C. Phillips, *An Introduction of Crystallography*, Wiley (1971).

● **Crystal Growth: Heterostructures**

- A. Y. Cho, *Journal of Vacuum Science Technology*, vol. 8, 531 (1971).
- F. Rosenberger, *Fundamentals of Crystal Growth*, Springer (1979).
- G. Haas and M. H. Francombe, C. E. C. Wood in *Physics of Thin Films*, Academic Press (1980).

● **X-Ray Diffraction: General**

- N. W. Ashcroft and N. D. Mermin, *Solid-State Physics*, Holt, Rinehart and Winston, New York (1976).
- R. B. Leighton, *Principles of Modern Physics*, McGraw-Hill, New York (1959).
- Segmueller and A. E. Blakeslee, *J. Appl. Crystallogr.*, **6**, 19 (1973).
- B. E. Warren, *X-Ray Diffraction*, Addison-Wesley (1969).

● **Diffraction from Surfaces and Crystal Growth RHEED**

- P. L. Gourley, and R. M. Biefeld, *J. Vac. Sci. Technol.*, **B1**, 383 (1983).
- B. A. Joyce, *Rep. Prog. Phys*, **48**, 1637 (1985).
- W. T. Tsang, *Semiconductors and Semimetals*, eds. R. K. Willardson and A. C. Beer, Academic Press, New York, vol.22, Part A (1985).
- J. M. VanHove, P. R. Pukite, G. J. Whaley, A. M. Worchak, and P. I. Cohen, *J. Vac. Sci. Technol.*, **B3**, 1116 (1985).

● **Point Defects in Semiconductors**

- P.K. Bhattacharya and S. Dhar, *Deep Levels in III-V Compound Semiconductors Grown by Molecular Beam Epitaxy, Semiconductors and Semimetals*, eds. A.C. Willardson and C. Beer, Academic Press, New York, vol. 26 (1988).
- E.N. Economou, *Green's Functions in Quantum Physics*, Springer-Verlag, Berlin (1979).
- G.F. Foster and J.C. Slater, *Phys. Rev.*, **96**, 1208 (1954).
- H.F. Matare, *Defect Electronics in Semiconductors*, Wiley-Interscience, New York (1971).
- S. Pantelides, *Rev. Mod. Phys.*, **50**, 797 (1978).

● **Dislocations and Lattice Mismatched Epitaxy**

- S. Amelinckx, *Dislocations in Solids*, ed. F.R.N. Nabarro, North-Holland, New York (1988).

- C.A.B Ball and J.H. van der Merwe, *Dislocations in Solids*, ed. F.R.N. Nabarro, North-Holland, New York, vol. 5 (1983).

- H.F. Matare, *Defect Electronics in Semiconductors*, Wiley-Interscience, New York (1971).

- J.W. Mathews and A.E. Blakeslee, *J. Cryst. Growth*, **27**, 118 (1974).

- J.Y. Tsao, B.W. Dodson, S.T. Picraux, and D.M. Cornelison, *Phys. Rev. Lett.*, **59**, 2455 (1987).

- J.H. van der Merwe, *J. Appl. Phys.*, **34**, 117 (1963).

- J.H. van der Merwe, *J. Appl. Phys.*, **34**, 123 (1963).

- **Microelectronic Processing**

- J.W. Mayer and S.S. Lau, *Electronic Materials Science: For Integrated Circuits in Si and GaAs*, MacMillan, New York (1990).

- W.S. Ruska, *Microelectronic Processing: An Introduction to the Manufacture of Integrated Circuits*, McGraw-Hill, New York (1988).

- P.K.L. Yu and P.C. Chan, *Introduction to GaAs Technology*, ed. C. Wang, Wiley, New York (1988).

CHAPTER 2

PROPERTIES OF SEMICONDUCTORS: ELECTRONIC STATES

2.1   INTRODUCTION

In the previous chapter we have seen that in semiconductor crystals, there is a certain periodicity in the way in which atoms are arranged. This periodicity has a profound influence on the properties of electrons inside the semiconductor. In this chapter we will examine the electronic properties of semiconductors. The electronic properties are represented by what is called the bandstructure which defines the energy levels that an electron can have in the semiconductor. Essentially all the transport and optical properties of a semiconductor are determined by the bandstructure.

We will also examine approaches that can be exploited to modify the bandstructure and thus modify and optimize optoelectronic devices. These approaches, based on alloying (mixing) of semiconductors and upon the use of heterostructures, are widely used in modern optoelectronics.

2.2   ELECTRONS IN A PERIODIC POTENTIAL: BLOCH THEOREM

$\mathcal{R}$ The electrons are one of the elementary particles of nature with mass 0.91×10^{-30} kg and negative charge with magnitude 1.6×10^{-19} C or 4.8×10^{-10} esu. The mass of the electrons is, of course, a universal constant, and it determines the response of the electrons to external forces.

To understand semiconductor devices, we need to know what happens to electrons when they enter a semiconductor and how they respond to external forces. In a simple minded classical approach, it may appear that as electrons move through the semiconductor, they will suffer scattering from the fixed ions. While this seems intuitively reasonable, it turns out that *electrons moving in periodic structures do not scatter at all!* Only the presence of imperfections causes scattering. *To understand the properties of electrons in semiconductors, we first study their properties in the perfectly periodic crystal. Then we allow for imperfections which cause scattering and affect their transport properties. In this chapter we will study the perfect system. In chapter 3 we will study the effect of imperfections.*

The description of the electron in the semiconductor has to be via the Schrödinger equation

$$\left[\frac{-\hbar^2}{2\,m_0}\nabla^2 + U(r)\right]\psi(r) = E\psi(r) \tag{2.1}$$

where $U(r)$ is the background potential seen by the electrons. Due to the crystalline nature of the material, the potential $U(r)$ has the same periodicity, R, as the lattice

$$U(r) = U(r + R) \tag{2.2}$$

As a result, we expect the *electron probability to be same in all unit cells of the crystal because each cell is identical.* If the potential was random, this would not be the case.

The Bloch theorem gives us the form of the electron wavefunction in a periodic structure and states that the eigenfunctions are the product of a plane wave $e^{i\mathbf{k}\cdot\mathbf{r}}$ times a function $u_k(\mathbf{r})$ which has the *same periodicity as the periodic potential*. Thus

$$\boxed{\psi_k(\mathbf{r}) = e^{i\mathbf{k}\cdot\mathbf{r}}u_k(\mathbf{r})} \tag{2.3}$$

is the form of the electronic function. The "cell" periodic part $u_k(r)$ has the same periodicity as the crystal, i.e.,

$$\boxed{u_k(r) = u_k(r + R)} \tag{2.4}$$

The wavefunction has the property

$$\begin{aligned}\psi_k(r + R) &= e^{ik\cdot(r+R)}u_k(r + R) = e^{ik\cdot r}u_k(r)e^{ik\cdot R} \\ &= e^{ik\cdot R}\psi_k(r)\end{aligned} \tag{2.5}$$

2.2.1 From Atomic Levels to Bands

The Bloch theorem tells us that the electron wavefunction in a crystal has a particularly simple form: it is the product of a cell periodic part and a plane wave. While this greatly simplifies the Schrödinger equation for the electron in the crystal, the problem is still quite complicated. However, the size of the problem is reduced greatly due to the periodicity. Methods like the tight binding method, orthogonalized plane wave method, pseudopotential method, $k \cdot p$ perturbation method, etc., are used to obtain the solution for the electronic levels. These solutions provide the energy of the electron as a function of the quantity k appearing in the Block theorem. This E vs. k relation is called the bandstructure of the semiconductor. In addition, one also obtains the central cell part of the wavefunction which is of great importance in optical transitions.

We will not discuss the derivation of the bandstructure of semiconductors; merely provide the student important facts regarding the bandstructure. To better understand the bandstructure and the makeup of the central cell part of the Bloch state, we start with a brief discussion of an isolated atom. In Fig. 2.1 we show the energy spectra of a typical isolated atom. The spectra is made up of two regions: i) bound states in which the electron is bound to the nucleus by Coulombic interactions, and ii) free state where the electron is free. These two regions are separated by the "vacuum energy" E_{vac} as shown, which is chosen as the zero of energy.

In the bound state, the electron energy is negative with respect to the vacuum level. The allowed energy levels are discrete as shown in Fig. 2.1 and are separated from each other by a region of "forbidden gap". The electron wavefunction in general is given by a form $\psi_{n\ell n}$ where the index n is a positive integer (1, 2, 3 ...), called the principle quantum number, the index ℓ is the angular momentum quantum number and gives the value of the orbital angular momentum of the electron's motion around the nucleus. Its values are $\hbar, 2\hbar, 3\hbar \ldots$. Finally, the index n gives the value of the projection of the angular momentum along the z-axis. It's values can be $0, \pm\hbar, \pm 2\hbar \ldots$.

From atomic physics we have the following:

n=1: The orbital angular momentum can only be zero, i.e., $\ell = 0$ and m = 0. The $\ell = 0$ state is also called an s-state. The s-state is a spherically symmetrical state in space as shown in Fig. 2.2. The n = 1, $\ell = 0$, m = 0 state is denoted by 1s in atomic physics notation.

n=2: ℓ can take a value of 0 or 1. The $\ell = 1$ state is called a p-state and p-states have a dumbbell shaped form in real space as shown in Fig. 2.2. Being a vector, the p-states can have components p_x, p_y and p_z as shown in Fig. 2.2. The n = 2, $\ell = 0$ state is called 2s and n = 2, $\ell = 1$ state is called 2p state.

Higher order states can have ℓ-values of 2 (d-states) or 3 (f-states), etc., For reasons that will become clear, the symmetry of the s-states and p-states is very important since selection rules for optical transitions depend upon these symmetries. We will exploit this information when we discuss optical selection rules and polarization properties of optoelectronic devices later.

If the atoms are brought together to form a crystal, the electron on a given atom starts to see the neighboring nuclei and the discrete levels start to broaden until eventually one starts to get bands as shown schematically in Fig. 2.1.

The allowed energy levels (i.e., the solutions of Eqn. 2.1) are not continuous in general as for the free electron case, but have *regions where there are no allowed energy values*. This means that the electron cannot exist in the semiconductor in these "gaps." To appreciate the presence of these "allowed bands" and "forbidden gaps" let us examine the atomic structure of some of the elements which make up various semiconductors.

IV Semiconductors

$$C \qquad 1s_2 \, \underbrace{2s_2 2p_2}$$

$$Si \qquad 1s_2 2s_2 2p_6 \, \underbrace{3s_2 3p_2}$$

$$Ge \qquad 1s_2 2s_2 2p_6 3s_2 3p_6 3d_{10} \, \underbrace{4s_2 4p_2}$$

III-V Semiconductors

$$Ga \qquad 1s_2 2s_2 2p_6 3s_2 3p_6 3d_{10} \, \underbrace{4s_2 4p_1}$$

$$As \qquad 1s_2 2s_2 2p_6 3s_2 3p_6 3d_{10} \, \underbrace{4s_2 4p_3}$$

$$etc.$$

The braces under each atomic structure represent the electrons in the outermost shell which are bound so loosely to the individual atom that when atoms are brought together to form the crystal, these outmost electrons form a series of allowed bands separated by a forbidden gap as shown in Fig. 2.1.

An important point to note from our discussion of the electronic levels of the atoms making up semiconductors is that the outermost electronic levels (in the valence shell) are all either s-type or p-type. Essentially all electronic and optoelectronic properties of semiconductors are determined by these outermost electrons since the core electrons do not participate in most processes (because they are so tightly bound to the nucleus). As a result, the central cell character of the Bloch functions in semiconductors

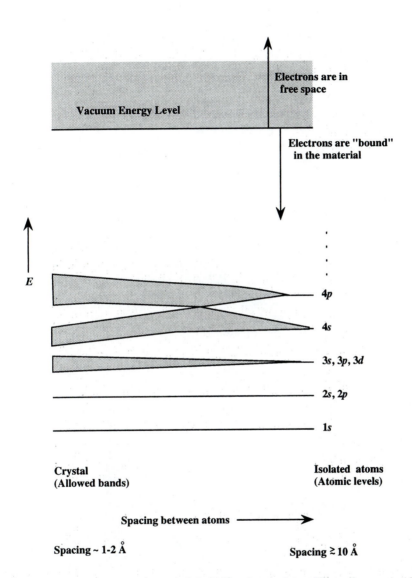

Figure 2.1: A schematic of how allowed and forbidden bands form. When the spacing between atoms is large, the allowed levels are discrete. As the atoms come closer, the electron energy levels interact with each other forming complicated bands. The deep core levels are relatively unaffected, but the higher levels broaden into bands.

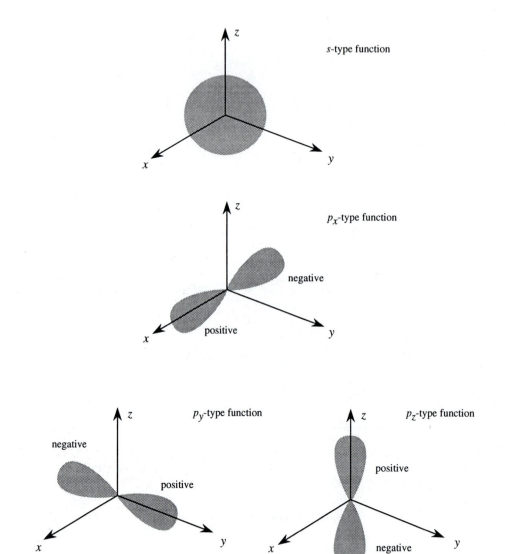

Figure 2.2: The spatial character of the s, p_x, p_y and p_z functions. The s function is isotropic while p_x, p_y, p_z functions have odd parity, i.e., they change sign when $x \to -x, y \to -y, z \to -z$.

are adequately described by *s*- and *p*-type functions. In general, the central cell part can be written for the electrons in a bulk semiconductor as

$$u_k(r) = \sum_{i=1}^{2} \sum_{n=1}^{4} a_{in}(k)\phi_{in} \tag{2.6}$$

where the first summation represents the sum over the two atoms that make up a unit cell of the diamond or zinc blende structure, and the second sum represents the 4 atomic-like functions s, p_x, p_y, p_z discussed above. The coefficients a_{in} represent the relative mixtures of the atomic-like functions in the central cell part of the electronic state. At certain special values of k, the central cell part becomes a pure combination of only *s*-type states or only *p*-type states as will be discussed later. A pure state has a well defined spatial symmetry which is important for "selection rules" in scattering as will be discussed later.

2.2.2 The Crystal Momentum

What is the significance of the wavevector k that appears in the electron wavefunction in the Bloch's theorem? For the free space electron, the k that appears in the electron wavefunction is related to the momentum of the electron by $\mathbf{p} = \hbar\mathbf{k}$. For electrons moving in free space there are two very important laws which are used to describe their properties: i) Newton's second law of motion which tells us how the electron's trajectory evolves in the presence of an external force, and ii) the law of conservation of momentum which allows us to determine the electron's trajectory when there is a collision. We are obviously interested in finding out what the analogous laws are when an electron is inside a crystal and not in free space.

The quantity ħk takes on exactly the same role for the electron in the perfect semiconductor as a "momentum." However, $\hbar k$ only reacts to the external forces such as an electric field *as if* it was the momentum of the electron. It is called crystal momentum because it includes the effect of the atoms in the crystal on the electrons. Let us examine the correspondence of the equations satisfied by the electron crystal momentum and the electron momentum in free space:

Electron momentum relations:

$$\text{Free space} \quad : \quad E = \frac{\hbar^2 k^2}{2\,m_0}$$

$$\text{Crystal} \quad : \quad E = E(k) \tag{2.7}$$

The relation between E and k in the material is the semiconductor bandstructure.

Equation of motion (in the absence of any collisions): The equation of motion

Important high symmetry points

Γ **point:** $k_x = 0 = k_y = k_z$

X point: $k_x = \frac{2\pi}{a}$; $k_y = k_z = 0$

L point: $k_x = k_y = k_z = \frac{\pi}{a}$

a = **lattice constant (cube edge)**

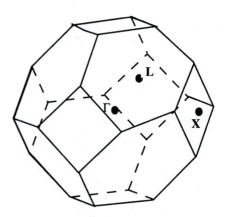

Figure 2.3: The k-vector for the electrons in a crystal is limited to a space called the Brillouin zone. The figure shows the Brillouin zone for the fcc lattice relevant for most semiconductors. The values and notations of certain important k-points are also shown. Most semiconductors have bandedges of allowed bands at one of these points.

is the same for the free electron k and the crystal momentum k

$$\frac{\hbar dk}{dt} = F_{ext} \tag{2.8}$$

Collisions of electrons: Momentum is conserved in collisions just as in free space. In reality, there are some rare scattering processes in semiconductors where momentum is not conserved. However, for most cases of interest, momentum is conserved.

Thus the only effect of the crystal is to modify the E vs. k relation which can be quite complicated for real semiconductors. The vector k for the electron is related to the wavelength λ of the electron wavefunction by

$$k = \frac{2\pi}{\lambda} \tag{2.9}$$

The shortest wavelength that is allowed in a crystal is governed by spacing between the lattice points in space. Thus the wavelength ranges from infinite to lattice spacing distances. The value of k correspondingly varies from 0 to some finite values determined by the lattice spacing. In the fcc lattice which is the lattice for most semiconductors, the k-vector values are thus confined to a volume called the Brillouin zone. This Brillouin zone is shown in Fig. 2.3. The origin of the Brillouin zone is $k = (0,0,0)$ and this point is called the Γ-point. There are other points in the Brillouin zone that occur at high symmetry points. These are all denoted by standard notations and are given in Fig. 2.3. These points are important because the bandedges of the allowed bands generally occur at k-values corresponding to these points.

2.3 METALS, SEMICONDUCTORS AND INSULATORS

$\mathcal{R}$ We have discussed in Section 2.2 that the solution of the electron problem in a crystal gives an E vs. k relationship which has regions of allowed bands separated by forbidden bandgaps. The question now arises: which of these allowed states are occupied by electrons and which are unoccupied? Two important situations arise when we examine the electron occupation of allowed bands: In one case we have a situation where an allowed band is completely filled with electrons while the next allowed band is separated in energy by a gap E_g and is completely empty at 0 K. In a second case, the highest occupied band is only half full (or partially full).

At this point a very important concept needs to be introduced. *When an allowed band is completely filled with electrons, the electrons in the band cannot conduct any current.* This important concept is central to the special properties of semiconductors.

The electrons being Fermions cannot carry any net current in a filled band since an electron can only move into an empty state. One can imagine a net cancellation of the motion of electrons moving one way and those moving the other. Because of this effect, when we have a material in which a band is completely filled, while the next allowed band is separated in energy and empty, the material has, in principle, infinite resistivity and is called an *insulator* or a *semiconductor*. The material in which a band is only half full with electrons has a very low resistivity and is called a *metal*.

The band that is normally filled with electrons at 0 K in semiconductors is called the valence band while the upper unfilled band is called the conduction band. The energy difference between the vacuum level and the highest occupied electronic state in a metal is called the metal work function. The energy between the vacuum level and the bottom of the conduction band is called the electron affinity.

Although in what follows, we are anticipating some results that are to be discussed later, it is important to summarize a very important difference between metals and semiconductors. *The metals have a very high conductivity because of the very large number of electrons that can participate in current transport. It is difficult to alter the conductivity of metals in any simple manner as a result of this. On the other hand, semiconductors have zero conductivity at 0 K and quite low conductivity at finite temperatures, but it is possible to alter their conductivity by orders of magnitude.* This is the key reason why semiconductors can be used for active devices.

We have already discussed that the solution of the Schrödinger equation leads to the bandstructure of the semiconductor. The top of the valence band of most semiconductors occurs at $k=0$, i.e., at effective momentum equal to zero. A typical bandstructure of a semiconductor near the top of the valence band is shown in Fig. 2.4.

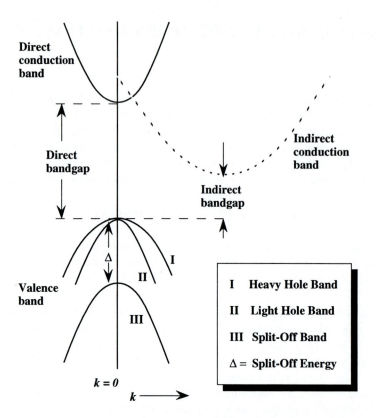

Figure 2.4: Schematic of the valence band, direct bandgap and indirect bandgap conduction bands. The conduction band of the direct gap semiconductor is shown in the solid line while the conduction band of the indirect semiconductor is shown in the dashed line. The curves I, II, III in the valence band are called heavy hole, light hole and split-off hole states, respectively. The reason why valence band states are called holes will be discussed in Section 2.6.

The bottom of the conduction band in some semiconductors occurs at $k=0$. Such semiconductors are called direct bandgap materials and, for reasons discussed in Chapter 4, are "optically active." Semiconductors such as GaAs, InP, InGaAs, etc., are direct bandgap semiconductors. In other semiconductors, the bottom of the conduction band does not occur at the $k=0$ point but at certain other points. Such semiconductors are called indirect semiconductors. Examples are Si, Ge, AlAs, etc. These materials have very weak interactions with light and cannot be used for efficient optical devices. The reasons are based on momentum conservation rules in optical transitions which make it difficult to have strong transitions in indirect semiconductors.

2.4 PROPERTIES OF CONDUCTION AND VALENCE BANDEDGE STATES

$\longrightarrow$ The Bloch theorem tells us that the wavefunction of the electron is made up of a plane wave part and a central cell part that is periodic. When the appropriate Schrödinger equation is solved, one obtains both the energy of the electron as a function of the wavevector k and the central cell part. The character of the central cell part is very important in optoelectronic devices. Important selection rules determining the strength of the photon-electron interactions depend upon the nature of the central cell function.

As noted in Section 2.2, the valence band and conduction bands arise from the outermost shell electron states of the atoms making up the semiconductor. In essentially all semiconductors, the outermost shell electron states of the atoms have an s-type or p-type character. As a result, the central cell part of the wavefunction is made up of a combination of s and p-type states.

In atomic physics terminology, the electron in the s-type state has zero orbital angular momentum and the function is isotropic in space. The p-type state, on the other hand, has an orbital angular momentum of $\hbar$ and is anisotropic in space as shown in Fig. 2.2. There are three independent p-functions: p_x, p_y, and p_z.

In addition to the orbital angular momentum of an electron state, one also has an angular momentum due to the spin of the electron. An electron has a spin $\hbar/2$ and can have a magnitude $+ \hbar/2$ or $- \hbar/2$. Often the spin is denoted by $+ 1/2$ or $- 1/2$, assuming that it is expressed in units of $\hbar$. It is useful to express the states of the electron not in terms of orbital angular momentum and spin but in terms of *total angular momentum*. The total angular momentum states are related to the orbital angular momentum and spin states by a simple transformation which is given by the following relation for the states with an orbital angular momentum of $\hbar$ (the p-states):

$$\Phi_{3/2,3/2} = \frac{-1}{\sqrt{2}} \left(|p_x> +i|p_y> \right) \uparrow$$

$$\Phi_{3/2,-3/2} = \frac{1}{\sqrt{2}} \left(|p_x> -i|p_y> \right) \downarrow \qquad (2.10)$$

$$\Phi_{3/2,1/2} = \frac{-1}{\sqrt{6}} \left[\left(|p_x> +i|p_y> \right) \downarrow -2|p_z> \uparrow \right]$$

$$\Phi_{3/2,-1/2} = \frac{1}{\sqrt{6}} \left[\left(|p_x> -i|p_y> \right) \uparrow +2|p_z> \downarrow \right] \qquad (2.11)$$

$$\Phi_{1/2,1/2} = \frac{-1}{\sqrt{3}} \left[\left(|p_x> +i|p_y> \right) \downarrow +|p_z> \uparrow \right]$$

$$\Phi_{1/2,-1/2} = \frac{-1}{\sqrt{3}} \left[\left(|p_x> -i|p_y> \right) \uparrow +|p_z> \downarrow \right] \qquad (2.12)$$

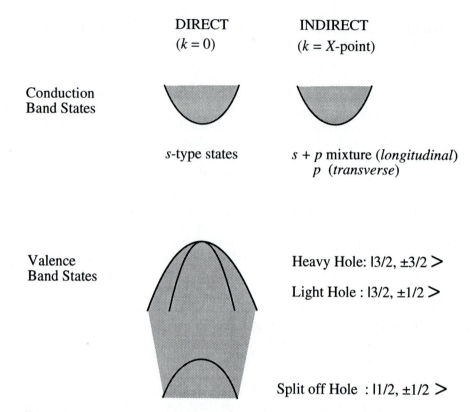

Figure 2.5: The character of the valence and conduction band states in a typical semiconductor.

Here the subscripts of the total angular momentum states represent the total angular momentum and its projection along the z-axis.

Having established the general character of electron states in semiconductors, we will now summarize the character of the central cell part of the electron states at the bottom of the conduction band and the top of the valence band. In most semiconductor devices, the device performance is controlled by these bandedge states. In Fig. 2.5 we show a generic form of the bandedge bandstructure.

The central cell part of the electrons at the bottom of the conduction band of direct bandgap materials is purely s-type. This is a very important feature of direct gap semiconductors and will be exploited in calculating the absorption and emission of photons from such materials. The central cell part of the conduction bandedge states in indirect gap semiconductors is a mixture of s and p states.

As shown in Fig. 2.5, the valence band has three branches near the top of the band. These are called the heavy hole, light hole and split-off branches. Counting the spin degeneracy (i.e, spin up (+ 1/2) and spin down (− 1/2) have the same energy), each branch is 2-fold degenerate. The central cell character of these states is p-type. In terms of the total angular momentum states, we use the following notation for these states:

Heavy Hole State: $\Phi_{(3/2,\pm3/2)}$

Light Hole State: $\Phi_{(3/2,\pm1/2)}$

Split Off State: $\Phi_{(1/2,\pm1/2)}$

where the makeup of these states in terms of the p-component has already been given.

Due to the simplicity of the conduction bandedge, the E vs. k relation near the conduction bandedge is given by simple parabolic relations that will be discussed for direct and indirect materials later in this section. However, the problem is more complicated for the valence band because of the very strong interaction of the heavy hole and light hole states. A very useful description of the bandstructure is given by the Kohn-Luttinger formulation according to which the bandstructure is given by the solution of the following equation:

$$
-\begin{bmatrix} H_{hh} & b & c & 0 \\ b^* & H_{lh} & 0 & c \\ c^* & 0 & H_{lh} & -b \\ 0 & c^* & -b^* & H_{hh} \end{bmatrix} \begin{bmatrix} \Phi_{(3/2,3/2)} \\ \Phi_{(3/2,1/2)} \\ \Phi_{(3/2,-1/2)} \\ \Phi_{(3/2,-3/2)} \end{bmatrix} = E \begin{bmatrix} \Phi_{(3/2,3/2)} \\ \Phi_{(3/2,1/2)} \\ \Phi_{(3/2,-1/2)} \\ \Phi_{(3/2,-3/2)} \end{bmatrix} \tag{2.13}
$$

where the coefficients H_{hh}, H_{lh}, c, b are

$$
H_{hh} = \frac{\hbar^2}{2\,m_0}\left[(\gamma_1 + \gamma_2)\left(k_x^2 + k_y^2\right) + (\gamma_1 - 2\gamma_2)\,k_z^2\right]
$$

$$
H_{lh} = \frac{\hbar^2}{2\,m_0}\left[(\gamma_1 - \gamma_2)\left(k_x^2 + k_y^2\right) + (\gamma_1 + 2\gamma_2)\,k_z^2\right]
$$

$$
b = \frac{-\sqrt{3}i\hbar^2}{m_0}\gamma_3\left(k_x - ik_y\right)k_z
$$

$$
c = \frac{\sqrt{3}\hbar^2}{2\,m_0}\left[\gamma_2\left(k_x^2 - k_y^2\right) - 2i\gamma_3 k_x k_y\right] \tag{2.14}
$$

The quantities γ_1, γ_2 and γ_3 are the Kohn-Luttinger parameters. These can be obtained by fitting experimentally obtained hole masses. The eigenstates of the Kohn-Luttinger Hamiltonian are the angular momentum states which are related to the p states. The Kohn-Luttinger matrix can be extended very easily to the quantum well problem in a manner to be discussed in Section 2.9. It may be noted that the 4 × 4 matrix equation can be reduced to two 2 × 2 matrix equations , and the E vs. k relations can be solved analytically (see Problem 13 and the paper by Broido and Sham (1985)).

Near the valence bandedge, the HH, LH states can be represented by an energy momentum relation of the following form

$$E_v(k) = \frac{-\hbar^2}{2\,m_0} \left\{ Ak^2 \pm \left[Bk^4 + C\left(k_x^2 k_y^2 + k_y^2 k_z^2 + k_z^2 k_x^2\right)\right]^{1/2}\right\} \tag{2.15}$$

where A, B, and C are dimensionless parameters. The upper sign is for HH and the lower one is for the LH states. As an example for Si, we have $A = 4.0$, $B = 1.1$, and $C = 4.1$ with $m_{hh}^* = 0.49\,m_0$, $m_{lh}^* = 0.16\,m_0$. For Ge we have $A = 13.1$, $B = 8.3$, and $C = 12.5$, giving $m_{hh}^* = 0.29\,m_0$, $m_{lh}^* = 0.047\,m_0$.

Near the conduction bandedges (occurring at $k = k_o$) it is usually possible to represent the bandstructure by a simple relation of the form

$$E(k, k_o) = E(k_o) + \hbar^2 \sum_{i=x,y,z} \frac{(k_i - k_{oi})^2}{2m_i^*} \tag{2.16}$$

where the index i represents the x, y, z components of k or k_o. For direct bandgap materials $k_o = (0,0,0)$.

If the bandstructure is isotropic as is the case for direct gap semiconductors, the relation becomes

$$E(k) = E_c + \frac{\hbar^2 k^2}{2m^*} \tag{2.17}$$

where E_c is the conduction bandedge, and the bandstructure is a simple parabola with equal energy surfaces being the surfaces of a sphere.

For indirect materials like Si, the bottom of the conduction band occurs at six equivalent points: $\frac{2\pi}{a}$ (0.85,0,0), $\frac{2\pi}{a}$ (0,0.85,0), $\frac{2\pi}{a}$ (0,0,0.85) and their inverses. The energy momentum relation has the form

$$E(k) = E_c + \frac{\hbar^2 k_l^2}{2m_l^*} + \frac{\hbar^2 k_t^2}{2m_t^*} \tag{2.18}$$

where k_l is the longitudinal part of k (i.e., parallel to the k-value at the bandedge) and k_t is the transverse part measured from the conduction bandedge k-value. The constant energy surface of the bandstructure is an ellipsoid.

Near the bandedges, the electrons in semiconductors behave as if they have a mass m^* which is called the effective mass. For direct gap semiconductors, the effective mass of the electrons at the conduction bandedge is given by the following approximate relation which results from detailed bandstructure calculations

$$\frac{1}{m^*} \cong \frac{1}{m_0} + \frac{2p_{cv}^2}{E_g} \tag{2.19}$$

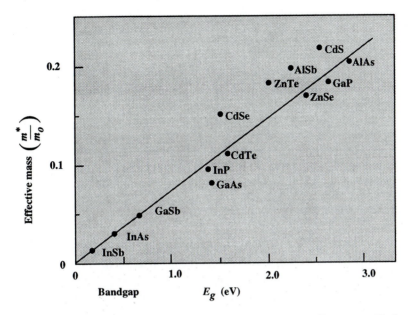

Figure 2.6: Electron effective mass, m* as a function of the lowest-direct gap E_g for various compounds semiconductors. It is interesting to note that the effective mass decreases as the bandgap decreases.

where, for most semiconductors

$$\frac{p_{cv}^2}{m_0} \cong 20.0 \ eV \qquad (2.20)$$

where m_0 is the free electron mass.

The conduction band effective mass thus decreases rapidly with decreasing bandgap as shown in Fig. 2.6.

Near the top of the valence band, as noted earlier (see Fig. 2.4), there are three important curves. The heavier mass band is called the heavy hole band (I in Fig. 2.4). The second lighter band is called light hole band (II in Fig. 2.4) and the third band, separated by an energy Δ, is called the split-off band (III in Fig. 2.4). The masses of the valence band electrons are usually much heavier than those in the conduction band and are also negative. *The reason we call the valence band states "holes" will be discussed in Section 2.6.* In general, the effective mass is defined by the relation

$$\frac{1}{m^*} = \frac{1}{\hbar^2} \frac{d^2E}{dk^2} \qquad (2.21)$$

EXAMPLE 2.1 An electron in GaAs is subjected to a 1 ps pulse of electric field of magnitude

10 kV/cm. Assuming that there is no scattering, calculate the energy of the electron at the end of the pulse.

The electrons obey the equation (in the absence of scattering)

$$\frac{\hbar dk}{dt} = eF$$

After a time t the change in the electron momentum is

$$\hbar k_f = eFt$$

and the change in energy is

$$
\begin{aligned}
\Delta E &= \frac{(\hbar k_f)^2}{2m^*} = \frac{(eFt)^2}{2m^*} \\
&= \frac{\left[1.6 \times 10^{-19} \ (C) \times 10^6 \ (V/m) \times 10^{-12} \ s\right]^2}{2 \times 0.067 \times 0.91 \times 10^{-30} \ kg} \\
&= 2.1 \times 10^{-19} \ J \\
&= \frac{2.1 \times 10^{-19}}{1.6 \times 10^{-19}} = 1.31 \ eV
\end{aligned}
$$

The scattering of electrons does not allow such high energy to be reached in steady state. In fact in steady state conditions, due to scattering, the energy of the electron on an average is ~ 0.3 eV.

2.5   DENSITY OF STATES

$\longrightarrow$ The electronic states in a semiconductor have a plane wave form modulated by the central cell part. Also, as we have seen near the bandedges, it is possible to write the energy momentum relation in the simple parabolic form

$$E = \frac{\hbar^2 k^2}{2m^*} \tag{2.22}$$

where the quantity m^* is the effective mass. If we consider a volume $L^3 = V$ of the material, suppressing the cell periodic part of the wavefunction, we have

$$\psi(r) = \frac{1}{\sqrt{V}} e^{\pm i k \cdot r} \tag{2.23}$$

where the factor $\frac{1}{\sqrt{V}}$ comes because we wish to have one electron per volume V or

$$\int_V d^3 r \mid \psi(r) \mid^2 = 1 \tag{2.24}$$

We assume that the volume V is a cube of side L.

To correlate with physical conditions that we may want to describe, there are two kinds of boundary conditions that are imposed on the wavefunction. In the first one the wavefunction is considered to go to zero at the boundaries of the volume. In this case, the wave solutions are of the form $\sin(k_x x)$ or $\cos(k_x x)$, etc., and k-values are restricted to the positive values,

$$k_x = \frac{\pi}{L}, \frac{2\pi}{L}, \frac{3\pi}{L} \ldots \tag{2.25}$$

The standing wave solution is often used to describe stationary electrons confined in finite regions such as quantum wells discussed earlier. For describing moving electrons, the boundary condition used is known as a periodic boundary condition. Even though we focus our attention on a finite volume V, the wave can be considered to spread in all space as we conceive the entire space was made up of identical cubes of sides L. Then

$$\begin{aligned}
\psi(x, y, z + L) &= \psi(x, y, z) \\
\psi(x, y + L, z) &= \psi(x, y, z) \\
\psi(x + L, y, z) &= \psi(x, y, z)
\end{aligned} \tag{2.26}$$

Because of the boundary conditions the allowed values of k are (n are integers— positive and negative)

$$k_x = \frac{2\pi n_x}{L}; k_y = \frac{2\pi n_y}{L}; k_z = \frac{2\pi n_z}{L} \tag{2.27}$$

If L is large, the spacing between the allowed k values is very small. It is useful to discuss the *volume in k-space that each electronic state occupies*. As can be seen from Fig. 2.7, this volume is (in 3-dimensions)

$$\left(\frac{2\pi}{L}\right)^3 = \frac{8\pi^3}{V} \tag{2.28}$$

If Ω is a volume of k-space, the number of electronic states in this volume is k-space

$$\boxed{\frac{\Omega V}{8\pi^3}} \tag{2.29}$$

2.5.1 Density of States for a 3-Dimensional System

We will now use the discussion of the previous subsection to derive the extremely important concept of density of states. Although we will use the periodic boundary conditions to obtain the density of states, the stationary conditions lead to the same result.

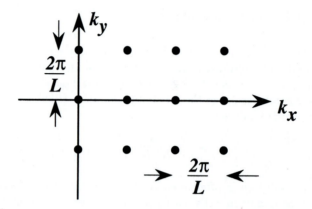

Figure 2.7: k-space volume of each electronic state. The separation between the various allowed components of the k-vector is $\frac{2\pi}{L}$.

The concept of density of states is extremely powerful, and important physical properties such as optical absorption, transport, etc., are intimately dependent upon this concept. Density of states is the number of available electronic states *per unit volume per unit energy* around an energy E. If we denote the density of states by $N(E)$, the number of states in a unit volume in an energy interval dE around an energy E is $N(E)dE$. To calculate the density of states, we need to know the dimensionality of the system and the energy vs. k relation that the electrons obey. We assume that we have a parabolic E-k relation:

$$E = \frac{\hbar^2 k^2}{2m^*} \tag{2.30}$$

The energies E and $E + dE$ are represented by surfaces of spheres with radii k and $k + dk$ as shown in Fig. 2.8. In a 3-dimensional system, the k-space volume between vector k and $k + dk$ is $4\pi k^2 dk$. We have shown in Eqn. 2.29 that the k-space volume per electron state is $(\frac{2\pi}{L})^3$. Therefore, the number of electron states in the region between k and $k + dk$ is

$$\frac{4\pi k^2 dk}{8\pi^3} V = \frac{k^2 dk}{2\pi^2} V \tag{2.31}$$

Denoting the energy and energy interval corresponding to k and dk as E and dE, we see that the number of electron states between E and $E + dE$ per unit volume are

$$N(E)dE = \frac{k^2 dk}{2\pi^2} \tag{2.32}$$

and since

$$E = \frac{\hbar^2 k^2}{2m^*} \tag{2.33}$$

$$k^2 dk = \frac{\sqrt{2}(m^*)^{3/2}E^{1/2}dE}{\hbar^3} \tag{2.34}$$

and

$$N(E)dE = \frac{(m^*)^{3/2}E^{1/2}dE}{\sqrt{2}\pi^2\hbar^3} \tag{2.35}$$

An electron is a "Fermion" and can have two possible spin states with a given energy. Accounting for spin, the density of states obtained above is simply multiplied by 2:

$$\boxed{N(E) = \frac{\sqrt{2}(m^*)^{3/2}E^{1/2}}{\pi^2\hbar^3}} \tag{2.36}$$

2.5.2 Density of States in Sub-3-Dimensional Systems

Let us now consider a 2D system, a concept that has become a reality with use of quantum wells. Similar arguments (see Fig. 2.8) show that the density of states for a parabolic band is (including spin)

$$\boxed{N(E) = \frac{m^*}{\pi\hbar^2}} \tag{2.37}$$

Finally, in a 1D system or a "quantum wire," the density of states is (including spin)

$$\boxed{N(E) = \frac{\sqrt{2}m^{*1/2}}{\pi\hbar}E^{-1/2}} \tag{2.38}$$

We note that as the dimensionality of the system changes, the energy dependence of the density of states also changes. As shown in Fig. 2.9, for a 3-dimensional system we have a $E^{1/2}$ dependence; for a 2-dimensional system we have no energy dependence; and for a 1-dimensional system we have an $E^{-1/2}$ dependence.

One may wonder why we are interested in 2- and 1-dimensional systems since our life is 3-dimensional. It is possible, indeed, to create semiconductor heterostructures where the electron feels it is in a 2-dimensional space or even in a 1-dimensional space.

For direct bandgap semiconductors, the density of states mass for the conduction band is the same as the effective mass discussed above

$$\boxed{m^*_{dos} = m^*} \tag{2.39}$$

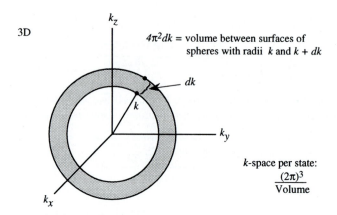

3D

$4\pi^2 dk$ = volume between surfaces of
spheres with radii k and $k + dk$

dk

k

k-space per state:
$$\frac{(2\pi)^3}{\text{Volume}}$$

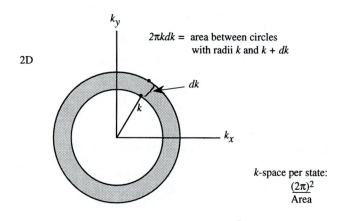

$2\pi k dk$ = area between circles
with radii k and $k + dk$

2D

dk

k

k-space per state:
$$\frac{(2\pi)^2}{\text{Area}}$$

1D

$2dk = k$ space between k and $k + dk$

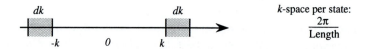

dk dk

$-k$ 0 k

k-space per state:
$$\frac{2\pi}{\text{Length}}$$

Figure 2.8: Geometry used to calculate density of states in 3-, 2-, and 1-dimensions. By finding the k-space volume in an energy interval between E and $E + dE$, one can find out how many allowed states there are.

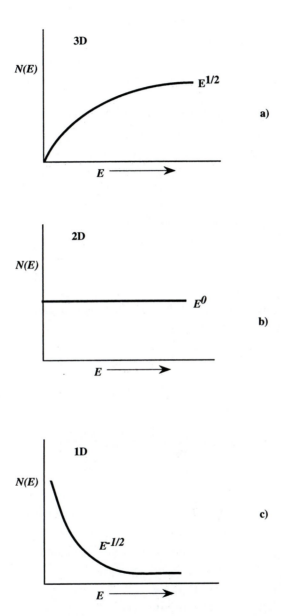

Figure 2.9: Variation in the energy dependence of the density of states in (a) 3-dimensional, (b) 2-dimensional, and (c) 1-dimensional systems. The energy dependence of the density of states is determined by the dimensionality of the system. This feature is exploited in advanced semiconductor devices.

For materials like silicon in which the effective mass is different in various directions, the density of states mass for one valley is

$$\boxed{m_{dos}^* = (m_1 m_2 m_3)^{1/3}} \tag{2.40}$$

where m_1, m_2 and m_3 are the effective masses in the three principle directions. For silicon

$$m_\ell^* = m_1 \; ; \; m_t^* = m_2 = m_3 \tag{2.41}$$

Since silicon has 6 conduction bandedge valleys, the density of states calculated for one valley must be multiplied by six to obtain the total density of states.

In the case of valance band masses, there is the heavy hole and light hole mass. One can define the effect of both these bands by an effective density of states mass given by (the mass for holes to be discussed next is positive)

$$m_{dos}^{*3/2} = \left(m_{\ell h}^{*3/2} + m_{hh}^{*3/2} \right) \tag{2.42}$$

EXAMPLE 2.2 Calculate the density of states of a 3D system and a 2D system at an energy of 0.1 eV, if the effective mass is m_0.

The density of states in a 3D system (including the spin of the electron) is given by (E is the energy in Joules)

$$
\begin{aligned}
N(E) &= \frac{\sqrt{2}(m_0)^{3/2} E^{1/2}}{\pi^2 \hbar^3} \\
&= \frac{\sqrt{2}(0.91 \times 10^{-30} kg)(E^{1/2})}{\pi^2 (1.05 \times 10^{-34} J - s)^3} \\
&= 1.07 \times 10^{56} E^{1/2} J^{-1} m^{-3}
\end{aligned}
$$

Expressing E in eV and the density of states in the commonly used units of eV^{-1} cm^{-3}, we get

$$
\begin{aligned}
N(E) &= 1.07 \times 10^{56} \times (1.6 \times 10^{-19})^{3/2} (1.0 \times 10^{-6}) \\
&= 6.8 \times 10^{21} E^{1/2} \; eV^{-1} \; cm^{-3} \\
\text{at } E &= 0.1 \; eV \text{ we get} \\
N(E) &= 2.15 \times 10^{21} eV^{-1} cm^{-3}
\end{aligned}
$$

For a 2D system the density of states is independent of energy and is

$$N(E) = \frac{m_0}{\pi \hbar^2} = 4.21 \times 10^{14} \; eV^{-1} \; cm^{-2}$$

EXAMPLE 2.3. Due to k-conservation rules, the optical transitions in semiconductors are "vertical," i.e., the initial electron k-value in the valence band and the final electron k-value in the conduction band are equal. How exact is this "vertical transition rule" in GaAs ($E_g = 1.5\ eV$), a HgCdTe alloy with bandgap of 0.1 eV and a "blue laser" CdSe material with bandgap of 2.8 eV?

In a scattering of an electron from the valence band to conduction band (or absorption of a photon), the energy and momentum have to be conserved. Since the photon energy is roughly equal to the bandgap, the corresponding k-vector of the photon is

$$k_{ph} = \frac{2\pi}{\lambda_{ph}} = \frac{(\hbar\omega)}{\hbar v} = \frac{E_g}{\hbar v}$$

where v is the velocity of light in the material. For GaAs the photon k-vector is

$$
\begin{aligned}
k_{ph} &\sim \frac{1.5 \times 1.6 \times 10^{-19}\ J}{10^8\ m/s \times 1.05 \times 10^{-34}\ J-s} \\
&= 2.29 \times 10^7\ m^{-1}
\end{aligned}
$$

This is essentially negligible on the scale of k used in the bandstructure which ranges from 0 to $\sim 10^{10}\ m^{-1}$. The transitions thus appear to be "vertical." The k_{ph} values for a HgCdTe alloy ($E_g = 0.1$ eV) is $1.53 \times 10^6 m^{-1}$ and for CdSe ($E_g = 2.8$ eV) is $4.27 \times 10^7 m^{-1}$.

2.6   HOLES IN SEMICONDUCTORS

$\mathcal{R}$ As noted in the previous section, semiconductors are defined as materials in which the valence band is full of electrons and the conduction band is empty at 0 K. At finite temperatures some of the electrons leave the valence band and occupy the conduction band. The valence band is then left with some unoccupied states. Let us consider the situation as shown in Fig. 2.10 where an electron with momentum $\mathbf{k}_e$ is missing from the valence band.

When all the valence band states are occupied, the sum over all wavevector states is zero, i.e.,

$$\sum \mathbf{k}_i = 0 = \sum_{\mathbf{k}_i \neq \mathbf{k}_e} \mathbf{k}_i + \mathbf{k}_e \tag{2.43}$$

This result is just an indication that there are as many positive k states occupied as negative. Now in the situation where the electron at wavevector $\mathbf{k}_e$ is missing, the total wavevector is

$$\sum_{\mathbf{k}_i \neq \mathbf{k}_e} \mathbf{k}_i = -\mathbf{k}_e \tag{2.44}$$

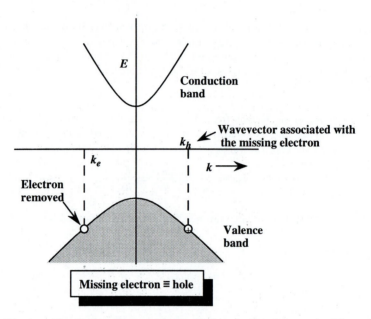

Figure 2.10: Diagram illustrating the wavevector of the missing electron k_e. The wavevector of the system with the missing electron is $-k_e$, which is associated with the hole.

The missing state is called a hole and the wavevector of the system $-\mathbf{k}_e$ is attributed to it. It is important to note that the electron is missing from the state $\mathbf{k}_e$ and the momentum associated with the hole is at $-\mathbf{k}_e$. The position of the hole is depicted as that of the missing electron. But in reality the hole wavevector $\mathbf{k}_h$ is $-\mathbf{k}_e$, as shown in Fig. 2.10.

$$\mathbf{k}_h = -\mathbf{k}_e \tag{2.45}$$

Note that the hole is a representation for the valence band with a missing electron. As discussed earlier, if the electron is not missing the valence band electrons cannot carry any current. However, if an electron is missing the current flow is allowed. If an electric field is applied, all the electrons move in the direction opposite to the electric field. This results in the unoccupied state moving in the field direction. *The hole thus responds as if it has a positive charge.* It therefore responds to external electric and magnetic fields F and B, respectively, according to the equation of motion

$$\hbar \frac{dk_h}{dt} = e\left[F + v_h \times B\right] \tag{2.46}$$

where $\hbar k_h$ and v_h are the momentum and velocity of the hole.

Thus, the equation of motion of holes is that of a particles with a *positive* charge e. The mass of the hole has a positive value, although the electron mass in its valence band is negative. For most semiconductors, the split off band energy shown in Fig.2.4 is quite large, therefore there is a negligible hole in this band. As a result it is sufficient to only consider the heavy hole and the light hole bands. It is important to point out that the terms heavy hole and light hole should not always be associated with the effective mass of the hole states. Under certain conditions (e.g. when the material is under strain — a topic we will discuss later) the valence band mass can be altered considerably. It is then possible that the light hole mass can be actually heavier than the heavy hole mass in some directions. The term heavy hole and light hole should in general be associated with the angular momentum state of the band as discussed earlier.

When we discuss the conduction band properties of semiconductors we refer to electrons, but when we discuss the valence band properties, we refer to holes. This is because in the valence band, only the missing electrons or holes lead to charge transport and current flow.

2.7 BANDSTRUCTURES OF SOME SEMICONDUCTORS

$\mathcal{R}$ We will now examine special features of some semiconductors. Of particular interest are the bandedge properties since they dominate the transport and optical properties. We will examine some important semiconductors in this section. The optoelectronic technology relies on both elemental and compound semiconductors. Thus while silicon is the material of choice for electronic applications, materials like GaAs, AlAs, InAs, InP, GaSb, HgTe, CdTe etc. all play an important role in optoelectronics.

Silicon
Silicon is the unchallenged material of choice for electronic products. The bandstructure of Si is shown in Fig. 2.11a. The bandgap is 1.1 eV with the bottom of the conduction band occurring at $k = (0.85\frac{2\pi}{a}, 0, 0)$ and the five other equivalent points, where a is the lattice constant (5.43 Å). The bandstructure near the conduction band minima is (k is measured from the bandedge value)

$$E(k) = \frac{\hbar^2 k_l^2}{2m_l^*} + \frac{\hbar^2 k_t^2}{2m_t^*} \tag{2.47}$$

where $m_l^* = 0.98\ m_0$ and $m_t^* = 0.19\ m_0$, and gives ellipsoid constant energy surfaces.

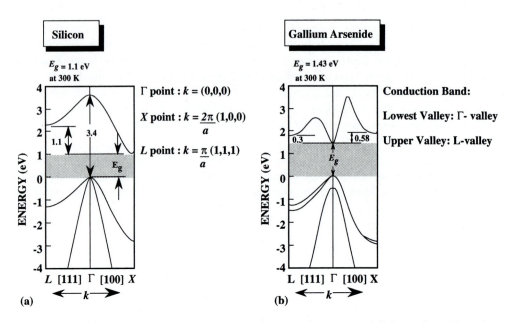

Figure 2.11: (a) Bandstructure of Si. Although the bandstructure of Si is far from ideal, having an indirect bandgap, high hole masses and small spin-orbit splitting, processing related advantages make Si the premier semiconductor for consumer electronics. (b) Bandstructure of GaAs.

Being an indirect material, Si has very poor optical properties and cannot be used to make lasers. The reason for this is the need for "vertical k" transitions in optical processes due to momentum conservations (see Example 2.3). The details of optical processes will be given in Chapter 4. The hole transport properties of Si are also quite poor, since the hole masses are quite large. However, for electronic devices, silicon is the material of choice because of its highly reliable processing technology.

GaAs
The bandstructure of GaAs is shown in Fig. 2.11b. The bandgap is direct which is the chief attraction of GaAs. The direct bandgap ensures excellent optical properties of GaAs as well as superior electron transport in the conduction band. The bandedge E vs. $\mathbf{k}$ relation is quite isotropic leading to spherical equal energy surfaces. The bandstructure can be represented by the relation

$$E = \frac{\hbar^2 k^2}{2m^*} \tag{2.48}$$

with $m^* = 0.067\ m_0$. A better relationship is the non-parabolic approximation

$$\frac{\hbar^2 k^2}{2m^*} = E(1 + \alpha E) \tag{2.49}$$

with $\alpha = 0.67\ eV^{-1}$. The values of the hole masses are $m^*_{hh} = 0.45\ m_0$; $m^*_{lh} = 0.1\ m_0$.

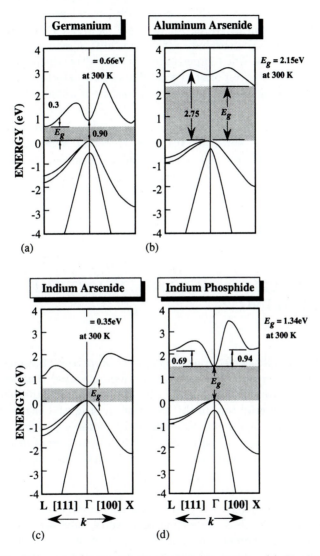

Figure 2.12: (a) Bandstructure of Ge. (b) Bandstructure of AlAs. (c) Bandstructure of InAs. Since no good substrate matches InAs directly, it is often used as an alloy (InGaAs, InAlAs, etc.,) for devices. (d) Bandstructure of InP. InP is a very important material for high speed devices as well as a substrate and barrier layer material for semiconductor lasers.

The bandstructure of several other semiconductors is shown in Fig. 2.12 along with brief comments about their important properties. It is also important to note that the bandgap of semiconductors, in general, decreases as the temperature increases. The temperature dependence of the bandgap for several semiconductors is given in Appendix B. The bandgap of GaAs, for example, is 1.51 eV at 0 K and 1.43 eV at room temperature. These changes have very important consequences for both electronic and optoelectronic devices. The temperature variation alters the laser frequency in solid state lasers, and alters the response of modulators and detectors. For most materials the decrease in the bandgap around room temperature is about $0.5\ meV/K$.

The bandstructure and bandgap picture we have discussed so far is called the single electron picture. It represents the case where the valence band is filled with electrons and the conduction band is empty. When electrons and holes are injected into the conduction and valence bands respectively, the bandgap is slightly reduced as discussed in Chapter 4.

2.8   MODIFICATION OF BANDSTRUCTURE BY ALLOYING

$\longrightarrow$ Since essentially all the electronic and optical properties of semiconductor devices are dependent upon the bandstructure, an obvious question that arises is—can the bandstructure of a material be changed? The ability to tailor the bandstructure can obviously become a powerful tool in the hands of an insightful engineer. Novel devices can be conceived and designed for superior and tailorable performance. The answer to the question above is an emphatic yes. The bandstructure of semiconductors can, indeed, be changed and over the last decade this has become one of the driving forces in semiconductor physics.

In principle, many physical phenomena can modify the electronic bandstructures but we will focus on three important ones, since these are widely used for band tailoring. These three phenomena involve: i) alloying of two or more semiconductors; ii) use of heterostructures to cause quantum confinement or formation of "superlattices;" iii) use of strained epitaxy. These three concepts are increasingly being used for improved performance electronic and optical devices and their importance is expected to become greater with each passing year. In this section we will briefly examine the effect of alloying on the bandstructure.

When two semiconductors A and B are mixed via an appropriate growth technique, one has the following properties of the alloy:

a) The crystalline structure of the lattice: In most semiconductors the two (or more) components of the alloy have the same crystal structure so that the final alloy also has the same crystalline structure. For the same lattice structure materials the lattice constant obeys the Vegard's law for the alloy $A_x B_{1-x}$

$$\boxed{a_{alloy} = xa_A + (1-x)a_B}$$ (2.50)

b) Bandstructures of alloys: The bandstructure of alloys is difficult to calculate in principle since alloys are *not perfect crystals even if they have a perfect lattice*. This is because the atoms are placed randomly and not in any periodic manner. A simple approach used for the problem is called the virtual crystal approximation according to which the bandstructure of the alloy $A_x B_{1-x}$ is simply the weighted bandstructure of the individual bandstructures of A and B. Thus the bandgap is given by

$$\boxed{E_g^{alloy} = xE_g^A + (1-x)E_g^B}$$ (2.51)

In most alloys, however, there is a bowing effect arising from the increasing disorder due to the alloying. One usually defines the bandgap described by the relation

$$E_g^{alloy} = a + bx + Cx^2$$ (2.52)

where C is the bowing parameter. In the Appendix B we show how the bandgaps of various materials combinations change as alloys are made. Most properties of alloys can be approximately obtained by a linear averaging of the individual properties of the components of the alloy.

The relations given above for the lattice constant and bandgap are strictly valid if there is a good "mixing" in the alloy formation process. Thus, if an alloy $A_x B_{1-x}$ is grown, the probability on an average that an A type atom is surrounded by a B type atom should be $(1-x)$ and the probability that a B type atom has an A type register should be x. If this is true, the alloy is called a random alloy. If, on the other hand, the probability that an A type atom is next to a B type atom is smaller than x, the alloy is clustered or phase separated.

It is essential that an alloy not be phase separated, otherwise the material is not very useful for optoelectronic devices and cannot be used in any reliable device process. Certain semiconductor combinations are not allowed by thermodynamics to produce a miscible (or random) alloy at some compositions. The key reason for this is that the sum of the bond energies between A type atoms and B type atoms is larger than twice the bond energies between A and B type bonds. Thus, in the lowest free energy the system prefers to be segregated. The growth of such alloys is quite difficult if a miscible state is to be reached and non-equilibrium growth approaches are used to overcome the "natural dislike" of such materials for each other.

EXAMPLE 2.4 Calculate the bandgap of $Al_{0.3}Ga_{0.7}As$ and $Al_{0.6}Ga_{0.4}As$. Use the virtual crystal approximation with the following values for the conduction band energies measured from the top of the valence band (at 300 K): 1.43 eV and 1.91 eV for the GaAs Γ-point and X-point respectively. For Al As the corresponding values are 2.75 eV and 2.15 eV. In the virtual crystal approximation one calculates the position of an energy point in an alloy by simply taking the *weighted average of the same energy points in the alloy components.*

For the $Al_{0.3}Ga_{0.7}As$ alloy, the Γ- and X-point energies are

$$
\begin{aligned}
E(\Gamma - \text{point}) &= 2.75(0.3) + (1.43)(0.7) = 1.826 \ eV \\
E(X - \text{point}) &= 2.15(0.3) + (1.91)(0.7) = 1.982 \ eV
\end{aligned}
$$

For the $Al_{0.6}Ga_{0.4}As$ system, the energies are

$$
\begin{aligned}
E(\Gamma - \text{point}) &= 2.75(0.6) + (1.43)(0.4) = 2.22 \ eV \\
E(X - \text{point}) &= 2.15(0.6) + (1.91)(0.4) = 2.05 \ eV
\end{aligned}
$$

We see that in the $Al_{0.3}Ga_{0.7}As$, the lowest conduction band point is at the Γ-point and the bandgap is direct. However, for the $Al_{0.6}Ga_{0.4}As$, the lowest conduction band is at the X-point and the material is indirect.

2.9   BANDSTRUCTURE MODIFICATION BY HETEROSTRUCTURES

$\longrightarrow$ In Chapter 1, we have discussed some important epitaxial crystal growth techniques, such as MBE and MOCVD. Using these epitaxial growth techniques, it is possible to grow a sequence of semiconductor layers so that a narrow bandgap material is surrounded by a larger bandgap material. In some semiconductors making up such a heterostructure, the bandgap of the narrow gap material is completely enclosed by the bandgap of the larger gap material. An important example is the GaAs and the $Al_xGa_{1-x}As$ system, which have a very good match of their lattice constants. In Fig. 2.13a we show schematically the change in the potential created when a heterostructure is formed. Quantum wells are formed in both the conduction band and the valence bands. As a result, the electrons and holes are unable to move freely in the crystal growth (confinement) direction. They can still move freely in the plane perpendicular to the growth direction.

An extremely important parameter in the quantum well problem is the band-edge discontinuity produced when two semiconductors are brought together. As shown in Fig. 2.13a, a part of the bandgap discontinuity $(Eg^A - Eg^B)$ of two semiconductors A and B would appear in the conduction band and a part would appear in the valence band. It may appear from a simple minded electron affinity rule based upon lining up the vacuum energy as shown in Fig. 2.13b that the conduction band discontinuity is simply

$$\Delta E_c = e(\chi^B - \chi^A) \tag{2.53}$$

Unfortunately, the problem is not so simple because of charge transfer that occurs when the two semiconductors are brought together. The interface bonds involve elements of different chemical properties and as a result, there is a charge sharing across these bonds which can be considered to produce an interface dipole. This produces a correction to the simple electron affinity model. A number of theoretical models have been proposed to calculate the bandedge discontinuities, but there is sufficient uncertainty that, other than simple trends, it is difficult to predict exact values. Experimental evaluations of the discontinuities are still being carried out, especially for new semiconductor combinations.

It is useful to point out that the band discontinuities obey the transitivity relation, i.e., if the discontinuity between A and B is ΔE_c^{AB} and that between B and C is ΔE_c^{BC}, then the discontinuity between A and C is $\Delta E_c^{AB} - \Delta E_c^{BC}$. Note that the important semiconductor system of GaAs and AlAs has a discontinuity ratio given by

$$\frac{\Delta E_c}{\Delta E_v} \cong \frac{60}{40} \text{to} \frac{65}{35} \tag{2.54}$$

i.e., 60 to 65% of the direct bandgap difference between AlGaAs and GaAs is in the conduction band.

Once the band discontinuity is known, the bandstructure of a quantum well structure can be calculated. The simplest way to do this to use the effective mass theory in which the electron in each region of the structure is represented by its effective mass. The effect of the quantum well is to impose a background confining potential.

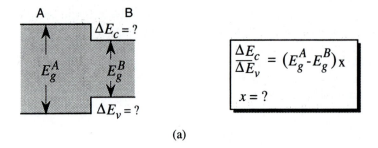

$$\frac{\Delta E_c}{\Delta E_v} = (E_g^A - E_g^B)x$$

$$x = ?$$

(a)

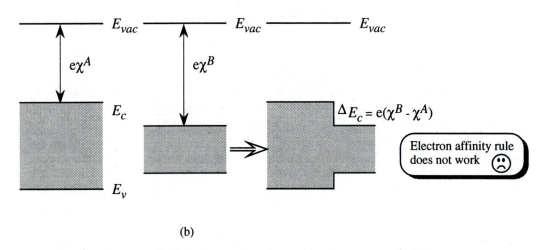

(b)

Figure 2.13: (a) An extremely important issue in the properties of a quantum well is the bandedge discontinuity ΔE_c (and ΔE_v), which defines the confining potential. (b) The electron affinity rule suggests that the conduction band discontinuity is determined by the electron affinity difference. Unfortunately, the rule does not work.

The simple quantum well structure such as shown in Fig. 2.14 is one of the most studied heterostructure. It's simple square well shape allows for easy solutions which are quite accurate and can be used to do first order comparisons with many experiments. The Schrödinger equation for the electron states in the quantum well can be written in a simple approximation as

$$\left[-\frac{\hbar^2 \nabla^2}{2m^*} + V(z)\right]\psi = E\psi \tag{2.55}$$

where m^* is the effective mass of the electron. The equation written above is strictly valid only for the conduction band of direct bandgap materials where the effective mass has a simple isotropic form. For the valence band to a first approximation, one can use the heavy hole mass m_{hh}^* and the light hole mass $m_{\ell h}^*$ to represent the problem. A more accurate equation will be discussed later.

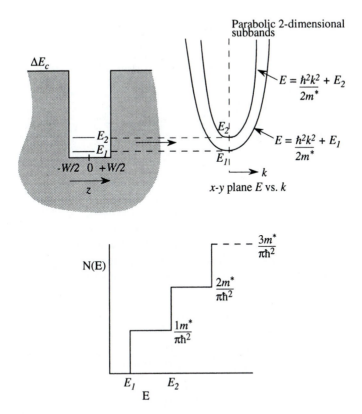

Figure 2.14: Schematic of a quantum well and the subband levels. In the x-y plane the subbands can be represented by parabolas. Note that in a semiconductor quantum well, one has a quantum well for the conduction band and one for the valance band. Subbands are produced in the conduction band and the valence band.

The Schrödinger equation for the quantum well in which the confinement is along the z-axis can be separated into equations in the x and y direction and in the z direction as follows:

$$\Psi = \psi_x \psi_y \psi_z \tag{2.56}$$

$$\frac{-\hbar^2}{2m^*} \frac{\partial^2 \psi_x}{\partial x^2} = E_x \psi_x$$

$$\frac{-\hbar^2}{2m^*} \frac{\partial^2 \psi_y}{\partial y^2} = E_y \psi_y$$

$$\left[\frac{-\hbar^2}{2m^*} \frac{\partial^2 \psi}{\partial z^2} + V(z)\right] \psi_z = E_z \psi_z \tag{2.57}$$

The solutions along the x,y plane in the quantum well are very simple and can be written as

$$\psi_x = \frac{1}{\sqrt{L_x}} e^{ik_x \cdot x}; \quad E_x = \frac{\hbar^2 k_x^2}{2m^*} \tag{2.58}$$

$$\psi_y = \frac{1}{\sqrt{L_y}} e^{ik_y y}; \quad E_y = \frac{\hbar^2 k_y^2}{2m^*} \tag{2.59}$$

where L_x, L_y represent the dimensions for normalization along the x and y direction. If the potential $V(z)$ is assumed to be zero in the quantum well and infinite outside the well of dimension W, the z-direction solutions are quite simple and are given by:

$$\psi_z = \frac{\sqrt{2}}{\sqrt{W}} \cos \frac{n\pi z}{W}, \quad \text{n is odd}$$

$$= \frac{\sqrt{2}}{\sqrt{W}} \sin \frac{n\pi z}{W}, \quad \text{n is even} \tag{2.60}$$

$$E_z = \frac{\pi^2 \hbar^2 n^2}{2m^* W^2} \tag{2.61}$$

The total energy of the electron is then (measured from the bulk material bandedge)

$$E(n, k_x, k_y) = \frac{\pi^2 \hbar^2 n^2}{2m^* W^2} + \frac{\hbar^2 k_x^2}{2m*} + \frac{\hbar^2 k_y^2}{2m^*} \tag{2.62}$$

This gives a series of subbands for $n = 1,2,3 \ldots$ as shown in Fig. 2.14. *In each subband, the electron behaves as if it is in a 2-dimensional world and the density of states thus have a 2-dimensional step like behavior, as shown in Fig. 2.14.* Such a step like density of states is exploited for optoelectronic devices as we will discuss later.

The energy values given above are valid only for an infinite quantum well potential as discussed above. For a finite quantum well potential, the results for the energy values are not analytical, but can be obtained by solving a simple set of equations given below:

$$\alpha \tan\left(\frac{\alpha W}{2}\right) = \beta \qquad (2.63)$$

or

$$\alpha \cot\left(\frac{\alpha W}{2}\right) = -\beta \qquad (2.64)$$

where

$$\alpha = \sqrt{\frac{2m^* E}{\hbar^2}}$$

$$\beta = \sqrt{\frac{2m^*(V_o - E)}{\hbar^2}}$$

Those values of energy which satisfy either of these sets of equations are allowed. The energy values one obtains by this more accurate treatment are somewhat smaller than the values given by Eqn. 2.61 and the difference is more pronounced for the higher subband energies.

The Valence Bands in a Quantum Well

↝ The valence bandstructure in a quantum well can be obtained simply if one assumes that one has two uncoupled bands—the heavy hole band and the light hole band. Thus one would get a subband structure for the heavy hole and one for the light hole. This is shown schematically in Fig. 2.15a. However, it can be seen from the discussions in Section 2.4 that there is a strong coupling between the light hole and heavy hole bands so that for an accurate treatment, one cannot use the uncoupled band Schrödinger equation. One has to solve a four-band equation of the form

$$[H - V(z)]\,\Psi_k = E\Psi_k \qquad (2.65)$$

where H is the 4-band $k \cdot p$ hamiltonian given by Eqns. 2.13 and 2.14, and Ψ is the valence band function which has the general form

$$\Psi_k = \sum_{\nu=1}^{4} g_\nu(z) u_\nu \, e^{ik \cdot \rho} \qquad (2.66)$$

where k is in the x-y plane, ρ is a spatial vector in the x-y plane, the u_ν represent the four states (two for HH, LH and two for their spin states), and $g_\nu(z)$ is the envelope function in the z-direction. The 4-band equation can be reduced to two band equations as discussed in Problem 2.13.

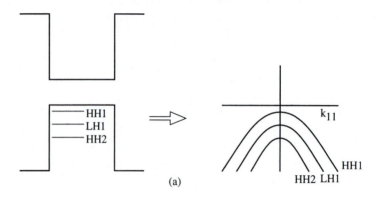

(a)

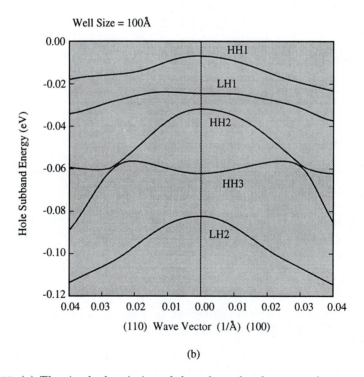

(b)

Figure 2.15: (a) The simple description of the valence bandstructure in a quantum well. The light hole and heavy hole bands are treated independently using parabolic bands. (b) The coupling between HH and LH bands causes the valence bandstructure to have strong non-parabolicity.

When the coupled HH, LH band problem is solved in the quantum well, the E vs. $k_{||}$ relations are no longer parabolic. In Fig. 2.15b we show a typical valence bandstructure in a GaAs/AlGaAs quantum well structure. We note that there is a splitting between the HH and LH states because of quantum confinement at $k_{||} = 0$. Also, the non-parabolicity of the bandstructure is quite clear from this figure. An important point to appreciate is also that the in-plane masses (say the density of states mass) depend upon the quantum well size.

The quantum confinement produces a situation where the HH state (the $|3/2, \pm 3/2 >$) state is at the top of the valence band in the usual quantum well. This has important effects on the polarization dependence of the optical transitions as will be discussed in later chapters. It is possible to incorporate strain in the quantum well and alter the ordering of the heavy hole and light hole states. This will be discussed next.

EXAMPLE 2.5 In the GaAs/AlGaAs heterostructure, 60% of the bandgap discontinuity is in the conduction band of the narrow gap material. Calculate the conduction band and valence band quantum well potentials for GaAs/Al$_{0.3}$Ga$_{0.7}$As.

The bandgap difference between GaAs and Al$_{0.3}$Ga$_{0.7}$As is (from Appendix B)

$$\Delta E_g = 1.247 \times 0.3 = 0.374 \ eV$$

Since 60% of this difference is in the conduction band, the discontinuity of the quantum well formed in the conduction band is (this is the barrier height)

$$\Delta E_c = 0.374 \times 0.6 = 0.224 \ eV$$

The barrier height for the valence band is

$$\Delta E_v = 0.15 \ eV$$

These barrier heights are not infinity and so our simple model is only an approximation to the real problem. However, the use of the infinite barrier problem is reasonable once the well size is $\sim$ 150 Å for this system.

EXAMPLE 2.6 Using a simple infinite barrier approximation, calculate the "effective bandgap" of a 100 Å GaAs/AlAs quantum well. If there is a one-monolayer fluctuation in the well size, how much will the effective bandgap change? This example gives an idea of how stringent the control has to be to exploit heterostructures.

The confinement of the electron ($m^* = 0.067 \ m_0$) pushes the effective conduction band up, and the confinement of electrons at the valence band push the effective edge down ($m_{hh}^* = 0.4 \ m_0$). The change in the electron ground state is (using $n = 1$) 55.77 meV. The shift (downwards) in the valence band is (why do we only need to worry about the shift of the HH band to find the effective gap?) 9.34 meV. The net shift is 65.11 meV.

The effective bandgap is thus 65.11 meV larger than the bulk GaAs bandgap.

If the well size changes by one monolayer (e.g., goes from 100 Å to 102.86 Å), the change in the electron level is

$$
\begin{aligned}
\Delta E_e &= E_e \left[1 - \frac{(100)^2}{(102.86)^2} \right] = E_e \times 0.055 \\
&= 3.06 \ meV
\end{aligned}
$$

The hole energy changes by

$$
\Delta E_{hh} = E_{hh} \times 0.055 = 0.51 meV.
$$

Thus, the bandgap changes by 3.56 meV for a one monolayer variation. In optical frequencies this represents a change of 0.86 Terrahertz, which is too large a shift for many optoelectronic applications.

2.10   BANDSTRUCTURE MODIFICATION BY STRAIN

In Chapter 1, Section 1.11, we have discussed how one can introduce a built-in strain in an epitaxial layer by growing it on a lattice mismatched substrate. As long as the mismatched epitaxial layer is below the critical thickness, the strain produced is uniform and no dislocations are produced. As a result of this uniform strain, the in-plane lattice constant of the epitaxial layer adjusts to *exactly* fit the in-plane lattice constant of the substrate. The out-of-plane lattice constant then adjusts according to the Poisson law to a new lattice constant. For example, if ϵ is the strain between an overlayer A with lattice constant a_A and a substrate with lattice constant a_S, and growth occurs in the (001) direction, the strain tensor produced is given by

$$
\begin{aligned}
\epsilon &= \frac{a_S - a_A}{a_A} \\
\epsilon_{xx} &= \epsilon \\
\epsilon_{yy} &= \epsilon \\
\epsilon_{zz} &= \frac{-2C_{12}}{C_{11}} \epsilon
\end{aligned}
\tag{2.67}
$$

where C_{12} and C_{11} are the force constants of the material. In the fcc based semiconductor crystals, there is no off-diagonal term by symmetry for the (001) growth. As a result of the strain, the cubic symmetry of the x, y, z is no longer present in the system. This cubic symmetry is the reason we have a degeneracy between the heavy hole state and the light hole state at the top of the valence band. The presence of strain where $\epsilon_{xx} = \epsilon_{yy} \neq \epsilon_{zz}$ produces a lifting of the HH, LH degeneracy which plays a very significant role in optoelectronic devices.

The strain tensor generated by (001) epitaxy produces a lifting of the HH, LH degeneracy by shifting these bands by an amount given by

$$\Delta E_{hh}(\epsilon) = \left[2a_d \left(\frac{C_{11} - C_{12}}{C_{11}} \right) + b_d \left(\frac{C_{11} + 2C_{12}}{C_{11}} \right) \right] \epsilon \qquad (2.68)$$

$$\Delta E_{\ell h}(\epsilon) = \left[2a_d \left(\frac{C_{11} - C_{12}}{C_{11}} \right) - b_d \left(\frac{C_{11} + 2C_{12}}{C_{11}} \right) \right] \epsilon \qquad (2.69)$$

where a_d and b_d are the deformation potentials. The deformation potential b_d is negative. The shifts given above are referenced to the conduction bandedge which is then assumed to have a fixed position.

As a result of the splitting between the HH and LH bands, we can have a situation where the HH is either on the top of the valence band (for an epilayer with lattice constant larger than that of the substrate) or the LH is on the top of the valence band. This situation is shown in Fig. 2.16. We will exploit these possibilities when designing semiconductor lasers for specific applications. An important consequence of the splitting between the HH and LH states is that the density of states mass in the valence band decreases dramatically as a function of strain. In Fig. 2.17 we show the effect of the strain on the near bandedge density of states mass for $In_x Ga_{1-x} As$ system grown on GaAs. We can see that the mass can be reduced by up to a factor of 3.

It is also interesting to note that while the normal quantum confinement due to a quantum well produces a situation where the HH state is at the top of the valence band, the strain can put the HH or LH state on the top. An interesting corollary is also that the strain can restore the degeneracy of the HH, LH splitting due to quantum confinement. This situation is produced as shown in Fig. 2.18 by introducing a small tensile strain in the quantum well. The merger of the HH, LH states produces a very high density of hole states at the bandedge which can be exploited for some optical modulation applications.

It is important to note that the magnitude of the splitting that can be produced between the HH and LH states can be ~100 meV which is quite large and can have a profound effect on optoelectronic properties.

EXAMPLE 2.7 Calculate the heavy hole-light hole splitting produced when $In_{0.2} Ga_{0.8} As$ is grown on a (001) GaAs substrate. Assume that the structure is pseudomorphic and the deformation potential is given by $b_d = -2.0$ eV. The force constants are $C_{11} = 11.5 \times 10^{11}$ dynes/cm^2; $C_{12} = 5.5 \times 10^{11}$ dynes/cm^2.

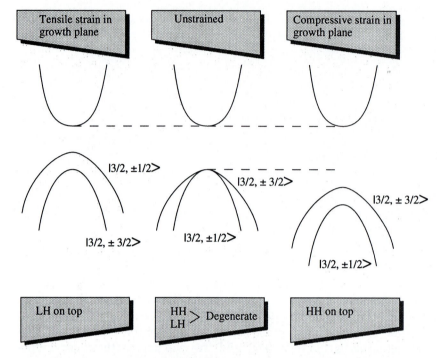

Figure 2.16: The consequence of pseudomorphic strain on the bandedges of a direct bandgap semiconductor. The valence band degeneracy at $k = 0$ is lifted as the HH and LH states are split as shown.

The splitting produced by a pseudomorphic strain is (in eV)

$$\Delta E_{hh} - \Delta E_{\ell h} = 2b_d \left(\frac{C_{11} + 2C_{12}}{C_{11}} \right) \epsilon$$
$$= -5.91\epsilon$$

The strain between $In_{0.2}Ga_{0.8}As$ and GaAs is

$$\epsilon = \frac{5.653 - 5.734}{5.653} = -0.014$$

Thus the splitting is

$$\Delta E_{hh} - \Delta E_{\ell h} = 0.085 \ eV$$

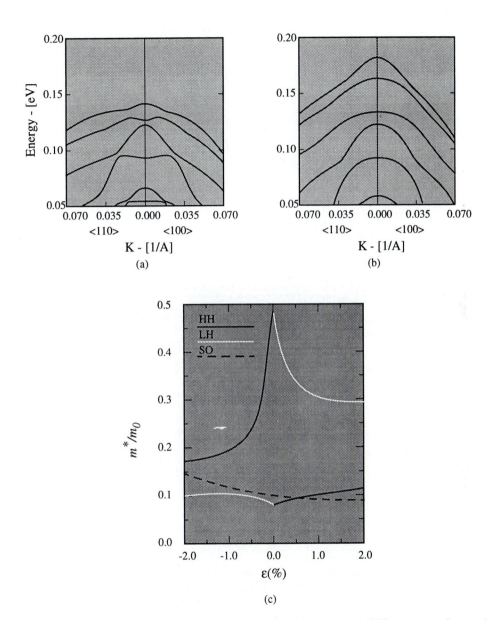

(a)

(b)

(c)

Figure 2.17: An important consequence of the splitting of the LH, HH states at $k = 0$ is that the bandedge density of states decreases. Hole dispersion in a 100 Å quantum well in (a) GaAs/Al$_{0.3}$Ga$_{0.7}$As and (b) In$_{0.1}$Ga$_{0.9}$As/Al$_{0.3}$Ga$_{0.7}$As; and (c) change in the density of states mass as a function of strain at the bandedge. (Figure provided by John Loehr.)

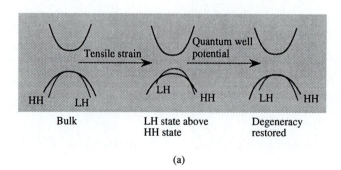

(a)

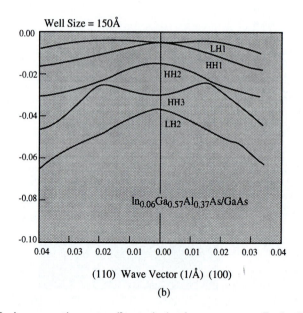

(b)

Figure 2.18: (a) By incorporating a tensile strain in the quantum well, the LH, HH degeneracy lifted by quantum confinement can be restored. (b) Calculated bandstructure of a 150 Å quantum well in which tensile strain is introduced to cause a merger of the heavy hole and light hole states. (Reproduced with permission from S. Hong, et. al., *Physical Review*, B, 37, 878 (1988).)

EXAMPLE 2.8 Estimate the pseudomorphic strain needed in a 100 Å quantum well to cause a merger of the HH and LH states at the top of the valence band. Assume the following parameters:

$$b_d = -2.0 \ eV; C_{11} = 12.0 \times 10^{11} dynes/cm^2; C_{12} = 6 \times 10^{11} dynes/cm^2;$$
$$m_{hh}^* = 0.45 \ m_0; m_{\ell h}^* = 0.1 \ m_0$$

In this problem we have to ensure that the HH-LH splitting due to strain exactly balances the splitting due to the quantum confinement. The quantum confinement splitting is given approximately by (note that the hole mass is taken as positive, while the electron mass in the valence band is negative)

$$
\begin{aligned}
\Delta E_{hh} - \Delta E_{\ell h} &= \frac{-\hbar^2 \pi^2}{2W^2} \left(\frac{1}{m_{hh}^*} - \frac{1}{m_{\ell h}^*} \right) \\
&= \frac{(1.05 \times 10^{-34} \ Js)^2 (\pi^2)}{2 \times (100 \times 10^{-10} \ m)^2} \left[\frac{1}{(0.45 \times 9.1 \times 10^{-31} \ kg)} - \frac{1}{(0.1 \times 9.1 \times 10^{-31} \ kg)} \right] \\
&= 4.64 \times 10^{-21} \ J \\
&= 29 \ meV
\end{aligned}
$$

The splitting produced by strain has to have an opposite effect, i.e.,

$$\Delta E_{\ell h} - \Delta E_{hh} = -2b_d \left(\frac{C_{11} + C_{12}}{C_{11}} \right) \epsilon = 0.029 \ eV$$

Thus a tensile strain of 0.005 is needed.

2.11   INTRINSIC CARRIER CONCENTRATION

$\longrightarrow$ We have discussed in the previous sections that a semiconductor is characterized by the fact that at zero Kelvin, the valence band is completely occupied while the conduction band is completely empty. We also saw that a completely full band does not conduct charge. Thus, at low temperatures the pure semiconductor offers an extremely high resistance to current transport. At finite temperatures, the occupation of electrons and holes is described by the Fermi distribution function given by

$$f(E) = \frac{1}{1 + exp\frac{E - E_F}{k_B T}} \tag{2.70}$$

As the temperature is raised, the Fermi distribution function smears and some electrons are emitted from the valence band into the conduction band. Now there are electrons in the conduction band and holes in the valence band which can carry current. However, such current carrying electrons produced by raising the temperature, known as intrinsic carriers, are not useful in semiconductor devices and often are a nuisance. The intrinsic carriers often are a source of limitation for high temperature operation of devices, since they cannot be controlled effectively by electric fields.

The intrinsic carrier concentration depends upon the bandgap and temperature as well as the details of the bandedge masses. We will assume that the bandedge density of states for electrons and holes originate from parabolic *E-k* relationships. The conduction and valence band density of states are shown in Fig. 2.19 along with the position of a Fermi level.

The concentration of electrons in the conduction band is

$$n = \int_{E_c}^{\infty} N_e(E) f(E) dE \tag{2.71}$$

where $N_e(E)$ is the electron density of states near the conduction bandedge and $f(E)$ is the Fermi function. Using the appropriate expressions for N_e and f we get, for a 3D system, (the conduction band density of states starts at $E = E_c$ as shown in Fig. 2.19)

$$n = \frac{1}{2\pi^2} \left(\frac{2m_e^*}{\hbar^2} \right)^{3/2} \int_{E_c}^{\infty} \frac{(E - E_c)^{1/2} dE}{exp(\frac{E - E_F}{k_B T}) + 1} \tag{2.72}$$

If the chemical potential is far from the bandedge, then the unity in the denominator can be neglected. *This approximation, called the Boltzmann approximation, is valid when n is small* ($\lesssim 10^{16}$ cm^{-3} *for most semiconductors*), and is usually valid for intrinsic concentrations. Then we get

$$
\begin{aligned}
n &= \frac{1}{2\pi^2} \left(\frac{2m_e^*}{\hbar^2} \right)^{3/2} exp\left(\frac{E_F}{k_B T} \right) \int_{E_c}^{\infty} (E - E_c)^{1/2} \, exp(-E/k_B T) dE \\
&= 2 \left(\frac{m_e^* k_B T}{2\pi\hbar^2} \right)^{3/2} exp[(E_F - E_c)/k_B T] = N_c \, exp[(E_F - E_c)/k_B T]
\end{aligned}
$$

where

$$N_c = 2 \left(\frac{m_e^* k_B T}{2\pi\hbar^2} \right)^{3/2} \tag{2.73}$$

N_c is known as the effective density of states at the conduction bandedge. Note that the units of the density of states N_e are eV^{-1} cm^{-3} while those of the effective density of states N_c are cm^{-3}.

The carrier concentration is known *when E_F is calculated.* To find the intrinsic carrier concentration, this requires finding the hole concentration p as well.

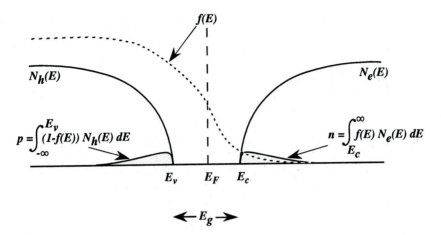

Figure 2.19: A schematic of the density of states of the conduction and valence band. N_e and N_h are the electron and hole density of states. Also shown is the Fermi function giving the occupation probability for the electrons. The resulting electron and hole concentrations are shown. For an intrinsic semiconductor $n = p$, since each electron produced in the conduction band leaves behind a hole in the valence band.

The hole distribution function f_h is given by (remember the hole is the absence of an electron)

$$f_h = 1 - f_e = 1 - \frac{1}{exp(\frac{E-E_F}{k_BT})+1} = \frac{1}{exp\left[\frac{(E_F-E)}{k_BT}\right]+1}$$

$$\cong exp-\left[\frac{(E_F-E)}{k_BT}\right] \tag{2.74}$$

The approximation is again based on our assumption that $E_F - E >> k_BT$, which is a good approximation for pure semiconductors. Carrying out the mathematics similar to that for electrons, we find that

$$p = 2\left(\frac{m_h^* k_BT}{2\pi\hbar^2}\right)^{3/2} exp\ [(E_v - E_F)/k_BT] \tag{2.75}$$

$$= N_v exp\ [(E_v - E_F)/k_BT] \tag{2.76}$$

where N_v is the effective density of states for the valence bandedge.

In the expressions above it is important to note that the relevant masses are the density of states masses. These were given in Eqns. 2.39 to 2.42.

One can define a total density of states mass for silicon by the expression

$$m_e^* = (6)^{2/3} \left(m_\ell^* \, m_t^{*2}\right)^{1/3} \tag{2.77}$$

This takes into account the presence of six valleys. If the effective mass given above is used, there is no further need to account for the six valleys.

In intrinsic semiconductors, the electron concentration is equal to the hole concentration since each electron in the conduction band leaves a hole in the valence band. If we multiply the electron and hole concentrations, we get

$$\boxed{np = 4 \left(\tfrac{k_B T}{2\pi\hbar^2}\right)^3 (m_e^* m_h^*)^{3/2} \, exp(-E_g/k_B T)} \tag{2.78}$$

and since for the intrinsic case $n = n_i = p = p_i$, we have from the square root of the equation above

$$\boxed{n_i = p_i = 2 \left(\tfrac{k_B T}{2\pi\hbar^2}\right)^{3/2} (m_e^* m_h^*)^{3/4} \, exp\left[-(E_c + E_v)/2k_B T\right]} \tag{2.79}$$

If we set $n = p$, we also obtain the Fermi level position measured from the valence bandedge using Eqns. 2.3 and 2.5. We denote the intrinsic Fermi level by E_{Fi}

$$exp\,(2E_{Fi}/k_B T) = (m_h^*/m_e^*)^{3/2} exp(E_g/k_B T) \tag{2.80}$$

or

$$\boxed{E_{Fi} = \tfrac{E_c + E_v}{2} + \tfrac{3}{4}k_B T \ell n\,(m_h^*/m_e^*)} \tag{2.81}$$

Thus the Fermi level of an intrinsic material lies close to the midgap.

We note that the carrier concentration increases exponentially as the bandgap decreases and has a strong temperature dependence. In electronic devices where current has to be modulated by some means, the concentration of intrinsic carriers is fixed by the temperature and therefore is detrimental to device performance. Once the intrinsic carrier concentration increases to $\sim 10^{15}$ cm^{-3}, the material becomes unsuitable for electronic devices. A growing interest in high bandgap semiconductors such as diamond (C), SiC etc., is partly due to the potential applications of these materials for high temperature devices where, due to their larger gap, the intrinsic carrier concentration remains low up to very high temperatures. *We note that the product np calculated above is independent of the Fermi level E_F. This is an expression of the law of mass action. This result is valid not only for the intrinsic case but also when we have dopants.* The assumption made for the particular derivation simply required that E_F is far from the bandedges. At room temperature, the np product is 2.25×10^{20} cm^{-6}, 5.76×10^{26} cm^{-6} and 3.24×10^{12} cm^{-6} for Si, Ge, and GaAs, respectively. The corresponding intrinsic carrier densities are 1.5×10^{10} cm^{-3}, 2.4×10^{13} cm^{-3} and 1.8×10^6 cm^{-3}, respectively.

It is quite clear from the discussion above that pure semiconductors have a very low concentration of carriers that can conduct current. One must compare the room temperature concentrations of $\sim 10^{11}$ cm^{-3} to the carrier concentrations of $\sim 10^{21}$ cm^{-3} for metals. Indeed, pure semiconductors would have little use by themselves.

2.12 DEFECT LEVELS IN SEMICONDUCTORS

$\mathcal{R}$ We have seen that in a perfectly periodic structure, the electronic spectra consists of regions of allowed energy band and regions where there is a bandgap. In a real crystal, as discussed in Chapter 1, Section 1.9, there are a number of defects that are present. These defects may be native defects (vacancies, anti-site defects, etc) or extrinsic defects (chemical impurities) which are intentionally introduced in the crystal or are unintentional.

The defects of key interest in semiconductors are the point defects which create a local disturbance in the crystal structure as discussed in Chapter 1. The effect of this crystal disturbance can be of two kinds as shown in Fig. 2.20:

i) The disturbance may create a potential profile which differs from the periodic potential only over one or a few unit cells. This potential is deep and localized and the defect is called a deep level defect.

ii) The disturbance may create a long range potential disturbance which may extend over tens or more unit cells. Such defects are called shallow level defects.

Associated with the defects are new electronic states that are called defect levels. These new electronic states can be produced in the regions of allowed bands (i.e., the conduction or valence band) in which case their effects are minimal. However, the new levels could be produced in the bandgap region in which case they can greatly alter the electronic and optical properties of the semiconductor.

The bandgap levels have associated with them a wavefunction which no longer has a Bloch form (i.e., a form having a plane wave part $\exp(ik \cdot r)$). However, in general, the defect level can be expressed in terms of the perfect crystal states $\psi_k(\nu)$. The deep defect has associated with it a wavefunction that is highly localized in space as shown in Fig. 2.21a, and as a consequence is made up of a large number of k-states. *Thus the Fourier transform of the deep level function has essentially all k-values in more or less equal proportion.*

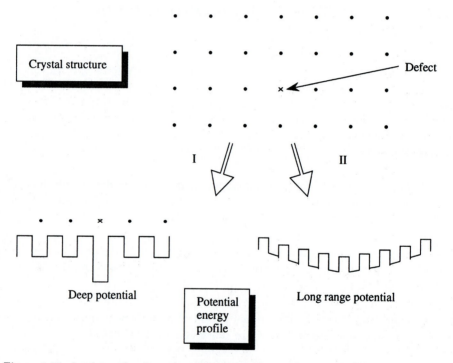

Figure 2.20: A schematic of a point defect in a crystal. The effect of the defect on the crystal background potential can be such that a "deep potential" variation is produced in a narrow spatial region or a long range disturbance is created.

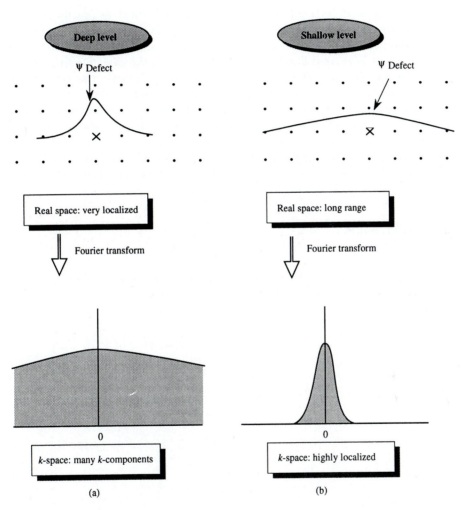

Figure 2.21: A schematic of the wavefunctions associated with (a) a deep short range potential defect and (b) a shallow long range potential defect.

The wavefunction associated with a shallow long range impurity is extended in real space as shown in Fig. 2.21b and is made up of only a few k-value functions arising from the bandedge states. The most important shallow defect levels are dopants that will be discussed in the next chapter.

Normally, deep level defects are to be avoided as much as possible in semiconductors. The key reason for this is that an electron in the defect level is "trapped" near the defect since the wavefunction of the electron is localized in space. Such electrons cannot participate in current flow very easily. However, in the case of indirect gap materials, deep level defects are often purposely introduced to increase the optical response of the material. The physical concepts involved are simple to understand. Consider an indirect material like Si. If a photon with energy just larger than the bandgap impinges on the material it cannot scatter an electron in the valence band to the conduction band since the photon momentum is not able to provide enough momentum to conserve momentum in the process. However, if a deep level is introduced, the wavefunction of the deep level has all k-values present in it and, therefore, an optical transition is possible.

2.13 SUMMARY

$\mathcal{R}$ In this chapter we have discussed the bandstructure of semiconductors and their heterostructures. Essentially all optical and transport properties of semiconductors are controlled by the bandstructure. We summarize the key findings of this chapter in Tables 2.1—2.4.

2.14 PROBLEMS

Sections 2.2-2.7

2.1 Consider electrons at the bottom of the conduction band in silicon. If an electric field pulse of 10 kV/cm is applied for 10ps along the (100) direction, what are the energies of the electrons at the end of the pulse if there is no scattering? Remember that Si has six equivalent conduction band minimas and, for the ($\bar{1}$00) and (100) valleys, the electron moves with its longitudinal mass while for the other 4 valleys the electrons move with transverse mass.

2.2 When a photon impinges upon a semiconductor, it can take an electron from the valence band to the conduction band. The momentum is conserved in such transitions. Calculate the momentum that a 2.0 eV photon carries. The electron in the valence band can go into the conduction band with the momentum change of the photon.

2.3 Use the data given in the Appendix B to identify semiconductors with bandgaps large enough to emit photons with wavelength less than 0.5 μm.

2.4 An electron at the top of the valence band in silicon is to be transferred in a momentum conserving scattering process to the bottom of the conduction band. Calculate how much momentum is needed to cause this scattering.

2.5 The effective mass of a conduction band electron in a semiconductor is 0.1 m_0. Calculate the energy of this electron if the k-vector is 0.3 Å^{-1}. If the electron affinity of this semiconductor is 10 eV, calculate the energy of the electron measured from the vacuum level.

2.6 Plot the conduction band and valence band density of states in Si, Ge and GaAs from the bandedges to 0.5 eV into the bands. Use the units eV^{-1} cm^{-3}. Use the following data:

$$Si : m_1^* = m_\ell^* = 0.98 \ m_0$$
$$m_2^* = m_3^* = m_t^* = 0.19 \ m_0$$
$$m_{hh}^* = 0.49 \ m_0$$
$$m_{\ell h}^* = 0.16 \ m_0$$

$$Ge : m_\ell^* = 1.64 \ m_0$$
$$m_t^* = 0.082 \ m_0$$
$$m_{hh}^* = 0.29 \ m_0$$
$$m_{\ell h}^* = 0.044 \ m_0$$

$$GaAs : m_e^* = m_{dos}^* = 0.067 \ m_0$$
$$m_{hh}^* = 0.5 \ m_0$$
$$m_{\ell h}^* = 0.08 \ m_0$$

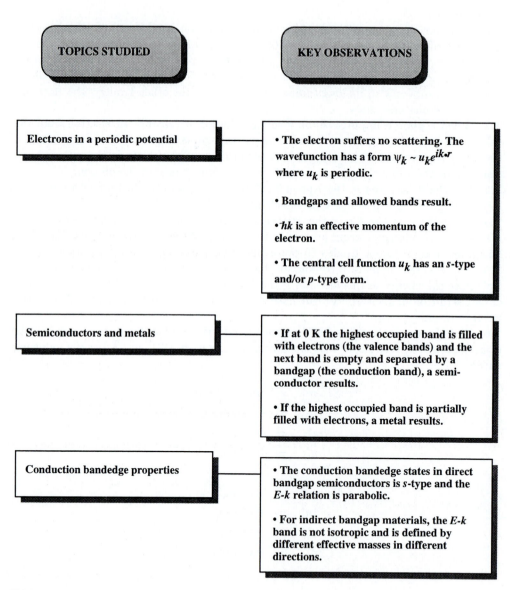

TOPICS STUDIED

KEY OBSERVATIONS

Electrons in a periodic potential

• The electron suffers no scattering. The wavefunction has a form $\psi_k \sim u_k e^{ik\cdot r}$ where u_k is periodic.

• Bandgaps and allowed bands result.

• $\hbar k$ is an effective momentum of the electron.

• The central cell function u_k has an s-type and/or p-type form.

Semiconductors and metals

• If at 0 K the highest occupied band is filled with electrons (the valence bands) and the next band is empty and separated by a bandgap (the conduction band), a semiconductor results.

• If the highest occupied band is partially filled with electrons, a metal results.

Conduction bandedge properties

• The conduction bandedge states in direct bandgap semiconductors is s-type and the E-k relation is parabolic.

• For indirect bandgap materials, the E-k band is not isotropic and is defined by different effective masses in different directions.

Table 2.1: Summary table.

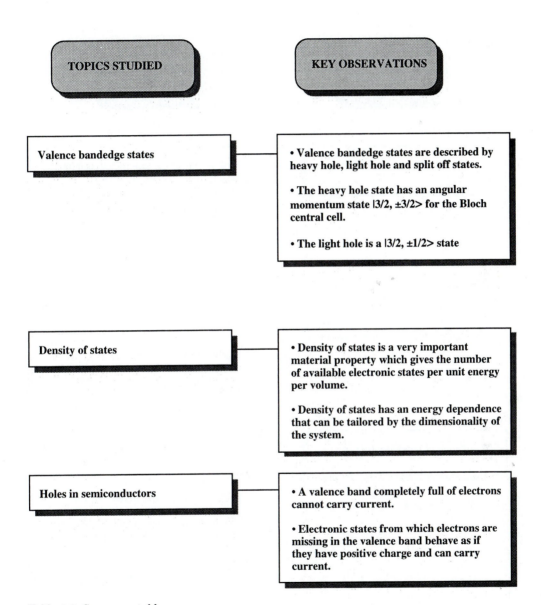

TOPICS STUDIED

KEY OBSERVATIONS

Valence bandedge states

• Valence bandedge states are described by heavy hole, light hole and split off states.

• The heavy hole state has an angular momentum state |3/2, ±3/2> for the Bloch central cell.

• The light hole is a |3/2, ±1/2> state

Density of states

• Density of states is a very important material property which gives the number of available electronic states per unit energy per volume.

• Density of states has an energy dependence that can be tailored by the dimensionality of the system.

Holes in semiconductors

• A valence band completely full of electrons cannot carry current.

• Electronic states from which electrons are missing in the valence band behave as if they have positive charge and can carry current.

Table 2.2: Summary table.

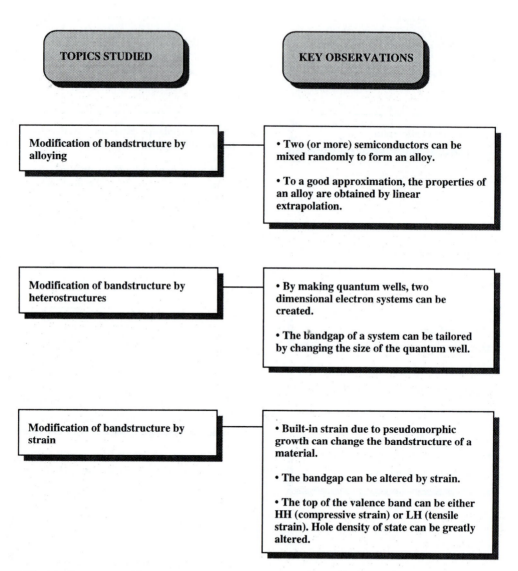

Table 2.3: Summary table.

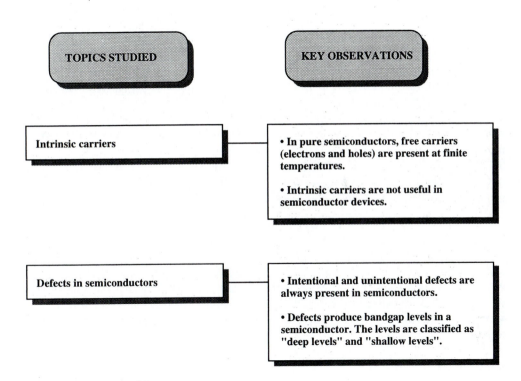

TOPICS STUDIED

KEY OBSERVATIONS

Intrinsic carriers

- **In pure semiconductors, free carriers (electrons and holes) are present at finite temperatures.**

- **Intrinsic carriers are not useful in semiconductor devices.**

Defects in semiconductors

- **Intentional and unintentional defects are always present in semiconductors.**

- **Defects produce bandgap levels in a semiconductor. The levels are classified as "deep levels" and "shallow levels".**

Table 2.4: Summary table.

2.7 The wavevector of a conduction band electron in GaAs is $k = (0.1, 0.1, 0.0)$ Å^{-1}. Calculate the energy of the electron measured from the conduction bandedge.

2.8 A conduction band electron in silicon is in the (100) valley and has a k- vector of $\frac{2\pi}{a}$ (1.0, 0.1, 0.1). Calculate the energy of the electron measured from the conduction bandedge. Here a is the lattice constant of silicon.

2.9 An electron in the Γ-valley of GaAs is to be transferred to the L-valley. Using the bandstructure of GaAs estimate the smallest k-vector change that is needed for this transition. The electron in the Γ-valley must have an energy equal to the position of the L-valley.

2.10 Calculate the energies of electrons in GaAs and InAs conduction band with k-vectors $(0.01, 0.01, 0.01)$ Å^{-1}. Refer the energies to the conduction bandedge values.

2.11 Consider an electron at the bottom of the conduction band in GaAs. An electric field of 10^4 V/cm is applied to the materials in the x-direction. Calculate the time it takes the electron to reach the Brillouin zone. Consult the bandstructure of GaAs to see what the electron energy is at this point. What happens to the electron after it has reached the Brillouin zone edge?

2.12 An electric field of 10^5 V/cm is applied for 0.1 ps to holes at the top of the valence band in GaAs. Calculate the hole energies at the end of the pulse. Note that holes will move in both the heavy-hole and light-hole bands.

2.13 Express the heavy hole and light hole masses along [001], [011], and [111] directions in terms of the Kohn-Luttinger parameters. Hint: The 4×4 $\mathbf{k} \cdot \mathbf{p}$ matrix can be reduced to two doubly degenerate (in the absence of external fields) 2×2 equations of the form

$$\begin{vmatrix} H_{hh} & b - ic \\ (b - ic)^* & H_{lh} \end{vmatrix} \psi = E\psi$$

where ψ is a 1×2 vector (Broido and Sham, 1985).

2.14 Assuming k conservation, what is the phonon wavevector which can take an electron in the GaAs Γ-valley to the L-valley? Also calculate the phonon vector which causes scattering of silicon electrons at the bottom of the conduction band to other equivalent valleys.

Section 2.8

2.15 According to the virtual crystal approximation for the electronic properties of alloys (discussed in the text), the high symmetry point (i.e., bandedges and other symmetry points like X point, L point, etc) energy of the alloy is simply given by a composition weighted averaging of the values for the individual semiconductors. Based on this, calculate the bandgap of $Si_x Ge_{1-x}$ alloy as x goes from 1.0 to 0. Note that the conduction bandedge is X-like in Si and L-like in Ge.

2.16 Using the Vegard's Law for the lattice constant of an alloy (i.e., lattice constant is the weighted average), find the bandgaps of alloys made in InAs, InP, GaAs, GaP which can be lattice matched to InP.

2.17 For long haul optical communication, the optical transmission losses in a fiber dictate that the optical beam must have a wavelength of either 1.3 μm or 1.55 μm. Which alloy combinations lattice matched to InP have a bandgap corresponding to these wavelengths?

2.18 Using the virtual crystal approximation upto what Al composition does the alloy $Al_x Ga_{1-x} As$ remain a direct gap semiconductor? What is the maximum bandgap achievable in the direct alloy? (see Example 2.4).

2.19 Consider the alloy system $Si_{1-x} Ge_x$. Using the virtual crystal approximation, calculate the positions of the Γ, X, and L point energies in the lowest conduction band. Use the top of the valence band as a reference. At what Ge composition does the conduction bandedge change from X-like to L-like?

2.20 Using the bandstructure of GaAs and AlAs, calculate the conduction band minima at Γ, X, L points in $Al_x Ga_{1-x} As$ alloy, as x varies from 0 to 1.

2.21 Calculate the composition of $Hg_x Cd_{1-x} Te$ which can be used for a night vision detector with bandgap corresponding to a photon energy of 0.1 eV. Bandgap of CdTe is 1.6 eV *and that of HgTe is −0.3 eV* at low tempertures around 4 K.

Section 2.9

2.22 At room temperature the bandgap of GaAs is 1.43 eV. Assuming an infinite barrier approximation, what is the well size of a GaAs/AlAs quantum well which can produce an effective bandgap of 1.50 eV?

2.23 If a 100 Å cubic dot of GaAs is embedded in AlAs, what is the approximate bandgap of the quantum dot? Assume an infinite barrier model. Use an effective mass of 0.067 m_0 and 0.4 m_0 for the electron and heavy holes, respectively.

2.24 In the $In_{0.53} Ga_{0.47} As$/InP system, 40% of the bandgap discontinuity is in the conduction band. Calculate the conduction and valence band discontinuities. Calculate the effective bandgap of a 100 Å quantum well. Use the infinite potential approximation and the finite potential approximation and compare the results.

2.25 Compare the electron and HH ground state energies in a GaAs/$Al_{0.3} Ga_{0.7} As$ well for infinite potential and finite potential models as a function of well size. The well size goes from 40 Å to 150 Å.

2.26 Consider the GaAs/$Al_x Ga_{1-x} As$ $(0 \leq x < 1)$. Assume that the band discontinuity distribution between the conduction and valence band is given by

$$\Delta E_c = 0.6 \Delta E_g \text{ (direct)}$$
$$\Delta E_v = 0.4 \Delta E_g \text{ (direct)}$$

where ΔE_g(direct) is the bandgap discontinuity between AlGaAs (*even if the material is indirect*) and GaAs. Calculate the valence and conduction band offsets as a function of Al composition. Notice that the conduction band offset starts to decrease as Al composition increases beyond ~ 0.4.

2.27 Calculate the first and second subband energy levels for the conduction band in a GaAs/Al$_{0.3}$Ga$_{0.7}$As quantum well as a function of well size. Assume a 60:40 rule for ΔE_c:ΔE_v. Also, calculate the energy levels if an infinite barrier approximation was being used.

2.28 Calculate the first two valence subband levels for heavy hole and light hole states in a GaAs/Al$_{0.3}$Ga$_{0.7}$As quantum well as a function of the well size. Note that to find the subband levels (at $k_{\parallel} = 0$), one does not need to include the coupling of the heavy hole and light hole states.

2.29 Using the Eigenvalue method described in appendix C.3, write the valence band 4×4 Schrödinger equation (using the Kohn-Luttinger formalism discussed in the text) as a difference equation. Write a computer program to solve this $4N \times 4N$ matrix equation where N is the number of grid points used to describe the region of the well and the barrier. Using the parameter values $\gamma_1 = 6.85$, $\gamma_2 = 2.1$, and $\gamma_3 = 2.9$, calculate the E vs. $k_{\parallel}$ diagram for the valence band in a 100 Å GaAs/Al$_{0.3}$Ga$_{0.7}$As quantum well. You will need to find a proper mathematical library through your computer center to diagonalize the large matrix.

Section 2.10

2.30 Calculate the splitting between the heavy hole and light hole states at the zone edge when In$_{0.2}$Ga$_{0.8}$As is grown lattice matched to GaAs.

2.31 Consider a 100 Å GaAsP/Al$_{0.3}$Ga$_{0.7}$As quantum well. At what composition of P will be zone center heavy hole and light hole states merge?

2.32 Consider a 100 Å In$_x$Ga$_{1-x}$As/Al$_{0.3}$GA$_{0.7}$As (001) quantum well structure. Calculate the heavy hole effective mass near the zone edge along the (100) and (110) direction as x goes from 0 to 0.3, by solving the Kohn-Luttinger equation. You may write the Kohn-Luttinger equation in the difference form and solve it by calling a matrix solving subroutine from your computer library.

2.33 A 150 Å In$_x$Ga$_{1-x}$As quantum well is to be designed so that the HH-LH separation is $\frac{3}{2}k_BT$ at 300 K. Calculate the In composition needed, assuming the infinite potential approximation.

Section 2.12

2.34 Calculate the effective density of states at the conduction and valence bands of Si and GaAs at 77 K, 300 K and 500 K.

2.35 Calculate the intrinsic carrier concentration of Si, Ge and GaAs as a function of temperature from 4 K to 600 K. Assume that the bandgap is given by

$$E_g(T) = E_g(0) - \frac{\alpha T^2}{T + \beta}$$

where $E_g(0)$, α, and β are given by

$$Si \quad : \quad E_g(0) = 1.17 \ eV; \alpha = 4.37 \times 10^{-4} \ eV \ K^{-1}; \beta = 636 \ K$$
$$Ge \quad : \quad E_g(0) = 0.74 \ eV; \alpha = 4.77 \times 10^{-4} \ eV \ K^{-1}; \beta = 235 \ K$$
$$GaAs \quad : \quad E_g(0) = 1.519 \ eV; \alpha = 5.4 \times 10^{-4} \ eV \ K^{-1}; \beta = 204 \ K$$

2.15 REFERENCES

- **General Bandstructure**

  - F. Bassani, in *Semiconductors and Semimetals*, eds. R. K. Willardson and A. C. Beer, Academic Press, New York, vol. 1, p. 21 (1966).
  - For general electronic properties a good reference is: H. C. Casey, Jr. and M. B. Panish, *Heterostructure Lasers, Part A, Fundamental Principles*, and *Part B, Materials and Operating Characteristics*, Academic Press, New York (1978).
  - J. R. Chelikowsky and M.L. Cohen, *Physical Review*, B, 14, 556 (1976).
  - G. C. Fletcher, *Electron Band Theory of Solids*, North-Holland, Amsterdam (1971).
  - W. A. Harrison, *Electronic Structure and Properties of Solids*, W. H. Freeman, San Francisco (1980).
  - E. O. Kane, *Semiconductors and Semimetals*, eds. R. K. Willardson and A. C. Beer, Academic Press, New York, vol. 1, p. 81 (1966).
  - J. C. Phillips, *Bonds and Bands in Semiconductors*, Academic Press, New York (1973).
  - K. Seeger, *Semiconductor Physics: An Introduction*, Springer, Berlin (1985).
  - H. F. Wolf, *Semiconductors*, Wiley-Interscience, New York (1971).

- **The $k.p$ Method**

  - D. A. Broido and L. J. Sham, *Phys. Rev.*, B, **31**, 888 (1985).
  - E. O. Kane, *J. Phys. Chem. Solids*, **8**, 38 (1959).
  - E. O. Kane, in *Semiconductors and Semimetals*, eds. R. K. Willardson and A. C. Beer, Academic Press, New York, vol. 1, p. 81 (1966).
  - P. Lawaetz, *Phys. Rev.*, B, **4**, 3460 (1971).
  - J. Luttinger, *Phys. Rev.*, **102**, 1030 (1956).
  - J. Luttinger and W. Kohn, *Phys. Rev.*, **97**, 869 (1955).

- **Semiconductor Alloys: Electronic Properties**

  - S. Adachi, *J. Appl. Phys.*, **58**, R-1 (1985).
  - For general alloy property values a good reference text is Casey, H. C., Jr. and M. B. Panish, *Heterostructure Lasers, Part A, Fundamental Principles, Part B, Materials and Operating Characteristics*, Academic Press, New York (1978).
  - For details of various methodologies beyond the virtual crystal approximation a good text is E. N. Economou, *Green's Functions in Quantum Physics*, Springer-Verlag, New York (1979).

- **Heterojunctions**

  - R. S. Bauer, P. Zurcher and H. W. Sang, *Appl. Phys. Lett.*, 43, 663 (1983).
  - W. A. Harrison, *J. Vac. Sci. Technol.*, **14**, 1016 (1977).
  - H. Kroemer, in *Molecular Beam Epitaxy and Heterostructures*, eds. L. L. Chang and K. Ploog, NATO ASI Series E, No. 87, Martinus Nijhoff, Dordrecht, Netherlands (1985).
  - A. G. Milnes and D. L. Feucht, *Heterojunctions and Metal Semiconductor Junctions*, Academic Press, New York (1972).
  - J. Tersoff, *Phys. Rev. Lett.*, **56**, 2755 (1986).
  - C. G. Van de Welle and R. M. Martin, *J. Vac. Sci. Technol.*, B, **4**, 1055 (1986).

- **Bandstructure in Quantum Wells**

  - For a simple discussion of electrons in quantum wells any book on basic quantum mechanics is adequate. An example is L. Schiff, *Quantum Mechanics*, McGraw-Hill, New York (1968).
  - H. Akera, S. Wakahana, and T. Ando, *Surf. Sci.*, **196**, 694 (1988).
  - E. Bangerk and G. Landwehr, *Superlattices and Microstructures*, **1**, 363 (1985).
  - D. A. Broido and L. J. Sham, *Phys. Rev.*, B, **31**, 888 (1985).
  - S. C. Hong, M. Jaffe, and J. Singh, *IEEE J. Quant. Electron.*, **QE-23**, 2181 (1987).
  - G. D. Sanders and Y. C. Chang, *Phys. Rev.*, B, **31**, 6892 (1985).
  - V. Sankaran and J. Singh, *Appl. Phys. Lett.*, **59**, 1963 (1991).
  - U. Ukenberg and M. Altarelli, *Phys. Rev.*, B, **30**, 3569 (1984).

- **Bandstructure Modification by Epitaxially Produced Strain**

  - J. M. Hinckley and J. Singh, *Phys. Rev.*, B, **42**, 3546 (1990).
  - H. Kato, N. Iguchi, S. Chika, M. Nakayama, and N. Sano, *Jap. J . Appl. Phys.*, **25**, 1327 (1986).
  - B. K., Laurich, K. Elcess, C. G. Fonstad, J. G. Berry, C. Mailhiot, and D. L. Smith, *Phys. Rev. Lett.*, **62**, 649 (1989).
  - C. Mailhiot and D. L. Smith, *J. Vac. Sci. Technol.*, A, **5**, 2060 (1987).
  - C. Mailhiot and D. L. Smith, *Phys. Rev.*, B, **35**, 1242 (1987).
  - E. P. O'Reilly, *Semicond. Sci. Technol.*, **4**, 121 (1989).
  - G. C. Osbourn, *J. Appl. Phys.*, **53**, 1586 (1982).
  - G. C. Osbourn, *J. Vac. Sci. Technol.*, B, **3**, 1586 (1985).
  - R. People, *IEEE J. Quant. Electron.*, **QE-22**, 1696 (1986).
  - J. Singh, *Physics of Semiconductors and Their Heterostructures*, McGraw-Hill, New York (1993).

CHAPTER
3

DOPING AND
CARRIER TRANSPORT

3.1 INTRODUCTION

In Chapter 2 we have examined the important electronic properties of semiconductors. We have seen that in pure semiconductors the density of electrons (holes) that can participate in current flow is extremely small. In fact, pure semiconductors are rarely used for device applications by themselves. Semiconductors become useful when one uses the concept of doping to alter, in a controllable manner, the density of carriers that can carry current. In this chapter we will examine the physics behind doping. An important consideration in electronic and optoelectronic devices is how electrons respond to electric fields and concentration gradients in carrier density. To understand this response we need to examine the transport theory which will also be discussed in this chapter.

Another important issue in optoelectronic devices is how "hot" electrons or "hot" holes (highly energetic carriers) injected into a semiconductor lose their energy. This energy relaxation process is extremely important in the dynamic performance of many optoelectronic devices and will be examined.

3.2 DOPING: DONORS AND ACCEPTORS

$\mathcal{R}$ We have seen that the free carriers that can carry current in pure semiconductors have a very low density. To increase the free carrier density, impurities known as dopants are introduced. The dopants are chosen from the periodic table so that they either have an extra electron in their outer shell compared to the host semiconductor, or have one less electron. The resulting dopant is called a donor or acceptor. The donor atom is treated as a shallow defect (see Section 2.12) and the electron-donor interaction is represented by a Coulombic potential. The problem is solved by a simple analogy to the hydrogen atom problem. The donor produces a shallow level whose energy is simply given by the hydrogen atom problem in quantum mechanics, except that the mass of the electron is replaced by the effective mass at the conduction bandedge:

$$E_d = E_c - \frac{e^4 m_e^*}{2(4\pi\epsilon)^2 \hbar^2} \frac{1}{n^2}, n = 1, 2, ... \tag{3.1}$$

A series of energy levels are produced, with the ground state energy level

$$
\begin{aligned}
E_d &= E_c - \frac{e^4 m_e^*}{2(4\pi\epsilon)^2 \hbar^2} \\
&= E_c - 13.6 \left(\frac{m^*}{m_o}\right) \left(\frac{\epsilon_o}{\epsilon}\right)^2 \ eV
\end{aligned}
\tag{3.2}
$$

The values of the ground state donor energy are shown for some semiconductors in Table 3.1.

At this point we introduce another effective mass called the conductivity effective mass m_σ^* which tells us how electrons respond to external potentials. This mass is used for donor energies as well as for charge transport in an electric field. For direct bandgap materials like GaAs, this is simply the effective mass. For materials like Si the conductivity mass is

$$m_\sigma^* = 3 \left(\frac{2}{m_t^*} + \frac{1}{m_\ell^*}\right)^{-1} \tag{3.3}$$

In indirect bandgap materials the density of states mass is different from the conductivity mass. The density of states mass represents the properties of the electrons at a constant energy surface in the bandstructure. The conductivity mass, on the other hand, gives the response of electrons to an external potential.

According to the simple picture of the donor impurity discussed above, the donor energy levels depend only upon the host crystal (through ϵ and m^*) and *not* on the nature of the dopant. According to Eqn. 3.1, the donor energies for Ge, Si, and GaAs should be 0.006 V, 0.025, and 0.007 eV, respectively. However, as can be seen from Table 3.1, there is a small deviation from these numbers, depending upon the nature of the dopant. This difference occurs because the real impurity potential perturbation is not simply the Coulombic potential as assumed by us, but has a short range correction

Semiconductor	Impurity (Donor)	Shallow Donor Energy (meV)	Impurity (Acceptor)	Shallow Acceptor Energy (meV)
GaAs	Si	5.8	C	26
	Ge	6.0	Be	28
	S	6.0	Mg	28
	Sn	6.0	Si	35
Si	Li	33	B	45
	Sb	39	Al	67
	P	45	Ga	72
	As	54	In	160
Ge	Li	9.3	B	10
	Sb	9.6	Al	10
	P	12.0	Ga	11
	As	13.0	In	11

Table 3.1: Shallow level energies in some semiconductors. The values are all referred to as the energy *below* the conduction bandedge (for donors) and *above* the valence bandedge (for acceptors).

which depends upon the dopant impurity atom. More accurate theories for the donor levels include this potential to get a better agreement with the experiments.

Another important class of intentional impurities are the acceptors. Just as donors are defect levels which are neutral when an electron occupies the defect level and positively charged when unoccupied, the acceptors are neutral when empty and negatively charged when occupied by an electron. The acceptor levels are produced when impurities which have a similar core potential as the atoms in the host lattice, but have one less electron in the outermost shell, are introduced in the crystal. Thus group III elements can form acceptors in Si or Ge while Si could be an acceptor if it replaces As in GaAs.

The acceptor impurity potential could now be considered to be equivalent to a host atom potential together with a *negatively* charged Coulombic potential. The "hole" (i.e., the absence of an electron) can then bind to the acceptor potential. The effective mass equation can again be used since only the top of the valence band contributes to the acceptor level. The valence band problem is considerably more complex and requires the solution of multiband effective mass theory. However, acceptor level can be reasonably predicted by using the heavy hole mass. Due to the heavier hole masses, the Bohr radius for the acceptor levels is usually a factor of 2 to 3 smaller than that for donors.

EXAMPLE 3.1 Calculate the donor and acceptor level energies in GaAs and Si. The shallow level energies are in general given by

$$E_d = E_c - 13.6(eV) \times \frac{m^*/m_0}{(\epsilon/\epsilon_0)^2}$$

The conduction band effective mass in GaAs is 0.067 m_0 and $\epsilon = 13.2\epsilon_0$, and we get for the donor level

$$E_d(GaAs) = E_c - 5.5 \; meV$$

The problem in silicon is a bit more complicated since the effective mass is not so simple. For donors we need to use the conductivity mass which is given by

$$m_\sigma^* = \frac{3 \; m_0}{\left(\frac{1}{m_\ell^*} + \frac{2}{m_t^*} \right)}$$

where m_ℓ^* and m_t^* are the longitudinal and transverse masses for the silicon conduction band. Using $m_\ell^* = 0.98$ and $m_t^* = 0.2 \; m_0$, we get

$$m_\sigma^* = 0.26 \; m_0$$

Using $\epsilon = 11.9\epsilon_0$, we get

$$E_d(Si) = 25 meV$$

The acceptor problem is much more complicated due to the degeneracy of the heavy hole and light hole band. The simple hydrogen atom problem does not give very accurate results. However, a reasonable approximation is obtained by using the heavy hole mass ($\sim 0.45 \; m_0$ for GaAs, $\sim 0.5 \; m_0$ for Si), we get

$$E_a(Si) \quad = \quad 48 \; meV$$
$$E_a(GaAs) \quad \cong \quad 36 \; meV$$

This is the energy above the valence bandedge. However, it must be noted that the use of the heavy hole mass is not strictly valid.

3.2.1   Extrinsic Carrier Density

In the lowest energy state of the donor atom, the extra electron of the donor is trapped at the donor site and occupies the donor level E_d. Such an electron cannot carry any current and is not useful for changing the electronic properties of the semiconductor. At very low temperatures, the donor electrons are, indeed, tied to the donor sites and this effect is called carrier freeze out. At higher temperatures, however, the donor electron

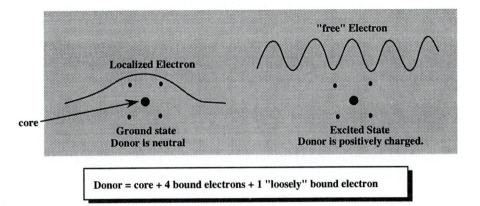

Donor = core + 4 bound electrons + 1 "loosely" bound electron

Figure 3.1: An electron bound to a donor does not contribute to charge conduction. However, if the donor is "ionized," the electron becomes free and can contribute to the charge transport.

is "ionized" and resides in the conduction band as a free electron as shown in Fig. 3.1. Such electrons can, of course, carry current and modify the electronic properties of the semiconductor. The ionized donor atom is positively charged.

In Section 2.11, we had calculated expressions for the free electron and hole densities and the Fermi level in a pure semiconductor. We will now generalize the results for the case where dopants are present in a semiconductor. Because of the doping (N_d is the donor density, N_a is the acceptor density), we no longer have the equality between electrons and holes, i.e.,

$$n - p = \Delta n \neq 0 \tag{3.4}$$

However, the law of mass action still holds (the value of the product changes only at high doping levels as will be discussed later)

$$np = \text{constant} = n_i^2 \tag{3.5}$$

If we eliminate p in these equations we get, after solving a quadratic equation

$$n = \frac{1}{2}\Delta n + \frac{1}{2}(\Delta n^2 + 4n_i^2)^{1/2} \tag{3.6}$$

Similarly, we have for the holes

$$p = -\frac{1}{2}\Delta n + \frac{1}{2}(\Delta n^2 + 4n_i^2)^{1/2} \tag{3.7}$$

Using the approximation, the distribution function can be replaced by a Boltzmann-like function $(f(E) = exp(-(E - E_F)/k_BT))$. We get (using Eqn. 2.73 for $n = n$ and

$n = n_i$)

$$\boxed{\frac{n}{n_i} = e^{(E_F - E_{Fi})/k_B T}} \tag{3.8}$$

and similarly from Eqn. 2.75 ($n_i = p_i$)

$$\boxed{\frac{p}{n_i} = e^{-(E_F - E_{Fi})/k_B T}} \tag{3.9}$$

where E_F is the Fermi level corresponding to the doped semiconductor. This gives us the simple equation

$$\frac{n - p}{n_i} = \frac{\Delta n}{n_i} = 2 \sinh\left(\frac{(E_F - E_{Fi})}{k_B T}\right) \tag{3.10}$$

When the semiconductor is doped n-type (p-type), the Fermi level moves towards the conduction (valence) bandedge. In this case, the Boltzmann approximation is not very good and the simple expressions relating the carrier concentration and the Fermi level are not very accurate. The carrier density is

$$n = \frac{1}{2\pi^2}\left(\frac{2m^*}{\hbar^2}\right)^{3/2} \int_{E_c}^{\infty} \frac{(E - E_c)^{1/2} dE}{exp\left(\frac{E - E_F}{k_B T}\right) + 1} \tag{3.11}$$

Defining

$$\eta = \frac{E - E_c}{k_B T} \, ; \eta_F = \frac{E_F - E_c}{k_B T} \tag{3.12}$$

$$n = \frac{1}{2\pi^2}\left(\frac{2m^* k_B T}{\hbar^2}\right)^{3/2} \int_0^{\infty} \frac{\eta^{1/2} d\eta}{exp\left(\eta - \eta_F\right) + 1} \tag{3.13}$$

The integral is called the Fermi-Dirac integral (or Fermi half integral)

$$F_{1/2}(\eta_F) = \int_0^{\infty} \frac{\eta^{1/2} d\eta}{exp\left(\eta - \eta_F\right) + 1} \tag{3.14}$$

and, using the definition of the effective density N_c

$$n = \frac{2}{\sqrt{\pi}} N_c F_{1/2}(\eta_F) \tag{3.15}$$

The value of this integral is plotted in Fig. 3.2 and can be used to calculate the carrier density for a given position of the Fermi level or η_F.

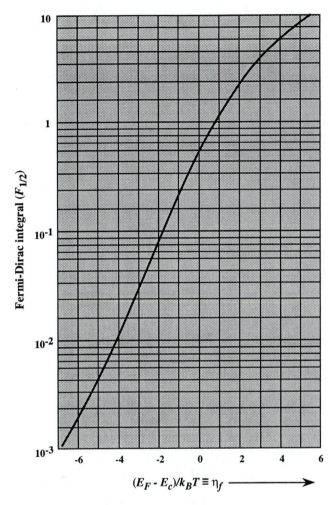

Figure 3.2: A plot of the Fermi-Dirac integral $F_{1/2}(\eta_F)$ as a function of η_F. The graph can be used to calculate an accurate value of the electron or hole densities for a given position of the Fermi level.

A useful expression for the relation between the carrier concentration and the Fermi level is the Joyce-Dixon approximation. According to this relation we have

$$E_F = E_c + k_B T \left[\ell n \frac{n}{N_c} + \frac{1}{\sqrt{8}} \frac{n}{N_c} \right] = E_v - k_B T \left[\ell n \frac{p}{N_v} + \frac{1}{\sqrt{8}} \frac{p}{N_v} \right] \quad (3.16)$$

This relation can be used to obtain the Fermi level if n is specified. Or else, it can be used to obtain n if E_F is known by solving for n iteratively. *If the term* $(n/\sqrt{8}\ N_c)$ *is ignored the result corresponds to the Boltzmann approximation.*

At room temperature and above, the donors and acceptors are essentially all ionized. Three regions can be identified. At low temperatures, the electrons are localized at the donors so that the free carrier density is very low. This is the freezeout regime. At higher temperatures $(k_B T \gtrsim E_c - E_d)$, the donors are ionized and the free carrier density (electrons) is essentially equal to the donor density. This is the saturation regime. Finally, at very high temperatures, the intrinsic carrier density dominates. We will assume that we are in the saturation region where the free electron density is equal to the donor density for n-type materials and hole density is equal to acceptor density for p-type materials.

EXAMPLE 3.2 A sample of GaAs has a free electron density of 10^{17} cm^{-3}. Calculate the position of the Fermi level using the Boltzmann approximation and the Joyce-Dixon approximation at 300 K.

In the Boltzmann approximation, the carrier concentration and the Fermi level are related by the following equation (E_F is measured from the bandedge):

$$E_F = k_B T \left[\ell n \frac{n}{N_c} \right]$$

$$= 0.026 \left[\ell n \left(\frac{10^{17}}{4.45 \times 10^{17}} \right) \right] = -0.039 \ eV$$

The Fermi level is 39 meV below the conduction band. In the Joyce-Dixon approximation we have

$$E_F = k_B T \left[\ell n \left(\frac{n}{N_c} \right) + \frac{1}{\sqrt{8}} \frac{n}{N_c} \right]$$

$$= 0.026 \left[\ell n \left(\frac{10^{17}}{4.45 \times 10^{17}} \right) + \frac{10^{17}}{\sqrt{8}(4.45 \times 10^{17})} \right]$$

$$= -0.039 + 0.002 = -0.037 \ eV$$

The error produced by using the Boltzmann approximation (compared to the more accurate Joyce-Dixon approximation) is 2 meV.

EXAMPLE 3.3 Assume that the Fermi level in silicon coincides with the conduction band-edge at 300 K. Calculate the electron carrier concentration using the Boltzmann approximation, the Joyce-Dixon approximation, and the Fermi-Dirac integral.

In the Boltzmann approximation, the carrier density is simply

$$n = N_c = 2.78 \times 10^{19} \ cm^{-3}$$

According to the Joyce-Dixon approximation, the carrier density is obtained from the solution of the equation

$$E_F = 0 = k_B T \left[\ell n \frac{n}{N_c} + \frac{n}{\sqrt{8}N_c} \right]$$

This gives

$$\frac{n}{N_c} = 0.76 \text{ or } n = 2.11 \times 10^{19} cm^{-3}$$

From the Fermi-Dirac integral we see that at $\eta_F = 0$, $F_{1/2}(0) \cong 0.65$

$$n = \frac{2}{\sqrt{\pi}} N_c F_{1/2}(0) = 0.74 N_c$$

We see that the Joyce-Dixon result gives a very close match with the exact calculation using the Fermi-Dirac integral. However, in this case the Boltzmann approximation gives a higher charge density.

EXAMPLE 3.4 A semiconductor is said to be n-type degenerate when the probability that the conduction bandedge electronic levels are occupied by electrons is close to unity. Similarly one can define a p-type degenerate semiconductor. Assume a criterion that the Fermi level has to be $\sim 3k_B T$ into the band before the material can be called degenerate. What are the free electron densities in Si and GaAs before the semiconductors are n-type degenerate?

We will use the Joyce-Dixon approximation

$$E_F = -E_c = k_B T \left[\ell n \frac{n}{N_c} + \frac{1}{\sqrt{8}} \frac{n}{N_c} \right]$$

Using $E_F = -E_c = 3k_B T$ we get

$$\frac{n}{N_c} = 4.4$$

If we were to use the Fermi-Dirac integral, $F_{1/2}(3) = 4$, which gives

$$\frac{n}{N_c} = \frac{2 \times 4}{\sqrt{\pi}} = 4.51$$

The two results are quite similar. For Si this leads to a carrier density of

$$n = 4.4 \times 2.78 \times 10^{19} = 1.22 \times 10^{20} \ cm^{-3}$$

For GaAs the density for degeneracy is

$$n = 4.4 \times 4.45 \times 10^{17} = 1.96 \times 10^{18} \ cm^{-3}$$

It must be noted, however, that the condition for degeneracy used in this example is somewhat arbitrary.

3.2.2 Heavily Doped Semiconductors

$\mathcal{R}$ In the theory discussed so far, we have made several important assumptions which are valid only when the doping levels are low: i) we have assumed that the bandstructure of the host crystal is not seriously perturbed and the bandedge states are still described by simple parabolic bands; ii) the dopants are assumed to be independent of each other and their potential is thus a simple Coulombic potential. These assumptions become invalid as the doping levels become higher. The Bohr radius of the impurity states is of the order of 100 Å. Thus when the average spacing of the impurity atoms reaches this level, the potential seen by the impurity electron is influenced by the neighboring impurities. In a sense this is like the problem of electrons in atoms. When the atoms are far apart, we get discrete atomic levels. However, when the atomic separation reaches a few angstroms, as in a crystal, we get electronic bands. At high doping levels we get impurity bands. Several other important effects occur at high doping levels. All these effects require us to abandon our simple picture that works well for low doping levels. The many body effects which self-consistently include the effects of other free electrons present in the system at high doping levels are an important area of research. This is especially so for low dimensional systems such as quantum wells and quantum wires.

An important effect of heavy doping is the narrowing of the bandgap. This can have serious effects on performance of devices such as bipolar transistors. In silicon, if N_d is the donor density (cm^{-3}), the bandgap narrowing is given by a simple expression:

$$\Delta E_g \cong -22.5 \left(\frac{N_d}{10^{18}} \frac{300}{T(K)} \right)^{1/2} meV \qquad (3.17)$$

This expression gives reasonable agreement with experiments at low doping levels. At high doping levels it over-estimates the bandgap narrowing. However, due to its simplicity, we can use it to get an estimate of bandgap narrowing.

3.3 MODULATION DOPING

$\mathcal{R}$ We have already discussed how doping is an integral part of semiconductor devices. The purpose behind doping of semiconductors is to controllably change the free carrier density in the semiconductor. This requires that the dopant be ionized. When the donor is ionized, a positively charged ion is present in the crystal. This fixed charged center causes scattering for the free electron. As discussed in Section 3.5, ionized impurity scattering is an important scattering mechanism in carrier transport. *Scattering causes a deterioration in the transport properties of electrons as we will study later in this chapter.* An obvious question that arises is whether one can have a controllable free electron density *without* scattering. The answer to this question is yes and this is realized through the concept of modulation doping. Before addressing this concept, it is worth remarking that the modulation doping also overcomes another problem with doping—the carrier freeze out problem. As we have discussed in the previous section, at low temperatures, the electrons are localized at the donor sites, thus reducing the free carriers available for conduction. This effect can negate some of the benefits of operating devices at low temperatures. The concept of modulation doping is able to overcome this problem as well.

Modulation doping can be understood by examining Fig. 3.3. A heterostructure is grown (say, GaAs/AlGaAs) and the high bandgap material is doped. In equilibrium the electrons associated with the donors see lower lying energy states in the narrow bandgap material and thus transfer to the GaAs region. This spatial separation of the positively charged donors and negatively charged electrons produces an electric field profile governed by the Poisson equation which causes a band-bending as shown in Fig. 3.3. Usually the dopants are placed some distance away from the heterointerface by including an undoped "spacer" region. The ionized impurity scattering is essentially eliminated by this physical separation between the mobile electrons and the fixed ionized scattering centers. Also since the electrons are at energy positions lower than the localized ground state of the donor atoms, the electrons remain mobile even at the lowest temperatures provided the material quality is pure. Extremely high sheet charge density of electrons ($\gtrsim 10^{12}$ cm^{-2}) can thus be maintained at low temperatures. Transistors based on such concepts (Modulation Doped Field Effect Transistors—MODFETs) can operate at low temperatures and are often used for detection of very weak signals from space and in other applications where low noise devices are required.

When the donors are separated from the mobile electrons, an electric field is created due to the charge separation. The electric field causes the formation of a "quantum well" as shown in Fig. 3.3. As discussed in Chapter 2, the quantum well has subbands starting at energies E_1, E_2 etc., as shown. The density of states of the electrons in each subband was given in Chapter 2. Using the effective mass of the electrons in the narrow gap quantum well region, the density of states is

$$N_{2D}(E) = \frac{m^*}{\pi \hbar^2} \tag{3.18}$$

Using Eqn. 2.71 for the total carrier density and the energy independent density of states, it is straightforward to see that the Fermi level measured from the bottom of the subband, (E_1), is related to the electron density by (assuming that only the first subband is occupied)

$$n_{2D} = \frac{m^* k_B T}{\pi \hbar^2} \, \ell n \left[1 + exp \left(\frac{E_F - E_1}{k_B T} \right) \right] \tag{3.19}$$

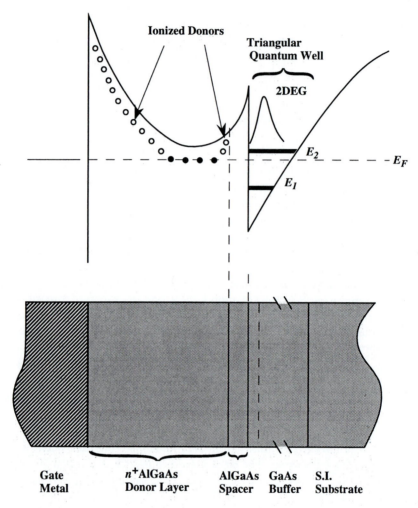

Figure 3.3: A modulation doped heterostructure showing the band profile and the layer sequence. The electrons are transferred to the narrow gap semiconductor where one has a 2-dimensional electron gas (2DEG).

or

$$E_F - E_1 = k_B T \, \ell n \left[exp \left(\frac{n_{2D} \pi \hbar^2}{m^* k_B T} \right) - 1 \right] \qquad (3.20)$$

EXAMPLE 3.5 The electrons in GaAs can be treated as if they have a mass of 0.067 m_0. Consider a 3D bulk GaAs where there are 10^{17} electrons per cubic centimeter, and a 2D GaAs quantum well which is 100 Å wide which also has the same electron density. Calculate the Fermi level position in each case at 300 K. Use the Joyce-Dixon approximation for the 3D system.

The effective density for the electrons in the 3D system is (using $m^* = 0.067\ m_0$)

$$
\begin{aligned}
N_c &= 2\left(\frac{m^*}{2\pi\hbar^2}\right)^{3/2}(k_BT)^{3/2} = 2\left[\frac{(0.067 \times 0.91 \times 10^{-30})(0.026 \times 1.6 \times 10^{-19})}{2\pi(1.05 \times 10^{-34})^2}\right]^{3/2} \\
&= 4.45 \times 10^{23}\ m^{-3} = 4.45 \times 10^{17}\ cm^{-3}
\end{aligned}
$$

The Fermi level is according to the Joyce-Dixon approximation,

$$
\begin{aligned}
E_F &= k_BT\left[\ell n\frac{n}{N_c} + \frac{1}{\sqrt{8}}\frac{n}{N_c}\right] = (0.026)\left[\ell n\left(\frac{10^{17}}{4.45 \times 10^{17}}\right) + \frac{10^{17}}{\sqrt{8}(4.45 \times 10^{17})}\right] \\
&= -36.7\ meV
\end{aligned}
$$

The Fermi level is 36.7 meV below the conduction band.

In the case of the quantum well the Fermi energy is determined by the surface density which, in our case, is

$$
n_{2D} = n_{3D} \times \text{well size} = 10^{17} \times 10^{-6} = 10^{11}\ cm^{-2}
$$

The Fermi energy position is given by (relative to the first subband)

$$
\begin{aligned}
E_F &= k_BT\ \ell n\left[exp\left(\frac{n_{2D}\pi\hbar^2}{m^*k_BT}\right) - 1\right] \\
&= 0.026(eV)\ell n\left[exp\left\{\frac{(10^{15}m^{-2})(\pi)(1.05 \times 10^{-34}J - s)^2}{(0.067)(0.91 \times 10^{-30}kg)(0.026 \times 1.6 \times 10^{-19})}\right\} - 1\right] \\
&= -77.8\ meV
\end{aligned}
$$

EXAMPLE 3.6 A GaAs/AlGaAs sample is modulation doped. This is done by putting a donor density of 2×10^{18} cm^{-3} over 100 Å of AlGaAs. Assume that all the donors spill their electrons into the GaAs region. If all these electrons go into the first subband of the quantum well, calculate the position of the Fermi level at 300 K.

The total 2D carrier density is

$$
n_{2D} = N_d \cdot d
$$

where d is the distance over which the dopants are placed. Thus,

$$
n_{2D} = 2 \times 10^{18} \times 10^{-6} = 2 \times 10^{12}\ cm^{-2}
$$

We have

$$
\begin{aligned}
E_F - E_1 &= k_BT\ \ell n\left[exp\left(\frac{n_{2D}\pi\hbar^2}{m^*k_BT}\right) - 1\right] \\
&= 0.026\ \ell n\left[exp\left(\frac{2 \times 10^{16} \times 3.1416 \times 1.1 \times 10^{-68}}{0.067 \times 0.91 \times 10^{-30} \times 0.026 \times 1.6 \times 10^{-19}}\right) - 1\right] \\
&= 0.069\ eV
\end{aligned}
$$

3.4   TRANSPORT IN SEMICONDUCTORS: A CONCEPTUAL PICTURE

$\mathcal{R}$ In Chapter 2 we discussed the Fermi-Dirac distribution which tells us how electrons are distributed in energy or momentum space at equilibrium. What happens if an electric field is applied or a concentration gradient in electron density is present so that the system is not in equilibrium? This is a complicated problem and involves the transport of electrons (holes) in a semiconductor.

The Bloch theorem tells us that in the perfect semiconductor the electron wavefunctions have the form (including the time dependent part)

$$\psi_k(r, t) = u_k \; exp \; i(k \cdot r - \omega t) \tag{3.21}$$

where $\omega = E/\hbar$ is the electron wave frequency. There is no scattering of the electron in the perfect system. Also, if an electric field F is applied, the electron behaves as a "free" space electron would, obeying the equation of the motion

$$\frac{\hbar dk}{dt} = F_{ext} = -eF \tag{3.22}$$

The electron would travel along a particular E-k band increasing and decreasing it's energy as shown schematically in Fig. 3.4. Such oscillations are called Bloch oscillations and scientists have been looking for them for decades since they could be used for very high frequency applications. Unfortunately, except at very low temperatures, in very special heterostructure samples, these oscillations do not occur because electrons scatter in real semiconductors.

In a real semiconductor, there are always imperfections which cause scattering of electrons so that the equation of motion of electrons is not given by Eqn. 3.22. A conceptual picture of electron transport can be developed where the electron moves in space for some time, then scatters and then again moves in space and again scatters. The process is shown schematically in Fig. 3.5. The average behavior of the ensemble of electrons will then represent the transport properties of the electron.

3.5   SCATTERING OF ELECTRONS

$\longrightarrow$ The key to understanding the non-equilibrium properties of electrons is the understanding of the scattering process of the electrons. The scattering problem in semiconductors is treated by using the perturbation theory in quantum mechanics (see Appendix

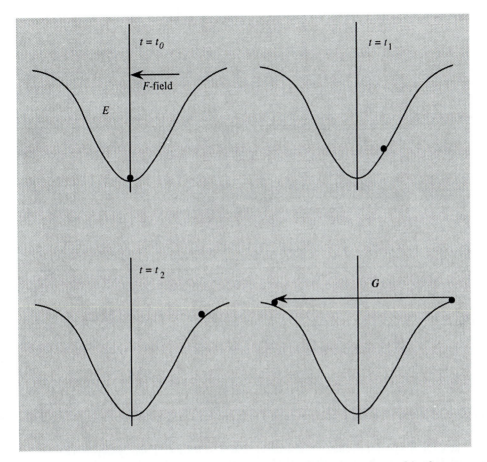

Figure 3.4: The motion of an electron in a band in absence of any scattering and in the presence of an electric field. The electron oscillates in k-space gaining and losing energy from the field.

C.2). We are interested in solving the quantum mechanics problem formally represented by

$$H\phi = E\phi \tag{3.23}$$

where H is the full hamiltonian (potential energy + kinetic energy operator) of the problem. This hamiltonian is, in our case, the sum of the hamiltonian of the perfect crystal H_o and the energy operator V corresponding to the imperfection causing scattering. Thus,

$$H = H_o + V \tag{3.24}$$

We know how to solve the problem

$$H_o\psi = E\psi \tag{3.25}$$

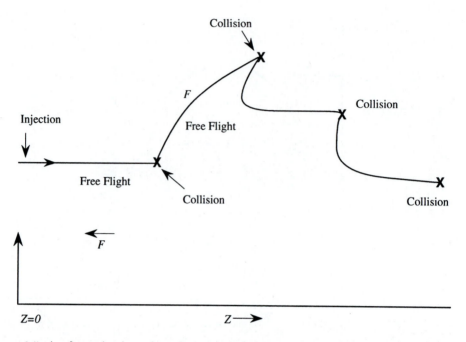

Figure 3.5: A schematic view of an electron as it moves under an electric field in a semiconductor. The electron suffers a scattering as it moves. In between scattering the electron moves
according to the "free" electron equation of motion.

which just gives us the bandstructure of the semiconductor. In the perturbation theory,
one uses the approach that the effect of the perturbation V is to cause scattering of
the electron from one perfect crystalline state to another. This theory works well if the
perturbation is small. The effect of the scattering is shown schematically in Fig. 3.6.
The rate of scattering for an electron initially in state i to a state f in the presence of
a perturbation of the form

$$V(r, t) = V(r)\, exp\,(i\omega \tau) \qquad (3.26)$$

is given by the Fermi golden rule (discussed in Appendix C)

$$W_{if} = \frac{2\pi}{\hbar} \mid M_{ij} \mid^2 \delta(E_i \pm \hbar\omega - E_f) \qquad (3.27)$$

where the various quantities in the equation represent the following:

$\frac{2\pi}{\hbar}$: this is a factor that appears from the details of the calculations.

$\mid M_{ij} \mid^2$: The quantity is called the matrix element of the scattering and is given
by

$$M_{ij} = \int \psi_f^* V(r)\psi_i d^3r \qquad (3.28)$$

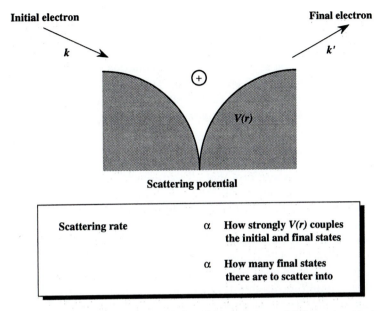

Figure 3.6: The scattering of an electron initially with momentum $\hbar k$ from a scattering potential $V(r)$. The final momentum is $\hbar k'$. The scattering process is assumed to be instantaneous. The scattering depends upon the coupling of the initial state to the final state by the scattering potential.

The matrix element tells us how the potential couples the initial and the final state.

$\delta(E_i \pm \hbar\omega - E_f)$: This δ-function is simply a representative of energy conservation. The process where

$$E_f = E_i + \hbar\omega \tag{3.29}$$

is called absorption, while the process

$$E_f = E_i - \hbar\omega \tag{3.30}$$

is called emission. Thus, both absorption or emission of energy can occur if the perturbation has a time dependence $\exp(i\omega t)$. If the potential is time independent, the scattering is elastic $(E_i = E_f)$.

In principle, the evaluation of the scattering rates is fairly straightforward since it simply involves the calculations of some integrals. In practice, the problem is complicated by the fact that the scattering potential $V(r)$ is not well defined and models have to be constructed to represent a defect by a proper potential. Thus, while it may be easy to describe the physical nature of the defect, it is quite difficult to represent the potential perturbation that the electron sees due to this defect. We will now briefly review some important scattering sources in semiconductors.

Phonon Scattering

In Chapter 1, we had discussed the crystalline structure in which atoms were at fixed periodic positions. In reality, the atoms in the crystal are vibrating. These lattice vibrations are called phonons and they satisfy an equation of motion similar to that of masses coupled to each other by springs. The properties of the lattice vibrations are represented by the relation between the vibration, amplitide, u, frequency, ω, and the wavevector q. The vibration of a particular atom is given by

$$u_i(q) = u_{oi} \; exp \; i(q \cdot r - \omega t) \tag{3.31}$$

which has the usual plane wave form that all solutions of periodic structures have. Recall that in a semiconductor there are two kinds of atoms in a basis. This results in a ω vs. k relation shown in Fig. 3.7. This relation which is for GaAs is typical of all compound semiconductors. One notices two kinds of lattice vibrations denoted by acoustic and optical. Additionally, there are two transverse and one longitudinal modes of vibration for each kind of vibration. The acoustic branch can be characterized by vibrations where the two atoms in the basis of a unit cell (i.e., Ga and As for GaAs) vibrate with the same sign of the amplitude as shown in Fig. 3.7b. In optical vibrations, the two atoms with opposing amplitudes are shown.

While the dispersion relations represent the allowed lattice vibration modes, an important question is how many such modes are actually being excited at a given temperature. In quantum mechanics the modes are called phonons and the number of phonons with frequency ω are given by

$$n_\omega = \frac{1}{exp \; \left(\frac{\hbar\omega}{k_B T} \right) - 1} \tag{3.32}$$

The lattice vibration problem is mathematically similar to the harmonic oscillator problem (discussed in Appendix C.1). The quantum mechanics of the harmonic oscillator problem tells us that the energy in the mode frequency ω is then

$$E_\omega = (n_\omega + \frac{1}{2})\hbar\omega \tag{3.33}$$

Note that even if there are no phonons in a particular mode, there is a finite "zero point" energy $\frac{1}{2}\hbar\omega$ in the mode.

The vibrations of the atoms produces three kinds of potential disturbances that result in the scattering of electrons. The phonon scattering is a key source of scattering and limits the performance of both electronic and optoelectronic devices. A schematic of the potential disturbance created by the vibrating atoms is shown in Fig. 3.8. In a simple physical picture, one can imagine the lattice vibrations causing spatial and temporal fluctuations in the conduction and valence band energies. The electrons (holes) then

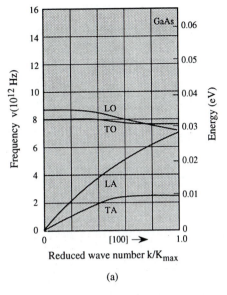

(a)

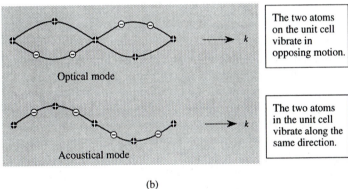

(b)

Figure 3.7: (a) Typical dispersion relations of a semiconductor (GaAs in this case). (b) The displacement of atoms in the optical and acoustic branches of the vibrations are shown. The motion of the atoms is shown for small k vibrations.

scatter form these disturbances. The acoustic phonons produce a strain field in the crystal and the electrons see a disturbance which produces a potential of the form

$$V_{AP} = D\frac{\partial u}{\partial x} \tag{3.34}$$

where D is called a deformation potential (units are eV) and $\frac{\partial u}{\partial x}$ is the amplitude gradient of the atomic vibrations.

The optical phonons produce a potential disturbance which is proportional to the atomic vibration amplitude, since in the optical vibrations the two atoms in the

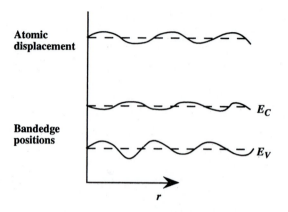

Figure 3.8: A schematic showing the effect of atomic displacement on bandedge energy levels in real space. The lattice vibrations cause spatial and time dependent variations in the bandedges from which the electrons scatter.

basis vibrate opposing each other

$$V_{op} = D_o u \tag{3.35}$$

where D_o (units are eV/cm) is the optical deformation potential.

In compound semiconductors where the two atoms in the basis of the crystal structure, an extremely important scattering potential arises additionally from optical phonons. Since the two atoms are different, there is an effective positive and negative charge e^* on each atom. When optical vibrations take place, the effective dipole in the unit cell vibrates causing polarization fields from which the electron scatters. This scattering, called polar optical phonon scattering has a scattering potential of the form

$$V_{po} \sim e^* u \tag{3.36}$$

The effective charge is related to the dc and high frequency dielectric constants of the semiconductor. In phonon scattering, the energy-momentum conservation laws require that the scattering primarily occurs with small q values, unless the electron moves from one valley to another. As a result, in acoustic phonon scattering, the electron energy does not change much after scattering (since $\omega \to 0$ at $q \to o$). However, for optical phonons, the scattering can increase or decrease the electron energy by $\hbar\omega_o$ depending upon whether absorption or emission of phonons has occurred.

The acoustic phonon scattering rate for an electron with energy E_k to any other state is given by

$$\boxed{W_{ac}(E_k) = \frac{2\pi D^2 k_B T N(E_k)}{\hbar \rho v_s^2}} \tag{3.37}$$

where $N(E_k)$ is the electron density of states, ρ is the density of the semiconductor, v_s is the sound velocity and T is the temperature.

In materials like GaAs, the dominant optical phonon scattering is polar optical phonon scattering and the scattering rate is given by (assuming the bandstructure is defined by a non-parabolic band; ϵ_∞ and ϵ_s are the high frequency and static dielectric constants of the semiconductor while ϵ_o is the free space dielectric constant)

$$W(k) = \frac{e^2 m^{*1/2} \omega_o}{4\pi\sqrt{2}\hbar} \left(\frac{\epsilon_o}{\epsilon_\infty} - \frac{\epsilon_o}{\epsilon_s} \right) \frac{1 + 2\alpha E'}{\gamma^{1/2}(E)} F_o(E, E')$$

$$\times \left\{ \begin{array}{ll} n(\omega_o) & \text{absorption} \\ n(\omega_o) + 1 & \text{emission} \end{array} \right. \tag{3.38}$$

where

$$
\begin{aligned}
E' &= E + \hbar\omega_o \text{ for absorption} \\
&= E - \hbar\omega_o \text{ for emission} \\
\gamma(E) &= E(1 + \alpha E) \\
F_o(E, E') &= C^{-1} \left(A\ell n \left| \frac{\gamma^{1/2}(E) + \gamma^{1/2}(E')}{\gamma^{1/2}(E) - \gamma^{1/2}(E')} \right| + B \right) \\
A &= [2(1 + \alpha E)(1 + \alpha E') + \alpha \{\gamma(E) + \gamma(E')\}]^2 \\
B &= -2\alpha\gamma^{1/2}(E)\gamma^{1/2}(E') \\
&= \times [4(1 + \alpha E)(1 + \alpha E') + \alpha \{\gamma(E) + \gamma(E')\}] \\
C &= 4(1 + \alpha E)(1 + \alpha E')(1 + 2\alpha E)(1 + 2\alpha E')
\end{aligned}
$$

The above equations assume a non-parabolic E-k relationship with non-parabolicity factor α (see Eqn. 2.49). It is important to examine typical values of scattering rates from these processes. The values for GaAs are shown in Fig. 3.9. *Note that the phonon emission process can start only after the electron has an energy equal to the phonon energy.* The emission rates are about 3 times as strong as the absorption rates at room temperature (the ratio between $n(\omega_o) + 1$ and $n(\omega_o)$).

Polar optical phonon scattering is the most important scattering mechanism for high field or high temperature transport of electrons. It is also responsible for relaxation of hot electrons injected into a semiconductor. The hot carrier relaxation is a key process in semiconductor laser performance and will be discussed later.

Ionized Impurity Scattering

In Section 3.2 we had discussed the importance of doping in semiconductors. When a dopant ionizes to produce an extra "free" electron, the electron scatters from the ion. The scattering potential is Coulombic in nature except that the potential is suppressed by screening effects. The screening is due to the presence of the other free electrons which form a cloud around the ion so the effect of the potential is short ranged. There are several models for the ionized impurity scattering potential. A popular form used is

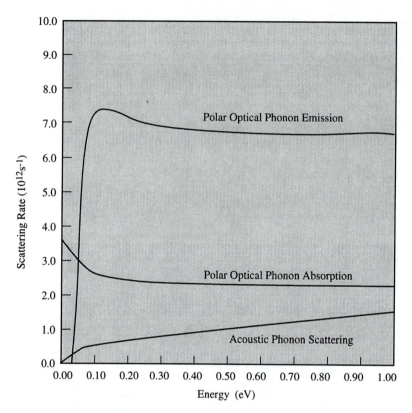

Figure 3.9: A comparison of the scattering rates due to acoustic and optical phonons for GaAs electrons at room temperature.

the screened Coulombic potential

$$V(r) = \frac{e^2}{\epsilon}\frac{e^{-\lambda r}}{r} \tag{3.39}$$

where

$$\lambda^2 = \frac{ne^2}{\epsilon k_B T} \tag{3.40}$$

with n the free electron density. The scattering rate then becomes, for an electron with energy E_k and momentum $\hbar k$

$$\boxed{\begin{aligned} W(k) &= 4\pi F\left(\frac{2k}{\lambda}\right)^2\left[\frac{1}{1+(\lambda/2k)^2}\right] \\ F &= \frac{1}{\hbar}\left(\frac{e^2}{\epsilon}\right)^2\frac{N(E_k)}{32k^4}N_I \end{aligned}} \tag{3.41}$$

where N_I is the ionized impurity density.

While the phonon and impurity scattering are the dominant scattering processes for most transport problems, electron-electron scattering, electron-hole scattering, alloy potential scattering, etc., can also play an important role.

EXAMPLE 3.7 Calculate the polar optical phonon emission rate for an electron in GaAs with energy 0.2 eV. Use the following parameters:

$$
\begin{aligned}
w_o &= 5.4 \times 10^{13}\,\mathrm{rad/s} \\
\hbar w_o &= 36\ meV \\
\frac{\epsilon_s}{\epsilon_o} &= 13.2 \\
\frac{\epsilon_\infty}{\epsilon_o} &= 10.9 \\
n(w_o) &= 0.33 \\
\alpha &= 0.6\ eV^{-1}
\end{aligned}
$$

The scattering rate is given by

$$
\begin{aligned}
W(k) &= \frac{(1.6 \times 10^{-19} C)^2 (0.067 \times 0.9 \times 10^{-30}\,kg)^{1/2}(5.4 \times 10^{13}\,\mathrm{rad/s})(1.33)}{4\Pi(1.414)(1.05 \times 10^{-34}\ J-s)(8.84 \times 10^{-12}\ F/m)} \\
&\quad \times \left(\frac{1}{10.9} - \frac{1}{13.2}\right)\left(\frac{1 + 2(0.6\ eV^{-1})(0.164\ eV)}{(0.224 \times 1.6 \times 10^{-19}\ J)^{1/2}}\right) F_o(E, E') \\
&= 2.43 \times 10^{12}\ F_o(E, E')\ s^{-1}
\end{aligned}
$$

The value of $F_o(E, E')$ can be found to be 2.87. Thus, the emission rate is

$$
W(k) = 7.0 \times 10^{12}\,s^{-1}
$$

EXAMPLE 3.8 Consider a semiconductor with effective mass $m^* = 0.26\ m_0$. The optical phonon energy is 50 meV. The carrier scattering relaxation time is 10^{-13} sec at 300 K. Calculate the electric field at which the electron can emit optical phonons on the average.

In this problem we have to remember that an electron can emit an optical phonon only if it's energy is equal to (or greater than) the phonon energy. According to the transport theory, the average energy of the electrons is (v_d is the drift velocity)

$$
E = \frac{3}{2}k_B T + \frac{1}{2}m^* v_d^2
$$

In our case, this has to be 50 meV at 300 K. Since $k_B T \sim 26$ meV at 300 K, we have

$$
\frac{1}{2}m^* v_d^2 = 50 - 39 = 11\ meV
$$

or

$$v_d^2 = \frac{2 \times (11 \times 10^{-3} \times 1.6 \times 10^{-12} erg)}{(0.91 \times 10^{-27} \times 0.26g)}$$

$$v_d = 1.22 \times 10^7 \ cm/s$$

Also we know that

$$v_d = \frac{e\tau F}{m^*}$$

Substituting for v_d, we get (for the average electrons)

$$F = \frac{(0.26 \times 0.91 \times 10^{-27} g)(1.22 \times 10^7 \ cm/s)}{(4.8 \times 10^{-10} \ esu)(10^{13} \ s)}$$

$$= 60.14 \ esu/cm = 18.04 kV/cm$$

The results discussed correspond approximately to silicon. Of course, since the distribution function has a spread, electrons start emitting optical phonons at a field lower than the one calculated above for the average electron.

3.6 MACROSCOPIC TRANSPORT PROPERTIES

$\mathcal{R}$ The scattering rates discussed in the previous section are only one of the ingredients of a transport theory. Note that the scattering rates are dependent upon the energy of the electron. What energy should be used to obtain transport properties? Clearly, an averaging must be carried out over the ensemble of the electrons in the semiconductor. However, this requires knowing the distribution function, which is only known at equilibrium. Since the scattering processes and the distribution function are inter-related at non-equilibrium, the problem is very complicated and various numerical and computer simulation techniques are developed to solve the problem.

Two important approaches to understanding transport in semiconductors are the solution of the Boltzmann transport equation using numerical methods and the Monte Carlo method using computer simulations. We will summarize the results of such theories by examining the drift velocity versus electric field relations in semiconductors.

3.6.1 Velocity-Electric Field Relations in Semiconductors

$\longrightarrow$ When an electron distribution is subjected to an electric field, the electrons tend to move in the field direction (opposite to the field **F**) and gain velocity from the field.

However, because of imperfections, they scatter in random directions. A steady state is established in which the electrons have some net drift velocity in the field direction. The response of the electrons to the field can be represented by a velocity-field relation. We will briefly discuss the velocity-field relationships at low electric fields and moderately high electric fields.

Low Field Response: Mobility

At low electric fields, the macroscopic transport properties of the material (mobility, conductivity) can be related to the microscopic properties (scattering rate or relaxation time) by simple arguments. We will not solve the Boltzmann transport equation, but we will use simple conceptual arguments to understand this relationship. We will follow an approach developed by Drude at the turn of the century, except that at the time Drude did not know about the concept of effective mass and the source of scattering. In this approach we make the following assumptions derived from modifications to the Drude approach:

i) The electrons in the semiconductor do not interact with each other. This approximation is called the independent electron approximation.

ii) Electrons suffer collisions from various scattering sources and the time τ_{sc} describes the mean time between successive collisions.

iii) The electrons move according to the free electron equation

$$\frac{\hbar dk}{dt} = \mathbf{F}_{ext} \tag{3.42}$$

in between collisions. After a collision, the electrons lose all their excess energy (on the average) so that the electron gas is essentially at thermal equilibrium. This assumption is really valid only at very low electric fields.

According to these assumptions, immediately after a collision the electron velocity is the same as that given by the thermal equilibrium conditions. This average velocity is thus zero, after collisions. The electron gains a velocity in-between collisions, i.e., only for the time τ_{sc}.

This average velocity gain is then that of an electron with mass m^* traveling in a field F for a time τ_{sc}:

$$v_{avg} = -\frac{eF\tau_{sc}}{m^*} = v_d \tag{3.43}$$

where v_d is the drift velocity. The current density is now

$$\mathbf{J} = -ne\mathbf{v}_d = \frac{ne^2\tau_{sc}}{m^*}\mathbf{F} \tag{3.44}$$

Comparing this with the Ohm's law result for conductivity σ

$$\mathbf{J} = \sigma\mathbf{F} \tag{3.45}$$

we have

$$\boxed{\sigma = \frac{ne^2\tau_{sc}}{m^*}} \tag{3.46}$$

The resistivity of the semiconductor is simply the inverse of the conductivity. From the definition of mobility μ, for electrons

$$\mathbf{v}_d = \mu\mathbf{F} \tag{3.47}$$

we have

$$\boxed{\mu = \frac{e\tau_{sc}}{m^*}} \tag{3.48}$$

Notice that the mobility has an explicit $\frac{1}{m^*}$ dependence in it. Additionally τ_{sc} decreases with m^* also as we had discussed in the previous section due to the dependence of scattering rates on the density of states. Thus the mobility has a strong dependence on the carrier mass. The effective mass to be used in these transport parameters is the conductivity mass. In Appendix B we show mobilities of some pure semiconductors.

The simple formulation above allows us to relate microscopic quantities such as effective mass and scattering time to a macroscopic property such as conductivity or mobility. The macroscopic quantity can be easily measured in the laboratory.

Low Field Mobility in Silicon
Si has a rather low electron mobility as compared to the mobilities in direct bandgap materials. Fortunately, in electronic devices high field transport dominates the device performance allowing Si devices to perform rather well. In Fig. 3.10 we show typical mobilities for Si as a function of temperature and doping. For high quality samples, the mobility continues to increase as the temperature decreases since ionized impurity scattering is absent and the lattice or phonon scattering decreases with temperature. However, for doped samples the mobility shows a peak at low temperatures and then decreases, since impurity scattering dominates transport at lower temperatures in doped materials.

Low Field Mobility in GaAs
The electrons in GaAs have a superior low field mobility as compared to the case in Si. This is mainly due to the lower density of states at the bandedge. Room temperature mobilities in high quality GaAs samples are ~ 8500 cm$^2V^{-1}s^{-1}$ compared to only

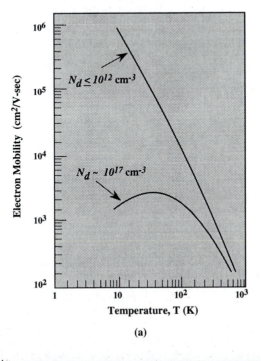

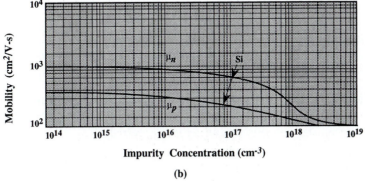

Figure 3.10: (a) Mobility of electrons in silicon as a function of doping and temperature; (b) mobility of electrons and holes in silicon as a function of doping at 300 K. (After S. Sze, *Physics of Semiconductor Devices*, John Wiley and Sons, New York (1981).)

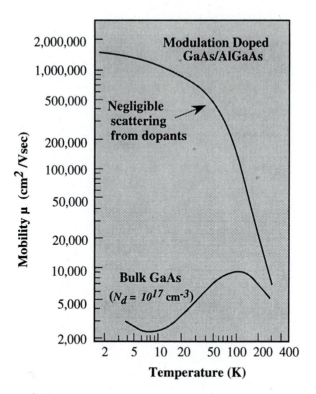

Figure 3.11: Typical mobility versus temperature for a modulation doped and uniformly doped (bulk GaAs) semiconductor structure. Mobilities as high as 10^7 cm^2/V-s have been achieved at 4 K in modulation doped structures, where scattering from dopants can be essentially removed by spatially separating the electrons from the dopants. (After the work of A.C. Gossard.)

~ 1500 cm$^2 V^{-1} s^{-1}$ for Si. Figure 3.11 shows a typical plot of mobility versus temperature for GaAs. Also shown is the mobility in a modulation doped GaAs/Al$_{0.3}$Ga$_{0.7}$As structure where ionized impurity scattering is suppressed.

Another semiconductor of considerable importance for high speed devices is In$_{0.53}$Ga$_{0.47}$As. The material has a very high room temperature mobility of $\sim$11,000 cm$^2 V^{-1} s^{-1}$. At low temperatures the mobility is dominated by alloy scattering effects, and is less than the mobility in pure GaAs. Low field mobilities of different materials are given in Appendix B.

EXAMPLE 3.9 The mobility of electrons in pure GaAs at 300 K is 8500 cm^2/V-s. Calculate the relaxation time. If the GaAs sample is doped at N$_d = 10^{17}$ cm^{-3}, the mobility decreases to 5000 cm^2/V-s. Calculate the relaxation time due to ionized impurity scattering.

The relaxation time is related to the mobility by

$$\tau_{sc}^{(1)} = \frac{m^*\mu}{e} = \frac{(0.067 \times 0.91 \times 10^{-30} kg)(8500 \times 10^{-4} m^2/V - s)}{1.6 \times 10^{-19} C}$$

$$= 3.24 \times 10^{-13} s$$

If the ionized impurities are present, the time is

$$\tau_{sc}^{(2)} = \frac{m^*\mu}{e} = 1.9 \times 10^{-13} s$$

According to Mathieson's rule, the impurity related time τ_{imp} is given by

$$\frac{1}{\tau_{sc}^{(2)}} = \frac{1}{\tau_{sc}^{(1)}} + \frac{1}{\tau_{sc}^{(imp)}}$$

which gives

$$\tau_{sc}^{(imp)} = 4.6 \times 10^{-13} s$$

EXAMPLE 3.10 The mobility of electrons in pure silicon is 1500 cm^2/V-s. Calculate the time between scattering events using the conductivity effective mass.

The conductivity mass is given by

$$m_\sigma^* = 3\left(\frac{2}{m_t^*} + \frac{1}{m_\ell^*}\right)^{-1}$$

$$= 3\left(\frac{2}{0.19\ m_0} + \frac{1}{0.98\ m_0}\right)^{-1} = 0.26\ m_0$$

The scattering time is then

$$\tau_{sc} = \frac{\mu m_\sigma^*}{e} = \frac{(0.26 \times 0.91 \times 10^{-30})(1500 \times 10^{-4})}{1.6 \times 10^{-19}}$$

$$= 2.2 \times 10^{-13} s$$

High Field Transport

In most electronic devices a significant portion of the electronic transport occurs under strong electric fields. This is especially true of field effect transistors. At such high fields ($F \sim 1$ - 100 kV/cm^{-1}) the electrons get "hot" and their temperature (electron temperature defined through their average energy) can be much higher than the lattice temperature. The extra energy comes due to the strong electric fields. The drift velocities are also quite high. The description of electrons at such high electric fields is quite complex and requires either numerical techniques (balance equation approach) or computer simulations (Monte Carlo method). In particular, the details of the bandstructure become very important. We will consider two generic cases: Si and GaAs which are representative of transport in indirect and direct bandgap materials. The velocity-field relations for several semiconductors are shown in Appendix B.

High Field Transport in Si
As discussed in Chapter 2, the bottom of conduction band in Si has six equivalent valleys. Up to electric fields of ~100 kV/cm^{-1}, the electrons remain in these valleys. As the field is increased, the velocity of the electrons increases steadily. However, gradually, as the electrons get hotter and occupy higher energy states, the scattering rates start to increase as well. This is mainly due to the increase in density of states with energy. *The velocity eventually saturates as shown in Fig. 3.12 at a value of $\sim 1.2 \times 10^7$ cm/s^{-1}.*

The monotonic increase and then saturation of velocity with electric field is generic of all indirect bandgap semiconductors (AlAs, Ge, etc.,) where electrons remain in the same lower valleys with a high effective mass.

High Field Transport in GaAs
Electron transport in GaAs has some very remarkable features which are exploited for microwave applications. These features are related to the negative differential resistance in the v-F relationships. In GaAs and other direct bandgap semiconductors, the conduction band minima are made up of a single valley with a very light mass $m^* = 0.067\ m_0$ (the so-called Γ- valley). Slightly higher up in energy, separated by ~0.3 eV, is a heavy mass valley (the L-valley) with $m^* = 0.22\ m_0$ as shown in Fig. 3.13a. At electric fields below ~3 kV/cm^{-1}, the electrons are primarily in the Γ-valley where they have low mass and a high mobility. The peak velocity in pure GaAs can reach a value of $\sim 2 \times 10^7$ cm/s^{-1} at ~2.5 kV/cm^{-1} as shown in Fig. 3.12. As the field increases, the fraction of electrons in the upper valley starts to increase as shown in Fig. 3.13b. Since the upper valley electrons are heavy, the *electron velocity decreases even though the field is increasing.* Eventually when most of the electrons are in the upper valley, the velocity saturates at $\sim 10^7$ cm/s^{-1}.

The results described for GaAs are typical of electron transport in most direct bandgap semiconductors. A material system that has become very important for high frequency microwave devices is In$_{0.53}$Ga$_{0.47}$As. This material has a very small carrier mass ($m^* = 0.04\ m_0$) and a large intervalley separation ($\Delta E_{L-\Gamma} \sim 0.55$ eV) which gives it a superior v-F relation than that of GaAs. The peak velocity in In$_{0.53}$Ga$_{0.47}$As is $\sim 2.4 \times 10^7$ cm/s^{-1}.

3.6.2   Transport of Holes

Hole transport is considerably more complex than electron transport, because of the complexity of the valence bandstructure. The density of states is very high for most

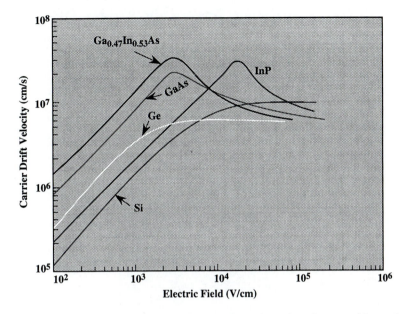

Figure 3.12: Velocity-field relation for electrons in several semiconductors. (A good source of velocity-field relations in several semiconductors is a chapter by S. Luryi, *High Speed Semiconductor Devices*, ed. S.M. Sze, John Wiley and Sons, New York (1990).)

semiconductors, the chief exception being Ge. In Fig. 3.14 we show velocity-field relations for Si and Ge.

Hole transport is extremely important in many electronic devices such as *p-n* junctions, bipolar transistors, and CMOS devices. It is also of importance in all optoelectronic devices where electrons and holes participate as partners in the device performance. Only in Ge is the hole transport comparable to that of the electrons.

EXAMPLE 3.11 The mobility of electrons in a semiconductor decreases as the electric field is increased. This is because the scattering rate increases as electrons become hotter due to the applied field. Calculate the relaxation time of electrons in silicon at 1 kV/cm and 100 kV/cm at 300 K.

The velocity of the silicon electrons at 1 kV/cm and 100 kV/cm is approximately 1.4×10^6 cm/s and 1.0×10^7 cm/s, respectively, from the *v-F* curves given in Appendix B. The mobilities are then

$$\mu(1kV/cm) = \frac{v}{F} = 1400 \ cm^2/V-s$$
$$\mu(100kV/cm) = 100 \ cm^2/V-s$$

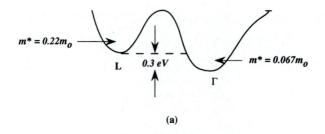

(a)

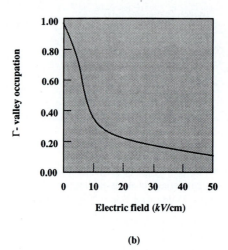

(b)

Figure 3.13: (a) The bandstructure of GaAs. At low fields, the electrons are all in the low mass Γ-valley. At high fields the electrons transfer by scattering into the high mass, low mobility L-valley. (b) The occupation of the Γ-valley in GaAs electron transport as a function of electric field.

The corresponding relaxation times are

$$\tau_{sc}(1kV/cm) = \frac{(0.26 \times 0.91 \times 10^{-30}\,kg)(1400 \times 10^{-4}\,m^2/V-s)}{1.6 \times 10^{-19}\,C} = 2.1 \times 10^{-13}\,s$$

$$\tau_{sc}(100kV/cm) = \frac{(0.26 \times 0.91 \times 10^{-30})(100 \times 10^{-4})}{1.6 \times 10^{-19}} = 1.48 \times 10^{-14}\,s$$

Thus the scattering rate has dramatically increased at the higher field.

EXAMPLE 3.12 The average electric field in a particular 2.0 μm GaAs device is 5 kV/cm. Calculate the transit time of an electron through the device a) if the low field mobility value of 8000 cm^2/V-s is used; b) if the saturation velocity value of 10^7 cm/s is used.

If the low field mobility is used, the average velocity of the electron is

$$v = \mu F = (8000\ cm^2/V-s) \times (5 \times 10^3\,V/cm) = 4 \times 10^7\ cm/s$$

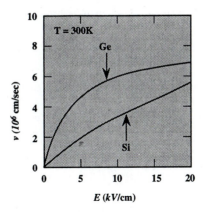

Figure 3.14: Si and Ge hole drift velocity vs. electric field relations. Ge has the best hole transport of all semiconductors. However, the material is not widely used because of processing related difficulties.

The transit time through the device becomes

$$\tau_{tr} = \frac{L}{v} = \frac{2.0 \times 10^{-4}\, cm}{4 \times 10^7}\ cm/s = 5ps$$

The transit time, if the saturation velocity (which is the correct velocity value) is used, is

$$\tau_{tr} = \frac{L}{v} = \frac{2 \times 10^{-4}}{10^7} = 20ps$$

3.7   VERY HIGH FIELD TRANSPORT: BREAKDOWN PHENOMENA

$\longrightarrow$ When the electric field becomes extremely high ($\gtrsim 100kV\,cm^{-1}$), the semiconductor suffers a "breakdown" in which the current has a "runaway" behavior. The breakdown occurs due to carrier multiplication which arises from the two sources discussed below. By carrier multiplication we mean that the number of electrons and holes that can participate in current flow increases. Of course, the total number of electrons is always conserved.

3.7.1   Impact Ionization or Avalanche Breakdown

In the transport considered in the previous sections, the electron (hole) remains in the same band during the transport. At very high electric fields, this does not hold true. In

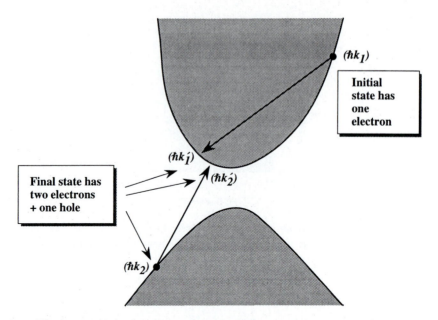

Figure 3.15: The impact ionization process where a high energy electron scatters from a valence band electron producing two conduction band electrons and a hole. Hot holes can undergo a similar process. Both energy and momentum conservation are required in such a process. Thus the initiating electron must have an energy above a minimum threshold.

the impact ionization process shown schematically in Fig. 3.15, an electron which is "very hot" scatters with an electron in the valence band via the Coulombic interaction and knocks it out into the conduction band. The initial electron must provide enough energy to bring the valence band electron up into the conduction band. Thus the *initial electron should have energy slightly larger than the bandgap* (measured from the conduction band minimum). In the final state we now have two electrons in the conduction band and one hole in the valence band. Thus the number of current carrying charges have multiplied, and the process is often called avalanching. Note that the same could happen to "hot holes" that could then trigger the avalanche.

In order for the impact ionization process to begin, the initiating electron with momentum $\hbar k_1$ must have enough energy to knock an electron from the valence band to the conduction band. Thus the initial electron must overcome a threshold energy before the carrier multiplication can begin. For most semiconductors, this threshold energy is essentially equal to the bandgap of the material

$$E_{k_1}(th) = E_g \tag{3.49}$$

where E_g is the material bandgap.

For the purpose of device applications, the current in the device, once avalanch-

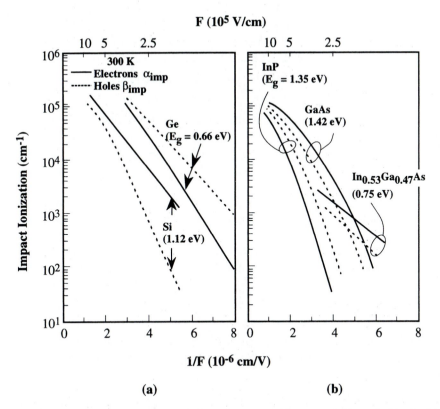

Figure 3.16: Ionization rates for electrons and holes at 300 K versus reciprocal electric field for Ge, Si, GaAs, In$_{0.53}$Ga$_{0.47}$ and InP. (Si, Ge results are after S.M. Sze, *Physics of Semiconductor Devices*, John Wiley and Sons (1981); InP, GaAs, InGaAs results are after G. Stillman, *Properties of Lattice Matched and Strained Indium Gallium Arsenide*, ed. P. Bhattacharya, INSPEC, London (1993).)

ing starts, is given by

$$\frac{dI(t)}{dt} = \alpha_t I \text{ or } \frac{dI(z)}{dz} = \alpha_z I \tag{3.50}$$

where I is the current and α_t or α_z represent the average rate of ionization per unit time or distance respectively.

The coefficients α_z (or simple α_{imp}) for electrons and β_z (or β_{imp}) for holes depend upon the bandgap of the material in a very strong manner. This is because, as discussed above, the process can only start if the initial electron has a kinetic energy equal to a certain threshold (roughly equal to the bandgap). This is achieved for lower electric fields in narrow gap materials. In Fig. 3.16 we show the impact ionization coefficients for some materials.

If the electric field is constant so that α_{imp} is constant, the number of times an initial electron will suffer impact ionization after travelling a distance x is

$$\boxed{N(x) = \frac{I(x)}{I(0)} = exp\,(\alpha_{imp}x)} \qquad (3.51)$$

Of course, the "daughter" electrons and holes produced will also ionize further.

A critical breakdown field F_{crit} is defined where α_{imp} or β_{imp} approaches 10^4 cm^{-1}. When α_{imp} (β_{imp}) approaches 10^4 cm^{-1}, there is about one impact ionization when a carrier travels a distance of one micron. Values of the critical field are given for several semiconductors in Appendix B. The avalanche process places an important limitation on the power output of devices. Once the process starts, the current rapidly increases due to carrier multiplication and the control over the device is lost. The push for high power devices is one of the reasons for research in large gap semiconductor devices. It must be noted that in devices such as avalanche photodetectors, the process is exploited for high gain detection. The avalanche photodetectors are important high performance photon detectors and will be discussed in Chapter 7.

EXAMPLE 3.13 Using the results in Fig. 3.16 for the impact ionization coefficients, calculate the number of secondary electrons generated by an electron travelling a distance of 1.0 μm in silicon at a field of 3×10^5 V/cm^{-1}.

The impact ionization coefficient for electrons is approximately 2×10^4 cm^{-1} at a field of 3×10^5 V cm^{-1} in silicon. The number of secondary electrons generated by a single initiating electron is

$$\begin{aligned} \frac{I(x)}{I(0)} &= exp\,(\alpha_{imp}W) = exp\,(2 \times 10^4 \times 10^{-4}) \\ &= 7.4 \end{aligned}$$

Thus a single electron causes 7.4 impact ionization events (on an average) as it moves a distance of 1.0 μm. Of course, these secondary electrons and holes will further generate more e- h pairs as they participate in the process.

EXAMPLE 3.14 According to one formalism for impact ionization scattering, only those electrons that are "lucky" enough to move in an electric field without scattering can gain enough energy to cause impact ionization. If electrons need to gain about a bandgap of energy to cause impact ionization and the scattering rates are $\sim 10^{13}$ s^{-1}, estimate the electric field at which impact ionization becomes important in GaAs.

The bandstructure in GaAs can be represented by the equation (for the conduction band)

$$E(1 + \alpha E(k)) = \frac{\hbar^2 k^2}{2m^*}$$

where $m^* = 0.067\, m_0$ and $\alpha\ (= 0.6\ eV^{-1})$ is called the non-parabolicity factor. If $\alpha = 0$, we have the usual parabolic E-k relation. The value of k at the point where the electron energy is 1.5 eV (= bandgap of GaAs) is given by

$$
\begin{aligned}
k^2 &= \frac{2m^* E(1 + \alpha E)}{\hbar^2} \\
&= \frac{2 \times (0.067 \times 0.91 \times 10^{-30}\, kg)(1.5 \times 1.6 \times 10^{-19}\, J)(1 + 0.6 \times 1.5)}{(1.05 \times 10^{-34}\, J - s)^2} \\
&= 5.04 \times 10^{18}\, m^{-2} \\
k &= 2.25 \times 10^9\, m^{-1}
\end{aligned}
$$

is

If an electron is moving without scattering starting at $k = 0$, after time τ_{sc}, its k-value

$$
k = \frac{eF\tau}{\hbar}
$$

where F is the applied field. Using $k = 2.25 \times 10^9\, m^{-1}$ and $\tau = 10^{-13}$ s, the field is

$$
\begin{aligned}
F &= \frac{\hbar k}{e\tau} = \frac{(1.05 \times 10^{-34}\, J - s)(2.25 \times 10^9\, m^{-1})}{(1.6 \times 10^{-19}\, C) \times (10^{-13}\, s)} \\
&= 1.47 \times 10^7\, V/m \\
&= 1.47 \times 10^5\, V/cm
\end{aligned}
$$

3.8 CARRIER TRANSPORT BY DIFFUSION

$\mathcal{R}$ So far we have studied carrier transport due to drift in an electric field. It is intuitively easy to see why the force due to an electric field would cause electrons to move a certain way. However, there is another important transport mechanism which does not involve such direct forces. This is the diffusion process. Whenever there is a gradient in the concentration of a species of mobile particles, the particles diffuse from the regions of high concentration to the regions of low concentration. This diffusion is due to the random motion of the particles which tends to increase the disorder or entropy of the system.

In the case of electrons (or holes), as the particles move they suffer random collisions as we had discussed earlier in the previous section. The collision process can be described by the mean free path ℓ and the mean collision time τ_{sc}. The mean free path is the average distance the electron (hole) travels in between successive collisions. These collisions are due to the various scattering processes that were discussed for the drift problem. In between the collisions the electrons move randomly with equal probability of moving in any direction (remember there is no electric field). We are interested in finding out how the electrons move (diffuse) when there is a concentration gradient in space.

The net flux of particles is given by

$$\phi_n(x,t) = -\frac{\ell^2}{2\tau_{sc}}\frac{dn(x,t)}{dx} = -D_n\frac{dn(x,t)}{dx} \qquad (3.52)$$

where D_n is the diffusion coefficient of the electron system and clearly depends upon the scattering processes which control ℓ and the τ_{sc}. *Since the mean free path is essentially* $v_{th}\tau_{sc}$, *where v_{th} is the mean thermal speed, the diffusion coefficient depends upon the temperature as well.*

Because of this electron and hole flux, a current can flow in the structure which in the absence of an electric field is given by (current is just charge multiplied by particle flux)

$$\begin{aligned} \mathbf{J}_{tot}(diff) &= \mathbf{J}_n(diff) + \mathbf{J}_p(diff) \\ &= eD_n\frac{\mathbf{d}n(x,t)}{\mathbf{d}x} - eD_p\frac{\mathbf{d}p(x,t)}{\mathbf{d}x} \end{aligned} \qquad (3.53)$$

While both electrons and holes move in the direction of less concentration of electrons and holes respectively, the currents they carry are opposite due to their charge difference.

In many electronic devices the charge moves under the combined influence of electric fields and concentration gradients. The current is then given by

$$\begin{aligned} \mathbf{J}_n(x) &= -e\mu_n n(x)\mathbf{F}(x) + eD_n\frac{\mathbf{d}n(x)}{\mathbf{d}x} \\ \mathbf{J}_p(x) &= e\mu_p p(x)\mathbf{F}(x) - eD_p\frac{\mathbf{d}p(x)}{\mathbf{d}x} \end{aligned} \qquad (3.54)$$

While using such a relation it is important to keep in mind that the electron and hole mobilities are not constant values. They are constant only at low electric fields. At high fields the mobilities decrease and the product μF reaches a constant corresponding to the saturation velocity. Thus the drift current saturates and becomes independent of the field.

There is a simple relation between the mobility and the diffusion coefficient given by

$$\frac{D_n}{\mu_n} = \frac{k_B T}{e} \qquad (3.55)$$

	D_n (cm^2/s)	D_p (cm^2/s)	μ_n $(cm^2V\text{-}s)$	μ_p $(cm^2/V\text{-}s)$
Ge	100	50	3900	1900
Si	35	12.5	1350	480
GaAs	220	10	8500	400

Table 3.2: Low field mobility and diffusion coefficients for several semiconductors at room temperature. The Einstein relation is satisfied quite well.

which is called the Einstein relation. A similar relation exists for the holes.

We list in Table 3.2 the mobilities and diffusion coefficients for a few semiconductors at room temperature. The Einstein relation is seen to be satisfied quite well.

At high electric fields the mobility decreases as an inverse of the electric field because of the decrease in the scattering time τ_{sc}. The diffusion coefficient also decreases at high fields. At high fields the Einstein relation does not hold as well as it does at low fields, since the electron distribution function is quite complicated.

EXAMPLE 3.15 In an n-type GaAs crystal at 300 K, the electron concentration varies as

$$n(x) = 10^{16} \ exp \left(-\frac{x}{L}\right) \ cm^{-3}. \qquad x > 0$$

where L is 1μm. Calculate the diffusion current density at $x = 0$ if the electron diffusion coefficient is 220 cm^2/s

The diffusion current density is

$$\begin{aligned}
J_n(diff) &= eD_n \left.\frac{dn}{dx}\right|_{x=0} \\
&= \left(1.6 \times 10^{-19} \ C\right)\left(220 \ cm^2/s\right)\left(\frac{10^{16} \ cm^{-3}}{10^{-4} \ cm}\right) \\
&= 3.5 \ kA/cm^2
\end{aligned}$$

Note that in this problem the diffusion current of electrons changes with the position in space. Since the total current is constant in the absence of any source or sink of current, some other current must be present to compensate for the spatial change in electron diffusion current.

EXAMPLE 3.16 Use the velocity-field relations for electrons in silicon to obtain the diffusion coefficient at an electric field of 1 kV/cm and 10 kV/cm at 300 K.

According to the v-F relations given in Appendix B, the velocity of electrons in silicon is $\sim 1.4 \times 10^6$ cm/s and $\sim 7 \times 10^6$ cm/s at 1 kV/cm and 10 kV/cm. Using the Einstein relation we have for the diffusion coefficient

$$D = \frac{\mu k_B T}{e} = \frac{v k_B T}{e F}$$

This gives

$$
\begin{aligned}
D(1kV/cm^{-1}) &= \frac{(1.4 \times 10^4 m/s)(0.026 \times 1.6 \times 10^{-19} J)}{(1.6 \times 10^{-19} C)(10^5 Vm^{-1})} \\
&= 3.64 \times 10^{-3} m^2/s = 36.4 \ cm^2/s \\
D(10kV/cm^{-1}) &= \frac{(7 \times 10^4 m/s)(0.026 \times 1.6 \times 10^{-19} J)}{(1.6 \times 10^{-19} C)(10^6 Vm^{-1})} \\
&= 1.82 \times 10^{-3} m^2/s = 18.2 \ cm^2/s
\end{aligned}
$$

The diffusion coefficient decreases with the field because of the higher scattering rate at higher fields.

3.9   CARRIER INJECTION AND THERMALIZATION

In both electronic and optoelectronic devices electrons and holes are injected into a semiconductor either by an external biasing circuit or by absorption of light, and usually these carriers are injected under non-equilibrium conditions. For example, in the optical absorption process to be discussed in the next chapter, electrons (holes) can be generated at energies much greater than the thermal energies ($\sim k_B T$). Similarly, in a Schottky diode or a p-n diode, carriers may be injected at high energies. These excess energy carriers are called hot electrons and, as shown in Fig. 3.17a, they lose their excess energies to reach energies corresponding to the conditions in the semiconductor.

If hole electrons are injected into a region of semiconductor where there is no electric field, the electrons will lose energy until they come in thermal equilibrium with the semiconductor lattice. In this state they will, on an average, have no net gain or loss of energy. If the hot electrons are injected in a region with an electric field, they will gain or lose energy until they are described by the distribution function of the electrons corresponding to the applied field. This process of carrier relaxation takes a time, say t_{relax}, and is shown schematically in Fig. 3.17b.

The carrier thermalization time is an extremely important parameter for the high speed performance of optoelectronic devices. This is especially true in semiconductor lasers and, as we shall discuss in Chapter 11, leads to important limitations. The carrier thermalization is due to the scattering processes that we discussed in the transport section of this chapter. Let us take a brief overview of the scattering processes and their role in the energy loss phenomenon:

Optical Phonon Emission and Absorption

Optical phonon emission is the most important energy loss mechanism and in each scattering event the electron loses energy equal to $\hbar\omega_o$ as discussed in Section 3.5. The rate of scattering from a state k to a state k' is given by (the *total scattering* rate to all possible states k' was given in Section 3.5.)

$$W(k, k') = \frac{2\pi}{\hbar} \frac{2\pi e^2 \hbar\omega_o}{V|k - k'|^2} \left(\frac{\epsilon_o}{\epsilon_\infty} - \frac{\epsilon_o}{\epsilon_s} \right)$$

$$\times \left\{ \begin{array}{ll} n(\omega_o)\delta(E(k') - E(k) - \hbar\omega_o) & \text{absorption} \\ (n(\omega_o) + 1)\delta(E(k') - E(k) + \hbar\omega_o) & \text{emission} \end{array} \right. \qquad (3.56)$$

where V is the volume of the system (rates are normalized to unit volume), and $(k-k') = q$ is the change in the wavevector of the electron due to the phonon with wavevector q. It is important to note the temperature dependence of the scattering rate via $n(\omega_o)$ and the dependence upon the wavevector change $(k-k') = q$. *The scattering is dominated by the small q scattering and decreases for large q values. This has important consequences for thermalization in low dimensional systems,* as will be discussed later.

The optical phonon absorption is another important energy exchange mechanism. The electron gains energy from the lattice by this process. The absorption rate is smaller than the emission rate by a factor $n(\omega_o)/(n(\omega_o) + 1)$, a factor that is about 1/3 for GaAs at 300 K (of course, if the electron energy is less than $\hbar\omega_o$, it cannot emit an optical phonon). For the hot carrier thermalization problem, there is a competition between energy loss by phonon emission and energy gain by phonon absorption. At low temperature, where $n(\omega_o)$ is small, the emission rate dominates, but as temperature increases, the rate $n(\omega_o)/(n(\omega_o)+1)$ approaches unity and it takes longer for the carriers to thermalize.

An important issue in regard to optical phonons is that if the electrons emit a high density of optical phonons, the phonon density $n(\omega_o)$ can become higher than the equilibrium value given by Eqn. 3.32. This is referred to as the hot phonon problem. As a result of this, the value of $n(\omega_o)$ can be quite large and the emission and absorption rates can approach unity. This causes the carrier thermalization time to be longer.

Acoustic Phonon Scattering

The acoustic phonon scattering rates are usually in the range of $\sim 10^9$ s^{-1} compared to optical phonon rates that can be $\sim 10^{12}$ s^{-1}. Thus, normally acoustic phonon scattering

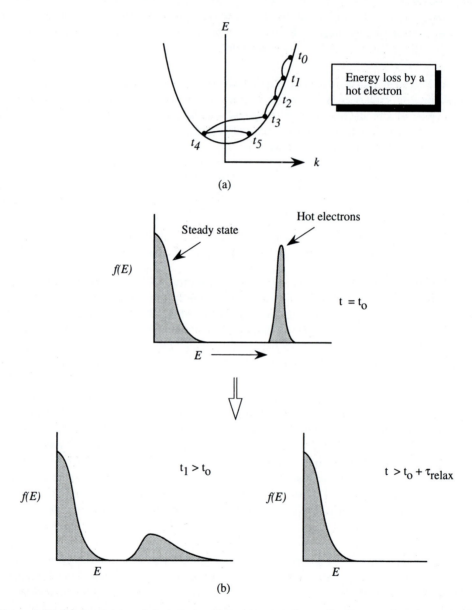

Figure 3.17: (a) An electron injected at a high energy at time t_o is shown to lose its energy by a series of scattering events. (b) A schematic of how a hot carrier distribution loses its energy to reach the thermal distribution function.

is not very important for energy loss. Also, since the acoustic phonon energies at small phonon wavevector is $\sim$ 1-2 meV, the energy loss is very small. However, if the electron is at an energy below the optical phonon energy $\hbar\omega_o$, the only way for it to lose its energy is by acoustic phonons.

Impurity Scattering

Impurity scattering, while very important for carrier mobility, does not cause energy loss since it is an elastic scattering. Other defect scattering processes such as alloy scattering also do not result in energy loss.

Electron-Electron Scattering

At high electron (hole) injection density the electrons can scatter from each other by Coulombic interaction. The scattering matrix element is quite similar to that in electron-ionized impurity scattering. When two electrons scatter from each other, the total energy does not change. Thus, the electron-electron scattering is only important for randomizing the electron energy, not for net energy loss.

Electron-Hole Scattering

Electron-hole scattering is another important Coulombic interaction mediated scattering process. This scattering process can use energy loss for electrons or for holes, although the total electron-hole energy is conserved in the scattering process. As we shall see later, this is an important process for carrier relaxation in quantum dot structures.

3.9.1 Carrier Relaxation in 3D Systems

In 3-dimensional systems, the polar optical phonon scattering is the dominant energy loss mechanism. The density of states is continuous so that the electrons with energy greater than $\hbar\omega_o$ can always find states into which they can scatter. As seen from Fig. 3.9, the polar optical emission rates are $10^{12}-10^{13}$ s^{-1}. As a result, the carrier relaxation takes 1-2 picoseconds.

At low temperatures, the hot phonon bottleneck discussed earlier can become an important issue since the hot phonons ensure that the emission and absorption rates become almost equal so that the electrons are unable to lose their energies.

The relaxation time of 1-2 ps for hot carriers in 3D systems is extremely important in determining the ultimate bandwidth of devices that depend upon carrier injection at high energies. These devices include semiconductor lasers, Schottky diodes and a wide variety of electronic devices.

3.9.2   Carrier Relaxation in 2D Systems

The key difference between the 3D system and the 2D system is the density of states. A typical quantum well has a series of subbands in, say, the conduction band, as shown schematically in Fig. 3.18. The quantum well configuration chosen in Fig. 3.18 corresponds to a typical graded index quantum well laser configuration. Consider a situation where an electron is injected into a high lying subband under non-equilibrium conditions. The scattering processes for the electron can be classified as intra-subband where the electron does not change its subband and inter-subband where the electron moves from one subband to another after scattering. It is obvious that if a hot electron injected into, say, the 3rd or 4th subband is to lose its energy, it must suffer inter-subband scattering.

The wavefunction of the electron in a quantum well has the form

$$\psi_n(z, k_{||}) = g_n(z) \; exp \; (ik_{||} \cdot \rho) \tag{3.57}$$

where the functions $g_n(z)$ have symmetries shown in Fig. 3.18. It is clear that for an intra-subband transition the value of $g_n(z)$ in the initial and final state is unchanged and the scattering potential simply couples the initial $k_{||}$ value to the final $k_{||}$ value. The scattering rates are quite similar to those in the 3D system except for the modification of the density of states.

In the inter-subband transitions the initial and final states have a different envelope function, say $g_n(z)$ and $g_{n'}(z)$ and the scattering potential must couple the two states. If the quantum well width is small, this corresponds to a large change in the wavelength of the electron in the z-direction. As noted above, polar optical phonon scattering has a strong dependence upon the wavelength change and as a result inter-subband scattering rates can be a bit smaller compared to the intra-subband rates. As a result, the carrier relaxation process can be slowed down.

In Fig. 3.19 we show some results from computer simulation calculations of carrier relaxation in quantum well structures. Typical relaxation times for 50-80Å wells are $\sim$ 10 ps which are somewhat larger than the times encountered in 3D systems.

An important point to note here is the effect of screening of potentials at high carrier density. At high carrier injection, the free carrier density reduces the scattering potential by modifying the dielectric response of the material. As a result the relaxation time increases.

The carrier relaxation time of $\sim$ 10 ps in quantum wells has important consequences in limiting the temporal performance of quantum well lasers. We will discuss these effects in Chapter 11.

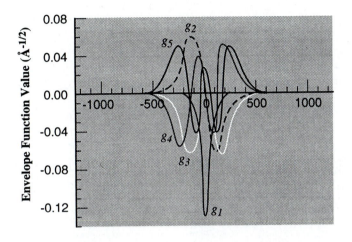

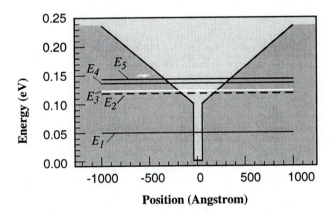

Figure 3.18: A schematic of the subband levels and electronic wavefunctions in a quantum well. A hot electron must suffer inter-subband scattering to lose its energy. The structure is a typical laser structure where electrons are "funneled" into a quantum well of narrow dimensions. The well shown here is 50 Å wide (after the work of Y. Lam and J. Singh.)

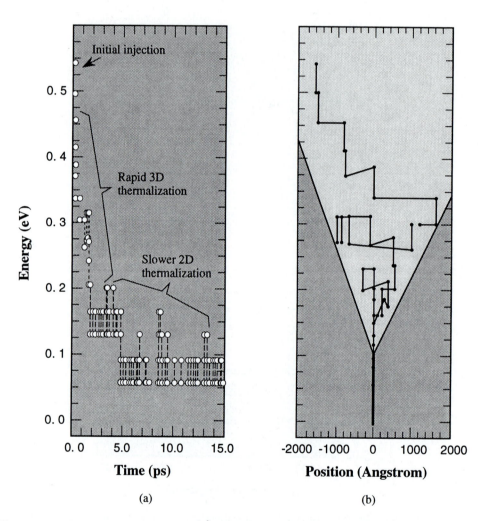

Figure 3.19: Carrier relaxation in a 50Å GaAs/Al$_{0.3}$Ga$_{0.7}$As quantum well. (a) The energy versus time of a typical hot electron injected into a 3D "funnel" from which it relaxes into the quantum well. (b) The trajectory followed by the electron (Y. Lam and J. Singh.)

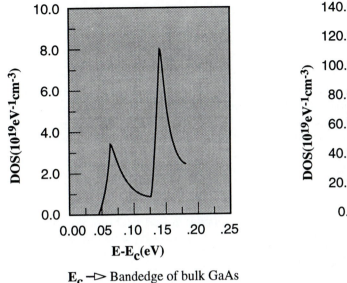
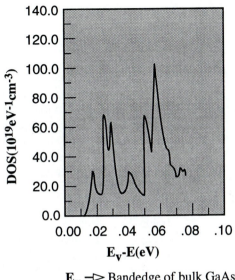

Figure 3.20: Density of states in the conduction and valence bands of a $100\text{Å} \times 100\text{Å GaAs}/\text{Al}_{0.3}\text{Ga}_{0.7}\text{As}$ quantum wire. A broadening function is used to broaden the density of states singularities. (After I. Vurgaftman and J. Singh.)

3.9.3 Carrier Relaxation in 1D Systems

In 1D systems (quantum wires) the density of states of the conduction band has singularities as shown in Fig. 3.20. It is important to point out that due to the strong valence band mixing and non-parabolic effects, the valence band density of states does not have the kind of sharp singularities seen in the conduction band, as shown in Fig. 3.20. For the conduction band, the singularities in the density of states produce singularities in the scattering rates as well. As a result of this, the relaxation times can be quite large in quantum wires, approaching 100 ps. These long times can have serious consequences for the dynamic performance of quantum wire lasers.

3.9.4 Carrier Relaxation in Quantum Dots

The density of states in a quantum dot system are a collection of discrete δ- functions. In reality, there is some broadening of these states due to the finite lifetime of the electrons in these states ($\Delta E \cdot \Delta t \sim \hbar$). For an infinite barrier case, the energy levels for a dot of dimensions $L_x \times L_y \times L_z$ are at (for a parabolic isotropic effective mass band)

$$E(\ell, m, n) = \frac{h^2}{2m^*} \left[\frac{\ell^2}{L_x^2} + \frac{m^2}{L_y^2} + \frac{n^2}{L_z^2} \right] \tag{3.58}$$

where ℓ, m, n takes the values 1, 2 3, ... Since the subband separation is, in general, different from the optical phonon energy $\hbar\omega_o$, it is not possible to couple different subband states. It may be noted that the phonon dispersion relation is different from the bulk phonon dispersion so that the phonon energies may have a greater spread for energies. Nevertheless, it is, in general, difficult to have available the phonon energies to allow efficient inter-subband energies. As a result, the hot electrons suffer from a "phonon bottleneck". If only phonon scattering were to occur, the carriers will take up to several nanoseconds to relax.

It has been seen experimentally that the relaxation into quantum dots is not limited strongly by the phonon bottleneck as expected. This is explained on the basis of the electron-hole scattering. The hole density of states is quite continuous, even in narrow quantum dots due to the greater mass of holes, and due to the strong non-parabolic effects. As a result, the holes can relax quite fast in the quantum dots. The hot electrons can scatter from holes and lose their energies to the holes and, as a result, the electrons can relax in several hundred picoseconds.

The hot carrier relaxation times discussed in this section are of great importance in the temporal dependence of optoelectronic devices and will be used in later chapters. As seen in this section, as the dimensionality of a structure becomes smaller, the time taken for carrier relaxation increases. However, there are other effects that benefit the device performance as the dimensions decrease.

3.10   CHARGE INJECTION AND QUASI-FERMI LEVELS

$\longrightarrow$ If electrons and holes are injected into a semiconductor, either by external contacts or by optical excitation, the system may not be in equilibrium, and the question then arises: What kind of distribution function describes the electron and hole occupation? We know that in equilibrium the electron and hole occupation is represented by the Fermi function. A new function is needed to describe the system when the electrons and holes are not in equilibrium.

3.10.1   Quasi-Fermi Levels

We know that in equilibrium the distribution of electrons and holes is given by the Fermi function which is defined once one knows the Fermi level. Also the product of electrons and holes, np is a constant. If excess electrons and holes are injected into the semiconductor, clearly the same function will not describe the occupation of states. Under certain assumptions the electron and hole occupation can be described by the use of *quasi-Fermi levels*. These assumptions are:

i) The electrons are essentially in thermal equilibrium in the conduction band and the excess holes are in equilibrium in the valence band. This means that the electrons are neither gaining nor losing energy from the crystal lattice atoms.

ii) The electron-hole recombination time is much larger than the time for the electrons and holes to reach equilibrium within the conduction and valence band, respectively.

As discussed in the previous section, in most problems of interest, the time to reach equilibrium in the same band is approximately a few picoseconds while the *e-h* recombination time is anywhere from a nanosecond to a microsecond. Thus, the above assumptions are usually met. In this case, the quasi-equilibrium electron and holes can be represented by an electron Fermi function f^e (with electron Fermi level) and a hole Fermi function f^h (with a *different* hole Fermi level). We now have

$$n = \int_{E_c}^{\infty} N_e(E) f^e(E) dE \tag{3.59}$$

$$p = \int_{-\infty}^{E_v} N_h(E) f^h(E) dE \tag{3.60}$$

where

$$f^e(E) = \frac{1}{exp\left(\frac{E-E_{Fn}}{k_BT}\right)+1} \tag{3.61}$$

$$f^h(E) = 1 - f^v(E) \quad = \quad 1 - \frac{1}{exp\left(\frac{E-E_{Fp}}{k_BT}\right)+1}$$

$$= \frac{1}{exp\left(\frac{E_{Fp}-E}{k_BT}\right)+1} \tag{3.62}$$

At equilibrium $E_{Fn} = E_{Fp}$. If excess electrons and holes are injected into the semiconductor, the electron Fermi level E_{Fn} moves towards the conduction band, while the hole Fermi level E_{Fp} moves towards the valence band. The ability to define quasi-Fermi levels E_{Fn} and E_{Fp} provides us a very powerful approach to solve non- equilibrium problems which are, of course, of greatest interest in devices.

By defining separate Fermi levels for the electrons and holes, one can study the properties of excess carriers using the same relationship between Fermi level and carrier density as we developed for the equilibrium problem. Thus, in the Boltzmann approximation we have

$$
\begin{aligned}
n &= N_c \, exp \left[\frac{(E_{Fn} - E_c)}{k_B T} \right] \\
p &= N_v \, exp \left[\frac{(E_v - E_{Fp})}{k_B T} \right]
\end{aligned}
\tag{3.63}
$$

In the more accurate Joyce-Dixon approximation we have

$$
(E_{Fn} - E_c) = k_B T \left[ln \, \frac{n}{N_c} + \frac{n}{\sqrt{8} N_c} \right]
\tag{3.64}
$$

with a similar expression for $E_v - E_{Fp}$.

EXAMPLE 3.17 Using Boltzmann statistics calculate the position of the electron and hole quasi-Fermi levels when an *e-h* density of 10^{17} cm^{-3} is injected into pure (undoped) silicon at 300 K.

The electron and hole densities are related to the quasi-Fermi levels by the equations

$$
\begin{aligned}
n_c &= N_c \, exp \, [(E_{Fn} - E_c)/k_B T] \\
p_v &= N_v \, exp \, [(E_v - E_{Fp})/k_B T]
\end{aligned}
$$

Recall that when we were discussing carrier concentrations in *semiconductors in equilibrium*, we used similar equations, except E_{Fn} and E_{Fp} were the same. Here, since we are not at equilibrium, the two are different.

At room temperature for Si we have

$$
\begin{aligned}
N_c &= 2.8 \times 10^{19} \; cm^{-3} \\
N_v &= 1.04 \times 10^{19} \; cm^{-3}
\end{aligned}
$$

If $n_c = p_v = 10^{17}$ cm^{-3}, we obtain ($k_B T = 0.026$ eV)

$$
\begin{aligned}
E_{Fn} &= k_B T \, ln \left[\frac{n}{N_c} \right] + E_c \\
&= E_c - 0.146 \; eV \\
E_{Fp} &= E_v - k_B T \left[ln \, \frac{p}{N_v} \right] \\
&= E_v + 0.121 \; eV
\end{aligned}
$$

Since in Si, the bandgap is $E_c - E_v = 1.1$eV, we have

$$
\begin{aligned}
E_{Fn} - E_{Fp} &= (E_c - E_v) - (0.146 + 0.121) \\
&= 1.1 - 0.267 = 0.833 \; eV
\end{aligned}
$$

If we had injected only 10^{15} cm^{-3} electrons and holes, the difference in the quasi-Fermi levels would be

$$
\begin{aligned}
E_{Fn} - E_{Fp} &= (E_c - E_v) - (0.266 + 0.24) \\
&= 1.1 - 0.506 = 0.59 \ eV
\end{aligned}
$$

Thus as the carrier injection is increased, the separation increases.

3.11 CHAPTER SUMMARY

$\mathcal{R}$ In this chapter we have discussed the basic transport phenomena upon which electronic and optoelectronic devices are based. All devices involve some physical response to external perturbations. These perturbations are usually electric fields or electromagnetic fields. The device performance depends upon how electrons respond to these external stimuli. The summary tables (Tables 3.3-3.5) highlight the concepts discussed in this chapter.

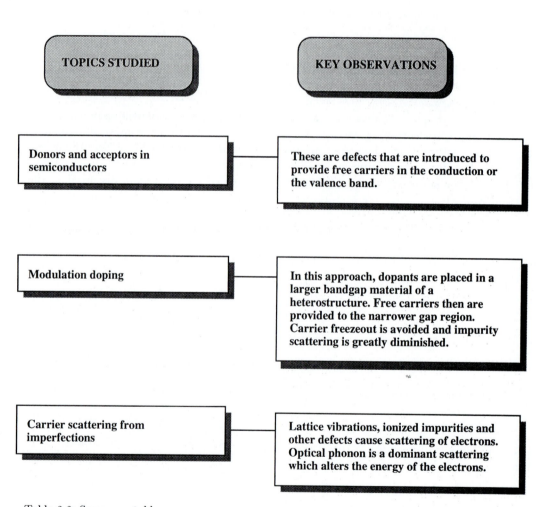

TOPICS STUDIED

KEY OBSERVATIONS

Donors and acceptors in semiconductors

These are defects that are introduced to provide free carriers in the conduction or the valence band.

Modulation doping

In this approach, dopants are placed in a larger bandgap material of a heterostructure. Free carriers then are provided to the narrower gap region. Carrier freezeout is avoided and impurity scattering is greatly diminished.

Carrier scattering from imperfections

Lattice vibrations, ionized impurities and other defects cause scattering of electrons. Optical phonon is a dominant scattering which alters the energy of the electrons.

Table 3.3: Summary table

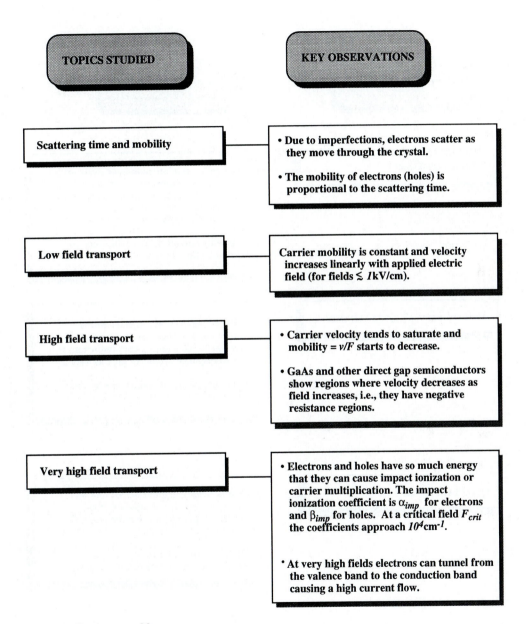

Table 3.4: Summary table

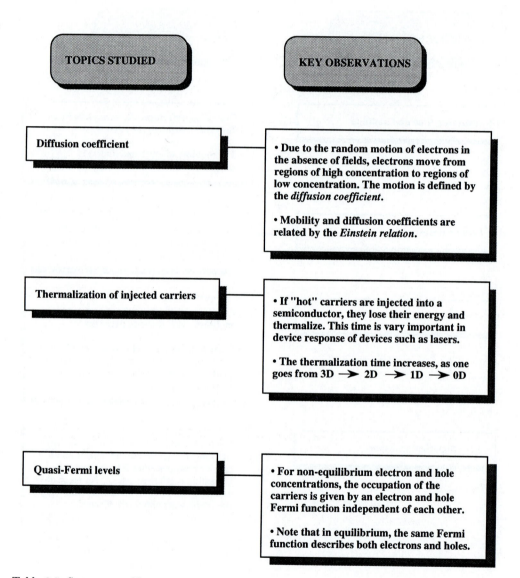

TOPICS STUDIED	KEY OBSERVATIONS
Diffusion coefficient	• Due to the random motion of electrons in the absence of fields, electrons move from regions of high concentration to regions of low concentration. The motion is defined by the *diffusion coefficient*. • Mobility and diffusion coefficients are related by the *Einstein relation*.
Thermalization of injected carriers	• If "hot" carriers are injected into a semiconductor, they lose their energy and thermalize. This time is vary important in device response of devices such as lasers. • The thermalization time increases, as one goes from 3D → 2D → 1D → 0D
Quasi-Fermi levels	• For non-equilibrium electron and hole concentrations, the occupation of the carriers is given by an electron and hole Fermi function independent of each other. • Note that in equilibrium, the same Fermi function describes both electrons and holes.

Table 3.5: Summary table

3.12   PROBLEMS

Sections 3.2-3.3

3.1 Calculate the density of electrons in silicon if the Fermi level is 0.2 eV below the conduction band at 300 K. Compare the results by using the Boltzmann approximation, the Joyce-Dixon approximation and the Fermi-Dirac integral.

3.2 In a GaAs sample at 300 K, the Fermi level coincides with the valence bandedge. Calculate the hole density using a) the Boltzmann approximation and b) the Joyce-Dixon approximation. Also calculate the electron density using the law of mass action.

3.3 The electron density in a silicon sample at 300 K is 10^{16} cm^{-3}. Calculate $E_c - E_F$ and the hole density using the Boltzmann approximation.

3.4 A GaAs sample is doped *n*-type at 5×10^{17} cm^{-3}. Assume that all the donors are ionized. What is the position of the Fermi level at 300 K?

3.5 The shrinking of bandgap with doping is a serious problem in bipolar junction transistors where one has to play an optimization game between band shrinkage and doping. Plot the band shrinkage in *n*-type Si as a function of doping up to a doping density of 5×10^{18} cm^{-3} at 300 K.

3.6 Calculate the position of the Fermi level with respect to the bottom of the first subband in a GaAs modulation doped structure that has an electron density of 10^{12} cm^{-2} at 300 K.

3.7 An InGaAs/InP modulation doped structure has 5×10^{12} cm^{-2} electrons in the InGaAs channel at 77 K. If the electron effective mass is 0.042 m_0, calculate the position of the Fermi level. Assume that a single subband is occupied.

Sections 3.4-3.6

3.8 Consider a sample of GaAs with electron effective mass of 0.067 m_0. If an electric field of 1 kV/cm is applied, what is the drift velocity produced if
a) $\tau_{sc} = 10^{-13}s$
b) $\tau_{sc} = 10^{-12}s$
c) $\tau_{sc} = 10^{-11}s$? How does the drift velocity compare to the average thermal speed of the electrons at room temperature?

3.9 Assume that at room temperature the electron mobility in Si is 1300 cm^2/V-s. If an electric field of 100 V/cm is applied, what is the excess energy of the electrons? How does it compare with the thermal energy? If you assume that the mobility is unchanged, how does the same comparison work out at a field of 5 V/cm? (Excess energy $= \frac{1}{2}m^* v_d^2$ where v_d is the drift velocity.)

3.10 Show that the average energy gained between collisions is

$$\delta E_{av} = \frac{1}{2}m^*(\mu F)^2$$

where F is the applied electric field. If the "optical phonon" energies in GaAs and Si are 36 meV and 47 meV, and the mobilities are 8000 cm^2/V-s and 1400 cm^2/V-s, respectively, what are the electric fields at which optical phonon emission can start?

3.11 The electron mobility of Si at 300 K is 1400 cm^2/V-s. Calculate the mean free path and the energy gained in a mean free path at an electric field of 1 kV/cm. Assume that the mean free path $= v_{th} \cdot \tau_{sc}$ where v_{th} is the thermal velocity of the electron ($v_{th} \sim 2.0 \times 10^7$ cm/s).

3.12 The mobility of electrons in the material InAs is $\sim$ 35,000 cm^2/V-s at 300K compared to a mobility of 1400 cm^2/V-s for silicon. Calculate the scattering times in the two semiconductors. Use Appendix B for carrier mass information.

3.13 Use the velocity-field relations for Si and GaAs to calculate the transit time of electrons in a 1.0 μm region for a field of 1 kV/cm^{-1} and 50 kV/cm^{-1}.

3.14 The velocity of electrons in silicon remains $\sim 1 \times 10^7$ cm s^{-1} between 50 kV/cm^{-1} and 200 kV/cm^{-1}. Estimate the scattering times at these two electric fields.

Section 3.7

3.15 The power output of a device depends upon a maximum voltage that the device can tolerate before impact ionization generated carriers become significant (say 10% excess carriers). Consider a device of length L over which a potential V drops uniformly. What is the maximum voltage that can be tolerated by an Si and a diamond device for $L = 2$ μm and $L = 0.5$ μm? Use the values of the critical fields given in the Appendix B.

3.16 An electron in a silicon device is injected in a region where the field is 500 kV cm^{-1}. The length of this region is 1.0 μm. Calculate the number of impact ionization events that occur for the incident electron.

3.17 In the previous problem, if a hole is injected under the same conditions, how many ionizing events will occur for the incident hole?

Sections 3.8, 3.9

3.18 In a silicon sample at 300 K, the electron concentration drops linearly from 10^{18} cm^{-3} to 10^{16} cm^{-3} over a length of 2.0 μm. Calculate the current density due to the electron diffusion current. Use the diffusion constant values given in this chapter.

3.19 In a GaAs sample, it is known that the electron concentration varies linearly. The diffusion current density at 300 K is found to be 100 A/cm^2. Calculate the slope of the electron concentration.

3.20 In a GaAs sample the electrons are moving under an electric field of 5 kV cm^{-1} and the carrier concentration is uniform at 10^{16} cm^{-3}. The electron velocity is the saturated velocity of 10^7 cm/s^{-1}. Calculate the drift current density. If a diffusion current has to have the same magnitude, calculate the concentration gradient needed. Assume a diffusion coefficient of 100 cm^2/s.

3.13 REFERENCES

- **General Doping and Transport**

  - R. A. Smith, *Semiconductors*, Cambridge University Press, London (1978).
  - R. B. Adler, A. C. Smith, and R. L. Longini, *Introduction to Semiconductor Physics*, Wiley, New York (1969).
  - J. R. Haynes and W. Shockley, *Phys. Rev.*, **81**, 835 (1951).
  - S. M. Sze, *Physics of Semiconductor Devices*, Wiley, New York (1981).
  - D. L. Rode, *Low Field Electron Transport in Semiconductors and Semimetals*, eds. R. K. Willardson and A. C. Beer, Academic Press, New York, vol. 10 (1975).

- **General High Field Transport**

  - M. Lundstrom, *Fundamentals of Carrier Transport*, Modular Series on Solid State Devices, eds. G. W. Neudeck and R. F. Pierret, Addison-Wesley, Reading, vol. X (1990).
  - J. Singh, *Physics of Semiconductors and Their Heterostructures*, McGraw-Hill, New York (1993).

- **Hot Carrier Thermalization Effects**

  - S.M. Goodnick and P. Lugli, *Phys. Rev.*, B, **37**, 2578 (1988).
  - C. Jacoboni and P. Lugli, *The Monte Carlo Method for Semiconductor Device Stimulation*, Springer, New York (1989).
  - Y. Lam and J. Singh, *Appl. Phys. Lett.*, **63**, 1874 (1993).
  - R. Mickevicius and V. Mitin, *Phys. Rev.*, B, **48**, 17184 (1993).
  - S. Morin, B. Deveaud, F. Clerot, K. Fugiwara, and K. Mitsunaga, *IEEE J. Quant. Electron.*, **QE-27**, 1669 (1991).
  - L. Rota, F. Rossi, S.M. Goodnick, P. Lugli, E. Molinari, and W. Porod, *Phys. Rev.*, B, **47**, 1632 (1993).
  - I. Vurgaftman and J. Singh, *Appl. Phys. Lett.*, **64**, 232 (1994).
  - S Weiss, J.M. Weisenfeld, D.S. Chemla, G. Raybon, G. Suchla, M. Wegener, G. Einsenstein, C.A. Burrus, A.G. Dentai, U. Koren, B.I. Miller, H. Temkin, R.A. Logan, and T. Tanbun-Ek, *Appl. Phys. Lett.*, **60**, 9 (1992).

CHAPTER 4

OPTICAL PROPERTIES OF SEMICONDUCTORS

4.1 INTRODUCTION

$\mathcal{R}$ This text deals with the physics and technology of the optoelectronic devices based on semiconductors. These devices depend upon the interactions of photons or electromagnetic fields with semiconductors. It is therefore essential to understand the interaction of photons with the electrons in a semiconductor. This interaction is described by the scattering theory in which the electrons scatter from one state to another by the photon field. In the previous chapter we have examined the effect of imperfections on scattering of electrons. Perturbations such as ionized impurities, phonons, etc., cause scattering of an electron from an initial Bloch state to another state. Electromagnetic radiation or photons cause a similar scattering process and the problem can be treated by using time dependent perturbation theory as we treated the electron-phonon scattering problem provided the photon intensities are not too high. The photons can introduce scattering within a band as well as between bands as shown in the Fig. 4.1. The interband scattering involving valence and conduction band states is, of course, most important for optical devices such as lasers and detectors. In addition to the band-to-band transitions, increasing interest has recently focussed on excitonic states especially in quantum well structures. The exciton-photon interaction in semiconductor structures contains important physics and is also of great technical interest for high speed modulation devices and optical switches.

We will briefly review some important concepts in electromagnetic theory and then discuss the interactions between electrons and photons. We will focus on the special aspects of this interaction for semiconductor electrons, especially those relating to

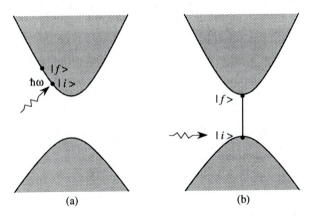

Figure 4.1: Intraband and interband scattering of an electron from an initial state k_i to a final state k_f.

selection rules.

We will first discuss a macroscopic theory for the optical effects in materials. In this formalism, the details of the Bloch states and the bandstructure are ignored and are instead represented by macroscopic quantities such as conductivity, dielectric constant, etc. The Kramers-Kronig relation will also be discussed. Next, we will discuss the one-electron picture of optical-semiconductor interactions where the details of the Bloch states and the bandstructure are critical. Band-to-band and intraband absorption coefficients will be discussed. Both 3-dimensional and 2-dimensional systems will be covered. We will also focus on the selection rules and "gain" in semiconductor structures, considerations which are extremely important in solid state lasers.

While the single electron picture which gives us the E vs. k relation provides a good description of optical processes for photon energies above the bandgap, just below the bandgap there are extremely important optical interactions involving the electron-hole system. These processes, known as excitonic processes, are becoming increasingly important for technological applications such as high speed switches and modulators. We will discuss the exciton related effects in Chapter 5.

4.2 MAXWELL EQUATIONS AND VECTOR POTENTIAL

$\longrightarrow$ The properties of electromagnetic fields in a medium are described by the four Maxwell equations. Apart from the electric ($\boldsymbol{F}$) and magnetic ($\boldsymbol{B}$) fields and velocity of light, the effects of the material are represented by the dielectric constant, permeability (we will assume that the permeability $\mu = \mu_o$), electrical conductivity, etc., We start with the four Maxwell equations

$$
\begin{aligned}
\nabla \times \boldsymbol{F} + \frac{\partial \boldsymbol{B}}{\partial t} &= 0 \\
\nabla \times \boldsymbol{H} - \frac{\partial \boldsymbol{D}}{\partial t} &= \boldsymbol{J} \\
\nabla \cdot \boldsymbol{D} &= \rho \\
\nabla \cdot \boldsymbol{B} &= 0
\end{aligned}
\tag{4.1}
$$

where $\boldsymbol{F}$ and $\boldsymbol{H}$ are the electric and magnetic fields, $\boldsymbol{D} = \epsilon \boldsymbol{F}$, $\boldsymbol{B} = \mu \boldsymbol{H}$, $\boldsymbol{J}$, and ρ are the current and charge densities. In dealing with the electron-photon interactions, it is convenient to work with the vector and scalar potentials $\boldsymbol{A}$ and ϕ respectively, which are defined through the equations

$$
\begin{aligned}
\boldsymbol{F} &= -\frac{\partial \boldsymbol{A}}{\partial t} - \nabla \phi \\
\boldsymbol{B} &= \nabla \times \boldsymbol{A}
\end{aligned}
\tag{4.2}
$$

The first and fourth Maxwell equations are automatically satisfied by these definitions. The potentials $\boldsymbol{A}$ and ϕ are not unique, but can be replaced by a new set of potentials $\boldsymbol{A}'$ and ϕ' given by

$$
\begin{aligned}
\boldsymbol{A}' &= \boldsymbol{A} + \nabla \chi \\
\phi' &= \phi - \frac{\partial \chi}{\partial t}
\end{aligned}
\tag{4.3}
$$

The new choice of potentials does not have any effect on the physical fields $\boldsymbol{F}$ and $\boldsymbol{B}$. To at least partially remove the arbitrariness of the potentials $\boldsymbol{A}$ and ϕ, we define certain transformations called the gauge transformations which restrict them. Before considering these transformations let us rewrite the Maxwell equations in terms of $\boldsymbol{A}$ and ϕ.

The second and third Maxwell equations become

$$
\begin{aligned}
\frac{1}{\mu_o} \nabla \times \nabla \times \boldsymbol{A} + \epsilon \frac{\partial^2 \boldsymbol{A}}{\partial t^2} + \epsilon \nabla \frac{\partial \phi}{\partial t} &= \boldsymbol{J} \\
\frac{\partial}{\partial t} \nabla \cdot \boldsymbol{A} + \nabla^2 \phi &= \frac{\rho}{\epsilon}
\end{aligned}
\tag{4.4}
$$

Now

$$
\nabla \times \nabla \times \boldsymbol{A} = \nabla(\nabla \cdot \boldsymbol{A}) - \nabla^2 \boldsymbol{A}
\tag{4.5}
$$

giving us

$$
\nabla \left(\frac{1}{\mu_o} \nabla \cdot \boldsymbol{A} + \epsilon \frac{\partial \phi}{\partial t} \right) - \frac{1}{\mu_o} \nabla^2 \boldsymbol{A} + \epsilon \frac{\partial^2 \boldsymbol{A}}{\partial t^2} = \boldsymbol{J}
\tag{4.6}
$$

$$\frac{\partial}{\partial t}(\nabla \cdot \boldsymbol{A}) + \nabla^2 \phi = -\frac{\rho}{\epsilon} \tag{4.7}$$

We will now impose certain restrictions on $\boldsymbol{A}$ and ϕ to further simplify these equations. Note that these restrictions have no implications on $\boldsymbol{F}$ and $\boldsymbol{B}$ fields as can be easily verified from their descriptions in terms of $\boldsymbol{A}$ and ϕ. The purpose of working in different gauges is mathematical elegance and simplicity.

Since the vector potential $\boldsymbol{A}$ is defined in terms of its curl: $\boldsymbol{B} = \nabla \times \boldsymbol{A}$, its divergence $(\nabla \cdot \boldsymbol{A})$ is arbitrary. Choice of a particular gauge is equivalent to the choice of the the value of $\nabla \cdot \boldsymbol{A}$.

In a gauge known as the Lorentz gauge, widely used in relativistic electrodynamics, we choose

$$\nabla \cdot \boldsymbol{A}' + \frac{\partial \phi'}{\partial t} = 0 \tag{4.8}$$

This is equivalent to imposing the following restriction on the arbitrary quantity χ

$$\nabla^2 \chi - \frac{\partial^2 \chi}{\partial t^2} = -\left(\nabla \cdot \boldsymbol{A} + \frac{\partial \phi}{\partial t}\right) \tag{4.9}$$

With this choice of χ we get for the Maxwell equations

$$\frac{1}{\mu_o}\nabla^2 \boldsymbol{A} - \epsilon\frac{\partial^2 \boldsymbol{A}}{\partial t^2} = \boldsymbol{J}$$

$$\nabla^2 \phi - \frac{\partial^2 \phi}{\partial t^2} = -\frac{\rho}{\epsilon} \tag{4.10}$$

This form of Maxwell's equations is extremely useful for generalizing to relativistic electrodynamics. In dealing with electron-photon interactions, a most useful gauge is the radiation or Coulomb gauge. If $\boldsymbol{J} = 0$ and $\rho = 0$, we can choose the constant background potential $\phi' = 0$. In addition we can choose $\nabla \cdot \boldsymbol{A}' = 0$. In this case the solutions for the vector potential are represented by plane wave transverse electromagnetic waves. We will be working in this gauge. It is useful to establish the relation between the vector potential $\boldsymbol{A}$ and the photon density which represents the optical power. The time dependent solution for the vector potential solution of Eqn. 4.10 with $\boldsymbol{J} = 0$ is

$$A(\boldsymbol{r},t) = \boldsymbol{A}_0 \left\{\exp\left[i(\boldsymbol{k} \cdot \boldsymbol{r} - \omega t)\right] + \text{ c.c.}\right\} \tag{4.11}$$

with

$$k^2 = \epsilon\mu_o\omega^2$$

Note that in the MKS units $(\epsilon_o\mu_o)^{1/2}$ is the velocity of light c $(3 \times 10^8 \text{ ms}^{-1})$. The electric and magnetic fields are

$$\boldsymbol{F} = \frac{\partial \boldsymbol{A}}{\partial t}$$

$$= -2\omega \boldsymbol{A}_0 \sin(\boldsymbol{k} \cdot \boldsymbol{r} - \omega t)$$

$$\boldsymbol{B} = \nabla \times \boldsymbol{A}$$

$$= -2\boldsymbol{k} \times \boldsymbol{A}_0 \sin(\boldsymbol{k} \cdot \boldsymbol{r} - \omega t) \tag{4.12}$$

The Poynting vector $\boldsymbol{S}$ representing the optical power is

$$\boldsymbol{S} = (\boldsymbol{F} \times \boldsymbol{H})$$

$$= \frac{4}{\mu_o} v k^2 \left|\boldsymbol{A}_0\right|^2 \sin^2(\boldsymbol{k} \cdot \boldsymbol{r} - \omega t)\hat{k} \tag{4.13}$$

where v is the velocity of light in the medium $(= c/\sqrt{\tilde{\epsilon}})$ and $\hat{k}$ is a unit vector in the direction of $\boldsymbol{k}$. Here $\tilde{\epsilon}$ is the relative dielectric constant. The time averaged value of the power is

$$<\boldsymbol{S}>_{\text{time}} = \hat{k}\frac{2vk^2 \left|\boldsymbol{A}_0\right|^2}{\mu_o}$$

$$= 2v\epsilon\mu_o\omega^2 \left|\boldsymbol{A}_0\right|^2 \hat{k} \tag{4.14}$$

since

$$|\boldsymbol{k}| = \omega/v \tag{4.15}$$

The energy density is then

$$\left|\frac{\boldsymbol{S}}{v}\right| = \frac{2\epsilon\,\omega^2 \left|\boldsymbol{A}_0\right|^2}{c^2} \tag{4.16}$$

Also, if the photon mode occupation is n_{ph}, the energy density is (for a volume V)

$$\frac{n_{ph}\hbar\omega}{V} \tag{4.17}$$

Equating these two, we get

$$\boxed{|\boldsymbol{A}_0|^2 = \frac{n_{ph}\hbar}{2\epsilon\omega V}} \tag{4.18}$$

The phonon number density n_{ph}/V is a physically measurable quantity which according to this relation tells us the strength of the vector potential. Eqn 4.18 is extremely useful since it allows us to relate the photon number density to the vector potential $\boldsymbol{A}$ which appears in the electron-photon interaction as we will see later in this chapter.

Having established these basic equations for the electromagnetic field, we will now develop a macroscopic picture of the photon-material interactions. Going back to the Maxwell's equations and writing $\boldsymbol{J} = \sigma\boldsymbol{F}$, we get the wave equation for the electric field (after eliminating the $\boldsymbol{B}$-field)

$$\nabla^2\boldsymbol{F} = \epsilon\mu_o\frac{\partial^2 \boldsymbol{F}}{\partial t^2} + \sigma\mu_o\frac{\partial \boldsymbol{F}}{\partial t} \tag{4.19}$$

This represents a wave propagating with dissipation. The general solution can be chosen to be of the form

$$\boldsymbol{F} = \boldsymbol{F}_0 \exp\left\{i(\boldsymbol{k} \cdot \boldsymbol{r} - \omega t)\right\} \tag{4.20}$$

so that k is given by

$$- k^2 = -\epsilon \mu_o \omega^2 - \sigma \mu_o i \omega \tag{4.21}$$

or $\left(c = (\epsilon_o \mu_o)^{-1/2}\right)$

$$k = \frac{\omega}{c} \left(\tilde{\epsilon} + \frac{\sigma \mu_o i}{\omega}\right)^{1/2} \tag{4.22}$$

In general, k is a complex number. In free space where $\sigma = 0$ we simply have $(\tilde{\epsilon} = 1)$

$$k = \omega/c \tag{4.23}$$

In a medium, the phase velocity is modified by dividing c by a complex refractive index given by

$$n_r = \left(\tilde{\epsilon} + \frac{\sigma \mu_o i}{\omega}\right)^{1/2} \tag{4.24}$$

We can write the complex refractive index in terms of its real and imaginary parts

$$n_r = n_r' + i n_r'' \tag{4.25}$$

so that

$$k = \frac{n_r' \omega}{c} + i n_r'' \frac{\omega}{c} \tag{4.26}$$

The electric field wave Eqn. 4.20 now becomes (for propagations in the +z direction)

$$\boldsymbol{F} = \boldsymbol{F}_0 \exp\left\{i\omega\left(\frac{n_r' z}{c} - t\right)\right\} \exp\left(\frac{-n_r'' \omega z}{c}\right) \tag{4.27}$$

The velocity of the wave is reduced by n_r' to c/n_r' and its amplitude is damped exponentially by a fraction $\exp\left(-2\pi n_r''/n_r'\right)$ per wavelength. The damping of the wave is associated with the absorption of the electromagnetic energy. The absorption coefficient α is described by the absorption of the intensity (i.e., square of Eqn. 4.27)

$$\alpha = \frac{2 n_r'' \omega}{c} \tag{4.28}$$

Note that in the absence of absorption, $n_r' = n_r$, and the refraction index will simply be denoted by n_r. The absorption coefficient can be measured for any material system and it provides information on n_r''. Another quantity which can be easily measured is

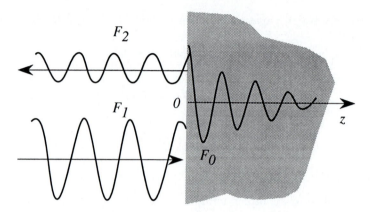

Figure 4.2: Schematic showing the incident, reflected, and transmitted waves.

the reflection coefficient R. If we consider a normally incident wave as shown in Fig. 4.2, impinging on a medium at $z = 0$, we have a reflected wave. For $z > 0$ we can write

$$\boldsymbol{F}_x = \boldsymbol{F}_0 \exp\left\{ i\omega \left(\frac{n_r z}{c} - t \right) \right\} \tag{4.29}$$

and for $z < 0$ we have

$$\boldsymbol{F}_x = \boldsymbol{F}_1 \exp\left\{ i\omega \left(\frac{z}{c} - t \right) \right\} + \boldsymbol{F}_2 \exp\left\{ -i\omega \left(\frac{z}{c} + t \right) \right\} \tag{4.30}$$

corresponding to a wave with amplitude $\boldsymbol{F}_1$ traveling to the right and a reflected wave to the left. Matching the boundary conditions we have

$$\boldsymbol{F}_0 = \boldsymbol{F}_1 + \boldsymbol{F}_2 \tag{4.31}$$

and for B_y, we get using Maxwell's equation

$$- n_r \boldsymbol{F}_0 = \boldsymbol{F}_2 - \boldsymbol{F}_1 \tag{4.32}$$

These give, for the reflection coefficient

$$
\begin{aligned}
R &= \left| \frac{F_2}{F_1} \right|^2 \\
&= \left| \frac{1 - n_r}{1 + n_r} \right|^2 \\
&= \frac{(n_r' - 1)^2 + n_r''^2}{(n_r' + 1)^2 + n_r''^2}
\end{aligned} \tag{4.33}
$$

The measurement of the reflection coefficient and absorption coefficient give us the optical constants n_r' and n_r'' of the material. Of course, we have not calculated n_r' and n_r'' from any microscopic information on the semiconductor as yet.

In the above formalism we have seen that the response of the medium to the outside world is represented by a complex dielectric constant or a complex refractive index. The real and imaginary parts of the relative dielectric constants are related to the refractive index components by the following relations (from Eqns. 4.24 and 4.25)

$$\tilde{\epsilon}_1 = n_r'^2 - n_r''^2$$
$$\tilde{\epsilon}_2 = 2n_r' n_r'' \tag{4.34}$$

where $\tilde{\epsilon}_1$ and $\tilde{\epsilon}_2$ are the real and imaginary parts of the relative dielectric constant. An important relation is satisfied by the real and imaginary parts of all response functions provided the causality principle is obeyed (effect follows the cause in time). The relation is given by an expression known as the Kramers Kronig relation which when applied to the refractive index and absorption coefficient gives the relation

$$n(\omega_o) - 1 = \frac{c}{\pi} P \int_o^\infty \frac{\alpha(\omega) d\omega}{\omega^2 - \omega_o^2} \tag{4.35}$$

This relation is extremely useful since it allows one to calculate the refractive index if the absorption coefficient is known. In particular, if the absorption spectra is modulated by some external means (say, an electric field), the effect on the refractive index can be calculated.

We will now proceed to derive a microscopic theory for the optical processes in semiconductors. We will focus on the photon energies close to the bandgap of the semiconductor since most optoelectronic devices cater to this energy range. Thus, our formalism, while general, will focus on the physics behind devices such as solid state detectors and lasers. The approach we use will be based on the perturbation theory in which we evaluate the scattering rate for electrons to scatter among various Bloch states. The transitions could be interband or intraband.

EXAMPLE 4.1 An electromagnetic radiation with a power density of 1 $\mu W/m^2$ impinges upon a receiver. Calculate the electric field amplitude of this radiation if the photon energy is 0.8 eV.

The power density is extremely small, but low noise detectors available in today's technology can detect such levels.

The photon density is related to the power density by

$$\text{Power density} = n_{ph} \hbar w c$$

Thus,

$$\frac{n_{ph}}{V} = \frac{(10^{-6} \ Wm^{-2})}{(0.8 \times 1.6 \times 10^{-19} J)(3 \times 10^8 \ m/s)}$$

$$= \quad 2.6 \times 10^4 \; m^{-3}$$

From Eqn. 4.18 we have

$$|A_o|^2 \quad = \quad \frac{(2.6 \times 10^4 \; m^{-3})(1.05 \times 10^{-34} \; J-s)}{2 \times (8.84 \times 10^{-12} \; F/m)(1.22 \times 10^{15} \; rad/s)}$$

The electric field amplitude is

$$|F_o| = 2\omega A_o \quad = \quad 2.745 \times 10^{-2} \; V/m$$
$$= \quad 2.745 \times 10^{-4} \; V/cm$$

At the other extreme of optical powers, femtosecond lasers can now produce power densities of 10^{16} W/m^2! At such high intensities, a material can suffer impact ionization due to the high electric field present.

4.3   ELECTRONS IN AN ELECTROMAGNETIC FIELD

$\rightsquigarrow$ The electron carries a negative charge which interacts with the electric and magnetic fields of the electromagnetic radiation. The force on the electron due to the electric field is simply the charge times the field and that due to the magnetic field is the Lorentz force determined by the cross product of the electron velocity and the field. The energy associated with this interaction determines the interaction Hamiltonian that is responsible for electron scattering. The Hamiltonian describing the interactions between a charge, e, and the electromagnetic field is

$$
\begin{aligned}
H \quad &= \quad \frac{1}{2m_0}\left(\boldsymbol{p} - e\boldsymbol{A}\right)^2 + e\phi + V(\boldsymbol{r}) \\
&= \quad \frac{p^2}{2m_0} - \frac{e}{2m_0}(\boldsymbol{p}\cdot\boldsymbol{A} + \boldsymbol{A}\cdot\boldsymbol{p}) + \frac{e^2}{2m_0}A^2 + e\phi + V(\boldsymbol{r}) \qquad (4.36)
\end{aligned}
$$

Here $\boldsymbol{A}$ is the vector potential and $V(\boldsymbol{r})$ is any background crystal potential.

We use the general commutation result for a space dependent function $f(r)$ and the momentum p

$$
\begin{aligned}
[f(\boldsymbol{r}), p] \quad &= \quad f(\boldsymbol{r})\boldsymbol{p} - \boldsymbol{p}f(\boldsymbol{r}) \\
&= \quad i\hbar\frac{\partial}{\partial\boldsymbol{r}}f(\boldsymbol{r}) \qquad (4.37)
\end{aligned}
$$

which can be explicitly verified by using $f(\boldsymbol{r}) = x; \; x^2$, etc., This then allows us to write

$$\frac{e}{2m_0}\boldsymbol{p}\cdot\boldsymbol{A} = \frac{e\boldsymbol{A}\cdot\boldsymbol{p}}{2m_0} - \frac{ie\hbar}{2m_0}\nabla\cdot\boldsymbol{A} \qquad (4.38)$$

The Hamiltonian becomes

$$H = \frac{p^2}{2m_0} - \frac{e}{m_0}\boldsymbol{A}\cdot\boldsymbol{p} + \frac{ie\hbar}{2m_0}\nabla\cdot\boldsymbol{A} + \frac{e^2}{2m_0}\boldsymbol{A}^2 + e\phi + V(\boldsymbol{r}) \qquad (4.39)$$

We will now use the perturbation theory to study the effect of the electromagnetic radiation on the electron. In the quantum theory of radiation, the electromagnetic field is written in terms of creation and destruction operators, in analog with the harmonic oscillator problem discussed in Appendix C. We will outline this approach, often called the second quantization approach.

The Schrödinger equation to be solved is

$$\begin{aligned}
i\hbar\frac{\partial\psi}{\partial t} &= \left[-\frac{\hbar^2}{2m_0}\nabla^2 + \frac{ie\hbar}{m_0 c}\boldsymbol{A}\cdot\nabla + \frac{ie\hbar}{2m_0}(\nabla\cdot\boldsymbol{A})\right.\\
&\quad + \left.\frac{e^2}{2m_0}A^2 + e\phi + V(\boldsymbol{r})\right]\psi
\end{aligned} \qquad (4.40)$$

We will work in the radiation gauge ($\nabla\cdot\boldsymbol{A} = \phi = 0$), and use the time dependent theory which results in scattering rates given by the Fermi golden rule. We assume that the optical power and consequently A is small, so that

$$\begin{aligned}
\left|\frac{ie\hbar}{m_0}\boldsymbol{A}\cdot\nabla\right| : \left|\frac{\hbar^2}{2m_0}\nabla^2\right| &\approx \left|\frac{e^2 A^2}{2m_0^2}\right| : \left|\frac{ie\hbar}{m}\boldsymbol{A}\cdot\nabla\right|\\
&\approx \frac{eA}{p}
\end{aligned} \qquad (4.41)$$

For an optical power of 1 W/cm^2 the photon density of a 1 eV energy beam is $\sim 10^9$ cm^{-3}. Using an electron velocity of 10^6 cm/s, one finds that

$$\frac{eA}{p} \sim 10^{-5} \qquad (4.42)$$

Thus, even for an optical beam carrying 1 MW/cm^2, the value of eA/p is small enough that perturbation theory can be used. We will, therefore, only retain the first order term in $\boldsymbol{A}$.

The Schrödinger equation is now written as

$$i\hbar\frac{\partial\psi}{\partial t} = (H_0 + H^{'})\psi \qquad (4.43)$$

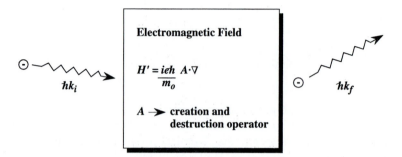

Figure 4.3: A schematic of the scattering of an electron by the electromagnetic field.

where

$$H_0 = -\frac{\hbar}{2m_0}\nabla^2 + V(\boldsymbol{r}) \tag{4.44}$$

and

$$H' = \frac{ie\hbar}{m_0}\boldsymbol{A}\cdot\nabla \tag{4.45}$$

In some cases, especially for forbidden transitions, the higher order term needs to be included. The scattering problem is schematically represented in Fig. 4.3 where the perturbation due to the electromagnetic field causes a scattering of the electron. The vector potential A represents the electromagnetic field which in the "second quantization" is represented by an operator. This operator can be written in terms of creation and destruction operators b^+ and b as discussed for the harmonic oscillator problem in Appendix C.1. The reader who is not familiar with the properties of the harmonic oscillator problem should go through this appendix.

The first order time dependent perturbation theory gives us the transition rates from the initial electron state $|i\rangle$ to the final state $|f\rangle$ by the golden rule

$$W(i) = \frac{2\pi}{\hbar}\sum_f \left|\langle f|H'|i\rangle\right|^2 \delta\left(E_f - E_i \mp \hbar\omega\right) \tag{4.46}$$

where the upper sign is for photon absorption and the lower one is for emission. For the scattering rate to have physical significance, we need to define the final state as having either a spread in the electronic states or a spread in the photonic states.

At this point we will use the results derived in Appendix C for the harmonic oscillator matrix elements. Note that the initial and final electron and photon states can be represented by the momentum states of the electron, along with the photon densities as shown in Fig. 4.4. The initial and final electron-photon states are represented by the following:

$$
\begin{aligned}
\text{Absorption:} \quad & |i> = |k_i, n_{ph}> \\
& |f> = |k_f, n_{ph} - 1> \\
\text{Emission:} \quad & |i> = |k_i, n_{ph}> \\
& |f> = |k_f, n_{ph} + 1>
\end{aligned}
$$

where k_i and k_f are the electron's initial and final wave vectors and n_{ph} is the photon density in the initial state. The vector potential is written as

$$
A_o = \sqrt{\frac{\hbar}{2\omega\epsilon V}}\,(b^+ + b) \tag{4.47}
$$

Here, as discussed in Appendix C.1, b^+ and b are the photon creation and destruction operators. Note that this choice gives the value for $|A_o|^2$ given in Eqn. 4.18.

The creation and destruction operations have the matrix elements (focusing only on the photon term of the matrix element; a is the polarization unit vector).

Absorption:

$$
\begin{aligned}
<i| \quad A\cdot\nabla\,|f> &\Rightarrow \sqrt{\frac{\hbar}{2\omega\epsilon}}<k_f, n_{ph} - 1\,|\,(b^+ + b)\cdot a\cdot\nabla\,|\,k_i, n_{ph}> \\
&= \sqrt{\frac{\hbar}{2\omega\epsilon}}(n_{ph})^{1/2}<k_f\,|\,a\cdot\nabla\,|\,k_i> \tag{4.48}
\end{aligned}
$$

$$
\tag{4.49}
$$

Emission:

$$
\begin{aligned}
<i| \quad A\cdot\nabla\,|f> &\Rightarrow \sqrt{\frac{\hbar}{2\omega\epsilon}}<k_f, n_{ph} + 1\,|\,(b^+ + b)a\cdot\nabla\,|\,k_i, n_{ph}> \\
&= \sqrt{\frac{\hbar}{2\omega\epsilon}}(n_{ph} + 1)^{1/2}<k_f\,|\,a\cdot\nabla\,|\,k_i> \tag{4.50}
\end{aligned}
$$

Notice the prefactors $(n_{ph})^{1/2}$ for the absorption process and $(n_{ph} + 1)^{1/2}$ for the emission process. This difference is extremely important, as will be discussed later.

Let us consider the photon absorption process where a photon with momentum $\hbar k_{ph}$ and energy $\hbar\omega$ is absorbed by an electron system. To calculate this rate we need to sum over all possible electron states which can allow such a process to occur. This is shown schematically in Fig. 4.5. The photon energy $\hbar\omega$ is transferred to the electron.

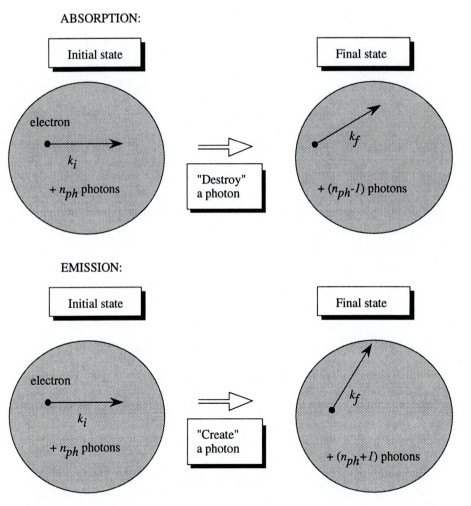

Figure 4.4: (a) A schematic of an absorption process where a photon is absorbed (destroyed) and the energy and momentum of the electron is altered; (b) the emission of a photon where a photon is created.

To find the total scattering rate we integrate over the final electron density of states

$$W_{abs} = \frac{2\pi}{\hbar} \frac{e^2}{m_0^2} \left(\frac{\hbar n_{ph}}{2\omega\epsilon} \right) \sum_{\text{final states}} | \int \psi_f^*(a \cdot p)e^{ik_{ph} \cdot r}\psi_i d^3r |^2 \cdot \delta(E_i - E_f + \hbar\omega) \quad (4.51)$$

The sum over the final states essentially gives the density of states corresponding to either the electron states, or the photon states. In the above expression, if we are considering the rate at which a photon is absorbed, this involves summing over any possible electronic levels, as long as the energy conservation is obeyed. One may, instead, be interested in the problem where a single electron with momentum k_i is scattered into a final state with momentum k_f. In this case the final states sum would be over all photon states that can cause such transitions and would involve the photon density of states. These two approaches are illustrated in Fig. 4.6.

The rate of an electron $\hbar k_i$ to emit any photon and reach a state with momentum $\hbar k_f$ is similarly

$$W_{em} = \frac{2\pi}{\hbar} \frac{e^2}{m_o^2} \left(\frac{\hbar(n_{ph}+1)}{2\omega\epsilon} \right) \sum_{\text{final states}} | \int \psi_f^*(a \cdot p)e^{-ik_{ph} \cdot r}\psi_i d^3r |^2 \cdot \delta(E_i - E_f - \hbar\omega)$$

$$(4.52)$$

we notice that the emission term can be rewritten as stimulated and spontaneous emission terms

$$W_{em} = W_{st} + W_{spon} \quad (4.53)$$

where

$$W_{st} = \frac{2\pi}{\hbar} \frac{e^2}{m_0^2} \frac{\hbar n_{ph}}{2\omega\epsilon} \sum_{\text{final states}} | \int \psi_f^* e^{-ik_{ph} \cdot r}(a \cdot p)\psi_i d^3r |^2$$

$$\cdot \, \delta(E_i - E_f - \hbar\omega) \quad (4.54)$$

$$W_{spon} = \frac{2\pi}{\hbar} \frac{e^2}{m_0^2} \frac{\hbar}{2\omega\epsilon} \sum_{\text{final states}} | \int \psi_f^* e^{-ik_{ph} \cdot r}\psi_i d^3r |^2 \quad (4.55)$$

$$\cdot \, \delta(E_i - E_f - \hbar\omega)$$

The stimulated emission is due to the initial photons present in the system and the emitted photons maintain phase coherence with the initial photons. The spontaneous emission comes from the perturbations due to the vacuum state (i.e., $n_{ph} = 0$) energy fluctuations and the emitted photons are incoherent with no phase relationship. The difference between these two kinds of processes is the key to understanding the differences between a light emitting diode (to be discussed in Chapter 9) and the laser diode (Chapters 10 and 11).

Let us now focus on the semiconductor electronic states involved in the absorption or emission process. The photon momentum $\hbar k_{ph}$ for most energies of interest

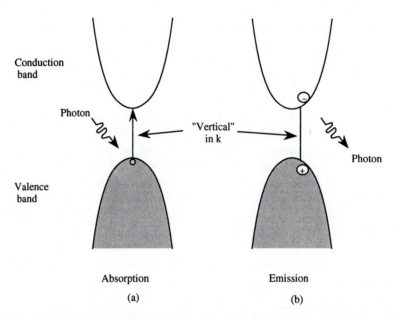

Figure 4.5: Band to band absorption and emission in semiconductors. In (a) an electron in the valence band absorbs a photon and moves into the conduction band; (b) In the reverse process an electron in the conduction band emits a photon and moves "vertically" down into the valence band.

in solid state devices ($\hbar\omega \sim 0.1$ - 2.0 eV) is extremely small compared to the electron momentum so that momentum conservation requires that

$$k_i = k_f$$

Thus, in first order perturbation theory, the electronic transitions due to photons are "vertical" in the $E\text{-}k$ description. An example for interband transitions is shown in Fig. 4.5. The approximation of neglecting k_{ph} is called the dipole approximation. In the dipole approximation the momentum matrix element, (the integral within the vertical bars in Eqns. 4.51 and 4.52) which we denote by p_{if} becomes quite simple.

Let us consider the initial and final states which have the Bloch function form. The momentum matrix element is in the dipole approximation

$$\boldsymbol{p}_{\text{if}} = -i\hbar \int \psi^*_{\mathbf{k}_f \ell'} \nabla \psi_{\mathbf{k}_i \ell} \, d^3 r \tag{4.56}$$

where we choose

$$
\begin{aligned}
|i\rangle &= \psi_{\mathbf{k}_i \ell} \\
&= e^{i\mathbf{k}_i \cdot \mathbf{r}} u_{\mathbf{k}_i \ell} \\
|f\rangle &= \psi_{\mathbf{k}_f \ell'} \\
&= e^{i\mathbf{k}_f \cdot \mathbf{r}} u_{\mathbf{k}_f \ell'}
\end{aligned}
$$

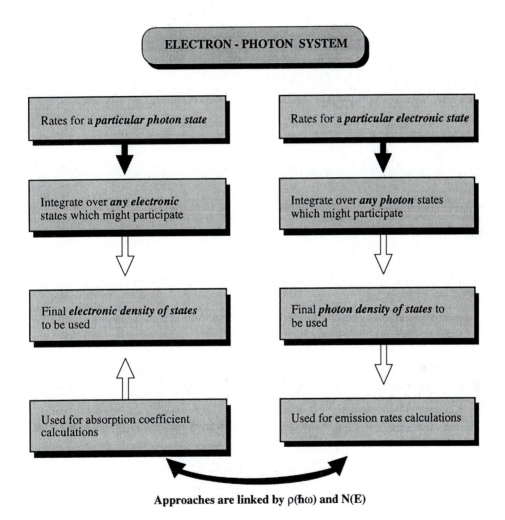

Figure 4.6: Final states used in scattering rates in the Fermi golden rule.

where $u_{\mathbf{k}\ell}$ is the cell periodic part of the Bloch state and ℓ, ℓ' are the band indices. Carrying out the differentiation we can write

$$\boldsymbol{p}_{\text{if}} = \hbar k_i \int \psi^*_{\mathbf{k}_f \ell'} \, \psi_{\mathbf{k}_i \ell} \, d^3 r - i\hbar \int u^*_{\mathbf{k}_f \ell'} \left(\nabla u_{\mathbf{k}_i \ell} \right) e^{i(\mathbf{k}_i - \mathbf{k}_f) \cdot \mathbf{r}} \, d^3 r \qquad (4.57)$$

In the next section, we will examine the selection rules explicitly, for semiconductor systems of interest.

4.4 INTERBAND TRANSITIONS

4.4.1 Interband Transitions in Bulk Semiconductors

⤳ Let us consider the selection rules for *band-to-band* transitions in direct gap semi-conductors as shown in Fig. 4.7. We will now use our understanding of the nature of conduction and valence band state central cell functions. The first term on the right-hand side of the momentum matrix element (Eqn. 4.57) is zero because of orthogonality of Bloch states. The second term requires ($u^*_{\mathbf{k}'\ell'} \nabla u_{\mathbf{k}\ell}$ is periodic)

$$\boldsymbol{k}_i - \boldsymbol{k}_f = 0$$

so that the interband transitions are "vertical" transitions. The interband matrix element is, therefore ($k_i = k_f = k$)

$$\langle u_{c\mathbf{k}} | \boldsymbol{p}_a | u_{v\mathbf{k}} \rangle \qquad (4.58)$$

where $u_{c\mathbf{k}}$ and $u_{v\mathbf{k}}$ represent the conduction band and valence band central cell states.

For near bandedge transitions we will assume that $u_{c\mathbf{k}}$ and $u_{v\mathbf{k}}$ are given by their zone center values. We remind ourselves that in this case the central cell states are (See Chapter 2 for a discussion on the central cell states in the conduction and valence band states),

- Conduction band:

$$u_{c0} = |s\rangle \qquad (4.59)$$

  where $|s\rangle$ is a spherically symmetric state.

- Valence band:

  Heavy hole states: $|3/2, 3/2\rangle = \dfrac{-1}{\sqrt{2}} \left(|p_x\rangle + i|p_y\rangle \right) \uparrow$

  $ |3/2, -3/2\rangle = \dfrac{1}{\sqrt{2}} \left(|p_x\rangle - i|p_y\rangle \right) \downarrow$

  Light hole states: $|3/2, 1/2\rangle = \dfrac{-1}{\sqrt{6}} \left[\left(|p_x\rangle + i|p_y\rangle \right) \downarrow \, -2|p_z\rangle \uparrow \right]$

  $ |3/2, -1/2\rangle = \dfrac{1}{\sqrt{6}} \left[\left(|p_x\rangle - i|p_y\rangle \right) \uparrow \, +2|p_z\rangle >\downarrow \right]$

  $\hfill (4.60)$

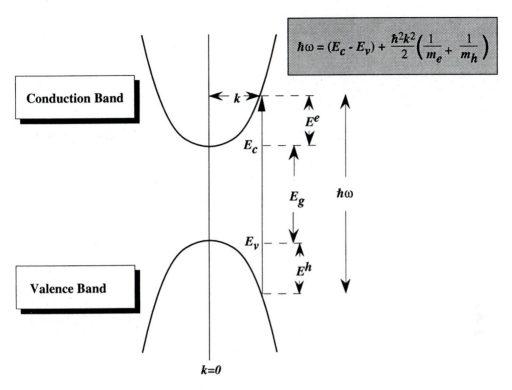

Figure 4.7: The positions of the electron and hole energies at vertical k-values. The electron and hole energies are determined by the photon energy and the carrier masses.

From symmetry we see that *only* the matrix elements of the form

$$\langle p_x | p_x | s \rangle = \langle p_y | p_y | s \rangle = \langle p_z | p_z | s \rangle = p_{cv}$$

are nonzero. Thus, for band-to-band transition, the only allowed transitions have the following matrix elements

$$\langle \text{HH} | p_x | s \rangle = \langle \text{HH} | p_y | s \rangle = \frac{1}{\sqrt{2}} \langle p_x | p_x | s \rangle \qquad (4.61)$$

$$\langle \text{LH} | p_x | s \rangle = \langle \text{LH} | p_y | s \rangle = \frac{1}{\sqrt{6}} \langle p_x | p_x | s \rangle \qquad (4.62)$$

and

$$\langle \text{LH} | p_z | s \rangle = \frac{2}{\sqrt{6}} \langle p_x | p_x | s \rangle = \frac{2}{\sqrt{6}} p_{cv} \qquad (4.63)$$

It is important to note that

$$\langle \text{HH} | p_z | s \rangle = 0 \qquad (4.64)$$

It is very useful to examine the matrix element square for light polarized along various orientations. This polarization dependence is accentuated in quantum wells because the HH and LH states are no longer degenerate in that case.

$$
\begin{aligned}
z\text{-polarized light:} \quad & \text{HH} \rightarrow \text{c-band:} \quad \text{No coupling} \\
& \text{LH} \rightarrow \text{c-band:} \quad |p_{\text{if}}|^2 = \tfrac{2}{3} |\langle p_x | p_x | s \rangle|^2 \\
x\text{-polarized light:} \quad & \text{HH} \rightarrow \text{c-band:} \quad |p_{\text{if}}|^2 = \tfrac{1}{2} |\langle p_x | p_x | s \rangle|^2 \\
& \text{LH} \rightarrow \text{c-band:} \quad |p_{\text{if}}|^2 = \tfrac{1}{6} |\langle p_x | p_x | s \rangle|^2 \\
y\text{-polarized light:} \quad & \text{HH} \rightarrow \text{c-band:} \quad |p_{\text{if}}|^2 = \tfrac{1}{2} |\langle p_x | p_x | s \rangle|^2 \\
& \text{LH} \rightarrow \text{c-band:} \quad |p_{\text{if}}|^2 = \tfrac{1}{6} |\langle p_x | p_x | s \rangle|^2
\end{aligned}
\tag{4.65}
$$

We see that the z-polarized light has no coupling to the HH states. Of course, the states have the pure form only at $k = 0$. Away from $k = 0$, the HH state and LH states have some mixture. For x-y polarized light the light couples three times as strongly to the HH states as to the LH states. In quantum well structures where the HH, LH degeneracy is lifted, the selection holes have important consequences for lasers and detectors and their polarization dependent properties.

It is convenient to define a quantity

$$
E_p = \frac{2}{m_0} |\langle p_x | p_x | s \rangle|^2
\tag{4.66}
$$

The values of E_p for several semiconductors are given in Table 4.1.

As a result of the vertical transitions we have, as shown in Fig. 4.7, the equality (using the parabolic band approximation)

$$
\begin{aligned}
\hbar\omega - E_g &= \frac{\hbar^2 k^2}{2} \left(\frac{1}{m_e^*} + \frac{1}{m_h^*} \right) \\
&= \frac{\hbar^2 k^2}{2 m_r^*}
\end{aligned}
\tag{4.67}
$$

where m_r^* is the reduced mass of the e-h system. The final state density of states in the summation are thus the reduced density of states given by

$$
N_{cv}(\hbar\omega) = \sqrt{2} \frac{(m_r^*)^{3/2} (\hbar\omega - E_g)^{1/2}}{\pi^2 \hbar^3}
\tag{4.68}
$$

and we have from Eqn. 4.51

$$
\boxed{W_{abs} = \frac{\pi e^2 \hbar n_{ph}}{\epsilon m_o^2 \hbar\omega} | (a \cdot p)_{cv} |^2 N_{cv}(\hbar\omega)}
\tag{4.69}
$$

Semiconductor	E_p (eV)
GaAs	25.7
InP	20.9
InAs	22.2
CdTe	20.7

Table 4.1: Values of E_p for different semiconductors. (After P. Lawaetz, *Physical Review*, B, **4**, 3460 (1971).)

In bulk semiconductors the expression for unpolarized light becomes (from Eqns. 4.69 and 4.65)

$$W_{abs} = \frac{\pi e^2 \hbar n_{ph}}{2\epsilon m_0 \hbar \omega} \left(\frac{2p_{cv}^2}{m_0} \right) \frac{2}{3} N_{cv}(\hbar \omega) \qquad (4.70)$$

4.4.2 Interband Transitions in Quantum Wells

⤳ The formalism developed so far can be extended in a straightforward manner to the case of the quantum well structures. The central cell functions in the quantum wells are relatively unaffected by the presence of the confining potential. The two changes that occur are the nature of the wavefunctions, which for the low lying states are confined to the well region, and the density of states which have the usual step-like form for parabolic 2-dimensional bands.

The absorption rates are calculated only for the well region since the barrier material has a higher bandgap and does not participate in the optical process till the photon energies are much higher. The quantum well conduction and valence band states are given in the envelope function approximation by

$$
\begin{aligned}
\psi_c^n &= \frac{1}{\sqrt{AW}} e^{i\mathbf{k}_e \cdot \rho} g_c^n(z) u_{c\mathbf{k}_e}^n \\
\psi_v^m &= \frac{1}{\sqrt{AW}} e^{i\mathbf{k}_h \cdot \rho} \sum_\nu g_v^{\nu m}(z) u_{v\mathbf{k}_h}^{\nu m}
\end{aligned}
\qquad (4.71)
$$

Here W is the well size and A is the area considered and we have used (as discussed in Chapter 2) the scalar description of the conduction band states and a multiband (indexed by ν) description of the valence band. The envelope functions g_c^n and $g_v^{\nu m}$ correspond to the n and m subband levels in the conduction and valence bands respectively. If we ignore the band mixing effects between the HH and LH states, i.e., ignore the off-diagonal terms in the Kohn-Luttinger Hamiltonian, we obtain simple analytic results for the absorption and emission rates. Note that the momentum matrix element now undergoes the following change when we go from our 3-dimensional calculation to a quasi-2-dimensional one

$$p_{\text{if}}^{3D} = \frac{1}{V} \int e^{i(\mathbf{k}_e - \mathbf{k}_h) \cdot \mathbf{r}} \langle u_v^\nu | p_a | u_c \rangle \, d^3 r$$

$$\rightarrow p_{\text{if}}^{2D} = \frac{1}{AW} \sum_\nu \langle g_v^{\nu m} | g_c^n \rangle \int e^{i(\mathbf{k}_e - \mathbf{k}_h) \cdot \rho} \langle u_v^{\nu m} | p_a | u_c \rangle \, d^2\rho \qquad (4.72)$$

where $\langle g_v^{\nu m} | g_c^n \rangle$ denotes the overlap between the z-dependent envelope functions of the conduction and valence bands. For symmetrical potentials one has the approximate condition that

$$\sum_\nu \langle g_v^{\nu m} | g_c^n \rangle \approx \delta_{nm} \qquad (4.73)$$

This condition is not exact and can be changed if there is any asymmetry present (e.g., if there is a transverse electric field present).

When expressing the photon absorption rate in terms of the final electronic density of states, the changes discussed above can be represented by a modified density of states which for parabolic bands in a 3D system given by

$$N_{cv}^{3D}(\hbar\omega) = \frac{\sqrt{2} \, m_r^{*3/2} (\hbar\omega - E_g)^{1/2}}{\pi \hbar^3}$$

are to be replaced in a 2D system by

$$\frac{N_{cv}^{2D}(\hbar\omega)}{W} = \frac{m_r^*}{\pi \hbar^2 W} \sum_{nm} \langle g_v^m | g_c^n \rangle \, \theta(E_{nm} - \hbar\omega) \qquad (4.74)$$

and

$$E_{nm} = E_{\text{gap}} + E_c^n + E_v^m \qquad (4.75)$$

Here m_r is the reduced electron hole mass. The θ-function is the Heaviside step function. In the emission process, if we are interested in the recombination of an electron and hole (for well-defined $\mathbf{k}_e = \mathbf{k}_h$), we integrate over the final density of photon states. In general, one can alter the photon density of states as well in heterostructures. If this is done, the recombination rate can change in quantum well structures. In fact, considerable work is being carried out to modify photon density of states by using specially tailored heterostructures. In this chapter we will assume that the photons are still 3-dimensional, with the dispersion relation $\omega = vk$.

Before summarizing and comparing the results for the optical absorption and electron-hole recombination in bulk and quantum well structures, we will briefly relate the absorption coefficient to the absorption rate. It is useful to talk about absorption coefficient rather than the rate at which a photon is absorbed. If we consider a beam of photons traveling along the x-axis, we can write the continuity equation for the photon density

$$\frac{dn_{ph}}{dt} = \frac{\partial n_{ph}}{\partial t}\bigg|_{\text{vol}} + \frac{\partial(vn_{ph})}{\partial x} \tag{4.76}$$

where the first term represents the absorption rate of photons and the second term represents the photons leaving due to the photon current. Here v is the velocity of light. In steady state we have, in general

$$n_{ph}(x) = n_0 \exp(-\alpha x) \tag{4.77}$$

which defines the absorption coefficient. Also

$$\frac{\partial n_{ph}}{\partial t} = W_{\text{abs}} \tag{4.78}$$

and in steady state we have

$$W_{\text{abs}} = \alpha v n_{ph}$$

or

$$\boxed{\alpha = \frac{W_{\text{abs}}}{vn_{ph}}} \tag{4.79}$$

Let us now summarize the absorption coefficients for interband and inter-subband transitions discussed above.

Bulk Semiconductors: Interband Transitions

$$\boxed{\alpha(\hbar\omega) = \frac{\pi e^2 \hbar}{m_o^2 c n_r \epsilon_o} \frac{1}{(\hbar\omega)} |\boldsymbol{a} \cdot \boldsymbol{p}_{\text{if}}|^2 N_{cv}(\hbar\omega)} \tag{4.80}$$

with

$$N_{cv}(\hbar\omega) = \frac{\sqrt{2}\,(m_r^*)^{3/2}\,(\hbar\omega - E_g)^{1/2}}{\pi^2 \hbar^3}$$

for parabolic electron-hole bands. The polarization dependence of the absorption coefficient is contained in the matrix element $|\boldsymbol{a} \cdot \boldsymbol{p}_{\text{if}}|^2$, n_r is the refractive index of the material.

Quantum Well: Interband Transitions

$$\boxed{\alpha(\hbar\omega) = \frac{\pi e^2 \hbar}{m_o^2 c n_r \epsilon_o} \frac{1}{(\hbar\omega)} |\boldsymbol{a} \cdot \boldsymbol{p}_{\text{if}}|^2 \frac{N_{2D}(\hbar\omega)}{W} \sum_{n,m} f_{nm}\, \theta(E_{nm} - \hbar\omega)} \tag{4.81}$$

Here, for parabolic bands, we have

$$N_{2D} = \frac{m_r^*}{\pi \hbar^2} \tag{4.82}$$

and f_{nm} represents the overlap between the n and m subband overlap functions. This function is close to unity for $n = m$ and close to zero, otherwise.

As discussed earlier, if we are interested in calculating the recombination rate of an electron with a hole at the same k value, we integrate over all possible photon states into which emission could occur. For total emission, we have

$$\boxed{W_{em} = \frac{\pi e^2 \hbar}{m_o^2 \hbar \omega \epsilon} (n_{ph} + 1) |a \cdot p_{if}|^2 \rho_a(\hbar \omega)} \tag{4.83}$$

where ρ_a is the photon density of states for the polarization a. The total photon density of states is given by (there are 2 transverse modes for each k value)

$$\boxed{\rho(\hbar \omega) = \frac{2\omega^2}{2\pi^2 \hbar v^3}} \tag{4.84}$$

for photons emitted in the 3-dimensional space. For a given polarization, the photon density of states is one third of the value in Eqn. 4.84. For quantum well structures the rate is simply modified by the overlap of the envelope functions of the electron and hole states. If this envelope function is unity, the emission rates are the same as in bulk, unless the structure is designed to change the photon density of states.

For $n_{ph} = 0$, the emission rate is called the spontaneous emission rate W_{spon} and it's inverse is the e-h recombination time τ_o. The time τ_o represents the time taken by an electron in a state k to recombine with an available hole in the state k.

It is useful to examine some numerical values of the absorption coefficient and the recombination time for, say, a common system like GaAs. In Fig. 4.8 and Fig. 4.9 we show the absorption coefficient for GaAs and a 100 Å GaAs /Al$_{0.3}$Ga$_{0.7}$As quantum well structure. In the bulk semiconductor the absorption coefficient starts at $\hbar \omega = E_g$, with a zero value and initially increases as $(\hbar \omega - E_g)^{1/2}$. It also has a $1/\hbar \omega$ behavior which only influences the absorption coefficients at high energies where the density of states is not parabolic anymore. Because of the degeneracy of the HH and LH states, there is no polarization dependence of the absorption coefficient near the bandgap region.

The absorption coefficient in the quantum well structure is quite distinct from the bulk case mainly because of the density of states function. Another difference arises because of the lifting of the HH, LH degeneracy which makes the absorption coefficient strongly polarization dependent as discussed earlier.

In quantum wells, the $1/W$ dependence of the absorption coefficient is quite interesting and somewhat misleading. We note that this dependence came from our

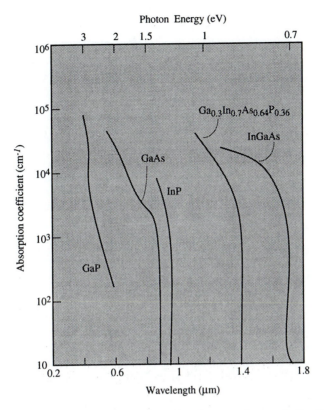

Figure 4.8: Absorption coefficient for several semiconductors. In a real situation, the effects of the *e-h* interaction discussed in Chapter 5 causes the absorption coefficient to be modified at the bandedge.

assumption that the wavefunction is localized a distance equal to the well size. The $1/W$ dependence suggests that the absorption coefficient can be increased indefinitely by decreasing W, the well size. This is, however, not true. As shown schematically in Fig. 4.10, as the well size is narrowed, the wavefunctions of the electron and hole no longer are confined to the well size. The region of interest for the electrons starts increasing beyond W. Also, because of the different masses of the electron and hole, the electron function starts spreading beyond the well at a larger well size making the overlap from less than unity. Thus, depending upon the material, the optimum well size is ~ 50 Å (for GaAs) to 80 Å (for $In_{0.53}Ga_{0.47}As$).

4.5 INDIRECT INTERBAND TRANSITIONS

⤳ In the previous section we discussed the interactions between photons and electrons

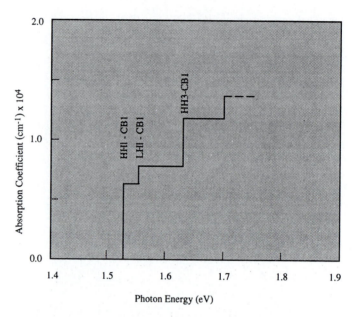

Figure 4.9: Absorption coefficient in a 100 Å GaAs/Al$_{0.3}$Ga$_{0.7}$As quantum well structure for in-plane polarized light. The HH transition is about three times stronger than the LH transition in this polarization. The higher subbands become closer spaced and eventually one gets the 3D absorption coefficient.

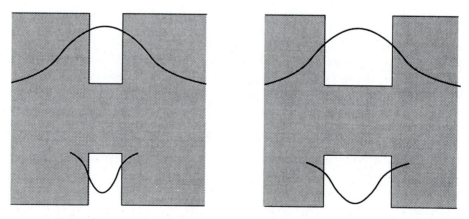

Figure 4.10: Schematic representation of the effect of well size on electron and hole wavefunctions and the associated overlap integrals. The overlap integral decreases at very narrow quantum well sizes.

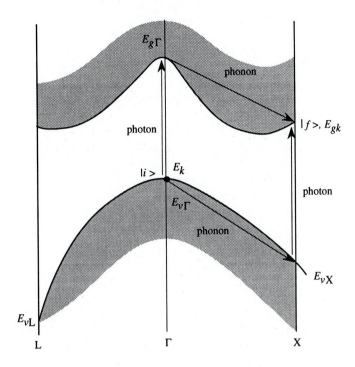

Figure 4.11: Typical processes responsible for optical absorption in indirect bandgap semiconductors. The photon energy need not be equal to the vertical energy, since the intermediate transitions are "virtual," i.e., the electron does not reside there for any length of time.

in direct bandgap material where vertical transitions in k-space are allowed. However, photons also cause interband transitions in indirect semiconductors where the bandedges are at different k-points. Thus, optical absorption is observed in Si and Ge although the absorption rate is far weaker than for GaAs.

Transitions between electron states that are not vertical in k-space are called indirect transitions and they can be mediated by a phonon interaction. The phonons are required to ensure momentum conservation in the process.

A typical process in the interband transition is shown in Fig. 4.11. The process is second order in which the electron is first scattered by a photon to the direct band conserving momentum and then scattering to the indirect band by a phonon. While momentum is conserved in the intermediate process, *the energy is not conserved since this process is virtual and the time-energy uncertainty ensures that there is no energy conservation requirement.* The overall process, however, does conserve energy. The scattering rate is again given by the Fermi golden rule except that the matrix element is a second order matrix element

$$W = \frac{2\pi}{\hbar} \int \left| \frac{\sum \langle f|H_{\text{per}}|n\rangle \langle n|H_{\text{per}}|i\rangle}{E_i - E_n} \right|^2 \delta(E_f - E_i) \frac{d^3k}{(2\pi)^3} \qquad (4.85)$$

where $|n\rangle$ is an intermediate state, and the perturbation is

$$H_{\text{per}} = H_{\text{ph}} + H_{\text{ep}} \qquad (4.86)$$

where H_{ph} is the electron-photon interaction discussed so far and H_{ep} is the electron-phonon interaction discussed in Chapter 3. In general, the processes shown in Fig. 4.11 can contribute to the scattering process. However, the processes which involve first a photon interaction are stronger since the denominator is smaller for them (denominator is $\sim$ direct bandgap), than for the processes involving the phonon first (denominator is $\sim |E_{v\Gamma} - E_{vX}|$ or $|E_{v\Gamma} - E_{vL}|$ which are larger than the direct bandgap). The scattering rate is then

$$W(\boldsymbol{k}) = \frac{2\pi}{\hbar} \int_f \left\{ |M_{\text{em}}|^2 + |M_{\text{abs}}|^2 \right\} \delta(E_f - E_i) \frac{d^3k}{(2\pi)^3} \qquad (4.87)$$

The matrix elements M_{em} and M_{abs} correspond to the cases where first a photon is absorbed and then a phonon is either emitted or absorbed. Note that the photon energy $\hbar\omega$ is smaller than the direct bandgap, but the intermediate transition can occur since energy need not be conserved. The form of the matrix elements is

$$M_{\text{abs}} = \frac{\left| \langle c, \boldsymbol{k} + \boldsymbol{q}|H_{\text{ep}}^{\text{abs}}|c, \boldsymbol{k}\rangle \right|^2 \left| \langle c, \boldsymbol{k}|H_{\text{ph}}^{\text{abs}}|v, \boldsymbol{k}\rangle \right|^2}{(E_{g\Gamma} - \hbar\omega)^2}$$

$$M_{\text{em}} = \frac{\left| \langle c, \boldsymbol{k} - \boldsymbol{q}|H_{\text{ep}}^{\text{em}}|c, \boldsymbol{k}\rangle \right|^2 \left| \langle c, \boldsymbol{k}|H_{\text{ph}}^{\text{em}}|v, \boldsymbol{k}\rangle \right|^2}{(E_{g\Gamma} - \hbar\omega)^2} \qquad (4.88)$$

The phonon scattering is due to the optical phonon intervalley scattering. Its matrix element is

$$M_q^2 = \frac{\hbar D_{ij}^2}{2\rho V \omega_{ij}} \left\{ \begin{array}{c} n(\omega_{ij}) \\ n(\omega_{ij}) + 1 \end{array} \right\} \qquad (4.89)$$

for the absorption and emission processes respectively. Here D_{ij} is the deformation potential, ρ is the mass density, and ω_{ij} is the phonon frequency which connects the Γ valley to the zone edge valley. It is useful to point out that due to the *indirect* nature of the transition, the rates calculated earlier for direct gap semiconductors are essentially *lowered* by a factor equal to

$$\boxed{\frac{M_q^2}{(E_{g\Gamma} - \hbar\omega)^2}} \qquad (4.90)$$

This factor is typically 10^{-2} to 10^{-3} and has a temperature dependence due to the temperature dependence of the phonon occupation $n(\omega_{ij})$. For direct gap transitions, for a given initial state $|i\rangle$, there was only one final state $|f\rangle$ which had the same k-value. For a given photon energy one then had the states $N_{cv}(\hbar\omega)d(\hbar\omega)$ in energy interval $\hbar\omega$ to $\hbar(\omega+d\omega)$ which contributed with N_{cv} given by Eqn. 4.68. In case of indirect transitions, for a given initial states $|v,k\rangle$, there is a spread in the final states due to the phonon scattering. The scattering rate sums over this spread, giving

$$
\begin{aligned}
W(k) &= \frac{2\pi}{\hbar} \frac{M_{\text{ph}}^2}{(E_{g\Gamma}-\hbar\omega)^2} \frac{\hbar D_{ij}^2}{2\rho\omega_{ij}} J_v \\
&\times\ [n(\omega_{ij})\,N_c(E_1+\hbar\omega_{ij}) \\
&+\ \{n(\omega_{ij})+1\}\,N_c(E_1-\hbar\omega_{ij})]
\end{aligned}
\tag{4.91}
$$

where J_v is the number of equivalent valleys, N_c is the density of states for a given spin in a valley, and

$$
E_1 = \hbar\omega - E_{gk'} - E_k
\tag{4.92}
$$

where $E_{gk'}$ is the *indirect gap* and the E_k is the energy of the initial electron measured from the top of the valence band.

To find the absorption coefficient we need to sum the above rate over all possible starting states which could absorb a photon with energy $\hbar\omega$. This means we must sum over all possible initial states from $E_k = 0$ to E_{kmax} where

$$
\begin{aligned}
E_{kmax} &= \hbar\omega - E_{gk'} + \hbar\omega_{ij} \text{ (phonon absorption)} \\
E_{kmax} &= \hbar\omega - E_{gk'} - \hbar\omega_{ij} \text{ (phonon emission)}
\end{aligned}
\tag{4.93}
$$

We multiply $W(k)$ by $2VN_v(E_k)dE_k$ where N_v is the single spin density of states in the valence band and iterate from 0 to E_{kmax}. For parabolic bands the integral is simple and we get

$$
\begin{aligned}
W_{\text{abs}}(\hbar\omega) &= \frac{M_{\text{ph}}^2\,D_{ij}^2\,J_v\,(m_c m_v)^{3/2}}{8\pi^2(E_{g\Gamma}-\hbar\omega)^2\,\hbar^6\,\rho\,\omega_{ij}} \\
&\times\ \left[n(\omega_{ij})\left(\hbar\omega - E_{gk'} + \hbar\omega_{ij}\right)^2 \right. \\
&+\ \left. \{n(\omega_{ij})+1\}\left(\hbar\omega - E_{gk'} - \hbar\omega_{ij}\right)^2\right]
\end{aligned}
\tag{4.94}
$$

and from our earlier calculations

$$
M_{\text{ph}}^2 = \frac{e^2\hbar n_{\text{ph}}\,|a\cdot p_{\text{if}}|^2}{2m_o^2\epsilon\omega}
\tag{4.95}
$$

The absorption coefficient is then given by $W_{\text{abs}}/(n_{\text{ph}}\,v_{\text{ph}})$ (see Eqn. 4.79). We note that once the threshold photon energy is reached the absorption coefficient

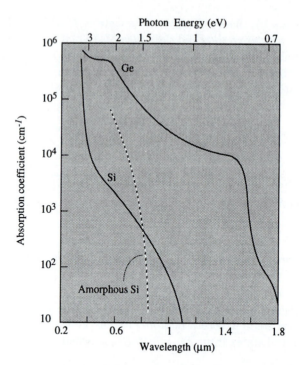

Figure 4.12: Absorption coefficient of Si. For the direct gap material, the absorption coefficient is very strong once the photon energy exceeds the bandgap. For indirect materials the absorption coefficient rises much more gradually. Note that amorphous silicon has an absorption coefficient almost like a direct gap semiconductor, since k-selection is not applicable.

increases as $(\hbar\omega - E_{\rm th})^2$ in contrast to the direct gap case where the energy dependence was $(\hbar\omega - E_g)^{1/2}$.

In Fig. 4.12 we show typical absorption measurements for Si, Ge, and amorphous silicon. We note the low absorption coefficient at near the bandgap in Si when compared to the results for a direct gap material like GaAs (Fig. 4.8). Once the photon energies reach the direct gap region, the absorption coefficient increases rapidly since direct transitions are possible. Notice that here we have only considered phonons as a scattering process. Other scattering processes such as alloy scattering, impurity scattering, etc., can also cause optical absorption in indirect semiconductors. As a result, "poor" quality indirect semiconductors have a better absorption coefficient than pure indirect materials. An example of absorption in amorphous silicon (a-Si) is shown in Fig. 4.12. As can be seen in this amorphous material, the absorption coefficient is quite strong.

EXAMPLE 4.2 A 1.6 eV photon is absorbed by a valence band electron in GaAs. If the bandgap of GaAs is 1.41 eV, calculate the energy of the electron and heavy hole produced by the photon absorption.

The electron, heavy-hole, and reduced mass of GaAs are $0.067m_0$, $0.45m_0$, and $0.058m_0$, respectively. The electron and the hole generated by photon absorption have the same momentum. The energy of the electron is

$$E^e = E_c + \frac{m_r^*}{m_e^*}(\hbar\omega - E_g)$$

$$E^e - E_c = \frac{0.058}{0.067}(1.6 - 1.41) = 0.164 \; eV$$

The hole energy is

$$E^h - E_v = -\frac{m_r^*}{m_h^*}(\hbar\omega - E_g) = -\frac{0.058}{0.45}(1.6 - 1.41)$$

$$= -0.025 \; eV$$

The electron by virtue of its lower mass is created with a much greater energy than the hole.

EXAMPLE 4.3 In silicon, an electron from the top of the valence band is taken to the bottom of the conduction band by photon absorption. Calculate the change in the electron momentum. Can this momentum difference be provided by a photon?

The conduction band minima for silicon are at a k-value of $\frac{2\pi}{a}$ (0.85, 0, 0). There are five other similar bandedges. The top of the valence band has a k-value of 0. The change in the momentum is thus

$$\hbar\Delta k = \hbar\frac{2\pi}{a}(0.85) = (1.05 \times 10^{-34})\left(\frac{2\pi}{5.43 \times 10^{-10}}\right)(0.85)$$

$$= 1.03 \times 10^{-24} \; kg \; m \; s^{-1}$$

A photon which has an energy equal to the silicon bandgap can only provide a momentum of

$$\hbar k_{ph} = \hbar \cdot \frac{2\pi}{\lambda}$$

The λ for silicon bandgap is 1.06 μm and thus the photon momentum is about a factor of 600 too small to balance the momentum needed for the momentum conservation. The lattice vibrations produced by thermal vibration are needed for the process.

EXAMPLE 4.4 The absorption coefficient near the bandedges of GaAs and Si are $\sim 10^4$ cm^{-1} and 10^3 cm^{-1} respectively. What is the minimum thickness of a sample in each case which can absorb 90% of the incident light?

The light absorbed in a sample of length L is

$$\frac{I_{abs}}{I_{inc}} = 1 - exp\,(-\alpha L)$$

or

$$L = = \frac{1}{\alpha}\,\ell n\,\left(1 - \frac{I_{abs}}{I_{inc}}\right)$$

Using $\frac{I_{abs}}{I_{inc}}$ equal to 0.9, we get

$$L(GaAs) = -\frac{1}{10^4}\,\ell n\,(0.1) \quad = \quad 2.3 \times 10^{-4}\,cm$$

$$= \quad 2.3\,\mu m$$

$$L(Si) = -\frac{1}{10^3}\,\ell n\,(0.1) \quad = \quad 23\,\mu m$$

Thus an Si detector requires a very thick active absorption layer to function.

EXAMPLE 4.5 Calculate the absorption coefficient of GaAs as a function of photon frequency.

The joint density of states for GaAs is (using a reduced mass of $0.065m_0$)

$$N_{cv}(E) \quad = \quad \frac{\sqrt{2}(m_r^*)^{3/2}(E - E_g)^{1/2}}{\pi^2\hbar^3}$$

$$= \quad \frac{1.414 \times (0.065 \times 0.91 \times 10^{-30}\ kg)^{3/2}(E - E_g)^{1/2}}{9.87 \times (1.05 \times 10^{-34})^3}$$

$$= \quad 1.78 \times 10^{54}(E - \hbar\omega)^{1/2}\ J^{-1}\ m^{-3}$$

The absorption coefficient is for unpolarized light

$$\alpha(\hbar\omega) = \frac{\pi e^2\hbar}{2n_r c\epsilon_0 m_0}\left(\frac{2p_{cv}^2}{m_0}\right)\frac{N_{cv}(\hbar\omega)}{\hbar\omega} \cdot \frac{2}{3}$$

The term $\frac{2p_{cv}^2}{m_0}$ is $\sim$23.0 eV for GaAs. This gives

$$\alpha(\hbar\omega) \quad = \quad \frac{3.1416 \times (1.6 \times 10^{-19}\ C)^2(1.05 \times 10^{-34}\ J-s)}{2 \times 3.4 \times (3 \times 10^8\ m/s)(8.84 \times 10^{-12}\ (F/m)^2)}$$

$$\cdot\ \frac{(23.0 \times 1.6 \times 10^{-19}\ J)}{(0.91 \times 10^{-30}\ kg)}\frac{(\hbar\omega - E_g)^{1/2}}{\hbar\omega} \times 1.78 \times 10^{54}\ \times\ \frac{2}{3}$$

$$\alpha(\hbar\omega) \quad = \quad 2.25 \times 10^{-3}\frac{(\hbar\omega - E_g)^{1/2}}{\hbar\omega}\ m^{-1}$$

Here the energy and $\hbar\omega$ are in units of Joules. It is usual to express the energy in eV, and the absorption coefficient in cm^{-1}. This is obtained by multiplying the result by

$$\left[\frac{1}{(1.6 \times 10^{-19})^{1/2}} \times \frac{1}{100}\right]$$

$$\alpha(\hbar\omega) = 5.6 \times 10^4\frac{(\hbar\omega - E_g)^{1/2}}{\hbar\omega}\ cm^{-1}$$

For GaAs the bandgap is 1.5 eV at low temperatures and 1.43 eV at room temperatures. From the value of α, we can see that a few microns of GaAs are adequate to absorb a significant fraction of light above the bandgap.

EXAMPLE 4.6 Calculate the electron-hole recombination time in GaAs.

The recombination rate is given by

$$W_{em} = \frac{e^2 n_r}{6\pi\epsilon_o m_0 c^3 \hbar^2} \left(\frac{2p_{cv}^2}{m_0}\right) \hbar\omega$$

with $\frac{2p_{cv}^2}{m_0}$ being 23 eV for GaAs.

$$
\begin{aligned}
W_{em} &= \frac{(1.6 \times 10^{-19} \ C)^2 \times 3.4 \times (23 \times 1.6 \times 10^{-19} \ J)\hbar\omega}{6 \times 3.1416 \times (8.84 \times 10^{-12} \ F/m) \times (0.91 \times 10^{-30} \ kg)} \\
&\quad \cdot \frac{1}{(3 \times 10^8 \ m/s)^3 \times (1.05 \times 10^{-34} \ J - s)^2} \\
&= 7.1 \times 10^{27} \hbar\omega \ s^{-1}
\end{aligned}
$$

If we require the value of $\hbar\omega$ in eV instead of Joules we get

$$
\begin{aligned}
W_{em} &= 7.1 \times 10^{27} \times (1.6 \times 10^{-19})\hbar\omega \ s^{-1} \\
&= 1.14 \times 10^9 \ \hbar\omega \ s^{-1}
\end{aligned}
$$

For GaAs, $\hbar\omega \sim 1.5$ eV so that

$$W_{em} = 1.71 \times 10^9 \ s^{-1}$$

The corresponding recombination time is

$$\tau_o = \frac{1}{W_{em}} = 0.58 \ ns$$

Remember that this is the recombination time when an electron can find a hole to recombine with. This happens when there is a high concentration of electrons and holes, i.e., at high injection of electrons and holes or when a minority carrier is injected into a heavily doped majority carrier region.

4.6 INTRABAND TRANSITIONS

In direct bandgap materials it is possible to satisfy both momentum and energy conservation laws for an optical interband transition to occur in first order. This is not possible for intraband transitions in bulk semiconductors. However, as we shall discuss shortly, in quantum well structures, it is possible to have strong intraband transitions making quantum well structures quite exciting for long wavelength optical devices.

4.6.1 Intraband Transitions in Bulk Semiconductors

⤳ For bulk semiconductors, the intraband transitions must involve a phonon or some other scattering mechanism (ionized impurity, defects, etc.) discussed in Chapter 3 to ensure momentum conservation. The second order process is essentially similar to the one we dealt with for indirect processes. The free carrier absorption, as absorption due to intraband transitions is called, is quite important particularly for lasers since it is responsible for losses in the cladding layers of the laser. We will not calculate the free carrier absorption rates in detail here. A good source for their evaluation is B. K. Ridley (1982). We will only note that for acoustic phonon assisted processes the absorption coefficient is

$$\alpha(\hbar\omega) \approx \frac{128\ \alpha_f\ e\ \hbar\ n_e}{g\ n_r\ (m_c^*)^2\ \omega^2\ \mu_{ac}} \tag{4.96}$$

where α_f is the fine structure constant ($\approx 1/137$), n_e is the free carrier density, n_r the refractive index, and μ_{ac} the mobility due to acoustic phonon scattering. The absorption coefficient is inversely proportional to the square of the frequency.

4.6.2 Intraband Transitions in Quantum Wells

⤳ As we have seen in this chapter, quantum well structures can produce remarkable changes in the optical properties of semiconductor structures. Nowhere are the changes more impressive than in the area of intraband transitions. As we have seen, the intraband transitions in bulk semiconductors are forbidden in the first order. This is because it is not possible to obey both energy and momentum conservation in such transitions. The momentum conservation is due to the plane wave part of the Bloch states. In quantum well structures, the electronic states are no longer of the plane wave form in the growth direction making it possible to have intraband transitions for certain polarizations of light. Since the intraband (or inter-subband) transition energy can be easily varied by changing the well size, these transitions have great importance for far infrared detectors and modulators.

Let us consider the intraband (inter-subband) transitions for the quantum well case. Due to the confinement in the z-direction, the subband functions can be written as (say, the first two functions)

$$
\begin{aligned}
\psi^1(\boldsymbol{k}, z) &= g^1(z)\, e^{i\boldsymbol{k}\cdot\rho}\, u^1_{n\boldsymbol{k}}(\boldsymbol{r}) \\
\psi^2(\boldsymbol{k}, z) &= g^2(z)\, e^{i\boldsymbol{k}\cdot\rho}\, u^2_{n\boldsymbol{k}}(\boldsymbol{r})
\end{aligned}
\tag{4.97}
$$

We are once again interested in vertical transitions since the photon momentum is very small. The functions g^1 and g^2 are orthogonal and to a good approximation the central cell functions are same for the different subbands (this especially true for the conduction band). Thus, the momentum matrix element is given by

$$\boldsymbol{p}_{\text{if}} = -\frac{i\hbar}{W} \int g^{2*}(z)\, e^{-i\mathbf{k}\cdot\boldsymbol{\rho}}\, \boldsymbol{a} \cdot \nabla g^1(z)\, e^{i\mathbf{k}\cdot\boldsymbol{\rho}}\, d^2\rho\, dz \qquad (4.98)$$

where W is the well width.

As in the case of the 3-dimensional system, the momentum matrix element is zero if the polarization vector (or the ∇ function) is in the ρ-plane. Thus, for the x-y polarized light, the *transitions rate is still zero*. (Note that if there is strong mixing of the central cell functions as in the valence bands, this condition can be relaxed.) However, if the light is z-polarized we get

$$\boldsymbol{p}_{\text{if}} = \frac{-i\hbar}{W} \int g^{2*}(z)\, \hat{z}\, \frac{\partial}{\partial z} g^1(z)\, dz \qquad (4.99)$$

Since $g^1(z)$ and $g^2(z)$ have even and odd parities respectively, as shown in Fig. 4.13, $g^2(z)$ and $\partial g^1(z)/\partial z$ both have odd parity. The momentum matrix element for z-polarization is then approximately

$$|\boldsymbol{p}_{\text{if}}| \approx \frac{\hbar}{W} \qquad (4.100)$$

This result is reasonably accurate if both the ground and excited states are confined to the well size. Often the excited state may have less confinement than the ground state in which case one has to explicitly evaluate the integral in Eqn. 4.99.

If we make a simple parabolic approximation, we see that the dispersion relations of the two subbands are essentially parallel and shifted by the subband energy levels difference $E_2 - E_1$. In principle, therefore, the joint density of states is a δ-function with infinite density of states at the transition energy. However, at this point we must include the statistics and broadening into the problem. For interband transitions it was reasonable to assume that the initial state (in the valence band) is occupied and the final state (in the conduction band) is empty. However, in intraband transitions discussed above (the inter-subband transitions), the electrons have to be introduced by doping the material so that the electrons are present in the first subband and hopefully not present in the second subband. Introducing the Fermi factors, we get for the absorption process

$$W_{\text{abs}} = \frac{\pi e^2 n_{ph}}{m_o^2 \omega \epsilon} \frac{1}{W} \sum_f |p_{\text{if}}|^2\, \delta\left(E_f - E_i - \hbar\omega\right)\, f(E_i)\left[1 - f(E_f)\right] \qquad (4.101)$$

with

$$|p_{\text{if}}| \approx \frac{\pi\hbar}{W}$$

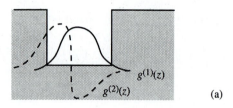

(a)

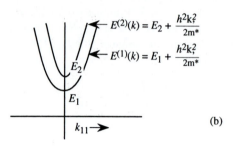

(b)

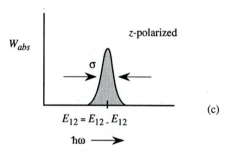

(c)

Figure 4.13: A schematic presentation of the (a) envelope functions, (b) bandstructure, and (c) absorption rate for z-polarized light in a quantum well.

If we assume that the second subband is empty, the sum over the final states is zero except at resonance where we have

$$\sum_f \delta(E_f - E_i - \hbar\omega) f(E_i) = N_c \qquad (4.102)$$

where N_c is the electron concentration in the first subband. The density of states at resonance is infinite for the simple parabolic band giving an infinite absorption rate. However, in reality the nonparabolicity and scattering mechanisms introduce some broadening in the density of states. If we assume a Gaussian broadening, the two-dimensional density of states becomes

$$N(E) = \frac{N_c \exp\left(-\frac{(E - E_{12})^2}{1.44\sigma^2}\right)}{\sqrt{1.44\pi}\sigma} \qquad (4.103)$$

where σ is the linewidth of the transition. The linewidth may have contributions from both homogeneous (phonon related) and inhomogeneous (structural imperfections) effects.

The absorption coefficient for the z-polarized light now becomes

$$\alpha(\hbar\omega) = \frac{\pi e^2 \hbar}{m_o^2 c n_r \epsilon_o} \frac{|p_{if}|^2}{\hbar\omega} \frac{N_c \exp\left(-\frac{(\hbar\omega - E_{12})^2}{1.44\sigma^2}\right)}{\sqrt{1.44\pi}\sigma} \qquad (4.104)$$

with

$$|p_{if}|^2 \sim \frac{\pi^2 \hbar^2}{W^2}$$

The absorption coefficient increases rapidly with decreasing well size although it is important to note again that as the well size becomes very small, the electronic states are no longer confined to the well size as assumed by us. For a transition which is only 1–2 meV wide, the absorption coefficient could reach 10^4 cm^{-1}, making such transitions useful for detectors or modulators.

4.7 CHARGE INJECTION AND RADIATIVE RECOMBINATION

$\longrightarrow$ In the formalism for interband transitions discussed so far, we have assumed that the initial electron state is occupied with unity probability while the final state is empty. In actual experimental situations this may, of course, not be true. In this section we will discuss the important optical properties which play a key role in the performance of light emitting diodes and laser diodes.

Spontaneous Emission Rate

When electrons and holes are injected into the conduction and valence bands of a semiconductor, they recombine with each other as we have discussed earlier. In then absence of any photon density in the cavity (i.e., $n_{ph} = 0$), the emission rate is the spontaneous emission rate which has a value of $\sim 1/(0.5$ ns$)$, provided an electron is present in the state $\boldsymbol{k}$ and a hole is present in the same state $\boldsymbol{k}$ in the valence band. In reality, however, the rate depends upon the occupation probabilities of the electron and hole with the same k-value. Therefore, we have to include the distribution functions for electrons and holes and integrate over all possible electronic states. Thus, the recombination rate is (units are $cm^{-3}s^{-1}$)

$$R_{spon} = \frac{2}{3} \int d(\hbar\omega) \frac{2e^2 n_r \hbar\omega}{m_0^2 c^3 \hbar^2} \left[\int \frac{1}{(2\pi)^3} d^3k \, |p_{if}|^2\right.$$

$$\times \quad \delta(E^e(\boldsymbol{k}) - E^h(\boldsymbol{k}) - \hbar\omega)$$

$$\times \quad f_e(E^e(\boldsymbol{k}))f_h(E^h(\boldsymbol{k}))\bigg] \tag{4.105}$$

The integral over $d(\hbar\omega)$ is to find the rate for all photons emitted and the integration over d^3k is to get the rate for all the occupied electron and hole states. The prefactor 2/3 comes about since we are considering emission into any photon polarization so that we average the matrix element square $|\boldsymbol{a} \cdot \boldsymbol{p}_{\mathrm{if}}|^2$.

The extension to quantum well structures is the usual one giving (the rate here is per unit area)

$$R_{\mathrm{spon}} = \frac{2}{3} \int d(\hbar\omega) \frac{2e^2 n_r \hbar\omega}{m_0^2 c^3 \hbar^2} \sum_{nm} \bigg[\int \frac{d^2 k}{(2\pi)^2} |p_{\mathrm{if}}|^2$$

$$\times \quad \delta(E_n^e(\boldsymbol{k}) - E_m^h(\boldsymbol{k}) - \hbar\omega)$$

$$\times \quad f_e(E_n^e(\boldsymbol{k}))f_h(E_m^h(\boldsymbol{k}))\bigg] \tag{4.106}$$

Using the definition of the time τ_o we have ($\tau_o = \frac{1}{W_{spon}}$)

$$R_{spon} = \frac{1}{\tau_o} \int d(\hbar\omega) N_{cv}\{f^e(E^e)\}\{f^h(E^h)\} \tag{4.107}$$

The spontaneous recombination rate is quite important for both electronic and optoelectronic devices. It is important to examine the rate for several important cases. We will give results for the electron hole recombination for the following cases:

i) **Minority carrier injection:** If $n \gg p$ and the sample is heavily doped, we can assume that $f^e(E^e)$ is close to unity. We then have for the rate at which holes will recombine with electrons,

$$R_{spon} \cong \frac{1}{\tau_o} \int d(\hbar\omega) N_{cv} f^h(E^h) \cong \frac{1}{\tau_o} \int d(\hbar\omega) N_h f^h(E^h) \left(\frac{m_r^*}{m_h^*}\right)^{3/2}$$

$$\cong \frac{1}{\tau_o} \left(\frac{m_r^*}{m_h^*}\right)^{3/2} p \tag{4.108}$$

Thus the recombination rate is proportional to the minority carrier density (holes in this case). This condition is useful for *p-n* diodes and bipolar transistors which will be discussed later.

ii) **Strong injection:** This case is important when a high density of both electrons and holes is injected. We can now assume that both f^e and f^h are sharp step functions and we get approximately

$$R_{spon} = \frac{n}{\tau_o} = \frac{p}{\tau_o} \tag{4.109}$$

iii) **Weak injection:** In this case we can use the Boltzmann distribution to describe the Fermi functions. We have

$$f^e \cdot f^h \cong exp\left\{-\frac{(E_c - E_{Fn})}{k_B T}\right\} exp\left\{-\frac{(E_{Fp} - E_v)}{k_B T}\right\} \cdot exp\left\{-\frac{(\hbar\omega - E_g)}{k_B T}\right\} \tag{4.110}$$

The spontaneous emission rate now turns out to be

$$R_{spon} = \frac{1}{2\tau_o}\left(\frac{2\pi\hbar^2 m_r^*}{k_B T m_e^* m_h^*}\right)^{3/2} np \tag{4.111}$$

If we write the total charge as equilibrium charge plus excess charge,

$$n = n_o + \Delta n; p = p_o + \Delta n \tag{4.112}$$

we have for the excess carrier recombination (note that at equilibrium the rates of recombination and generation are equal)

$$R_{spon} \cong \frac{1}{2\tau_o}\left(\frac{2\pi\hbar^2 m_r^*}{k_B T m_e^* m_h^*}\right)^{3/2} (\Delta n p_o + \Delta p n_o) \tag{4.113}$$

If $\Delta n = \Delta p$, we can define the rate of a single excess carrier recombination as $\frac{\Delta n}{\tau_r}$, where

$$\frac{1}{\tau_r} = \frac{1}{2\tau_o}\left(\frac{2\pi\hbar^2 m_r^*}{k_B T m_e^* m_h^*}\right)(n_o + p_o) \tag{4.114}$$

At low injection τ_r is much larger than τ_o and is a result of the fact that at low injection, electrons have a low probability to find a hole with which to recombine.

iv) **Inversion condition:** Another useful approximation occurs when the electron and hole densities are such that $f^e + f^h = 1$. *This is the condition for inversion when the emission and absorption coefficients become equal.* If we assume in this case $f^e \sim f^h = 1/2$, we get the approximate relation

$$R_{spon} \cong \frac{n}{4\tau_o} \cong \frac{p}{4\tau_o} \tag{4.115}$$

The recombination lifetime is approximately $4\tau_o$ in this case. This is a useful result to estimate the threshold current of semiconductor lasers.

The gain and recombination processes discussed here are extremely important in both electronic and optoelectronic devices that will be discussed later. We point out from the above discussion that the recombination time for a single excess carrier can be written in many situations in the form

$$\tau_r = \frac{\Delta n}{R_{spon}} \qquad (4.116)$$

For minority carrier injection or strong injection $\tau_r \cong \tau_o$. In general, R_{spon} has a strong carrier density dependence as does τ_r. A typical curve showing the dependence of τ_r on the carrier density is shown in Fig. 4.14. Note that the radiative lifetime in GaAs can range from microseconds to nanoseconds, depending upon the injection density.

Gain in a Material

In the previous sections we have discussed how a photon impinging upon a semiconductor can be absorbed by moving an electron from the valence to conduction band. The photon beam thus decays as it moves through the semiconductor. The process of absorption requires the valence band states to have electrons present and conduction band states to have no electrons. What happens if the valence band has holes and the conduction band has electrons? Normally, such a situation does not occur, but if electrons are injected into the conduction band and holes into the valence band (as happens for light emitting devices discussed in Chapters 9 and 10), this could occur. *Under such conditions the electron-hole pairs could recombine and emit more photons than could be absorbed.* Thus one must talk about the emission coefficient minus the absorption coefficient. This term is called the gain of the material. If the gain is positive, an optical beam will grow as it moves through the material instead of decaying. In the simple parabolic bands we have the gain $g(\hbar\omega)$ given by the generalization of Eqn. 4.80 (gain = emission coefficient − absorption coefficient)

$$g(\hbar\omega) = \frac{\pi e^2 \hbar}{n_r c m_0^2 \epsilon_0(\hbar\omega)} \mid a \cdot p_{if} \mid^2 N_{cv}(\hbar\omega)[f^e(E^e) - (1 - f^h(E^h))] \qquad (4.117)$$

The term in the square brackets arises since the emission of photons is proportional to $f^e \cdot f^h$, while the absorption process is proportional to $(1 - f^e) \cdot (1 - f^h)$. The difference of these terms appears in Eqn. 4.117.

The energies E^e and E^h are (see Fig. 4.15) determined by noting the following (see Eqn. 4.67)

$$\hbar\omega - E_g = \frac{\hbar^2 k^2}{2m_r^*} \qquad (4.118)$$

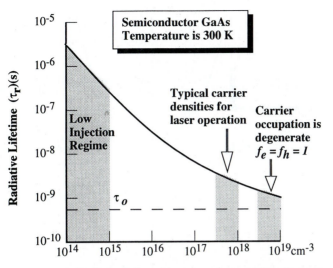

Figure 4.14: The dependence of the radiative lifetime in GaAs as a function of carrier injection $(n = p)$ or minority carrier injection into a doped region with doping density as shown.

$$E^e = E_c + \frac{\hbar^2 k^2}{2m_e^*} = E_c + \frac{m_r^*}{m_e^*}(\hbar\omega - E_g) \qquad (4.119)$$

$$E^h = E_v - \frac{\hbar^2 k^2}{2m_h^*} = E_v - \frac{m_r^*}{m_h^*}(\hbar\omega - E_g) \qquad (4.120)$$

If $f^e(E^e) = 0$ and $f^h(E^h) = 0$, i.e., if there are no electrons in the conduction band and no holes in the valence band, we see that the gain is simply $-\alpha(\hbar\omega)$ which we had discussed earlier. A positive value of gain occurs for a particular energy when

$$f^e(E^e) > 1 - f^h(E^h) \qquad (4.121)$$

a condition that is called inversion. In this case the light wave passing in the material has the spatial dependence

$$I(z) = I_o \, exp \, (gz) \qquad (4.122)$$

which grows with distance instead of diminishing as it usually does if $g(\hbar\omega)$ is negative. The gain in the optical intensity is the basis for the semiconductor laser.

EXAMPLE 4.7 According to the Joyce-Dixon approximation, the relation between the Fermi

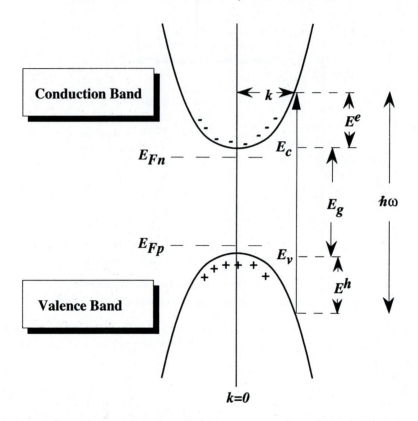

Figure 4.15: The positions of the quasi-Fermi levels, the electron and hole energies at vertical k-values. The electron and hole energies are determined by the photon energy and the carrier masses.

level and carrier concentration is given by

$$E_F - E_c = k_B T \left[\ell n \frac{n}{N_c} + \frac{1}{\sqrt{8}} \frac{n}{N_c} \right]$$

where N_c is the effective density of states for the band. Calculate the carrier density needed for the transparency condition in GaAs at 300 K and 77 K. The transparency condition is defined at the situation where the maximum gain is zero (i.e., the optical beam propagates without loss or gain).

At room temperature the valence and conduction band effective density of states is

$$N_v = 7 \times 10^{18} \, cm^{-3}$$
$$N_c = 4.7 \times 10^{17} \, cm^{-3}$$

The values at 77 K are

$$N_v = 0.91 \times 10^{18} \, cm^{-3}$$
$$N_c = 0.61 \times 10^{17} \, cm^{-3}$$

In the semiconductor laser, an equal number of electrons and holes are injected into the active region. We will look for the transparency conditions for photons with energy equal to the bandgap. The approach is very simple: i) choose a value of n or p; ii) calculate E_F from the Joyce-Dixon approximation; iii) calculate $f^e + f^h - 1$ and check if it is positive at the bandedge. The same approach can be used to find the gain as a function of $\hbar\omega$.

For 300 K we find that the material is transparent when $n \sim 1.1 \times 10^{18}$ cm^{-3} at 300 K and $n \sim 2.5 \times 10^{17}$ cm^{-3} at 77 K. Thus a significant decrease in the injected charge occurs as temperature is decreased.

4.8 CHARGE INJECTION: NON-RADIATIVE EFFECTS

In perfect semiconductors, there are no allowed electronic states in the bandgap region. However, in real semiconductors there are always intentional or unintentional impurities which produce electronic levels which are in the bandgap. These impurity levels can arise from chemical impurities or from native defects such as a vacancy or an anti-site defect (i.e., in compound semiconductors, an atom on the wrong sublattice—Ga on an As site, for example).

The bandgap levels are states in which the electron is "localized" in a finite space near the defect unlike the usual Bloch states which represent the valence and conduction band states, and which are extended in space. As the "free" electrons move in the allowed bands, they can be trapped by the defects (see Fig. 4.16). The defects can also allow the recombination of an electron and hole without emitting a photon as was the case in the previous section. This non-radiative recombination competes with radiative recombination, and can have a positive or negative impact depending upon the device. For example, in a laser the non-radiative recombination is not desirable, but it is purposely increased in p-n diodes to increase the speed. We will briefly discuss the non-radiative processes involving a midgap level with density N_t.

An empty state can be assigned a capture cross-section σ_e so that physically if an electron comes within this area around a trap, it will be captured. If v_{th} is the velocity of the electron and n is the concentration of the electrons, the capture rate is

$$r_n^c = N_t(1 - f(E_t))\sigma_e v_{th} n \tag{4.123}$$

where $f(E_t)$ is the occupation probability of the trap state at energy E_t. Once an electron is trapped at the defect site, it can be re-emitted with a rate P_n from each site. The total emission rate is then

$$r_n^e = N_t f(E_t) P_n \tag{4.124}$$

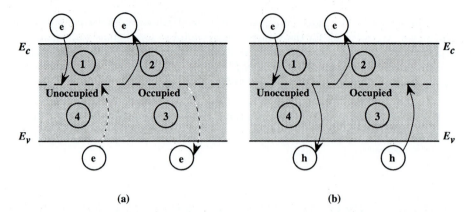

Figure 4.16: Various processes that lead to trapping and recombination via deep levels in the bandgap region (dashed line). The processes 1 and 2 in (a) represent trapping and emission of electrons while 3 and 4 represent the same for holes. The *e-h* recombination is shown in part (b).

If an electron is trapped by a defect site, the site can also trap a hole. If σ_h is the hole capture cross-section, and v_{th} the thermal velocity (assumed the same as for the electron), the hole capture rate is

$$r_p^c = N_t f(E_t)\sigma_h v_{th} p \tag{4.125}$$

As for the electron case, a vacant state may emit a hole to the valence band (physically capture an electron from the valence band). The rate is

$$r_p^e = N_t(1 - f(E_t))P_p \tag{4.126}$$

In steady state conditions, the capture and emission rates for both the individual carriers are equal

$$r_n^c = r_n^e \quad ; \quad r_p^c = r_p^e \tag{4.127}$$

We note that

$$n = n_i \, exp\left(\frac{E_F - E_{Fi}}{k_B T}\right) \tag{4.128}$$

$$f(E_t) = \frac{1}{1 + exp\left(\frac{E_t - E_F}{k_B T}\right)} \quad ; \quad 1 - f(E_t) = \frac{exp\left(\frac{E_t - E_F}{k_B T}\right)}{1 + exp\left(\frac{E_t - E_F}{k_B T}\right)} \tag{4.129}$$

so that we get, solving for P_n using Eqns. 4.123, 4.124, 4.127, 4.128 and 4.129

$$P_n = n_i v_{th}\sigma_n \, exp\left(\frac{E_t - E_{Fi}}{k_B T}\right) \tag{4.130}$$

Similarly

$$P_p = n_i v_{th} \sigma_p \; exp - \left(\frac{E_t - E_{Fi}}{k_B T} \right) \tag{4.131}$$

We are interested in the rate of change of the mobile electron and hole concentrations, which is now

$$\frac{-dn}{dt} = r_n^c - r_n^e = \frac{-dp}{dn} = r_p^c - r_p^e \tag{4.132}$$

Defining

$$\boxed{\tau_n = \frac{1}{N_t v_{th} \sigma_n} \; and \; \tau_p = \frac{1}{N_t v_{th} \sigma_p}} \tag{4.133}$$

we get for the recombination rate (after a little tedious mathematics)

$$R_t = \frac{-dn}{dt} = \frac{-dp}{dt} = \frac{np - n_i^2}{\tau_n \left[p + n_i \; exp \left(\frac{-E_t - E_{Fi}}{k_B T} \right) \right] + \tau_h \left[n + n_i \; exp \left(\frac{E_t - E_{Fi}}{k_B T} \right) \right]} \tag{4.134}$$

This expression is known as the Shockley, Read, Hall (SRH) equation for recombination rate via deep levels in the bandgap. To simplify this expression we make the following assumptions: i) $\tau_{nr} = \tau_n = \tau_p$; ii) $E_t = E_{Fi}$, i.e., the trap levels are essentially at midgap level; iii) $np \gg n_i^2$ under the injection conditions. This gives

$$\boxed{R_t = \frac{np}{\tau_{nr}(n + p)}} \tag{4.135}$$

The time constant τ_{nr} depends upon the impurity density, the cross-section associated with the defect and the electron thermal velocity as seen in Eqn. 4.133. Typically the cross-sections are in the range 10^{-13} to 10^{-15} cm^2.

We will see in Chapter 6 how the nonradiative recombination leads to non-ideal I-V characteristics for the *p-n* diode.

EXAMPLE 4.8 A silicon sample has an impurity level of 10^{15} cm^{-3}. These impurities create a midgap level with a cross-section of 10^{-14} cm^{-2}. Calculate the electron trapping time at 300 K and 77 K.

We can obtain the thermal velocities of the electrons by using the relation

$$\frac{1}{2} m^* v_{th}^2 = \frac{3}{2} k_B T$$

This gives

$$v_{th}(300 \ K) \ = \ 2 \times 10^7 \ cm/s$$
$$v_{th}(77 \ K) \ = \ 1 \times 10^7 \ cm/s$$

The electron trapping time is then

$$\tau_{nr}(300 \ K) \ = \ \frac{1}{10^{15} \times 2 \times 10^7 \times 10^{-14}} = 5 \times 10^{-9} \ s$$
$$\tau_{nr}(77 \ K) \ = \ 10^{15} \times 1 \times 10^7 \times 10^{-14} = 10^{-8} \ s$$

In silicon the *e-h* recombination by emission of photon is of the range of 1 ms to 1 μs so that the non-radiative (trap related) lifetime is much shorter.

4.9 NON-RADIATIVE RECOMBINATION: AUGER PROCESSES

In Chapter 3 we had discussed the impact ionization process in which a "hot" electron (hole) with energy greater than the bandgap can scatter from an electron (hole) in the valence band (conduction) to produce two electrons (holes) and a hole (electron). The impact ionization is mediated by the Coulombic interaction between charged particles. A reverse process can also occur in which an electron and hole recombine and the excess energy is transferred to either an electron or a hole as shown in Fig. 4.17. The important point to note is that no photons are produced in this process so that an electron-hole pair is lost without any photon output.

In Fig. 4.17a, we show a process called conduction-conduction-heavy hole-conduction or CCHC, the individual words representing the state of the carriers involved in the Auger process. After the scattering, an electron-hole pair is lost and one is left with a hot electron. The hot electron subsequently loses its excess energy by emitting phonons as discussed in Chapter 3.

In Fig. 4.17b we show a different process called CHHS (standing for conduction-heavy hole-heavy hole-split off) in which two holes and an electron interact to produce a hot hole in the split-off band.

To calculate the Auger rates we must discuss the occupation statistics of the various electrons and hole states involved in the process. We need to weigh the rate with the probability that state k_2 is full, k_1' is empty and k_1 is full. If we assume nondegenerate statistics, the occupation factor becomes

$$P(k_1, k_2, k_1') = f(k_1) \, f(k_2) \left(1 - f(k_1')\right) \tag{4.136}$$

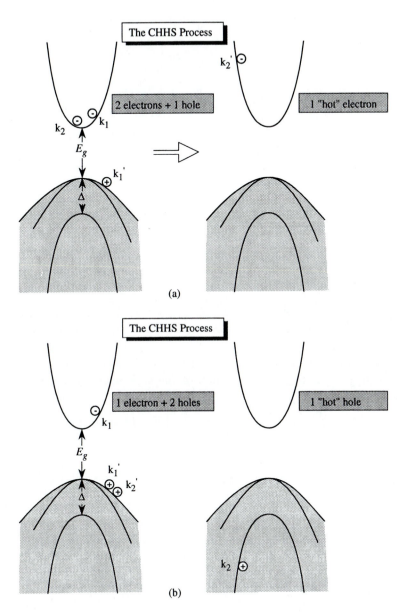

Figure 4.17: (a) The Auger process involving 2 electrons and 1 hole in the initial state and 1 hole electron after the scattering. (b) The process where two holes and an electron suffer an Auger process and give a hot hole in the split-off band.

where

$$f(\mathbf{k}_1) = \frac{n}{N_c} \exp\left(\frac{-E_{c\mathbf{k}_1}}{k_B T}\right)$$

$$f(\mathbf{k}_2) = \frac{n}{N_c} \exp\left(\frac{-E_{c\mathbf{k}_2}}{k_B T}\right)$$

$$1 - f(\mathbf{k}_1') = \frac{p}{N_v} \exp\left(\frac{-E_{v\mathbf{k}_1'}}{k_B T}\right) \tag{4.137}$$

Here n and p are the electron and hole carrier densities and N_c, N_v are the conduction and valence band effective density of states.

This gives the total probability factor

$$P(\mathbf{k}_1, \mathbf{k}_2, \mathbf{k}_1') = \frac{np}{N_c N_v} \frac{n}{N_c} \exp\left(-\frac{E_{c\mathbf{k}_2} + E_{v\mathbf{k}_1'} + E_{c\mathbf{k}_1}}{k_B T}\right)$$

$$\approx \frac{n}{N_c} \exp\left(-\frac{E_g + E_{c\mathbf{k}_2} + E_{v\mathbf{k}_1'} + E_{c\mathbf{k}_1}}{k_B T}\right) \tag{4.138}$$

We have assumed that the final state $\mathbf{k}_2'$ is always available, since it is a high energy electron state in the conduction band (or an electron state deep in the valence band).

It is useful to examine the energy at which the exponent of Eqn. 4.138 maximizes. This involves finding the extremum of the expression (in the parabolic band approximation)

$$\frac{k_1^2}{2m_c^*} + \frac{k_2^2}{2m_c^*} + \frac{k_1'^2}{2m_v^*} = \frac{1}{2m_c^*}\left[k_1^2 + k_2^2 + \mu k_1'^2\right] \tag{4.139}$$

where

$$\mu = \frac{m_c^*}{m_v^*} \tag{4.140}$$

The probability factor will maximize for the lowest energy values of $\mathbf{k}_1$, $\mathbf{k}_1'$ and $\mathbf{k}_2$ which are consistent with energy and momentum conservation. Since $\mathbf{k}_2'$ is the largest vector, we line up $\mathbf{k}_1$, $\mathbf{k}_1'$, and $\mathbf{k}_2$ with $\mathbf{k}_2'$ in the opposite direction as shown in Fig. 4.18. Thus we choose

$$\mathbf{k}_1 + \mathbf{k}_1' + \mathbf{k}_2 = -\mathbf{k}_2' \tag{4.141}$$

We also write

$$\mathbf{k}_1 = a\mathbf{k}_1'$$

$$\mathbf{k}_2 = b\mathbf{k}_1'. \tag{4.142}$$

We now have from conservation of energy

$$k_2'^2 = (a^2 + b^2 + \mu)k_1'^2 + K_g^2 \tag{4.143}$$

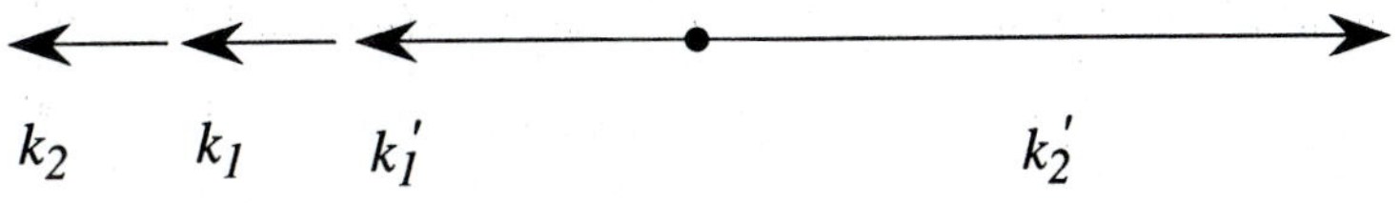

Figure 4.18: Procedure for finding the maximum in probability for Auger rates. This procedure is also used to find the threshold energy for the impact ionization process to occur. The threshold is reached when the high energy electron k_2' has its wavevector lined up opposite to those of the low energy electrons as shown.

where

$$E_g = \frac{\hbar^2 K_g^2}{2m_c^*} \qquad (4.144)$$

Also the conservation of momentum gives us

$$\boldsymbol{k}_2' = (a + b + 1)\boldsymbol{k}_2'$$

or

$$k_2'^2 = (a^2 + b^2 + 1 + 2ab + 2a + 2b)k_1'^2 \qquad (4.145)$$

Eliminating $k_2'^2$ from Eqn. 4.144 and Eqn. 4.146, we get

$$k_1'^2(1 + 2ab + 2a + 2b - \mu) = K_g^2 \qquad (4.146)$$

The quantity to be minimized for maximum Auger rate (or the impact ionization threshold discussed in Chapter 3, Section 3.7.1) is

$$k_1^2 + k_2^2 + \mu k_1'^2 = k_1'^2(a^2 + b^2 + \mu)$$

Substituting for $k_1'^2$ from Eqn. 4.146, we get

$$(a^2 + b^2 + \mu)k_1'^2 = \frac{a^2 + b^2 + \mu}{1 + 2ab + 2a + 2b - \mu}K_g^2 \qquad (4.147)$$

This quantity minimizes when

$$a = b = \mu \qquad (4.148)$$

This gives us the energy value for the initial state electrons

$$
\begin{aligned}
E_{ck_1} &= E_{ck_2} \\
&= \mu E_{vk_1'} \\
&= \left(\frac{\mu^2}{1 + 3\mu + 2\mu^2}\right) E_g \qquad (4.149)
\end{aligned}
$$

The maximum probability function is now

$$P(\boldsymbol{k}_1, \boldsymbol{k}_2, \boldsymbol{k}_1') = \frac{n}{N_c} \exp\left(-\frac{1 + 2\mu}{1 + \mu}\frac{E_g}{k_B T}\right) \qquad (4.150)$$

and the energy of the high energy electron is (this is the impact ionization threshold for direct gap semiconductors, where the initial electron is the "hot" carrier)

$$E_{ck_2'} = \frac{1 + 2\mu}{1 + \mu} E_g \tag{4.151}$$

If $\mu \ll 1$, we have the approximation

$$E_{ck_2'} \approx (1 + \mu) E_g \tag{4.152}$$

The higher the value of this threshold energy, the smaller the Auger rate, since the occupation factor decreases with increased energy. The threshold depends upon the bandgap, and the rate of the electron and hole masses. We had noted in Chapter 2 that the strain in a system can affect the hole masses and thus, the Auger rates.

To solve the general integral for the Auger rates (or the impact ionization process), one needs to evaluate the multiple integral (these are rates for a particular electron at k_1; the total rate over all electrons will involve a further integral over k_1, as well)

$$\begin{aligned}
W_{\text{Auger}}(k_1) &= 2 \left(\frac{2\pi}{\hbar} \right) \left(\frac{e^2}{\epsilon} \right)^2 \frac{1}{(2\pi)^6} \\
&\times \int d^3k_2 \int d^3k_1' \int d^3k_2' \, |M|^2 \, P(k_1, k_2, k_1') \\
&\times \delta(E_{ck_1} + E_{ck_2} - E_{vk_1'} - E_{ck_2'})
\end{aligned} \tag{4.153}$$

The matrix element $|M|^2$ has been mentioned before and is the screened Coulombic matrix element used for ionized impurity scattering. For most purposes it is adequate to use the screening length $\lambda = 0$. In general, one has to explicitly evaluate the overlap integrals by using an accurate bandstructure description. Typical results for such a calculation are shown in Fig. 4.19 where we show the Auger recombination rate for the narrow bandgap material $In_{0.53}Ga_{0.47}As$ ($E_g \approx 0.8$ eV) which is widely used for long-distance optical communication systems. The Auger rate is approximately proportional to n^3 (for the laser $n = p$), and is often written in the form

$$\begin{aligned}
W &= R_{\text{Auger}} \\
&= F n^3
\end{aligned} \tag{4.154}$$

where F is the Auger coefficient. This relation is, however, only approximate and breaks down at high injection where the statistics change from nondegenerate Boltzmann-like to degenerate Fermi-Dirac statistics.

An approximate treatment for the Auger rates can be made if one assumes parabolic bands and a simple form for the overlap integrals involving the central cell functions. The result is

$$W_{\text{recomb}} = \frac{e^4 m_c^* \, (k_B T)^{3/2} \left(\frac{m_c^*}{m} + \mu \right)}{4\pi^{5/2} \, \epsilon^2 \, \hbar^3 \, (1 + \mu)^{1/2} \, E_g^{3/2}} \exp \left\{ -\frac{(1 + \mu)E_g}{k_B T} \right\} \tag{4.155}$$

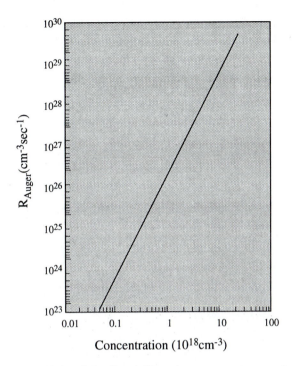

Figure 4.19: Auger rates calculated for $In_{0.53}Ga_{0.47}As$ at room temperature. The important process involves the final state with one hole in the split off band. The term CHHS (conduction-heavy hole-heavy hole-split off hole) is used for such events (figure provided by John Loehr).

The Auger rates increase exponentially as the bandgap is decreased. They also increase exponentially as the temperature increases. These are direct results of the energy and momentum conservation constraints and the carrier statistics. Auger processes are more or less unimportant in semiconductors with bandgaps larger than approximately 1.5 eV (e.g., GaAs, AlGaAs, InP). However, they become quite important in narrow bandgap materials such as $In_{0.53}Ga_{0.47}As$ ($E = 0.8$ eV) and HgCdTe ($E < 0.5$ eV), and are thus a serious hindrance for the development of long wavelength lasers. In Appendix B, we have given the values of Auger coefficients for some materials.

In general, Auger processes could be mediated by defects as well. As shown in Fig. 4.20, deep levels in the bandgap can be involved in the Auger processes. However, for high quality materials, these defect assisted processes are not important.

EXAMPLE 4.9 Consider an Auger process that involves the scattering shown in the Fig. 4.17b. This process denoted by CHHS is a dominant process for long distance communication lasers. The masses of the conduction, heavy hole and split off band are m_c, m_h, m_s, respectively, and the bands are parabolic. Calculate the threshold energy for the initial electron and hole states for such a transition using momentum and energy conservation.

Let k_1 and k_1' be the initial k-vectors for the electron and hole, respectively, and k_2 and k_2' be the vectors corresponding to the split-off hole and the heavy hole, respectively (as shown in Fig. 4.17b). Momentum conservation gives us

$$k_2 = k_1 + k_1' + k_2'$$

Energy conservation give us

$$E_{k_1} + E_g - E_{k_2} - \Delta = -E_{k_1'} - E_{k_2'}$$

We choose $k_1' = k_2'$, and define a variational parameter α, where

$$k_1' = \alpha k_1$$

We have

$$k_2 = k_1(2\alpha + 1)$$

Also, from energy conservation, we get

$$\frac{\hbar^2 k_1^2}{2m_c} + \frac{\alpha^2 \hbar^2 k_1^2}{m_h} + (E_g - \Delta) = \frac{\hbar^2 k_2^2}{2m_s} = \frac{\hbar^2 k_1^2}{2m_s}(1 + 4\alpha^2 + 4\alpha)$$

Solving for k_1^2, we get

$$k_1^2 = \frac{(E_g - \Delta)(2m_s/\hbar^2)}{1 + 4\alpha^2 + 4\alpha - \dfrac{m_s}{m_c} - \dfrac{2\alpha^2 m_s}{m_h}}$$

To minimize E_{k_2}' we must minimize

$$\frac{k_1^2}{2m_c} + \frac{k_2'^2}{m_h} = \frac{k_1^2}{2m_h}\left(2\alpha^2 + \frac{m_h}{m_c}\right)$$

Substituting for k_1^2, we find that the function to be maximized is

$$f(\alpha) = \frac{\left(2\alpha^2 + \dfrac{m_h}{m_c}\right)}{1 + 4\alpha^2 + 4\alpha - \dfrac{m_s}{m_c} - \dfrac{2\alpha^2 m_s}{m_h}}$$

Equating $\frac{\partial f}{\partial \alpha} = 0$, we get

$$\left(1 + 4\alpha^2 + 4\alpha - \frac{m_s}{m_c} - 2\alpha^2\frac{m_s}{m_h}\right)4\alpha = \left(2\alpha^2 + \frac{m_h}{m_c}\right)\left(8\alpha + 4 - 4\alpha\frac{m_s}{m_h}\right)$$

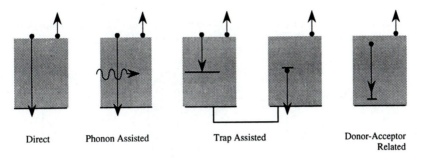

Figure 4.20: Various Auger processes that can occur in semiconductors.

so that

$$2\alpha^2 + \alpha \left(1 - \frac{2m_h}{m_c}\right) - \frac{m_h}{m_c} = 0$$

or,

$$\alpha = \frac{m_h}{m_c} = \mu$$

We can now calculate the energy of the initial electron

$$\frac{\hbar^2 k_1^2}{2m_c} = \frac{(E_g - \Delta)m_s/m_c}{1 + 4\mu^2 + 4\mu - \frac{m_s}{m_c} - 2\mu^2 \frac{m_s}{m_h}}$$

An important outcome of this result is that as E_g approaches Δ, the threshold for the Auger process goes to zero. This causes an extremely high Auger recombination in materials where the bandgap and the split-off energy are comparable.

4.10   THE CONTINUITY EQUATION: DIFFUSION LENGTH

$\mathcal{R}$ In our discussion on charge transport, we had considered the drift and diffusion processes through the semiconductor, without worrying about the electron-hole recombination. The recombination process removes the electrons and holes and thus alters the charge transport picture. To describe the transport and recombination of injected electrons and holes we develop a continuity equation for the problem. The carrier recombination process is critical for any device that involves both electron and hole flow (e.g., *p-n* diode, bipolar junction transistor).

If we consider a volume of space in which charge transport and recombination is taking place, we have the simple equality

Rate of particle flow = Particle flow rate due to current—Particle loss rate due to recombination

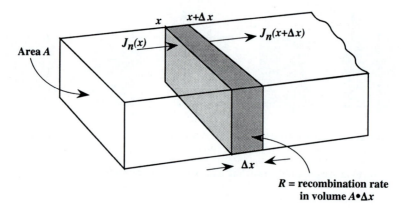

Figure 4.21: A conservation of particles applied to a volume $A \cdot \Delta x$. The difference in the particle currents has to equal the recombination rate.

This equation simply is a statement of conservations of particles. If δn is the excess carrier density in the region, the recombination rate in a volume $A \cdot \Delta x$ shown in Fig. 4.21 may be written approximately as

$$R = \frac{\delta n}{\tau_n} \cdot A \cdot \Delta x \tag{4.156}$$

where τ_n is the electron recombination time per excess particle and includes radiative and non- radiative components. The particle flow rate into the same volume due to the current J_n is given by

$$\left[\frac{-J_n(x)}{e} + \frac{J_n(x + \Delta x)}{e} \right] A \cong \frac{1}{e} \frac{\partial J_n(x)}{\partial x} \Delta x \cdot A \tag{4.157}$$

The total rate of electron build up in the volume $A \cdot \Delta x$ is then

$$A \cdot \Delta x \left[\frac{\partial n(x, t)}{\partial t} \equiv \frac{\partial \delta n}{\partial t} = \frac{1}{e} \frac{\partial J_n(x)}{\partial x} - \frac{\delta n}{\tau_n} \right] \tag{4.158}$$

where δn is the excess carrier density which is the only part which changes with time. The same considerations apply to holes. We thus have for the electrons and holes, the continuity equations

$$\frac{\partial \delta n}{\partial t} = \frac{1}{e} \frac{\partial J_n(x)}{\partial x} - \frac{\delta n}{\tau_n} \tag{4.159}$$

$$\frac{\partial \delta p}{\partial t} = -\frac{1}{e} \frac{\partial J_p(x)}{\partial x} - \frac{\delta p}{\tau_p} \tag{4.160}$$

It is interesting to examine these equations for the special case where the current is only being carried by the diffusion process. This occurs in the cases of *p-n* junction transports as well as in the bipolar transistors. Writing the diffusion currents as (see Eqn. 3.53)

$$J_n(diff) = eD_n \frac{\partial \delta n}{\partial x} \tag{4.161}$$

$$J_p(diff) = -eD_p \frac{\partial \delta p}{\partial x} \tag{4.162}$$

we have

$$\frac{\partial \delta n}{\partial t} = D_n \frac{\partial^2 \delta n}{\partial x^2} - \frac{\delta n}{\tau_n} \tag{4.163}$$

$$\frac{\partial \delta p}{\partial t} = D_p \frac{\partial^2 \delta p}{\partial x^2} - \frac{\delta p}{\tau_p} \tag{4.164}$$

These equations will be used when we discuss the transient time responses of the p-n diodes and bipolar transistors. These equations are also used to study the steady state charge profile in these devices. In steady state we have (the time derivative is zero)

$$\boxed{\begin{aligned} \frac{d^2 \delta n}{dx^2} &= \frac{\delta n}{D_n \tau_n} = \frac{\delta n}{L_n^2} \\ \frac{d^2 \delta p}{dx^2} &= \frac{\delta p}{D_p \tau_p} = \frac{\delta p}{L_p^2} \end{aligned}} \tag{4.165}$$

where $L_n(L_p)$ defined as $D_n\tau_n(D_p\tau_p)$ are called the diffusion lengths for reasons that will be clear below.

Consider first the case where due to some external injection mechanism an excess electron density $\delta n(0)$ is maintained at the semiconductor edge $x = 0$ as shown in Fig. 4.22. If n_o is the equilibrium density, we are interested in finding out how the excess density varies with position. The general solution of the second order differential Eqn. 4.165 is

$$\delta n(x) = A_1 e^{x/L_n} + A_2 e^{-x/L_n} \tag{4.166}$$

Since for large values of x, $\delta n(x)$ must go to zero, we must require that $A_1 = 0$. Note that we can only impose this condition if the sample is large. Since $\delta n(0)$ is known and fixed by the injection condition, we have

$$A_2 = \delta n(0) \tag{4.167}$$

The solution is then

$$\delta n(x) = \delta n(0)e^{-x/L_n} \tag{4.168}$$

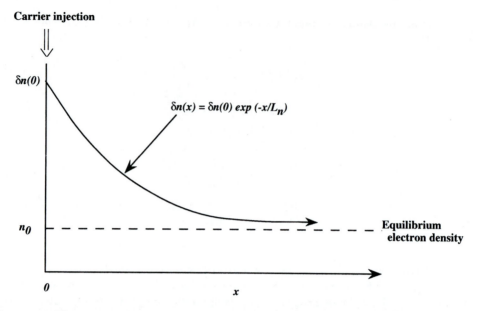

Figure 4.22: Electrons are injected at $x = 0$ into a sample. At $x = 0$, a fixed carrier concentration is maintained. The figure shows how the excess carriers decay into the semiconductor.

The diffusion length L_n represents the distance over which the injected carrier density falls to $1/e$ of its original value. It also represents the average distance an electron will diffuse before it recombines with a hole. This can be seen as follows.

The probability that an electron survives up to point x without recombination is

$$\boxed{\frac{\delta n(x)}{\delta n(0)} = e^{-x/L_n}} \qquad (4.169)$$

The probability that it recombines in a distance Δx is

$$\frac{\delta n(x) - \delta n(x + \Delta x)}{\delta n(x)} = -\frac{\Delta x}{\delta n(x)} \frac{d\delta n(x)}{dx} = \frac{1}{L_n}\Delta x \qquad (4.170)$$

where we have expanded $\delta n(x + \Delta x)$ in terms of $\delta n(x)$ and the first derivative of δn. Thus the probability that the electron survives up to a point x and then recombines is

$$P(x)\Delta x = \frac{1}{L_n}e^{-x/L_n}\Delta x \qquad (4.171)$$

Thus the average distance an electron can move and then recombine is

$$< x > = \int_o^\infty x P(x) dx = \int_0^\infty \frac{x e^{-x/L_n}}{L_n} dx$$
$$= L_n \qquad (4.172)$$

This average distance $(= \sqrt{D_n \tau_n})$ depends upon the recombination time and the diffusion constant in the material. In the derivations of this section, we used a simple form of recombination rate

$$R = \frac{\delta n}{\tau_n} \qquad (4.173)$$

where τ_n is given in terms of the radiative and non-radiative rates as

$$\frac{1}{\tau_n} = \frac{1}{\tau_r} + \frac{1}{\tau_{nr}} \qquad (4.174)$$

The simple $\delta n/\tau_n$ form is valid, for example, for minority carrier recombination (p ≫ n). These equations are therefore used widely to discuss minority carrier injection.

EXAMPLE 4.10 In a p-type GaAs sample, electrons are injected from a contact. If the minority carrier mobility is 4000 cm^2/V-s at 300 K, calculate the diffusion length for the electrons.

The diffusion constant is given by the Einstein relation

$$
\begin{aligned}
D_n &= \frac{\mu \, k_B T}{e} \\
&= \frac{(0.4 m^2/V - s) \times (1.38 \times 10^{-23} (J/K) \times 300 \, K)}{(1.6 \times 10^{-19} \, C)} \\
&= 1.04 \times 10^{-2} m^2/s
\end{aligned}
$$

The diffusion length is
$$L_n = \sqrt{D_n \tau}$$

If we use the recombination time corresponding to the radiation recombination time $(\tau \cong 0.6$ ns), we get

$$
\begin{aligned}
L_n &= 2.5 \times 10^{-6} \, m \\
&= 2.5 \, \mu m
\end{aligned}
$$

In silicon τ_n can be quite a bit larger unless some impurity levels are induced. Without the impurities τ_n can range from 10^{-3} to 10^{-6} s and L_n can be quite large.

4.11 CHARGE INJECTION AND BANDGAP RENORMALIZATION

In our discussions of the semiconductor bandstructure we have not discussed whether the E vs. k relation will change if one introduces extra electrons in the conduction band or holes in the valence band. In fact, without explicitly stating this, the bandstructure we have discussed so far is for the case where all the electrons are in the valence band and the conduction band is empty. What happens when the situation is changed, either by doping or by e-h injection? An e-h pair will have a Coulombic interaction with each other. A manifestation of this interaction is the exciton, which will be discussed in the next chapter. However, the presence of the excess electrons in the conduction band or holes in the valence band also shifts the bandgap energy, decreasing it slightly. This phenomenon is often called *bandgap renormalization.*

The bandgap renormalization effect has to be treated by the many body theory, since the energy seen by a carrier depends upon the presence of other electrons and holes. Extensive experimental and theoretical work has been done to study how the bandgap changes with carrier density for both bulk and quantum well systems. We will not discuss the many body treatment here, but simply provide an approximate result for the bandgap shrinkage. In bulk materials, the bandgap shrinkage is given by

$$\Delta E_g = -K(n^{1/3} + p^{1/3})$$

For GaAs, the constant K is such that we have (expressing n and p in cm^{-3})

$$\Delta E_g = -1.6 \times 10^{-8}(n^{1/3} + p^{1/3})\ eV$$

From this equation we see that if $n = p = 10^{18}$ cm^{-3}, the change in the bandgap is ~ -32 meV. While this is a small quantity, it can have an important effect on devices such as laser diodes, since the light emission will shift due to the bandgap change.

In quantum well systems, it is found that the bandgap change is proportional to $(n_{2D})^{1/3}$ at high carrier densities, and is proportional to $(n_{2D})^{1/2}$ at low carrier densities. Here, n_{2D} is the 2-dimensional electron density and is assumed to be the same as the hole density. Carrier density is considered high if it is greater than $\sim 5 \times 10^{11}$ cm^{-2}. At an injection of 10^{12} cm^{-2}, the shift in the bandgap is ~ -35 meV.

In addition to the bandgap renormalization, the presence of the Coulombic interaction between the e-h pairs also increases the absorption coefficient and gain near the bandedge. The basis for this is discussed in the next chapter. Away from the bandedge the effect is not important.

4.12   CHAPTER SUMMARY

$\mathcal{R}$ In this chapter we have discussed the basic physical phenomena upon which electronic and optoelectronic devices are based. All devices involve some physical response to external perturbations. These perturbations are usually electric fields or electromagnetic fields. The device performance depends upon how electrons respond to these external stimuli. The summary tables (Tables 4.2-4.5) highlight the concepts discussed in this chapter.

4.13   PROBLEMS

Sections 4.4-4.6

4.1 Identify the various semiconductors (including alloys) that can be used for light emission at 1.55 μm. Remember that light emission occurs at an energy near the bandgap.

4.2 Calculate and plot the optical absorption coefficients in GaAs, InAs, and InP as a function of photon energy assuming a parabolic density of states.

4.3 Consider a 100 Å GaAs/Al$_{0.3}$Ga$_{0.7}$As quantum well structure. Assuming that the problem can be treated as that with an infinite barrier, calculate the absorption spectra for in-plane and out-of-plane polarized light for $E \leq 100$ meV from the effective bandedge. Assume the simple uncoupled model for the HH and LH states.

4.4 Calculate and plot the overlap of an electron and heavy hole ground state envelope function in a GaAs/Al$_{0.3}$Ga$_{0.7}$As quantum well as a function of well size from 20 Å to 200 Å. Assume a 60:40 value for ΔE_c:ΔE_v. At what well size does the overlap significantly differ from unity?

4.5 Calculate the gain in a GaAs region as a function of injected carrier density at room temperature. Plot your results in the form of gain vs. energy.

4.6 Estimate the strength of the intraband transitions in a 100 Å GaAs/Al$_{0.3}$Ga$_{0.7}$As quantum well structure at an electron carrier concentration of 10^{12} cm^{-2}. Assume an infinite barrier model and a linewidth (full width at half maximum) of 1 meV for the transition.

4.7 Calculate the electron-hole recombination time τ_o for an HgCdTe alloy which has a bandgap of 0.1 eV. The momentum matrix element is the same as for GaAs.

Section 4.7

4.8 In a GaAs sample at 300 K, equal concentrations of electrons and holes are injected. If the carrier density is $n = p = 10^{17}$ cm^{-3}, calculate the electron and hole Fermi levels using the Boltzmann and Joyce-Dixon approximations.

4.9 In a p-type GaAs doped at $N_a = 10^{18}$ cm^{-3}, electrons are injected to produce a minority carrier concentration of 10^{15} cm^{-3}. What is the rate of photon emission assuming that all e-h recombination is due to photon emission ? What is the optical

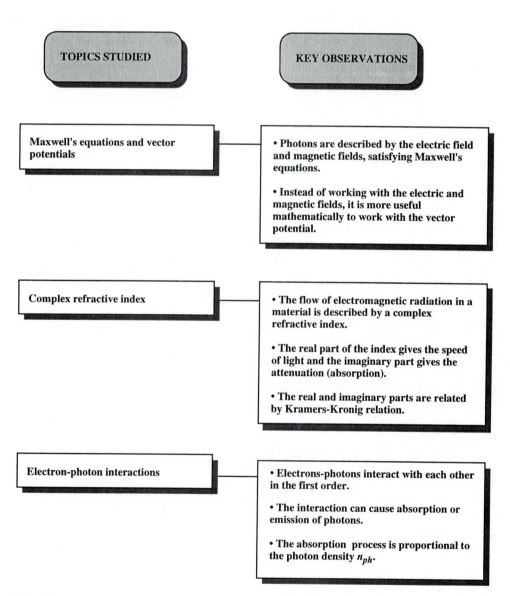

Table 4.2: Summary table.

TOPICS STUDIED	KEY OBSERVATIONS
Spontaneous and stimulated emission	**The emission of photons has two components:** **• spontaneous emission in which emitted photons are incoherent and the rate is independent of the photons in the system,** **• stimulated emission in which the photons emitted are coherent with the photons already present and the rate is proportional to the existing photon density.**
Interband transitions in semiconductors	**• In direct gap semiconductors, the electron transitions are "vertical" in k-space, due to momentum conservation.** **• Special polarization selection rates are applicable for LH$\rightarrow$ c-band and HH$\rightarrow$ c-band transitions.** **• In quantum wells, the step density of states alters the absorption profile.** **• In indirect semiconductors, phonons (or impurities) are needed for optical transitions; the rates are quite small, compared to the rates for direct gap semiconductors.**

Table 4.3: Summary table.

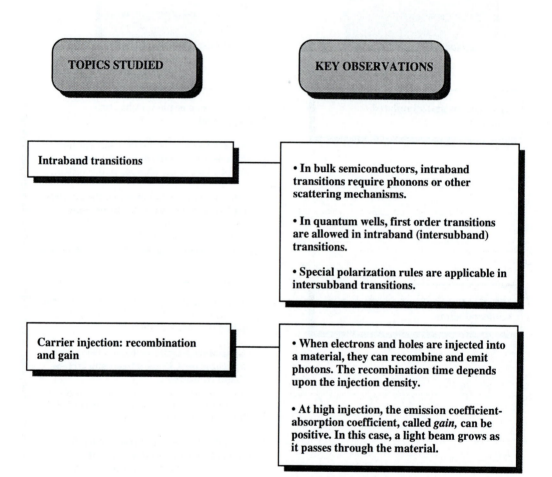

Table 4.4: Summary table.

TOPICS STUDIED	KEY OBSERVATIONS
Non-radiative recombination via traps	• Defect states produce bandgap states and electrons and holes can recombine via these states, without photon emission. • The recombination time is inversely proportional to the trap density, thermal velocity, and trap cross-section.
Non-radiative recombination by Auger processes	• The Auger process involves an electron and two holes (or two electrons and one hole) in which an *e-h* pair recombine and give their energy to the third particle. No photons are emitted in this process. • The Auger rate is proportional to np^2 or pn^2. The Auger coefficient is dependent upon the bandgap and temperature.
Diffusion length	An electron (hole) injected into a semiconductor moves a distance, called *diffusion length*, before recombining with a hole (electron).

Table 4.5: Summary table.

output power? The photon energy is $\hbar\omega = 1.41$ eV.

4.10 Calculate the electron carrier density needed to push the electron Fermi level to the conduction bandedge in GaAs. Also calculate the hole density needed to push the hole Fermi level to the valence bandedge. Calculate the results for 300 K and 77 K.

4.11 Calculate and plot the 300 K recombination rate in GaAs as a function of electron (hole) carrier density. Cover the electron (hole) density range from 10^{14} cm^{-3} to 10^{18} cm^{-3} in your calculations. Estimate the carrier density dependence of the recombination rate in the low carrier density and high carrier density regime.

Section 4.8

4.12 The radiative lifetime of a GaAs sample is 1.0 ns. The sample has a defect at the midgap with a capture cross-section of 10^{-15} cm^2. At what defect concentration does the non-radiative lifetime become equal to the radiative lifetime at i) 77 K and ii) 300 K?

Section 4.9

4.13 Electrons are injected into a *p*-type silicon sample at 300 K. The electron-hole radiative lifetime is 1 μs. The sample also has midgap traps with a cross-section of 10^{-15} cm^2 and a density of 10^{16} cm^{-3}. Calculate the diffusion length for the electrons if the diffusion coefficient is 30 cm^2/s.

4.14 REFERENCES

- **Maxwell's Equations; Gauge Transformation**

    - J. D. Jackson, *Classical Electrodynamics*, John Wiley and Sons, New York (1975).

- **Macroscopic Theory of Optical Processes**

    - J. D. Jackson, *Classical Electrodynamics*, John Wiley and Sons, New York (1975).

    - W. Jones and N. H. March, *Theoretical Solid State Physics*, Dover, New York, vol. 2 (1973).

    - E. Kreyszig, *Advanced Engineering Mathematics*, John Wiley and Sons, New York (1962).

    - F. Stern, "Elementary Theory of the Optical Properties of Solids," *Solid State Physics*, Academic Press, New York, vol. 15 (1963).

    - B.K. Ridley, *Quantum Processes in Semiconductors*, Oxford Press, New York (1982).

    - J. M. Ziman, *Principles of the Theory of Solids*, Cambridge (1972).

- **Microscopic Theory of Optical Processes in Semiconductors**

  - F. Bassani and G. P. Parravicini, *Electronic States and Optical Transitions in Solids*, Pergamon Press, New York (1975).
  - W. Jones and N. H. March, *Theoretical Solid State Physics*, Dover, New York, vol. 2 (1973).

- **Optical Processes in Quantum Wells**

  - S. Hong and J. Singh, *Superlattices and Microstructures*, **3**, 645 (1987).
  - R. C.Miller, D. A. Kleinman, W. A. Nordland, and A. C. Gossard, *Phys. Rev.*, B, **22**, 863 (1980).
  - G. D. Sanders and Y. C. Chang, *Phys. Rev.*, B, **31**, 6892 (1985).
  - J. S. Weiner, D. A. B. Miller, D. S. Chemla, T. C. Damen, C. A. Burrus, T. H. Wood, A. C. Gossard, and W. Wiegman, *Appl. Phys. Lett.*, **47**, 1148 (1985).

- **Optical Processes in Indirect Semiconductors**

  - F. Bassani and G. P. Parravicini, *Electronic States and Optical Transitions in Solids*, Pergamon, New York (1975).
  - W. C. Dash and R. Newman, *Phys. Rev.*, **99**, 1151 (1955).
  - H. R. Phillip and E. A. Taft, *Phys. Rev.*, **113**, 1002 (1959).
  - H. R. Phillip and E. A. Taft, *Phys. Rev. Lett.*, **8**, 13 (1962).

- **Intraband Transitions in Quantum Wells**

  - D. D. Coon and R. P. G. Karunasiri, *Appl. Phys. Lett.*, **45**, 649 (1984).
  - A. Harwitt and J. S. Harris, *Appl. Phys. Lett.*, **50**, 685 (1987).
  - B. F.Lavine , K. K. Choi, C. G. Bathea, J. Walker and R. J. Malik, *Appl. Phys. Lett.*, **50**, 1092 (1987).

- **Auger Processes**

  - A.R. Beattie, *J. Phys. Chem. Solids*, **49**, 589 (1988).
  - A.R. Beattie and P.T. Landsberg, *Proceedings of the Royal Society*, A 249, 16 (1959).
  - A. Haug, *J. Phys. Chem. Solids*, 49, 599 (1988).
  - Y. Jiang, M.C. Teich, and W.I. Wang, *Appl. Phys. Lett.*, **57**, 2922 (1990).
  - J.P. Loehr and J. Singh, *IEEE J. Quant. Electron.*, **QE-29**, 2583 (1993).
  - A. Sugimura, *IEEE J. Quant. Electron.*, **QE-18**, 352 (1982).

- **Bandgap Renormalization**

  - H.C. Casey and F. Stern, *J. Appl. Phys.*, **47**, 631 (1976).
  - S Schmitt-Rink, C. Ell, and H. Haug, *Phys. Rev.*, B, **38**, 3342 (1988).
  - D.A. Kleinmann and R.C. Miller, *Phys. Rev.*, B, **32**, 2266 (1985).

CHAPTER
5

EXCITONIC EFFECTS
AND MODULATION OF
OPTICAL PROPERTIES

5.1 INTRODUCTION

In our discussion of bandstructure,transport and optical properties of semiconductors
we made the independent electron approximation. The energy spectrum of the electrons is independent of the presence or absence of other electrons. The effect of other
electrons is only manifested through the occupation probabilities without altering the
bandstructure. This approach is usually valid for many important physical phenomena like transport in nondegenerate semiconductors or optical absorption away from
the bandedges. In reality, of course, there is a Coulombic interaction between an electron and another electron or hole. Some very important properties are modified by such
interactions. We already mentioned the effects of strong electron-charged impurity interactions in heavily doped semiconductors. The electron-hole interaction is also important
for bandgap narrowing in lasers and other devices with strong charge injection. The full
theory of the electron-electron interaction depends upon many body theory, which is
beyond the scope of this text. However, fortunately, there is one important problem,
that of excitonic effects in semiconductors, that can be addressed by simpler theoretical
techniques.

The exciton problem can be motivated by the simple schematic of Fig. 5.1. On
the left-hand side, we show the bandstructure of a semiconductor with a full valence
band and an empty conduction band. Under these conditions there are no allowed states
in the bandgap. Now consider the case where there is one electron in the conduction

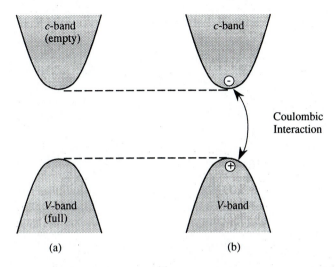

Figure 5.1: (a) The bandstructure in the independent electron picture and (b) the Coulombic interaction between the electron and hole which would modify the band picture.

band and one hole in the valence band. In this new configuration, the Hamiltonian describing the electronic system has changed. We now have an additional Coulombic interaction between the electron and the hole. The electronic bandstructure should thus be modified to reflect this change. In fact, the problem is somewhat similar to the case of a dopant where the electron-ionized impurity interaction modifies the electronic states by introducing impurity levels in the forbidden gap region. The electron-hole system, coupled through the Coulombic interaction, is called the exciton and will be the subject of this chapter.

In bulk semiconductors, excitons are usually observable optically, only in very high purity samples, due to their rather small binding energies. In poorer quality samples, the excitonic optical resonances often merge with the band-to-band transitions. These bulk transitions are, therefore, mainly studied from the point of view of material characterization. However, with the advent of heterostructure technology, the study of excitonic transitions in quantum wells has become an extremely important area, both from the point of view of new physics and of new technology. We will see in this chapter that due to quantum confinement, the exciton binding energy is greatly increased. This, and improved optical transition strength, allows one to observe extremely sharp resonances in quantum well optical spectra. Moreover, the energy, or strength, of these resonances can be controlled easily by simple electronics or optics. This ability allows one to use the excitonic transitions for high speed modulation of optical signals, as well as for optoelectronic switches, which could serve important functions in future information processing systems.

The modulation of optical properties is an essential ingredient for advanced

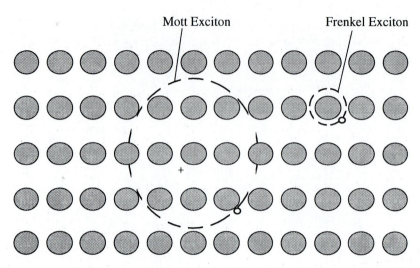

Figure 5.2: A conceptual picture of the periodic envelope function extent of the Frenkel and Mott excitons. The Frenkel exciton periodic function is of the extent of a few unit cells while the Mott exciton function extends over many unit cells.

optoelectronic systems. While a number of different stimulii can modulate the optical properties of a material (electric field, magnetic field, strain field) only electric field induced modulation can produce high speed operation. In the bulk form, most semiconductors do not have very good modulation properties. However, these properties can be greatly enhanced in quantum wells. In this chapter we will examine the physics behind various optical modulation approaches.

5.2   EXCITONIC STATES IN SEMICONDUCTORS

$\longrightarrow$ Two important classes of excitons exist depending upon the extent of the periodic envelope function. When the envelope function is confined to just a few unit cells, the excitons are classified as Frenkel excitons. Due to their restricted spatial extent, the Heisenberg uncertainty principle indicates that their treatment necessitates dealing with the full bandstructure of the semiconductors. On the other hand, if the envelope function extends over several hundred Angstroms, near bandedge electron and hole states can be used to describe them. Such excitons are called Mott excitons and are responsible for the excitonic physics in semiconductors (see Fig. 5.2). The effective mass theory can be used to describe these excitons and accordingly the problem is represented by the following Schrödinger equation

$$\left[-\frac{\hbar^2}{2m_e^*}\nabla_e^2 - \frac{\hbar^2}{2m_h^*}\nabla_h^2 - \frac{e^2}{4\pi\epsilon\,|r_e - r_h|}\right]\psi_{\text{ex}} = E\psi_{\text{ex}} \qquad (5.1)$$

Here m_e^* and m_h^* are the electron and hole effective masses and $|r_e - r_h|$ is the difference in coordinates defining the Coulombic interaction between the electron and the hole. We will shortly discuss in more detail the makeup of the exciton wavefunction ψ_{ex}. The problem is now the standard two-body problem, which can be written as a one-body problem, by using the following transformation

$$
\begin{aligned}
r &= r_e - r_h \\
k &= \frac{m_e^* k_e + m_h^* k_h}{m_e^* + m_h^*} \\
R &= \frac{m_e^* r_e + m_h^* r_h}{m_e^* + m_h^*} \\
K &= k_e - k_h
\end{aligned}
\tag{5.2}
$$

The Hamiltonian then becomes

$$
H = \frac{\hbar^2 K^2}{2(m_e^* + m_h^*)} + \left\{ \frac{\hbar^2 k^2}{2m_r^*} - \frac{e^2}{4\pi\epsilon \, |r|} \right\}
\tag{5.3}
$$

where m_r^* is the reduced mass of the electron-hole system. The Hamiltonian consists of two parts, the first term giving the description for the motion of center of mass of the electron-hole system, while the second term describing the relative motion of the electron-hole system. The first term gives a plane wave solution

$$
\psi_{\text{cm}} = e^{i\mathbf{K}\cdot\mathbf{R}}
\tag{5.4}
$$

while the solution to the second term satisfies

$$
\left(\frac{\hbar^2 k^2}{2m_r^*} - \frac{e^2}{4\pi\epsilon \, |r|} \right) F(r) = E F(r)
\tag{5.5}
$$

This is the usual hydrogen atom problem and $F(r)$ can be obtained from the mathematics of that problem. The general exciton solution is now (writing $\mathbf{K}_{\text{ex}} = \mathbf{K}$)

$$
\psi_{n\mathbf{K}_{\text{ex}}} = e^{i\mathbf{K}_{\text{ex}}\cdot\mathbf{R}} \, F_n(r) \, \phi_c(r_e) \, \phi_v(r_h)
\tag{5.6}
$$

where ϕ_c and ϕ_v represent the central cell nature of the electron and hole bandedge states used in the effective mass theory. The excitonic energy levels are then

$$
E_{n\mathbf{K}_{\text{ex}}} = E_n + \frac{\hbar^2}{2(m_e^* + m_h^*)} K_{\text{ex}}^2
\tag{5.7}
$$

with E_n being the eigenvalues of the hydrogen atom-like problem

$$
E_n = -\frac{m_r^* e^4}{2(4\pi\epsilon)^2 \hbar^2} \frac{1}{n^2}
\tag{5.8}
$$

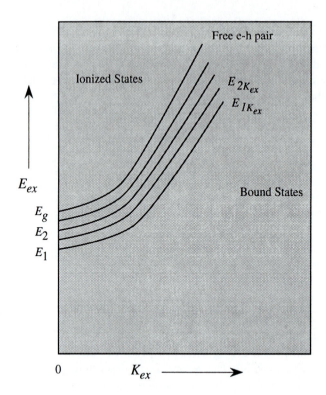

Figure 5.3: Dispersion curves for the electron-hole system in the exciton framework.

and the second term in Eqn. 5.7 represents the kinetic energy of the center of mass of the electron-hole pair.

The energy of the excitonic state is measured with respect to the energy of the state without the Coulomb interaction, i.e., the bandgap. Thus, excitonic levels appear slightly below the bandgap since typical values for E_1 are $\sim$2-6 meV for most semiconductors. The dispersion then looks as shown in Fig. 5.3.

This dispersion relation looks quite different from the usual E vs. $\boldsymbol{k}$ relation we are used to. This is because we are describing the system not in terms of the electron crystal momentum, but the electron-hole crystal momentum $\boldsymbol{K}_{ex}$. This is obviously the appropriate quantum number to describe the problem once the electron-hole Coulombic interaction is turned on. If the Coulombic interaction is turned off, the parabolas below the bandgap (bond states) all disappear and we simply have the free electron-hole dispersion which is the same as the bandstructure discussed in earlier chapters except that we plot the dispersion in a $\boldsymbol{k}_e - \boldsymbol{k}_h$ description.

We notice now that unlike the cases discussed in Chapter 4 where the electron-hole joint density of states started at the bandedge position, we now have a density of states below the bandedge energies. However, not all these states will couple to the photon because of momentum conservation.

↝ In order to examine the absorption spectra of excitonic transitions in semiconductors, it is useful to examine the problem in a little greater detail. As discussed earlier, the independent electron picture provides us the conduction and valence band states. The Coulombic interaction of the electron-hole pairs will now be treated as a perturbation, and the new wavefunction can be expressed in terms of the independent electron wavefunction basis. The general form of the excitonic problem is given by the Hamiltonian

$$H_e = H_0 + \frac{1}{2} \sum_{i \neq j} \frac{e^2}{4\pi\epsilon \, |r_i - r_j|} \tag{5.9}$$

where H_0 is the independent electron Hamiltonian giving rise to the usual bandstructure. The indices i and j represent the different electron pairs, with the factor $1/2$ to prevent double counting.

Since the Hamiltonian has the symmetry of the crystal, the Bloch theorem applies to the wavefunction, which must satisfy the condition

$$\psi_{\text{ex}}(r_1 + R, r_2 + R, r_3 + R, \ldots) = e^{iK_{\text{ex}} \cdot R} \, \psi_{\text{ex}}(r_1, r_2, r_3, \ldots) \tag{5.10}$$

where R is a lattice vector of the crystal.

The exciton state can be written in terms of a basis function $\Phi_{c,k_e,S_e;v,k_h,S_h}$ which represents a state where an electron, with momentum k_e and spin S_e, is in the conduction band and a hole, with momentum and spin k_h and S_h, is in the valence band, as shown in Fig 5.3. The difference $k_e - k_h$ represents the momentum of the exciton state. The exciton state is made up of a proper expansion of the Φ states. However, because of the Bloch theorem, the combination $k_e - k_h$ in the expansion must be constant for any given excitonic state. This greatly simplifies our exciton wavefunction, which can now be written as

$$\psi_{\text{ex}}^{n\ell m} = \sum_k A_{n\ell m}(k) \, \Phi_{c,k+K_{\text{ex}}/2,S_e;v,k-K_{\text{ex}}/2,S_h}^{n\ell m} \tag{5.11}$$

Here n is the energy eigenvalue index, ℓ and m are angular momentum indices representing the multiplicity of the excitonic state and $A_{n\ell m}(k)$ are the expansion coefficients. The exciton solution is given with the determination of $A_{n\ell m}(k)$. Since we are dealing with large envelope functions (order of ~ 100 Å), the coefficients $A_{n\ell m}(k)$ are expected to be localized sharply in k-space. We can define the Fourier transform of $A_{n\ell m}(k)$ as

$$F_{n\ell m}(r) = \sum_k A_{n\ell m}(k) \, e^{ik \cdot r} \tag{5.12}$$

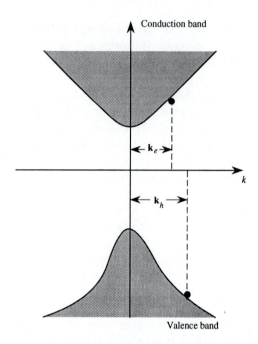

Figure 5.4: Schematic picture of an exciton in the Bloch representation. The state $\Phi_{c,\mathbf{k}_e,S_e;v,\mathbf{k}_h,S_h}$ represents an extra electron of wave vector k_e and spin S_e in the conduction band and a hole of wave vector k_h and spin S_h in the valence band.

This real space envelope function $F_{n\ell m}(\mathbf{r})$ is the same as we introduced in our simple derivation earlier and which obeys the hydrogen atom-like equation (ignoring exchange interactions):

$$\left[E_{\text{cv}}(-i\nabla, \mathbf{K}_{\text{ex}}) - \frac{e^2}{4\pi\epsilon r} \right] F_{n\ell m}(\mathbf{r}) = E_{\text{ex}} F_{n\ell m}(\mathbf{r}) \tag{5.13}$$

Here $E_{\text{cv}}(-i\nabla, \mathbf{K}_{\text{ex}})$ represents the operator obtained by expanding $E_c(\mathbf{k} + \mathbf{K}_{\text{ex}}/2) - E_v(\mathbf{k} - \mathbf{K}_{\text{ex}}/2)$ in powers of $\mathbf{k}$ and replacing $\mathbf{k}$ by $-i\nabla$. The exchange term is usually very small and will be ignored. The dielectric constant, in general, can be quite complicated, especially if the free carrier density is large. At low carrier densities ($n < 10^{14}$ cm^{-3}), the static dielectric constant is a good approximation to ϵ. It is important to note that this effective mass like equation is valid only if $F_{n\ell m}(\mathbf{r})$ is extended in space, i.e., $A_{n\ell m}(\mathbf{k})$ is peaked in k-space.

For a simple parabolic band, we have already discussed the solution of the exciton problem. The exciton energy levels are

$$E_n^{\text{ex}} = E_g - \frac{m_r^* e^4}{2\hbar^2 (4\pi\epsilon)^2} \frac{1}{n^2}$$

$$= E_g - \frac{R_{\mathrm{ex}}}{n^2} \qquad (5.14)$$

where R_{ex} denotes the exciton Rydberg. The kinetic energy of the electron-hole pair is to be added to Eqn. 5.14 for the total exciton energy.

The exciton envelope functions are the hydrogen atom-like functions, e.g., the ground state is

$$F_{100}(\boldsymbol{r}) = \frac{1}{\sqrt{\pi a_{\mathrm{ex}}^3}} e^{-r/a_{\mathrm{ex}}} \qquad (5.15)$$

with $a_{\mathrm{ex}} = (\epsilon m_0/\epsilon_0 m_r^*)\, a_B$ (a_B = Bohr radius = 0.529 Å). The exciton radius a_{ex} is ~100 Å for most semiconductors. Thus, the exciton is spread over a large number of unit cells, and the use of the effective mass equation is justified.

5.3 OPTICAL PROPERTIES WITH INCLUSION OF EXCITONIC EFFECTS

⤳ Excitonic effects have very dramatic consequences for the optical properties of semiconductors, especially near the bandedges. Below the bandedge, there is a strong and sharp excitonic absorption/emission transition. Also just above the bandgap, there is a strong enhancement of the absorption process especially in 3D systems.

As discussed in Chapter 4, in the absence of excitonic effects, the absorption coefficient can be written as

$$\alpha(\hbar\omega) = \frac{\pi e^2}{m_0^2 c n_r \epsilon_0} \frac{\hbar}{\hbar\omega} \int \frac{2\, d^3k}{(8\pi^3)} |\boldsymbol{a} \cdot \boldsymbol{p}_{\mathrm{if}}(\boldsymbol{k})|^2 \ \delta(E_c(\boldsymbol{k}) - E_v(\boldsymbol{k}) - \hbar\omega) \qquad (5.16)$$

For allowed transitions we can assume that $\boldsymbol{p}_{\mathrm{if}}$ is independent of $\boldsymbol{k}$ giving us the absorption coefficient

$$\begin{aligned}
\alpha(\hbar\omega) &= 0 & \text{if } \hbar\omega < E_g \\
&= \frac{\pi e^2}{m^2 c n_r \epsilon_0} \frac{\hbar}{\hbar\omega} |\boldsymbol{a} \cdot \boldsymbol{p}_{\mathrm{if}}|^2 \cdot N_{\mathrm{cv}}(\hbar\omega) & \text{if } \hbar\omega \geq E_g
\end{aligned} \qquad (5.17)$$

where $N_{\mathrm{cv}}(\hbar\omega)$ is the joint density of states.

If the excitonic effects are accounted for, these expressions are modified. We will again work in the dipole approximation and consider a transition from the ground state (all electrons are in the valence band) to the excited exciton state. This transition rate is, according to the Fermi golden rule

$$W(\psi_0 \rightarrow \psi_{\mathbf{K}_{\mathrm{ex}}}) = \frac{2\pi}{\hbar} \left(\frac{eA}{m_0}\right)^2 \delta_{\mathbf{K}_{\mathrm{ex}}} \left| \sum_{\boldsymbol{k}} A(\boldsymbol{k})\, \boldsymbol{a} \cdot \boldsymbol{p}_{\mathrm{cv}}(\boldsymbol{k}) \right|^2 \delta(E_{\mathrm{ex}} - E_0 - \hbar\omega) \qquad (5.18)$$

where E_0 is the energy corresponding to the ground state.

Once again, if we assume $p_{cv}(k)$ is independent of k

$$W(\psi_0 \to \psi_{K_{ex}}) = \frac{2\pi}{\hbar} \left(\frac{eA}{m_0}\right)^2 \delta_{K_{ex}} |a \cdot p_{if}(0)|^2 \left|\sum_k A(k)\right|^2 \delta(E_{ex} - E_0 - \hbar\omega) \quad (5.19)$$

From the definition of the Fourier transform $F_{n\ell m}$, we see that from Eqn. 5.12

$$F_{n\ell m}(0) = \sum_k A_{n\ell m}(k) \quad (5.20)$$

We also know from the theory of the hydrogen atom problem that $F_{n\ell m}(0)$ is nonzero only for s-type states, and, in general

$$F_{n\ell m}(0) = \frac{1}{\sqrt{\pi a_{ex}^3 n^3}} \delta_{\ell,0}\delta_{m,0} \quad (5.21)$$

Thus, the absorption rate is given by

$$W(\psi_0 \to \psi_{K_{ex}}) = \frac{2\pi}{\hbar} \left(\frac{eA_0}{m_0}\right)^2 \delta_{K_{ex}} |a \cdot p_{if}(0)|^2 \frac{\delta(E_{ex}^n - E_0 - \hbar\omega)}{\pi a_{ex}^3 n^3} \quad (5.22)$$

Comparing this result with the case for free band-to-band transitions, we note that the density of states in the free case is replaced by the term

$$N_{cv}(\hbar\omega) \to \frac{\delta(E_{ex}^n - E_0 - \hbar\omega)}{\pi a_{ex}^3 n^3} \quad (5.23)$$

with the δ-function eventually being replaced by a broadening function. If, for example, we assume a Gaussian broadening, we have (σ is the half width)

$$\delta(\hbar\omega - E) \to \frac{1}{\sqrt{1.44\pi}\,\sigma} \exp\left(\frac{-(\hbar\omega - E)^2}{1.44\,\sigma^2}\right)$$

For the ground state $n = 1$ exciton, the absorption coefficient becomes, after using the Gaussian form for the δ-function,

$$\alpha(\hbar\omega) = \frac{\pi e^2 \hbar}{2n_r \epsilon_0 c m_0(\hbar\omega)} \frac{2|p_{cv}|^2}{m_0} a_p \left(\frac{1}{\sqrt{1.44\pi}} \frac{1}{\sigma} \frac{1}{\pi a_{ex}^3} exp\left(\frac{-(\hbar\omega - E_{ex})^2}{1.44\sigma^2}\right)\right)$$

$$(5.24)$$

This result suggests that the excitonic transitions occur *as if* each exciton has a spatial extent of $1/(\pi a_{ex}^3 n^3)$. However, we note that this is not really the correct picture, since

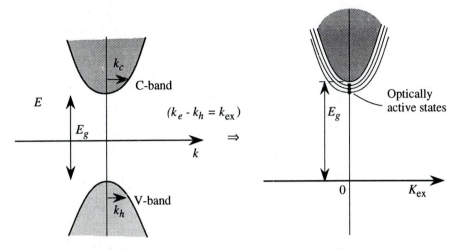

Independent electron picture + *e-h* interact ⇒ excitonic bandstructure

Figure 5.5: The effect of the electron-hole Coulombic interaction is to create exciton bands as shown. Only $K_{ex} = 0$ states are optically active.

the excitons are extended states. We also note that since only the $K_{ex} = 0$ state of the exciton is optically active, the transitions are *discrete*, even though the exciton density of states is *continuous*. The strength of the successive transitions decreases as $1/n^3$, so that the $n = 2$ resonance has one-eighth the strength of the $n = 1$ transition.

It is useful to examine again the independent electron picture and the exciton picture. This is done in Fig. 5.5. On the left-hand side we show the usual bandstructure with the valence and conduction band. The presence of the Coulomb interaction causes us to use $k_e - k_h = K_{ex}$ as an appropriate quantum number for the description. As discussed earlier, this leads to the exciton bands below the bandgap and free states above the bandgap. Due to the momentum conservation, only the $K_{ex} = k_e - k_h = 0$ states are optically active. Above the bandgap, these states are just the ones we considered earlier, in the band-to-band transitions. However, below the bandgap, these are the discrete excitonic resonances. Strong changes in the optical properties occur near the bandedge.

As one approaches the bandedge, the exciton lines become closer and closer and (see Eqn. 5.8), even though each transition becomes weaker, the absorption over an infinitesimal energy range reaches a finite value. In fact, the concept of the density of states of $K_{ex} = 0$ states becomes a meaningful concept. This density of states is from Eqn. 5.8

$$D_{ex}(E) = 2\frac{\partial n}{\partial E}$$

$$= \frac{n^3}{R_{\text{ex}}} \tag{5.25}$$

Extending the expression for the transition rates (Eqn. 5.22), by including a final state density of excitonic states we get

$$W(\psi_0 \rightarrow \psi_{\text{ex}}) = \frac{2\pi}{\hbar} \left(\frac{eA_0}{m_0} \right)^2 \delta_{\mathbf{K}_{\text{ex}}} |\mathbf{a} \cdot \mathbf{p}_{\text{if}}(0)|^2$$

$$\times \sum_n \left| \sum_{\mathbf{k}} A(\mathbf{k}) \right|^2 \delta \left(E_{\text{ex}}^n - E_0 - \hbar\omega \right)$$

$$= \frac{2\pi}{\hbar} \left(\frac{eA_0}{m_0} \right)^2 |\mathbf{a} \cdot \mathbf{p}_{\text{if}}(0)|^2 \frac{1}{\pi a_{\text{ex}}^3} \frac{1}{R_{\text{ex}}} \tag{5.26}$$

This expression is valid near the bandedge. If we compare this expression with the free electron-hole absorption rate near the bandedge, we see that the difference is that the density of states has been replaced by $1/(\pi a_{\text{ex}}^3 R_{\text{ex}})$ or, near the bandedge, the absorption coefficient is

$$\boxed{\alpha_{\text{ex}}(\hbar\omega \approx E_g) = \alpha_F \cdot \frac{2\pi \, R_{\text{ex}}^{1/2}}{(\hbar\omega - E_g)^{1/2}}} \tag{5.27}$$

where α_F is the absorption without excitonic effects. Thus, instead of α going to zero at the bandedge, it *becomes a constant*. By examining the nature of the "free" hydrogen atom-like states for the exciton *above* the bandedge, it can be shown that the absorption coefficient is given by the relation

$$\alpha_{\text{ex}}(\hbar\omega > E_g) = \alpha_F \cdot \frac{\pi x \, e^{\pi x}}{\sinh \pi x} \tag{5.28}$$

where

$$x = \frac{R_{\text{ex}}}{(\hbar\omega - E_g)^{1/2}}$$

When $E_g - \hbar\omega \gg R_{\text{ex}}$, the results reduce to the band-to-band transitions calculated in Chapter 4. These effects are shown in Fig. 5.6.

From these discussions it is clear that the excitonic transitions greatly modify the independent electron absorption spectra, especially near the bandedges. In Fig. 5.7, we show a low temperature measurement of the excitonic and band-to-band absorption in GaAs. The excitonic peak is clearly resolved here. The binding energy of the exciton in GaAs is ~ 4 meV. Since the exciton line is broadened by background impurity potential fluctuations, as well as phonons, it is possible to see such transitions only in high purity semiconductors, at low temperatures. In fact, the observation of excitons in bulk semiconductors is a good indication of the quality of the sample. For narrow bandgap semiconductors like $In_{0.53}Ga_{0.47}As$, the excitons are difficult to observe, because

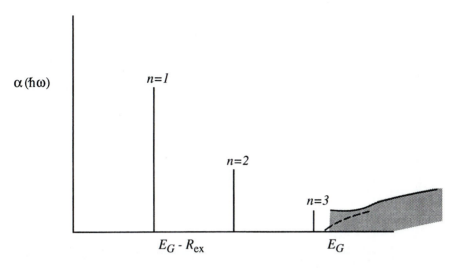

Figure 5.6: A schematic picture of the absorption spectra with (solid line) and without (dashed line) excitonic effects.

of the even small binding energy. In such materials, the exciton-related transitions are seen better in low temperature luminescence experiments. In these experiments, carriers are pumped into the material with a laser. The carriers relax into the low lying excitonic states in extremely short time scales (picoseconds) and then recombine, emitting photons with appropriate energies. At low temperatures, the band-to-band transitions are eliminated, since the electrons and holes thermalize into the excitonic states, from where they recombine. In such a spectrum, one sees not just the "free" exciton, which arises from the electron-hole system discussed above, but also various kinds of "bound" excitons. These bound excitons do not have the $\exp(i\boldsymbol{K}_{ex} \cdot \boldsymbol{R})$ dependence in their wavefunctions and are bound to local impurities such as donors or acceptors.

EXAMPLE 5.1 Consider a ground state exciton in GaAs having a halfwidth of $\sigma = 1.0$ meV. Calculate the peak absorption coefficient for the excitonic resonance.

The optical absorption coefficient is given for excitonic transitions by

$$\alpha(\hbar\omega) = \frac{\pi e^2 \hbar}{2 n_r \epsilon_0 c m_0 (\hbar\omega)} \frac{2 |p_{cv}|^2}{m_0} a_p \left(\frac{1}{\sqrt{1.44\pi}} \frac{1}{\sigma} \frac{1}{\pi a_{ex}^3} exp \left(\frac{-(\hbar\omega - E_{ex})^2}{1.44\sigma^2} \right) \right)$$

We have for GaAs (we will express σ in units of meV for convenience)

$$\frac{2p_{cv}^2}{m_0} \cong 25 \ eV$$

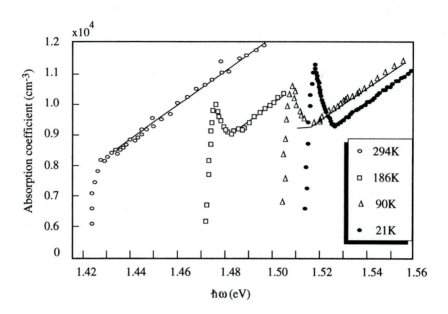

Figure 5.7: Measured excitonic and band-to-band spectra in GaAs. As can be seen, the excitonic peak essentially merges with the band-to-band absorption onset at room temperature. (Reproduced with permission, M. D. Sturge, *Phys. Rev.*, **127**, 768 (1962).)

Using $a_p = \frac{2}{3}$, we get ($a_{ex} = 120$ Å)

$$
\begin{aligned}
\alpha(\hbar\omega) &= \frac{\pi(1.6 \times 10^{-19}C)^2(1.05 \times 10^{-34}Js)}{2 \times 3.4(8.85 \times 10^{-12}F/m)(3 \times 10^8 m/s)(0.91 \times 10^{-30}kg)} \left(\frac{25}{1.5}\right)\left(\frac{2}{3}\right) \\
&\quad \cdot \frac{1}{\sqrt{1.44\pi}} \frac{1}{\sigma(meV)(10^{-3} \times 1.6 \times 10^{-19}J)} \frac{1}{(\pi \times 120 \times 10^{-10}m)^3} \\
&\quad exp\left(\frac{-(\hbar\omega - E_{ex})^2}{1.44\sigma^2}\right) \\
&= \frac{1.45 \times 10^6}{\sigma(meV)} exp\left(\frac{-(\hbar\omega - E_{ex})^2}{1.44\sigma^2}\right) \ m^{-1}
\end{aligned}
$$

For our case, where $\sigma = 1 \ meV$, we get, for the peak absorption coefficient

$$
\begin{aligned}
\alpha(\text{peak}) &= 1.45 \times 10^6 m^{-1} \\
&= 1.45 \times 10^4 cm^{-1}
\end{aligned}
$$

The strong dependence upon σ is quite evident and significant, since it is very difficult to control σ.

5.4 EXCITONIC STATES IN QUANTUM WELLS

↝ The ability to fabricate quantum well structures, where the electrons and holes can be strongly confined in the growth direction, has allowed excitonic resonances to assume an important technological aspect. The motivation for exciton studies is based equally between material characterization, pure physics, and "optical information processing." The highly controllable nature of the excitonic resonances lends itself to many versatile devices. The main reason for the interest in excitonic resonances in quantum well structures is the enhanced binding energy of the confined electron-hole system. Simple variational calculations show that the binding energy of a 2-dimensional electron-hole system with Coulombic interaction is four times that of the 3-dimensional system. In reality, of course, a quantum well system is not a 2-dimensional system, but is a quasi-2-dimensional system. The actual binding energy is, therefore, somewhat smaller than the $4 R_{ex}$ value. Nevertheless, the increased binding energy allows the excitonic transitions to persist up to high temperatures.

We address the exciton problem in the quantum well in the same manner as we addressed the 3-dimensional problem. The exciton state is again a Bloch state, with K_{ex} being a 2-dimensional quantum number in the plane of the quantum well. The kinetic energy of the exciton has the same form as for the 3D problem. The relative coordinate problem is then given by the Hamiltonian

$$
\begin{aligned}
H = & \frac{-\hbar^2}{2m_r^*}\left(\frac{1}{\rho}\frac{\partial}{\partial p}\rho\frac{\partial}{\partial p} + \frac{1}{\rho^2}\frac{\partial^2}{\partial\phi^2}\right) - \frac{\hbar^2}{2m_e}\frac{\partial^2}{\partial z_e^2} - \frac{\hbar^2}{2m_h}\frac{\partial^2}{\partial z_h^2} \\
& - \frac{e^2}{4\pi\epsilon\left|\boldsymbol{r}_e - \boldsymbol{r}_h\right|} + V_{ew}(z_e) + V_{hw}(z_h)
\end{aligned}
\qquad (5.29)
$$

Here V_{ew}, V_{hw} are the confining potentials, m_e and m_h are the in-plane effective masses of the electron and the hole, and μ is the reduced mass. The first term is the kinetic energy operator in the plane of the quantum well. The dielectric constant ϵ is the static dielectric constant, if there are no free carriers present. Otherwise, it is the screened dielectric constant, of the form used in the ionized impurity scattering problem. We will return to the screening problem later.

Unlike the 3D problem, the Hamiltonian of Eqn. 5.29 has no simple analytical solution, even if one assumes that the electron and hole states have a parabolic bandstructure. While the parabolic approximation is a good approximation for the electrons, it is a very poor approximation for the hole states, as can be seen from the in-plane valence bandstructure in a 100 ÅGaAs/Al$_{0.3}$Ga$_{0.7}$As quantum well structure shown in Fig. 5.8. As can be seen, the valence subbands are highly parabolic and the light hole state (LH1) even has a "negative" curvature at the zone center making the zone center reduced mass very large. It is possible to include these nonparabolic effects by using simple expressions of the form

$$
E_h(\boldsymbol{k}) = \alpha k^2 + \beta k^4
\qquad (5.30)
$$

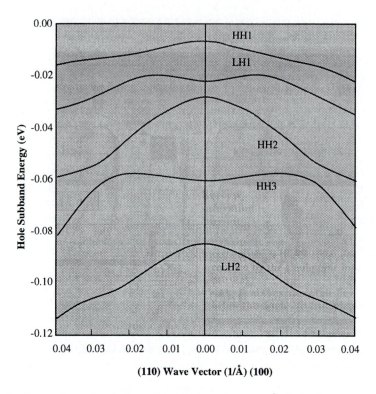

Figure 5.8: In plane valence band dispersion relations for a 100 Å GaAs /Al$_{0.3}$Ga$_{0.7}$As quantum well. Notice the strong nonparabolicity of the LH1 state.

where α and β can be fit to a calculated subband structure and then identifying $\boldsymbol{k}$ by $-i\nabla_\rho$. The function $F(\rho)$, describing the relative motion envelope, cannot be directly obtained. This function is usually obtained by assuming its form to be an exponential, or Gaussian or some combination of Gaussian functions, etc., with some variational constants which are then adjusted to minimize the energy

$$E = \frac{\int \psi_{ex}^* \, H \, \psi_{ex} \, dz_e \, dz_n \, \rho \, d\rho \, d\phi}{\int \psi_{ex}^* \, \psi_{ex} \, dz_e \, dz_n \, \rho \, d\rho \, d\phi} \tag{5.31}$$

This approach gives quite reliable results, with the effects of choosing different forms of variational functions being no more than ∼10% of the exciton binding energies. In Fig. 5.9, we show the effect of well size on the ground state excitonic states in GaAs/Al$_{0.3}$Ga$_{0.7}$As quantum well structures. As can be seen, the exciton binding energies can increase by up to a factor of ∼2.5 in optimally designed quantum well structures.

A useful approach to solve the exciton problem is to work in $\boldsymbol{k}$-space instead of the real space. This approach is more appropriate, since then the hole kinetic energy

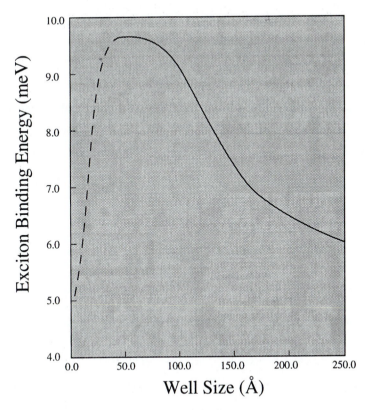

Figure 5.9: Variation of the heavy-hole exciton binding energy as a function of well size in GaAs/Al$_{0.3}$Ga$_{0.7}$As wells. The binding energy of the infinite barrier well should approach 16 meV as the well size goes to zero. However, in a real structure the binding energy starts to decrease below 50 Å well size.

need not be represented by any complicated real space differential operator. Instead, the bandstructure could be used with all nonparabolic effects included. This approach is also more useful when screening effects are to be included in the exciton problem. The important consideration in casting the exciton problem in k-space is to write the Coulombic interaction in the k-space basis of the free electron and hole states. If these states were simply the plane wave states, this would simply involve Fourier transforming the $1/r$ potential. The presence of the envelope functions and the central cell functions modifies the result slightly. Writing the electron and hole functions as

$$|n, k\rangle = f_n(z_e)\, e^{i k \cdot \rho_e}$$
$$|m, k\rangle = \sum_v g_m^v(k, z_h)\, e^{i k \cdot \rho_h} \qquad (5.32)$$

respectively, we need to evaluate the matrix elements of the Coulombic potential

$$V(\rho_e, z_e; \rho_h, z_h) = \frac{-e^2}{\epsilon \left[|\rho_e - \rho_h|^2 + (z_e - z_h)^2 \right]^{1/2}} \tag{5.33}$$

in the electron-hole product state basis. The k-space representation can be shown to have the form

$$
\begin{aligned}
V(\mathbf{k} - \mathbf{k}'; z_e, z_h) &= \frac{1}{(2\pi)^2} \int d^2\rho_e \, e^{-i(\mathbf{k}-\mathbf{k}')\cdot\rho_e} \, V(\rho_e, z_e; 0, z_h) \\
&= \frac{-e^2}{2\pi\epsilon_r |\mathbf{k} - \mathbf{k}'|} e^{|\mathbf{k}-\mathbf{k}'| \, |z_e - z_h|}
\end{aligned}
\tag{5.34}
$$

The full Hamiltonian in k-space is now

$$H = H_e^0 + H_h^0 + V \tag{5.35}$$

where H_e^0 and H_h^0 simply give the E vs. k relations for the electron and hole subbands. The exciton wavefunction then obeys the equation

$$H|\psi^{\text{ex}}\rangle = E^{\text{ex}}|\psi^{\text{ex}}\rangle \tag{5.36}$$

The optically active exciton function (i.e., $\mathbf{K}_{\text{ex}} = 0$), is expanded as usual, in the electron-hole basis with expansion coefficients $G_{nm}(\mathbf{k})$, the indices n,m representing various subband levels

$$|\psi^{\text{ex}}\rangle = \sum_n \sum_m \int d^2k \, |n, \mathbf{k}\rangle \, |m, -\mathbf{k}\rangle \, G_{nm}(\mathbf{k}) \tag{5.37}$$

An eigenvalue equation can now be set up for the $G_{nm}(\mathbf{k})$ by taking matrix elements of Eqn.5.29 between states $|n, \mathbf{k}\rangle|m, -\mathbf{k}\rangle$ and $|n'\mathbf{k}'\rangle|m', -\mathbf{k}'\rangle$, using Eqn.5.34 for the Coulombic potential matrix elements. Additionally, for narrow quantum wells ($W \approx 150$ Å) it is also reasonable to assume that the intersubband coupling is weak, i.e.,

$$\langle n, \mathbf{k}| \, \langle m, -\mathbf{k}|V|n'\mathbf{k}'\rangle \, |m', -\mathbf{k}'\rangle = 0 \tag{5.38}$$

unless $m = m'$, $n = n'$. This results in an eigenvalue equation

$$
\begin{aligned}
&\left[E^{\text{ex}} - E_n^e(\mathbf{k}) + E_m^h(-\mathbf{k}) \right] G_{nm}(\mathbf{k}) = \\
&\int d^2k' \int dz_e \int dz_h \, f_n^*(z_e) \, f_n(z_e) \\
&\sum_v g_m^{*v}(-\mathbf{k}, z_h) \, g_m^v(-\mathbf{k}', z_h) \, V(\mathbf{k} - \mathbf{k}'; z_e, z_h) \, G_{nm}(\mathbf{k}')
\end{aligned}
\tag{5.39}
$$

This equation can be solved numerically, by discretizing k-space and using the fact that the exciton function has a finite extent in k-space (~ 0.01 Å^{-1}). Alternately, the problem can be solved by variational techniques. Once the coefficients $G_{nm}(k)$ are known, the real-space dependence of the exciton function can be obtained by Fourier transformation, i.e.,

$$\psi_{nm}^{ex}(\rho, z_e, z_h) = \frac{1}{2\pi} f_n(z_e) \int d^2k \; e^{i\boldsymbol{k}\cdot\rho} \sum_v g_m^v(-\boldsymbol{k}, z_h) \, G_{nm}(\boldsymbol{k}) \qquad (5.40)$$

In Fig. 5.10, we show the CB1-HH1 (ground state) exciton wavefunction. Both the k-space and real-space functions are shown. We note that $G(\rho)$ has an exponential dependence $\psi^{ex} \sim \exp(-\rho/a_{ex})$, where a_{ex} can be used to define the exciton Bohr radius. In Fig. 5.10, we show that as free carriers (n_e) are injected into the material, the exciton radius increases. This effect will be discussed later in Section 5.9 and results from the screening of the e-h Coulombic interaction. It is interesting to note that the exciton radius decreases steadily as the well size is decreased. This is consistent with the increasing binding energy of the exciton.

5.5 EXCITONIC ABSORPTION IN QUANTUM WELLS

↝ The absorption spectra in quantum wells is given by the same formalism that was used for the case of bulk semiconductors. We simply need to represent the exciton envelope function by the function appropriate for the quantum well. The absorption coefficient is given by

$$\alpha_{nm}^{ex}(\hbar\omega) = \frac{\pi e^2 \hbar}{n_r m_0^2 c W \hbar\omega} \left| \sum_{\boldsymbol{k},n,m} G_{nm}(\boldsymbol{k}) \, \boldsymbol{a} \cdot \boldsymbol{p}_{nm}(\boldsymbol{k}) \right|^2 \delta\left(\hbar\omega - E_{nm}^{ex}\right) \qquad (5.41)$$

The matrix elements $\boldsymbol{p}_{nm}(\boldsymbol{k})$ are given by the central cell part used for the 3-dimensional problem, as well as the overlap of the envelope function as used in the Chapter 4 for band-to-band absorption

$$\boldsymbol{p}_{nm}(\boldsymbol{k}) = \sum_{\nu,\mu} \int d^2r \; U_0^\nu(\boldsymbol{r}) \, \boldsymbol{p} \, U_0^\mu(\boldsymbol{r}) \int dz \; f_n^\mu(z) \, g_m^\nu(k, z)$$

The selection rules and polarization dependencies discussed in Chapter 3 for band to band transitions, still hold for the excitonic transitions.

The absorption coefficient is often also written in terms of oscillator

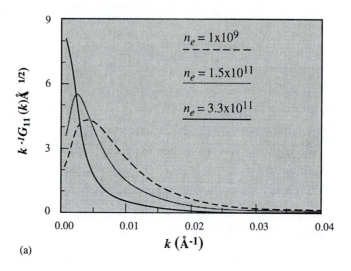

(a)

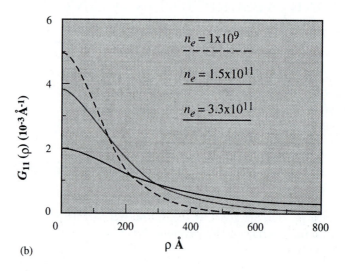

(b)

Figure 5.10: The (a) k-space and (b) real space exciton wavefunction defined by $\sqrt{k}G_{11}(k)$ and $G_{11}(\rho)$ for the CB1-HH1 exciton in a 100 Å GaAs/Al$_{0.3}$Ga$_{0.7}$As quantum well structure. (After J. Loehr and J. Singh, *Phys. Rev.*, **B 42**, 7154 (1990).)

strengths f_{nm} as

$$\alpha_{nm}(\hbar\omega) = \sum_{nm} \frac{\pi e^2 \hbar}{n_r \epsilon_0 m_0 cW} f_{nm} \, \delta\left(E_{nm}^{ex} - \hbar\omega\right) \tag{5.42}$$

where the oscillator strength per unit area is defined by $(\hbar\omega = E_{nm}^{ex})$

$$f_{nm} = \frac{2}{m_0 E_{nm}^{ex} (2\pi)^2} \left| \int d^2k \, G_{nm}(\boldsymbol{k}) \, \boldsymbol{a} \cdot \boldsymbol{p}_{nm}(\boldsymbol{k}) \right|^2 \tag{5.43}$$

The oscillator strength is a better measure of the excitonic absorption because it does not involve the δ-function. The δ-function will eventually be replaced by a broadening function whose value will be sample and temperature dependent as we will see below. Since quantum confinement decreases the spatial extent of the exciton function, the oscillator strength increases as the exciton is confined.

If the 2D exciton function can be represented by an exciton radius a_{ex}, then as in the 3D case, the absorption coefficient becomes

$$\alpha(\hbar\omega) = \frac{\pi e^2 \hbar}{2 n_r \epsilon_0 c m_0 \hbar\omega} \left(\frac{2|p_{cv}|^2}{m_0} \right) a_p \left(\frac{1}{\sqrt{1.44\pi}} \frac{1}{\sigma} \frac{1}{W\pi a_{ex}^2} \exp\left(\frac{-(\hbar\omega - E_{ex})^2}{1.44\sigma^2} \right) \right) \tag{5.44}$$

EXAMPLE 5.2 Consider a 100 Å GaAs/Al$_0$.3Ga$_0$.7As quantum well structure grown along (001) with a halfwidth of 1 meV. The exciton radius is $\frac{2}{3}$ the bulk value in GaAs. Calculate the peak of the HH ground state exciton resonance for x-polarized light.

The excitonic absorption coefficient is

$$\alpha(\hbar\omega) = \frac{\pi e^2 \hbar}{2 n_r \epsilon_0 c m_0 \hbar\omega} \left(\frac{2|p_{cv}|^2}{m_0} \right) a_p \left(\frac{1}{\sqrt{1.44\pi}} \frac{1}{\sigma} \frac{1}{W\pi a_{ex}^2} \exp\left(\frac{-(\hbar\omega - E_{ex})^2}{1.44\sigma^2} \right) \right)$$

Notice from Example 5.1 that the effect of quantization has been essentially the replacement

$$\frac{1}{\pi a_{ex}(3D)^3} \longrightarrow \frac{1}{W\pi a_{ex}(QW)^3}$$

Thus, the excitonic absorption becomes $\left(\text{using} \, a_{ex}(QW) \cong \frac{2}{3} \, a_{ex}(3D) \right)$ for light polarized along the x-axis where $a_p = \frac{1}{2}$

$$\alpha(\hbar\omega) = \frac{2.9 \times 10^6}{\sigma(meV)} \, exp\left(\frac{-(\hbar\omega - E_{ex})^2}{1.44\sigma^2} \right)$$

For $\sigma = 1 \, meV$, we get the peak value of

$$\begin{aligned} \alpha(\text{peak}) &= 2.9 \times 10^6 \, m^{-1} \\ &= 2.9 \times 10^4 \, cm^{-1} \end{aligned}$$

What would the peak value be for a z-polarized light?

5.6   EXCITON BROADENING EFFECTS

↝ We have seen from the previous discussions that the exciton absorption resonance includes a δ-function. In real systems the δ-function is always broadened. This line broadening or linewidth is extremely important from both a physics and a technological point of view. In general, the peak of the exciton absorption is inversely proportional to the linewidth of the resonance. This puts a tremendous premium on reducing the exciton linewidth (in a reliable manner). Unfortunately, small variations in growth of quantum wells can change the exciton linewidth by as much as 100%! In fact, the exciton linewidth is one of the serious hurdles in using exciton based devices in reliable technology.

The excitonic transitions are broadened by two important kinds of fluctuations: inhomogeneous and homogeneous. The inhomogeneous linewidth is due to local potential fluctuations which arise due to imperfections in the structure. Homogeneous broadening, on the other hand, is due to fluctuations which are extended in nature and influence the entire excitonic spectra. The main inhomogeneous broadening sources are:

1. Interface roughness.

2. Alloy potential fluctuations.

3. Well to well fluctuations in multiquantum well spectra.

4. Background impurity broadening.

In high purity materials the impurity broadening is usually negligible (if background density $< 10^{15} \mathrm{cm}^{-3}$). The homogeneous broadening mechanisms are:

1. Acoustic phonon scattering.

2. Optical phonon scattering.

3. Other mechanisms which may affect the exciton lifetime such as tunneling, recombination, etc.

The treatment of inhomogeneous broadening usually follows the approach developed to study electronic states in disordered materials. Line broadening arises due to spatially localized fluctuations which are capable of shifting the excitonic emission energy. In general, one may describe these fluctuations by concentration fluctuations in the mean compositions C_A^0 and C_B^0 of the structure. For example C_A^0 and C_B^0 may represent the fraction of the islands and the valleys representing interface roughness in a quantum well, or the mean composition of the alloy in the treatment of alloy broadening. The width of the concentration fluctuation occurring over a region β (which is an area for interface roughness treatment and a volume for the treatment of alloy broadening) is given by (Lifshitz, 1969):

$$\delta P = 2\sqrt{\frac{1.4\, C_A^0\, C_B^0\, \alpha}{\beta}} \tag{5.45}$$

where α is the smallest extent over which the fluctuation can take place. Again for the treatment of interface roughness, α is the average area of the two-dimensional islands representing the interface roughness and for alloy broadening, α is the volume of the smallest cluster in the alloy. For a perfectly random alloy, this is the volume per cation.

The shift in the excitonic energy due to this fluctuation is then

$$2\sigma(\beta) = \delta P\, |\Psi|_\beta^2\, \frac{\partial E_{\mathrm{ex}}}{\partial C} \tag{5.46}$$

where $|\Psi|_\beta^2$ is the fraction of the exciton sensing the region β, and $\partial E_{\mathrm{ex}}/\partial C$ represents the rate of change in the exciton energy with change in the concentration. To obtain the linewidth of the absorption, one needs to identify the volume (or area) β. For interface roughness broadening, this area is $\sim 3r_{\mathrm{ex}}^2$ and the linewidth is given by

$$2\sigma_{\mathrm{IR}} = 2\sqrt{\frac{1.4\, C_A^0\, C_B^0\, a_{2\mathrm{D}}}{3r_{\mathrm{ex}}^2 \pi}} \cdot \left.\frac{\partial E_{\mathrm{ex}}}{\partial W}\right|_{W_0} \cdot \delta_0 \tag{5.47}$$

where $a_{2\mathrm{D}}$ is the real extent of the two-dimensional islands representing the interface roughness, δ_0 their height, and r_{ex} is the exciton Bohr radius in the lateral direction parallel to the interface. $\partial E_{\mathrm{ex}}/\partial W$ is the change of the exciton energy as a function of well size.

Apart from interface roughness fluctuations in the same well, often one has to contend with intra-well fluctuations. Most optical devices using excitonic transitions use more than one well, i.e., use multiquantum wells that are produced by opening and shutting off fluxes of one or other chemical species. There is invariably a mono-layer or so of variation in the well sizes across a stack of quantum wells. For a 100 Å GaAs/Al$_{0.3}$Ga$_{0.7}$As quantum well, a monolayer fluctuation can produce ~ 1.5 meV change in the exciton resonance energy. This change arises almost entirely from the changes in subband levels since the *exciton binding energy* does not vary significantly with a monolayer change in the well size. In high quality GaAs/AlGaAs multiquantum well samples, an absorption linewidth of 3–4 meV is achievable.

In the GaAs based system, the well material is a binary material with no alloy scattering. In cases like In$_{0.53}$Ga$_{0.47}$As/InP or In$_{0.53}$Ga$_{0.47}$As/In$_{0.52}$Al$_{0.48}$As, one additionally has alloy broadening effects. These effects can also be of the order of 3–5 meV and are given by similar treatments. If C_A^0 and C_B^0 are the mean concentrations of the two alloy components and V_C is the average alloy cluster size ($=$ unit cell for random alloy), the linewidth is given approximately by

$$2\sigma_{\text{alloy}}^{\text{internal}} = 2\sqrt{\frac{1.4\, C_A^0\, C_B^0\, V_C}{3\pi r_{\text{ex}}^2 W_0}} \cdot \left.\frac{\partial E_{\text{ex}}}{\partial C_A}\right|_{C_A^0} \tag{5.48}$$

where $\partial E_{\text{ex}}/\partial C_A$ represents the change in the exciton resonance with changes in composition. If the barrier material is also an alloy, there is a contribution to the linewidth from the barrier as well

$$2\sigma_{\text{alloy}}^{\text{external}} = 2\sqrt{\frac{1.4\, C_A^0\, C_B^0\, V_C}{3\pi r_{\text{ex}}^2 L_{\text{eff}}}} \cdot \left.\frac{\partial E_{\text{ex}}}{\partial C_A}\right|_{C_A^0} \tag{5.49}$$

where L_{eff} is an effective length to which the exciton wavefunction penetrates in the barrier and has to be calculated numerically. L_{eff} is approximately equal to twice the distance over which the exciton wavefunction falls to $1/e$ of its initial value. Typically this is on the order of a few monolayers. The well size dependence of the exciton linewidth is quite apparent from the results of Fig. 5.11 and are due to the increase in $\partial E_{\text{ex}}/\partial W$ with decreasing well size.

Finally, to examine the homogeneous broadening note that the exciton state is optically created at the $\boldsymbol{K}_{\text{ex}} = 0$ state. The acoustic and optical phonons interact with the exciton causing either a transition to a $\boldsymbol{K}_{\text{ex}} \neq 0$ state in the same quantum level or even ionization of the exciton. This homogeneous broadening effect can be calculated in a manner similar to the one used for electron-phonon scattering. The linewidth is given approximately by

$$2\sigma = \frac{1}{W(\boldsymbol{K}_{\text{ex}} = 0)} \tag{5.50}$$

and has the form

$$2\sigma = \alpha T + \frac{\beta}{\exp\left(\hbar\omega_0/k_B T\right) - 1} \tag{5.51}$$

where the first term is due to acoustic phonon scattering and the second is due to optical phonons. At room temperature, the homogeneous linewidth due to phonons is $\sim$3–4 meV, implying that the exciton lifetime is ~ 0.2 ps.

The consequences of the exciton linewidth on devices cannot be emphasized enough. While the homogeneous linewidths are controlled by the temperature or other controllable effects, the inhomogeneous effects are dependent upon sample quality. This places a heavy burden on the crystal grower. In this context, it is useful to compare the effect of, say, one-monolayer fluctuation in the interface quality on an MODFET versus an exciton based device. The effect on room temperature mobility of a MODFET is less than 1%, while the effect on exciton linewidth at room temperature is $\sim$30%!

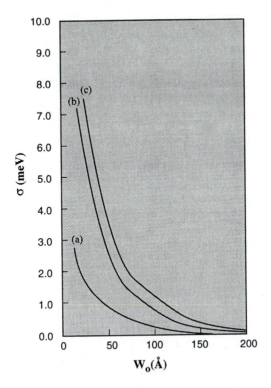

Figure 5.11: Variation of exciton linewidth as a function of well size for a one monolayer interface roughness. The half width σ is given for island size extents of (a) 20 Å; (b) 80 Å; (c) 100 Å for the GaAs/Al$_{0.3}$Ga$_{0.7}$As system.

In Fig. 5.12 we show the comparison of the excitonic spectra in a 100 Å GaAs/Al$_{0.3}$Ga$_{0.7}$As and 100 Å In$_{0.53}$Ga$_{0.47}$As/In$_{0.52}$Al$_{0.48}$As quantum well structure. The large reduction in the InGaAs spectra is primarily due to the alloy broadening of the exciton peak.

EXAMPLE 5.3 A multiquantum well (MQW) stack is used in an exciton absorption measurement. The nominal thickness of the GaAs/Al$_{0.3}$Ga$_{0.7}$As wells is 100 Å. Consider two cases: i) There is no structural disorder and the exciton halfwidth is given by a homogeneous width $\sigma = 1$ meV; ii) In addition to the above broadening, there is a structural disorder in the MQW stack described by a one monolayer rms fluctuation in the well size. Calculate the peak absorption coefficients in the two cases.

A one monolayer charge in the well size at 100 Å produces a shift in the HH exciton energy of ~ 1.5 meV, due to the shift in the subband energies (there is essentially no change in the binding energies). For simplicity, we assume that the linewidths from the homogeneous and inhomogeneous contributions simply add (in reality, the homogeneous broadening may be better described by a Lorentzian function, rather than a Gaussian function; also one has to use

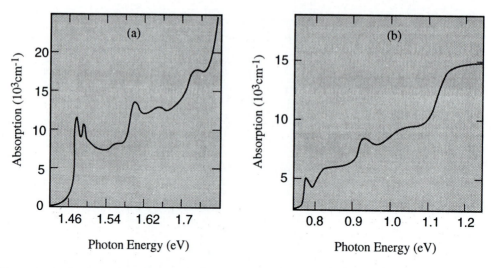

Figure 5.12: A comparison of the absorption spectra in (a) 100 Å GaAs /Al$_{0.3}$Ga$_{0.7}$As and (b) In$_{0.53}$Ga$_{0.47}$As /In$_{0.52}$Al$_{0.48}$As quantum wells. The excitons in InGaAs suffer alloy broadening which reduces their sharpness. (After D.S. Chemla, *Nonlinear Optics: Materials and Devices*, eds. C. Flytzanis and J.L. Oudar, Springer-Verlag, New York (1986).)

a convoluted linewidth).

From Example 5.2 we get, for the two cases,

i) α(peak) $= 2.9 \times 10^4$ cm^{-1}

i) α(peak) $= 1.2 \times 10^4$ cm^{-1}

This example shows the extreme sensitivity of excitonic resonances on minute structural imperfections.

5.7   MODULATION OF OPTICAL PROPERTIES IN 3D SYSTEMS

$\longrightarrow$ The modulation of the optical properties of a semiconductor can be exploited for a number of "intelligent" optoelectronic devices such as switches, logic gates, memory elements, etc. In Chapter 4 we had briefly discussed the complex refractive index of a material. The real part of this complex index determines the velocity of light in the medium and the imaginary part determines the attenuation of the light. In principle, any modification in either of these quantities can be used to modulate an optical signal propagating in the medium. Of course, as we know from the Kramer's-Kronig relation the real and imaginary part of the refractive index are related to each other.

When the refractive index of a material is modified, the effect on a light beam propagating in the material can be classified into two categories, depending upon the photon energy. As shown in Fig. 5.13, if the photon energy is in a region where the absorption coefficient is zero, the effect of the modification of the refractive index is to alter the velocity of propagation of light. On the other hand, if the photon energy is in a region where the absorption coefficient is altered, the intensity of light emerging from the sample will be altered. These two approaches for the modification of the optical properties by a applied electric field are called the electro-optic and the electro-absorption approaches, respectively.

In the electro-optic effect, an applied is used to alter the phase velocity of a propagating signal and this effect can be exploited in an interference scheme to alter the polarization or intensity of the light. We will first discuss this approach.

5.7.1 The Electro-Optic Effect

⤳ The electro-optic effect depends upon the modification of the refractive index of a material by an applied electric field. Before discussing the electro-optic effect, let us review some important optical properties of crystalline materials.

We had discussed in Chapter 4, section 4.2, how the Maxwell's equations and the material properties (ϵ, μ) determine the propagation of light in a material. In general, the medium is anisotropic and the material properties are described by the relation

$$
\begin{aligned}
D_x &= \epsilon_{xx} F_x + \epsilon_{xy} F_y + \epsilon_{xz} F_z \\
D_y &= \epsilon_{yx} F_x + \epsilon_{yy} F_y + \epsilon_{yz} F_z \\
D_z &= \epsilon_{zx} F_x + \epsilon_{zy} F_y + \epsilon_{zz} F_z
\end{aligned}
\tag{5.52}
$$

or in the short form

$$
D_i = \sum_j \epsilon_{ij} F_j
\tag{5.53}
$$

If one examines the energy density of the optical wave and imposes the conditions that the energy must be independent under the transformation $i \rightarrow -i$, we get

$$
\epsilon_{ij} = \epsilon_{ji}
\tag{5.54}
$$

Further reduction in the number of the independent ϵ_{ij} can be achieved, depending upon the symmetry of the crystal structure. A most useful way to describe the propagation of light in the medium is via the index ellipsoid as discussed in Appendix E. The equation

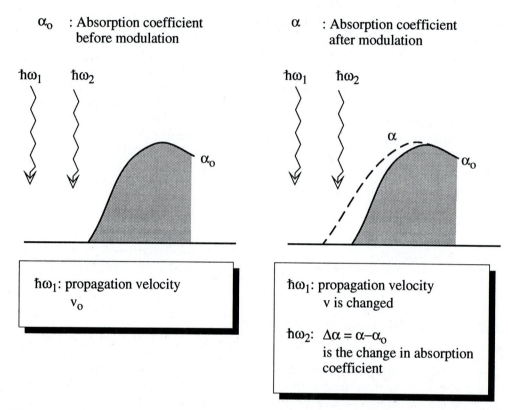

α_0 : Absorption coefficient before modulation

α : Absorption coefficient after modulation

$\hbar\omega_1$: propagation velocity v_0

$\hbar\omega_1$: propagation velocity v is changed

$\hbar\omega_2$: $\Delta\alpha = \alpha - \alpha_0$ is the change in absorption coefficient

Figure 5.13: A schematic of the effect of a change in optical properties of a material on an optical beam. For energy $\hbar\omega_1$, the main effect of the change in the optical properties is a change in propagation velocity. For $\hbar\omega_2$, the effect is a change in intensity.

for the index ellipsoid is

$$\frac{x^2}{\epsilon_x} + \frac{y^2}{\epsilon_y} + \frac{z^2}{\epsilon_z} = 1 \tag{5.55}$$

or in terms of the refractive indices,

$$\sum_{i=1}^{3} \frac{x_i^2}{n_i^2} = 1 \tag{5.56}$$

Here ϵ_x, ϵ_y *and* ϵ_z are the principal dielectric constants, expressed along the principle axes of the ellipsoid.

Now consider a situation where an electric field is applied to the crystal. The applied electric field modifies the bandstructure of the semiconductor through a number of interactions. These interactions may involve:

i) Strain: In a piezoelectric material where there is no inversion symmetry (e.g.,

GaAs, CdTe, etc.) the two atoms in the basis of the crystal have different charges. The electric field may cause a distortion in the lattice and, as a result, the bandstructure may change. This may cause a change in the refractive index.

ii) Distortion of the excitonic features: In the previous sections we discussed the optical properties of the exciton. The presence of an electric field can modify the excitonic spectra and thus alter the electronic spectra and, hence, the optical spectra of the material.

In general, the change in the refractive index may be written as

$$n_{ij}(F) - n_{ij}(0) = \Delta n_{ij} = r_{ijk} F_k + s_{ijk\ell} F_k F_\ell \tag{5.57}$$

where F_i is the applied electric field component along the direction i and r_{ijk} and $s_{ijk\ell}$ are the components of the electro-optic tensor. In materials like GaAs where the inversion symmetry is missing, r_{ijk} is non-zero and one has a linear term in the electro-optic effect. The linear effect is called the Pockel effect. In materials like Si where one has inversion symmetry $r_{ijk} = 0$ and the lowest order effect is due to the quadratic effect (known as the Kerr effect).

In general, r_{ijk} has 27 elements, but since the tensor is invariant under the exchange of i and j, there are only 18 independent terms. It is common to the use of contracted notation $r_{\ell m}$ where $\ell = 1, \ldots 6$ and m = 1, 2, 3. The standard contraction arises from the identification of $i, j = 1,1; 2,2; 3,3; 2,3; 3,1;1,2$ by $\ell = 1, 2, 3, 4, 5, 6$, respectively. The 18 coefficients are further reduced by the symmetry of the crystals. In semiconductors such as GaAs, it turns out that the only non-zero coefficients are

$$
\begin{aligned}
r_{41} & \\
r_{52} &= r_{41} \\
r_{63} &= r_{41}
\end{aligned}
\tag{5.58}
$$

Thus, a single parameter describes the linear electro-optic effect. The value of GaAs is $r_{41} = 1.2 \times 10^{-12}$ m/V. An important class of electro-optic materials are ferroelectric peroskites such as $LiNbO_3$ and $LiTaO_3$ which have trigonal symmetry and materials such as KDP (potassium dihydrogen phosphate) which have tetragonal symmetry). For the trigonal materials the non-zero tensor components are

$$
\begin{aligned}
r_{22} &= -r_{12} = -r_{61} \\
r_{51} &= r_{42} = r_{33} \\
r_{13} &= r_{23}
\end{aligned}
\tag{5.59}
$$

For KDP the non-zero elements are

$$
\begin{aligned}
r_{41} &= r_{52} \\
r_{63} &
\end{aligned}
\tag{5.60}
$$

Material	$\lambda(\mu m)$	$r_{ij}(10^{-12}m/V)$
GaAs	0.8 -10	$r_{41} = 1.2$
Quartz	0.6	$r_{11} = -0.47$ $r_{41} = -0.2$
LiNbO$_3$	0.5	$r_{13} = 9.0; r_{22} = 6.6$ $r_{42} = 30$
KDP	0.5	$r_{41} = 8.6; r_{63} = 9.5$

Table 5.1: Electro-optic coefficients for some materials. (After P.K Cheo, *Fiber Optics, Devices and Systems,* Prentice-Hall, New Jersey (1985).)

In Table 5.1, we give the values of the electro-optic coefficients for some materials. The second order electro-optic coefficients s_{ijkl} are usually not important for materials unless the optical energy $\hbar\omega$ is very close to the bandgap. In materials like GaAs, the second order coefficients that are non-zero from symmetry considerations are in the contracted form $s_{pq}, p = 1\ldots6, q = 1\ldots6$,

$$
\begin{aligned}
s_{11} &= s_{22} = s_{33} \\
s_{12} &= s_{13} \\
s_{44} &= s_{55} = s_{66}
\end{aligned}
\tag{5.61}
$$

The electro-optic effect is used to create a modulation in the frequency, intensity or polarization of an optical beam. To see how this occurs, consider a material like GaAs.

Let us first consider the linear electro-optic effect where a field F is applied to the crystal. The index ellipsoid is

$$
\left(\frac{1}{n_x^2} + r_{1k}F_k\right) x^2 \quad + \quad \left(\frac{1}{n_y^2} + r_{2k}F_k\right) y^2 + \left(\frac{1}{n_z^2} + r_{3k}F_k\right) z^2
$$
$$
+ \quad 2yzr_{4k}F_k + 2zxr_{5k}F_k + 2xyr_{6k}F_k = 1
\tag{5.62}
$$

where F_k (k = 1,2,3) is the component of the electric field in the x, y, and z directions.

Using the elements of the electro-optic tensor for GaAs, we get

$$
\frac{x^2}{n_x^2} + \frac{y^2}{n_y^2} + \frac{z^2}{n_z^2} + 2yzr_{41}F_x + 2zxr_{41}F_y + 2xyr_{41}F_z = 1
\tag{5.63}
$$

Let us now simplify the problem by assuming that the electric field is along the $< 001 >$ direction

$$F_x = F_y = 0, \qquad F_z = F \tag{5.64}$$

We now rotate the axes by 45° so that the new principal axes are

$$
\begin{aligned}
x' &= \frac{x}{\sqrt{2}} - \frac{y}{\sqrt{2}} \\
y' &= \frac{x}{\sqrt{2}} + \frac{y}{\sqrt{2}} \\
z' &= z
\end{aligned} \tag{5.65}
$$

In terms of this new set of axes, the index of ellipsoid is written as

$$\frac{x'^2}{n_x'^2} + \frac{y'^2}{n_y'^2} + \frac{z'^2}{n_z'^2} = 1 \tag{5.66}$$

where the new indices are

$$
\begin{aligned}
n_x' &= n_o + \frac{1}{2} n_o^3 r_{41} F \\
n_y' &= n_o - \frac{1}{2} n_o^3 r_{41} F \\
n_z' &= n_o
\end{aligned} \tag{5.67}
$$

where n_o is the index in absence of the field $(= n_x = n_y = n_z)$. As a result of this change in the indices along the x' and y' axes, for light along $< 01\bar{1} > (x')$ and $< 011 > (y')$ directions, a phase retardation occurs due to the field. The phase retardation for a wave that travels a distance L is, $(n_z' = n_o)$

$$
\begin{aligned}
\Delta\phi(x') &= \frac{\omega}{c}\left(n_z' - n_x'\right) L\xi_1 = -\frac{\pi}{\lambda} n_o^3 r_{41} F L\xi_1 \\
\Delta\phi(y') &= \frac{\omega}{c}\left(n_z' - n_y'\right) L\xi_1 = \frac{\pi}{\lambda} n_o^3 r_{41} F L\xi_1
\end{aligned} \tag{5.68}
$$

The quantity ξ_1 represents the overlap of the optical wave with the region where the electric field is present:

$$\xi_1 = \frac{1}{F} \int \int F \mid F_{\text{photon}} \mid^2 dA \tag{5.69}$$

where F_{photon} is the photon field. For bulk devices $\xi_1 \sim 1$.

Let us now extend our study to the second order term in the elctro-optic effect.

Using the contracted notiation, for quadratic electro-optic coefficients, the index ellipsoid can be written as

$$\left(\frac{1}{n_x^2} + s_{11}F_x^2 + s_{12}F_y^2 + s_{12}F_z^2\right) x^2$$

$$+ \left(\frac{1}{n_y^2} + s_{12}F_x^2 + s_{11}F_y^2 + s_{12}F_z^2\right) y^2$$

$$+ \left(\frac{1}{n_z^2} + s_{12}F_x^2 + s_{12}F_y^2 + s_{11}F_z^2\right) z^2$$

$$+2yz(2s_{44}F_yF_z) \quad + \quad 2zx(2s_{44}F_xF_z) + 2xy(2s_{44}F_yF_x) = 1 \qquad (5.70)$$

In the presence of an electric field, F, in the z direction, Eqn. 5.70 can be rewritten as

$$\left(\frac{1}{n_x^2} + s_{12}F^2\right) x^2 \quad + \quad \left(\frac{1}{n_y^2} + s_{12}F^2\right) y^2$$

$$+ \left(\frac{1}{n_z^2} + s_{11}F^2\right) z^2 = 1 \qquad (5.71)$$

This index ellipsoid can be rewritten as

$$\frac{x^2 + y^2}{n_o^2} + \frac{z^2}{n_e^2} = 1 \qquad (5.72)$$

with

$$n_o = n - \frac{1}{2}n^3 s_{12}F^2 \qquad (5.73)$$

and

$$n_e = n - \frac{1}{2}n^3 s_{11}F^2 \qquad (5.74)$$

The phase retardation due to the applied field is thus given by

$$\Delta\Phi = \frac{\omega}{c}(n_e - n_o)L\xi_2 = \frac{\pi}{\lambda}n^3(s_{12} - s_{11})F^2 L\xi_2$$

where ξ_2 is an overlap of the square of the electric field and the optical field given by

$$\xi_2 = \frac{1}{F^2}\int\int F^2 \mid F_{\text{photon}} \mid^2 dA \qquad (5.75)$$

The total phase change between waves travelling along x' and y' then becomes after adding the effects of the linear and quadratic electro-optic effects

$$\boxed{\begin{aligned} \Delta\phi(x') &= -\frac{\pi L}{\lambda}n_o^3\left[r_{41}F\xi_1 + (s_{12} - s_{11})F^2\xi_2\right] \\[2mm] \Delta\phi(y') &= \frac{\pi L}{\lambda}n_o^3\left[r_{41}F\xi_1 + (s_{12} - s_{11})F^2\xi_2\right] \end{aligned}} \qquad (5.76)$$

The phase charges produced by the electric field can be exploited for a number of important switching or modulation devices. These devices will be discussed in Chapter 12.

EXAMPLE 5.4 A bulk GaAs device is used as an electro-optic modulator. The device dimension is 1 mm and a phase change of 90° is obtained between light polarized along $< 01\bar{1} >$ and $< 011 >$. The wavelength of the light is 1.5 μm. Calculate the electric field needed if $\xi_1 = 1$.

The phase change produced is ($\xi = 1$)

$$\Delta\phi = \frac{2\pi}{\lambda} n_o^3 r_{41} F L = \frac{\pi}{4}$$

$$F = \frac{\lambda}{8 n_o^3 r_{41} L}$$

$$= \frac{(1.5 \times 10^{-6} \ m)}{8(3.3)^3 (1.2 \times 10^{-12} \ m/V)(10^{-3} \ m)}$$

$$= 4.35 \times 10^6 \ V/m$$

If the field is across a 10 μm thickness, the voltage needed is 4.35 V.

5.8 MODULATION OF EXCITONIC TRANSITIONS: QUANTUM CONFINED STARK EFFECT

$\rightsquigarrow$ We have seen that by confining the excitonic wavefunction in a quantum well, the binding energy greatly increases along with the oscillator strength. This has allowed excitonic transitions to persist up to high temperatures without merging into the band-to-band transitions. When static electric fields are applied to a 3D semiconductor, the energy term

$$eFz_e - eFz_h$$

must be added to the exciton Hamiltonian to solve the problem. The electric field ionizes the exciton state by pulling apart the electron hole pair. This causes a broadening of the exciton peak and also shifts the peak somewhat by the Stark effect. This effect is known as Franz-Keldysh effect and was first discovered theoretically in 1958. The shift of the absorption to lower energies can be understood as due to tunneling of an electron between the top of the valence band and bottom of the conduction band due to a photon. In fact, all energies are in principle possible for these transitions along the electric field since the energy gap between the valence and conduction band is a triangular potential well whose height is the energy gap and the width is $z = E_g/(eF)$ where F is the electric field. In presence of a photon of energy $\hbar\omega$, the barrier height becomes $(E_g - \hbar\omega)$ and the tunneling probability depends exponentially on this parameter. Thus, even when the photon energy approaches zero, there is a finite probability of such transitions.

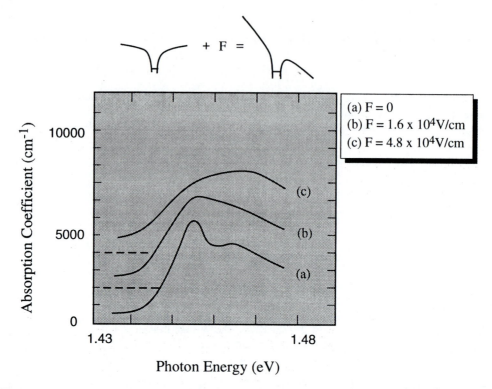

Figure 5.14: Absorption spectra at various electric fields for a parallel field . The insert shows schematically the distortion of the Coulomb potential of electron and applied field. The zeroes are displaced as shown by the dashed line clarity. (After D. S. Chemla, *Nonlinear Optics: Materials and Devices*, Springer-Verlag, New York (1986).)

We will focus our attention to the absorption spectra in quantum wells in presence of an electric field. The electric field can be either longitudinal (in the plane of the well) or transverse (in the growth direction). The longitudinal field problem is similar to the bulk problem and the excitonic transition essentially disappears at at fairly low field ($\leq$ 10 kV/cm) as shown in Fig. 5.14. The absorption edge shifts to lower energy as in the case of the bulk problem.

The transverse field problem is of great interest, since the exciton does not field ionize, since the electron and hole states are confined, because of the high barriers of the potential well. As a result, exciton transitions can persist up to electric fields of greater than 100 kV/cm. This effect is known as Quantum Confined Stark Effect (QCSE) and has been used to design new optoelectronic devices ranging from modulators to switches. The QCSE can be understood on the basis of the same formalism as the one discussed for the exciton and band to band transitions in absence of the electric field as long as one can assume that the quantum well subband levels are reasonably confined states. In principle, the quantum well states are quasi-bound states in presence of the field with

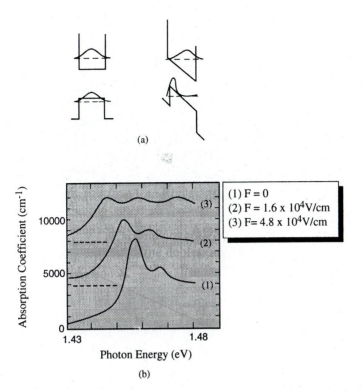

Figure 5.15: Absorption spectra at various electric fields for parallel fields. The insert shows schematically the distortion of the Coulomb potential of electron and applied field. The zeroes are displaced as shown by the dashed line clarity. (After D. S. Chemla, *Nonlinear Optics: Materials and Devices*, Springer-Verlag, New York (1986).)

the wavefunction primarily peaked in the quantum well region. Electrons and holes created in the form of a wavepacket in the well will eventually tunnel out of the well. In the addressing exciton problem, one assumes that the subband states are localized in the well and the exciton can be made up of only the confined states. There are several effects that occur in the presence of the transverse electric field:

1. The intersubband separations change. The field pushes the electron and hole functions to opposite sides making the ground state intersubband separation smaller. This effect is the dominant term in changing the exciton resonance energy.

2. Due to the separation of the electron and hole wavefunction, the binding energy of the exciton decreases. This effect is shown in Fig. 5.17.

In Figs. 5.16 and 5.17, we show the variation in the intersubband separation and exciton binding energies with applied fields. As can be seen from these figures, the change in the

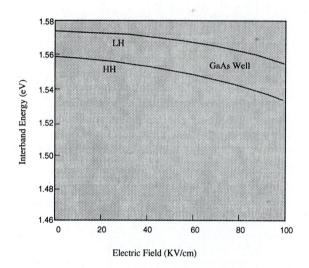

Figure 5.16: Variation of the ground state HH and LH (to conduction band ground state) intersubband transition energies as a function of electric field. The well is a 100 Å GaAs/Al$_{0.3}$Ga$_{0.7}$As.

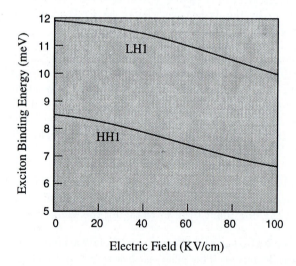

Figure 5.17: The variation in the exciton binding energy in a 100 Å GaAs/Al$_{0.3}$Ga$_{0.7}$As quantum well as a function of electric field. The HH and LH exciton results are shown.

exciton binding energy is only $\sim$2–3 meV while the intersubband energies are altered by up to 20 meV. The QCSE is, therefore, primarily determined by the intersubband effect.

While the exact calculation of the intersubband separation requires numerical techniques, one can estimate these changes by using perturbation theory. This approach gives reasonable results for low electric fields. The problem can be defined by the Hamiltonian

$$H = H_0 + eFz \tag{5.77}$$

where H_0 is the usual quantum well Hamiltonian. The eigenfunctions of H_0 are ψ_n and in the square quantum well, the ground state function ψ_1 has an even parity so that the first order correction to the subband energy is

$$\Delta E^{(1)} = \langle \psi_1 | eFz | \psi_1 \rangle \tag{5.78}$$

which is zero. Thus, one has to calculate the second order perturbation by using the usual approach. For the infinite barrier quantum well, the states ψ_n are known and it is, therefore, possible to calculate the second order correction. If the field is small enough such that

$$|eFW| \ll \frac{\hbar^2 \pi^2}{2m^* W^2} \tag{5.79}$$

i.e., the perturbation is small compared to the ground state energy then it can be shown that the ground state energy changes by

$$\Delta E_1^{(2)} = \frac{1}{24\pi^2} \left(\frac{15}{\pi^2} - 1 \right) \frac{m^* e^2 F^2 W^4}{\hbar^2} \tag{5.80}$$

One sees that the second order effect increases with m^* and has a strong well size dependence. This would suggest that for best modulation one should use a wide well. However, in wide wells the exciton absorption decreases and also the HH, LH separation becomes small. Optimum well sizes are of the order of $\sim$100 Å for modulators.

A most important consequence of QCSE is that an optical beam with a frequency near the exciton resonance can be modulated by applying an electric field across, say, a p-i(MQW)-n region. This is an obvious use of the QCSE. In fact, a number of devices can also be designed on this basis and will be discussed in Chapter 12. We note that the matrix element in the excitonic absorption obeys the same polarization selection rules as the ones we discussed in Chapter 4 for band to band transitions. In Fig. 5.18, we show the TE and TM absorption spectra calculated for a 100 Å GaAs/Al$_{0.3}$Ga$_{0.7}$As quantum well. As can be seen, in the TE mode, the LH exciton strength is approximately a third of the HH exciton strength. In the TM mode on the other hand, there is no HH transition allowed in accordance with our discussions in Chapter 4. Measured dependent spectra in similar structures is shown in Fig. 5.19.

In the context of polarization dependence of the absorption spectra, it is very interesting to see the effect of strain on the spectra. We noted in Chapter 2 that epitaxial

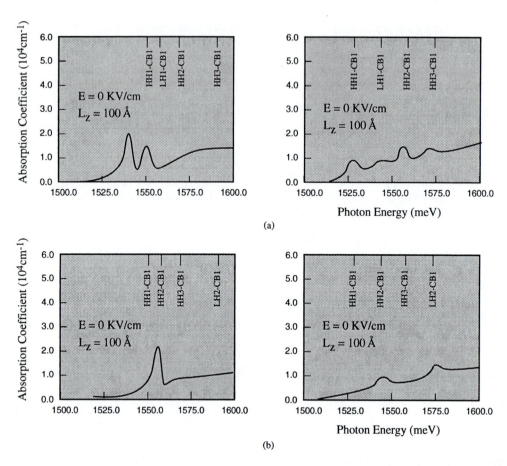

Figure 5.18: Calculated absorption coefficients for TE and TM modes in a 100 Å GaAs/AlGaAs quantum well structure at (a) $F = 0$ and (b) $F = 70$ kV/cm. (After S. Hong and J. Singh, *Superlattices and Microstructures*, 3, 645 (1987).)

strain can lift the degeneracy between the HH and LH states. In fact, a compressive in-plane strain pushes the LH below the HH state while a tensile strain does the opposite. Now the quantum confinement also pushes the LH state below the HH state. Thus, in principle a properly chosen tensile strain can cause a *merger* of the HH and LH states at the zone center in a quantum well. In Fig. 5.20, we show a schematic sketch of this possibility along with the valence bandstructure in a 150 Å GaAs/In$_{0.06}$Ga$_{0.57}$Al$_{0.37}$As well where we assume that the GaAs well and the barrier share the lattice mismatch to put the well under tensile strain. As can be seen, the LH and HH states are degenerate at the zone center with a very high hole density of states at the bandedge.

This HH, LH merger can produce a large enhancement of the absorption peak in quantum wells. The merger of HH and LH excitonic resonances by strain has been

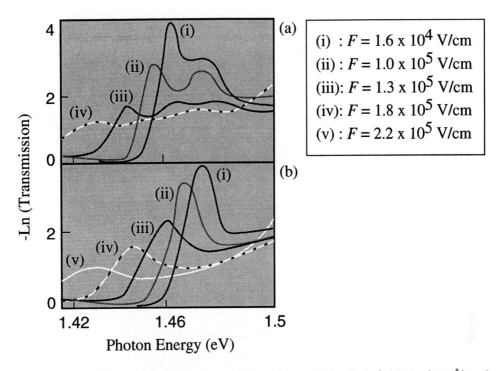

(a)

(i)	: $F = 1.6 \times 10^4$ V/cm
(ii)	: $F = 1.0 \times 10^5$ V/cm
(iii)	: $F = 1.3 \times 10^5$ V/cm
(iv)	: $F = 1.8 \times 10^5$ V/cm
(v)	: $F = 2.2 \times 10^5$ V/cm

Figure 5.19: Measured polarization dependent transmittances in GaAs/AlGaAs (100 Å) multiquantum well structures when light is coming in the waveguide geometry. (a) Incident polarization parallel to the plane of the layers . (b) Incident polarization perpendicular to the plane of the layers. (After D.A.B. Miller, et al., *IEEE J. Quantum Electronics*, QE-22, 1816 (1986).)

verified experimentally in several kinds of structures. In the GaAsP/AlGaAs system grown on GaAs, the tensile strain can be directly produced in the well, since GaAsP has a smaller lattice constant than GaAs.

EXAMPLE 5.5 A 100 Å GaAs/Al$_{0.3}$Ga$_{0.5}$As MQW structure has the HH exciton energy peak at 1.51 eV. A transverse bias of 80 kV/cm is applied to the MQW. Calculate the change in the transmitted beam intensity (there is no substrate absorption) if the total width of the wells is 1.0 μm and the exciton linewidth is $\sigma = 2.5$ meV. The photon energy is $\hbar\omega = 1.49$ eV.

The transmitted light is

$$I = I_o \, exp \, (-\alpha d)$$

At zero bias, we have (see Example 5.2)

$$\alpha(V = 0) = \frac{2.9 \times 10^4}{2.5} \, exp \left(\frac{-(1.49 - 1.51)^2}{1.44(2.5 \times 10^{-3})^2} \right)$$

$$\sim \quad 0$$

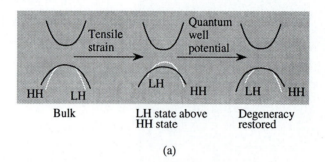

(a)

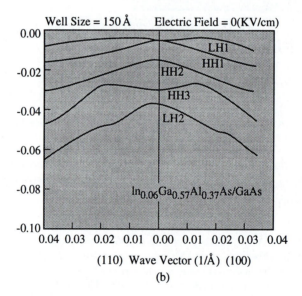

(b)

Figure 5.20: Biaxial tensile strain can be adjusted to produce a HH, LH degeneracy in a quantum well. The valence bandedge mass is extremely high in this case as shown by the dispersion relation.

At a bias of 80 kV/cm, the exciton peak shifts by ~20 meV, as can be seen from Fig. 5.17. The absorption coefficient is

$$\alpha(F = 80 kV/cm) = \frac{2.9 \times 10^4}{2.5} \exp\left(\frac{-(1.49 - 1.49)^2}{1.44(2.5 \times 10^{-3})^2}\right)$$
$$= 1.2 \times 10^4 \ cm^{-1}$$

The ratio of the transmitted intensity is

$$\frac{I(F = 80 \ kV/cm)}{I(F = 0)} = 0.3$$

Thus, an ON/OFF ratio is 3.3:1 for this modulator.

EXAMPLE 5.6 A small tensile strain is used to cause the HH and LH excitonic states to merge in a 100 Å GaAs/Al$_{0.3}$Ga$_{0.7}$As well. The quantum well is grown along the (001) direction. Calculate the absorption coefficient for light polarized along the z-direction and along the x (or y) direction. The exciton linewidth is $\sigma = 1.0$ meV.

When the HH and LH states are coincident, the total coupling of the z-polarized light gives, for the momentum matrix element, (see Eqn. 4.66)

$$\frac{2}{3} \ P_{cv}^2$$

and for the x-polarized light,

$$\frac{1}{2}P_{cv}^2 + \frac{1}{6}P_{cv}^2 = \frac{2}{3} \ P_{cv}^2$$

In Example 5.2, we calculated the excitonic transition strength using $a_p = \frac{1}{2}$ for x-polarized light, coupling only to the HH state. The absorption strength thus increases to 3.9 $\times 10^4$ cm^{-1} for the peak value.

5.9   EXCITON QUENCHING

⤳   In the previous section we discussed an extremely powerful approach used to modulate the excitonic transitions in quantum wells. There are several other approaches which can also modulate the exciton spectra and have been used to design various devices. Two important categories of these techniques are:

1. Quenching of the exciton by free carriers.

2. Quenching of the exciton by creation of a high density of excitons by high optical intensity.

In the first approach, a high density of free electrons (or holes) is introduced into the quantum well. The free carriers screen the Coulombic interaction between the electron and hole, weakening the exciton binding energy and reducing the exciton oscillator strength. The physics behind the second approach is quite complex and a number of important phenomenon including bandgap renormalization, exciton phase space filling, and screening effects participate to cause the modulation of the exciton resonance. Since we are not fully equipped to treat these phenomenon which require many body theory we will only summarize the current state of knowledge.

The screening of the exciton is relatively simple to understand and has been used to design high speed modulators. An example is the field effect transistor optical modulator (FETOM) in which the optical beam passes under the gate of a MODFET and has an energy equal to the exciton resonance energy in the undoped well of the device. The gate is used to inject free electrons into the channel where they screen the electron-hole interaction thus quenching the exciton transition .

In the simple theory of the effect of free carriers on Coulombic interaction, it can be shown that the Coulombic interaction is modified as

$$V(\boldsymbol{r}) = \frac{e^2}{\epsilon r} e^{-\lambda r} \tag{5.81}$$

where for a nondegenerate electron gas with doping density n,

$$\lambda^2 = \frac{ne^2}{\epsilon k_B T}$$

and $N(E_F)$ is the density of states at the Fermi energy E_F.

The screening comes about due to the dielectric response of the material especially due to the mobile carriers that can adjust their positions and reduce the effect of the Coulombic interaction. In a quantum well, one has to carry out a proper 2D treatment of the dielectric response, but the overall effect on the excitons can be understood on physical grounds.

As the background carrier density increases, the electron-hole attractive potential decreases due to screening and the exciton binding energy decreases. Another manifestation of this is that the exciton radius increases and the absorption coefficient decreases.

In Fig. 5.21, we show how the exciton binding energy changes as a function of carrier density for a 100 Å GaAs/Al$_{0.3}$Ga$_{0.7}$As quantum well structure. Also shown in Fig. 5.21 is the exciton radius as a function of carrier density for the same structure. We can see that the exciton essentially "disappears" once the background carrier density approaches 2–3 × 10^{11}cm^{-2}. This is a fairly low density and can be easily injected at high speeds into a quantum well. Since the injected charge removes the exciton peak, the

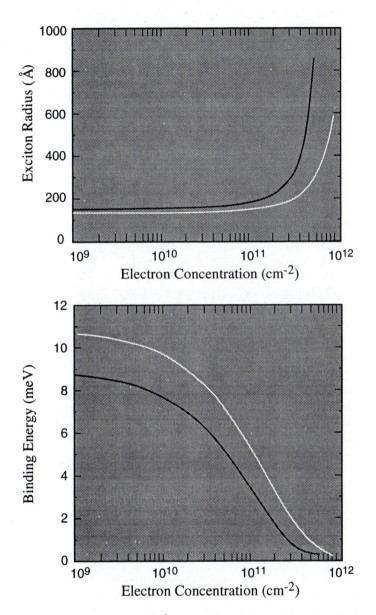

Figure 5.21: (a) Exciton radii a_{ex} in a 100 Å GaAs/Al$_{0.3}$Ga$_{0.7}$As quantum well verses electron concentration n_e at $T = 300$ K. The dark curve is for the CB1-HH1 exciton; the open curve is for the CB1-LH1 exciton. (b) Also shown is the exciton binding energy as a function of the carrier concentration. (After J. Loehr and J. Singh, *Physical Review, B 42*, 7154 (1990).)

absorption coefficient can be modulated rapidly by this process. We note that in addition
to the effect noted above, the injected carriers also renormalize the bandgap by shrinking
the gap somewhat. At the carrier concentrations of $\sim 10^{11} \text{cm}^{-2}$, the screening effects
are, however, more dominant. This phenomenon has been used in creating electronic
devices which can modulate an optical signal by charge injection.

Another area of intense research in exciton physics is the area of nonlinear
effects. At low optical intensity, the optical constants of the material are essentially
independent of the optical intensity. However, as the optical intensity is increased, excess
carriers are generated and these carriers affect the optical constants causing them to be
highly dependent upon the intensity. A number of devices have been proposed on the
basis of such effects. In this text we will not attempt to describe the physics behind the
optical nonlinearities. We will simply review a few cases.

When the optical intensity generates electron-hole pairs, the optical absorption
is affected by phase space filling. In the case of the free electron-hole pairs, this effect
is simply due to the band filling effects manifested in the product term $f^e(E) \cdot f^h(E)$
which we discussed when discussing the gain in a laser. While the free electron-hole
pairs, which are Fermions, clearly obey the exclusion principle, the situation for the
excitonic states, which are bound states, is not as clear. These composite particles are
essentially bosons when their concentration is small, but obey Fermi statistics when
their concentration is high. Thus, phase space filling effects occur in the case of excitons
once the exciton density starts approaching $\sim 10^{12} \text{cm}^{-2}$.

In addition to the phase space filling effects, the free electron-hole pairs, as
well as excitons, cause the screening of the Coulombic interaction as discussed already
and thus reduce the absorption coefficient. As discussed earlier, this causes the exciton
wavefunction to increase in real space causing a decrease in the oscillator strength.

5.10   REFRACTIVE INDEX MODULATION DUE TO EXCITON MODULATION

$\rightsquigarrow$ So far in this chapter we considered how the exciton transitions affect the absorption coefficient. We also addressed the issue of how the exciton absorption can be
modulated. While this provides an excellent means to modulate an optical signal with
energy at the exciton resonance, the signal is modulated by absorption. In many applications, the absorption of the optical signal is not desirable. Particular examples are
the "directional coupler" and the Mach-Zender interferometer shown in Fig. 5.22, which
will be discussed in Chapter 12. In the directional coupler, for example, an optical beam
is fed into a waveguide and is brought close to another waveguide which is close enough
that the evanescent wave can couple with the other guide. The picture is very similar to

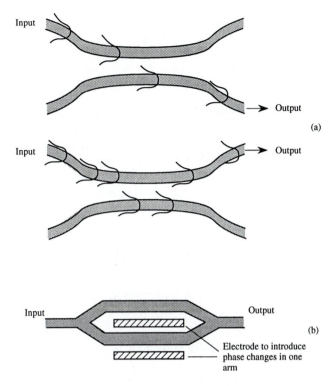

Figure 5.22: A schematic of optical devices which rely upon changes produced in refractive index and phase for switching optical signals. (a) A directional coupler; (b) the Mach-Zender interferometer.

two coupled pendulums. The optical energy sloshes back and forth between the waveguides and can come out from the other end via either of the two guides depending upon the coupling coefficient, the distance traveled, and the variation in the refractive index of the two guides.

In bulk semiconductors, the electro-optic effect described by the electric fields required to alter the relative phase enough to cause switching are extremely high, making them unable to compete with materials like Lithium Niobate for waveguide switching. However, in quantum well structures, since the exciton peak is so strong and can be modulated easily, large changes in the refractive index can be produced. Additionally, since there are strong selection rules for TE and TM modes in quantum wells and, in particular, for strained quantum wells, one can exploit these for altering the polarization of the transmitted light. In quantum well structures, the electro-optic effect is essentially controlled by the Stark effect shift of the excitonic and band-to-band transitions. The effect has been studied by a number of workers and utilized for devices. To determine the effect of excitonic and band-to-band absorption on the refractive index we use the

Kramers-Kronig transform yielding the result

$$n(\omega_0) - 1 = \frac{c}{\pi} P \int_0^\infty \frac{\alpha(\omega)\, d\omega}{\omega^2 - \omega_0^2} \tag{5.82}$$

It has been observed experimentally that the total area under the absorption curve (for energies slightly above the bandgap) remains constant with the application of electric field. This reflects the sum rules that hold for excitonic and band-to-band absorption: as the electric field breaks the reflection symmetry of the potential, the parity-allowed transitions have their absorption strengths decreased and the parity-forbidden transitions start increasing in strength causing the total area to remain approximately constant.

The effect of the electric field can be studied assuming that the excitonic broadening is a Lorentzian function with half-width Γ, centered at ω'

$$\Delta(\omega - \omega') = \frac{\Gamma}{\pi \left[(\omega - \omega')^2 + \Gamma^2 \right]} \tag{5.83}$$

The Kramers-Kronig transform is then carried out by using the result

$$P \int_0^\infty d\omega\, \frac{\Delta(\omega - \omega')}{\omega^2 - \omega_0^2} = \frac{-\left[\omega_0^2 - \omega'^2 + \Gamma^2 \right]}{\left[\omega_0^2 + \omega'^2 + \Gamma^2 \right]^2 - 4\omega_0^2 \omega'^2} \tag{5.84}$$

obtained by contour integration. The change in the refractive index can now be calculated from the shift of the exciton peak. The effect is quadratic for square quantum wells, since QCSE is a quadratic effect for such wells. A number of groups have examined the electro-optic effect including the effect of strain. One can see a strong quadratic effect corresponding to the quadratic Stark effect. The inclusion of strain separates the LH state making the effect stronger.

It is important to note that while the exciton linewidth is extremely important in electro-absorption effects where the exciton resonance is used to create *e-h* pairs, it is not as important in electro-optic effects where the exciton resonance is at a higher energy than the energy of the photon.

5.11   EXCITONIC EFFECTS IN SUB 2D SYSTEMS

$\mathcal{R}$ In this chapter we have seen that as one goes from a 3D system to a 2D system, the excitonic features become more prominent and show remarkable modulation characteristics. The question naturally arises: Would this trend of imporved properties persist, as one goes towards 1D and 2D systems? A number of theoretical studies have shown that, indeed, the excitonic binding energy becomes larger in quantum wires or dots. Also, the electro-optic effects are expected to greatly improve.

It must be pointed out, however, that the fabrication technology for quantum wires and dots is rather immature at present. Thus, structural imperfections are present that tend to mar the optical effects possible. Thus, while a number of groups have observed excitonic spectra in quantum wires and dots, the expected superior excitonic properties have not been demonstrated. However, as technology matures, improved effects are expected.

An area which has recently drawn some attention is the microcavity structure in which the electron states are confined by a quantum well and the photon states are also confined by an optical cavity. In this case, there is a strong photon-exciton coupling which promises very strong non-linear optical effects.

5.12 SUMMARY

In this chapter we have examined some of the physics in the excitonic phenomenon in bulk and quantum well structures. The enhanced excitonic binding energy and the modulation techniques that can be applied in quantum well structures have made this area of research very exciting. The physics of excitons in low-dimensional systems is by no means exhausted. In fact, exciton physics in quantum wires and quantum dots is at a very nascent stage at present. In these structures, the exciton effects are expected to be even stronger and more versatile than for the quantum well structures. In Tables 5.2-5.4, we present the key findings of this chapter.

5.13 PROBLEMS

5.1 Assume a simple parabolic density of states mass for the electrons and holes and calculate the exciton binding energies in GaAs, $In_{0.53}Ga_{0.47}As$ and InAs. What is the exciton Bohr radius in each case?

5.2 Assume a Gaussian exciton linewidth of 1 meV half-width at half maximum and plot the absorption coefficient due to the ground state exciton resonance in each of the three semiconductors of problem 1.

5.3 Assume a hydrogenic form of the ground state exciton function. Using the variational method, show that a 2D exciton has four times the binding energy of the 3D exciton. Also calculate the in-plane Bohr radius of the 2D exciton.

5.4 Assume that the exciton radius a_{ex} scales roughly as inverse of the exciton binding

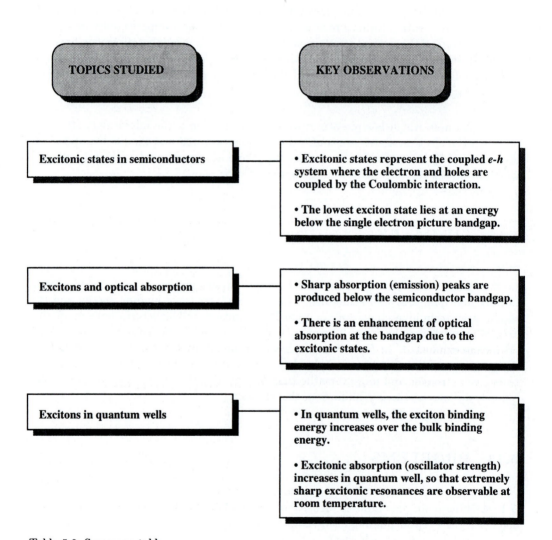

Table 5.2: Summary table.

TOPICS STUDIED	KEY OBSERVATIONS
Linewidth of excitonic resonances	• Excitonic linewidth is controlled by homogeneous and inhomogeneous broadening. • Homogeneous linewidth is due to exciton-phonon scattering effects. • Inhomogeneous effects are due to structural imperfections in the system.
The electro-optic effect	• The application of an electric field can alter the refractive index of a semiconductor. This causes light propagating in different directions to develop a relative phase shift which can be controlled by the field. • In bulk semiconductors, the electro-optic effect is linear with the field. • In quantum wells, the electro-optic effect has a strong quadratic component, especially near the exciton resonance.

Table 5.3: Summary table.

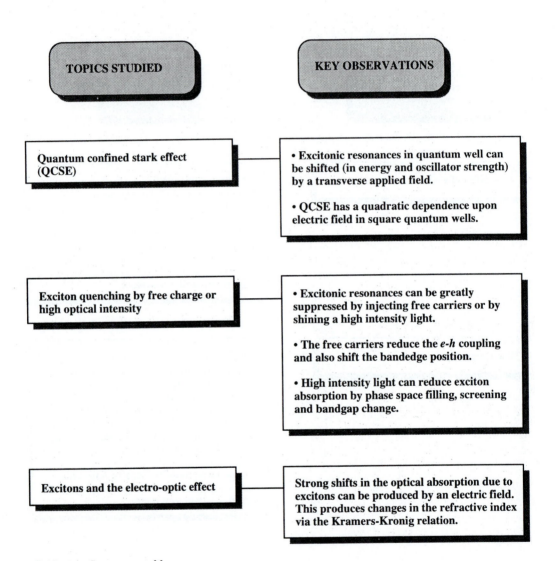

Table 5.4: Summary table.

energy. Using the in-plane dependence of the exciton envelope function as $\exp(-\rho/a_{ex})$, use the results of Fig. 5.10 to plot the well size dependence of the exciton peak absorption strength of the heavy hole ground state exciton. Use a 4 meV (HWHM) Gaussian width. In general, the linewidth may increase with decreasing well size.

5.5 Using the results of the simple perturbation theory (Eqn.5.80) calculate the electric field dependence of the HH exciton emission energy in a 100 Å GaAs quantum well. How does the position of the LH exciton (ground state) vary?

5.6 Design a GaAs quantum well (you may assume an infinite barrier) so that the ground state exciton resonance is at 1.5 eV. Calculate the peak absorption coefficient at this energy. If the quantum well width decreases by one monolayer, what is the absorption coefficient at 1.5 eV? Assume that:

exciton radius	$a_{ex} = 80$ Å
electron mass	$m_e^* = 0.067\ m_0$
hole mass	$m_h^* = 0.45\ m_0$
exciton halfwidth	$\sigma = 2.0\ meV$
GaAs bandgap	$E_g = 1.43\ eV$
1 monolayer	$d_{ml} = 2.86$ Å

Also assume that the exciton binding energy is 8.0 meV.

5.7 Consider a bulk GaAs electro-optic device on which an electric field is applied in the z-direction. The field is switched between 0 and 10^5 V/cm. Calculate the length of the device needed to produce a phase change of $\pi/2$ between the light waves polarized along $< 01\bar{1} >$ and $< 0\bar{1}1 >$. The wavelength of the light is 1.3 μm.

5.8 Consider a 100 Å GaAs quantum well in which the heavy-hole exciton resonance has a halfwidth of $\sigma = 3$ meV. A transverse electric field is switched between 0 and 60 kV/cm. Calculate the change in the absorption coefficient at a photon energy coincident with the peak of the exciton at zero field. If the total length of the active region is 2.0 μm, calculate the modulation depth that can be achieved (modulation depth $= I_{ph}(F = 60 \text{ kV/cm})/I_{ph}(F = 0)$. You may ignore the effect of the light hole exciton.

5.9 Consider a 100 Å In$_{0.53}$Ga$_{0.47}$As quantum well device in which the exciton halfwidth is $\sigma = 5$ meV. An optical beam impinges on the device with an energy 10 meV below the zero field exciton resonance. The active length of the device is 3.0 μm. Calculate the minimum electric field needed to produce an optical intensity modulation of 2:1 relative to the zero field transmission intensity. Exciton radius may be taken as 120 Å in the well.

5.10 Assume that the absorption coefficient due to excitonic transitions is given adequately by the exciton radius. Calculate the change in the peak of the absorption coefficient as a function of free carriers in a 100 Å quantum well using the information provided in this chapter. The exciton halfwidth is $\sigma = 3$ meV.

5.11 In a normal incidence GaAs/Al GaAs quantum well modulator based on Stark effect, the substrate (GaAs) has to be removed. Discuss the reasons for this. Calculate the minimum composition of In in a 100 Å InGaAs/AlGaAs quantum well so that the substrate removal is not necessary.

5.12 Consider a 100 Å In$_{0.2}$Ga$_{0.8}$As quantum well (grown lattice matched on a GaAs substrate) in which the ground state HH exciton at zero applied field is at 1.4 eV.

Estimate the position of the ground LH exciton. Note that the strain induced splitting between HH and LH is $6\,\epsilon$ eV where ϵ is the lattice mismatch.

5.13 In problem 5.12, a transverse electric field of 10^5 V/cm is applied. Estimate the change in the refractive index for the TE and TM polarized light at $\hbar\omega = 1.3$ eV. The exciton halfwidth is $\sigma = 3$ meV and the exciton radius can be assumed to be 100 Å for both the HH and LH excitons.

5.14 A 100 Å GaAs quantum well has a heavy-hole exciton resonance at 1.5 eV with a halfwidth of 3 meV. Calculate the change in the refractive index produced by a transverse electric field of 10^5 V/cm at a photon energy of 1.4 eV. Use the Kramers Kronig relation and assume that the change in solely due to the change in the excitonic effects. Consider both HH and LH excitons and assume that $a_{ex} = 100$ Å for both cases. Consider the TE and TM polarized light separately.

5.15 In non-linear exciton "bleaching," one needs approximately 5×10^{11} cm^{-2} *e-h* pairs to cause the bleaching of the exciton. If the *e-h* recombination time is 1.0 ns, what is the optical power density needed to cause bleaching in 100 Å quantum wells? The absorption coefficient is 10^4 cm^{-1} for the impinging optical beam with energy 1.5 eV.

5.14 REFERENCES

- **General**

  - F. Bassani and G. P. Parravicini, *Electronic States and Optical Transitions in Solids*, Pergamon Press, New York (1975).

  - J. O. Dimmock, in *Semiconductors and Semimetals*, eds. R. K. Willardson and A. C. Beer, Academic Press, New York, vol. 3 (1967).

  - R. J. Elliott, in *Polarons and Excitons*, eds. C. G. Kuper and G. D. Whitfield, Plenum Press, Englewood Cliffs (1963).

  - "Theory of Excitons," by R. S. Knox, *Solid State Physics*, Academic Press, New York, vol. Suppl. 5 (1963).

  - Y. Onodera and Y. Toyozawa, *J. Phys. Soc. Japan*, **22**, 833 (1967).

  - D. C. Reynolds, C. W. Litton, E. B. Smith, and K. K. Bajaj, *Sol. St. Comm.*, **44**, 47 (1982).

  - M. D. Sturge, *Phys. Rev.*, **127**, 768 (1962).

- **Excitons in Quantum Wells**

  - G. Bastard, E. E. Mendez, L. L. Chang, and L. Esaki, *Phys. Rev. B*, **26**, 1974 (1982).

  - "Nonlinear Interactions and Excitonic Effects in Semiconductor Quantum Wells," by D. S. Chemla, in *Nonlinear Optics: Materials and Devices*, eds. C. Flytzanis and J. L. Oudar, Springer-Verlag (1986).

  - R. L. Green and K. K. Bajaj, *J. Vac. Sci. Technol.*, **B1**, 391 (1983).

- J. P. Loehr and J. Singh, *Phys. Rev. B*, **42**, 7154 (1990).

- R. C. Miller, D. A. Kleinman, W. A. Nordland, and A. C. Gossard, *Phys. Rev. B*, **22**, 863 (1980).

- G. D. Sanders and Y. C. Chang, *Phys. Rev. B*, **31**, 6892 (1985).

● **Exciton Broadening Effects in Quantum Wells**

- J. Lee, E. S. Koteles, and M. O. Vassell, *Phys. Rev. B*, **32**, 5512 (1986).

- I. M. Lifshitz, *Adv. Phys.*, **13**, 483 (1969).

- J. Singh and K. K. Bajaj, *Appl. Phys. Lett.*, **48**, 1077 (1986).

- C. Weisbuch, R. Dingle, A. C. Gossard, and W. Wiegmann, *J. Vac. Sci. Technol.*, **17**, 1128 (1980).

● **Quantum Confined Stark Effect**

- J. D. Dow and D. Redfield, *Phys. Rev. B*, **1**, 3358 (1970).

- E. E. Mendez, G. Bastard, L. L. Chang, and L. Esaki, *Phys. Rev. B*, **26**, 7101 (1982).

- D. A. B. Miller, D. S. Chemla, T. C. Damen, A. C. Gossard, W.Wiegmann, T. H. Wood, and C. A. Burrus, *Phys. Rev. Lett.*, **53**, 2173 (1984).

- D. A. B. Miller, J. S. Weiner, and D. S. Chemla, *IEEE J. Quant. Electron.*, **QE-22**, 1816 (1986).

● **Exciton Quenching**

- C. Flytzanis and J. L. Oudar, editors, *Nonlinear Optics: Materials and Devices*, Springer-Verlag, New York (1985).

- H. Haug and S. Schmitt-Rink, *Prog. Quant. Electron.*, **9**, 3 (1984).

- W. H. Knox, R. F. Fork, M. C. Downer, D. A. B. Miller, D. S. Chemla, and C. V. Shank, in *Ultrafast Phenomenon IV*, eds. D. H. Anston and K. B. Eisenthal, Springer-Verlag (1984).

● **Electro-Optic Effect**

- J. Pamulapati, J. P. Loehr, J. Singh, and P. K. Bhattacharya, *J. Appl. Phys.*, **69**, 4071 (1991).

- J. S. Weiner, D. A. B. Miller, D. S. Chemla, T, C. Damen, C.A.Burrus, T. H. Wood, A. C. Gossard, and W. Wiegmann, *Appl. Phys. Lett.*, **47**, 1148 (1985).

- J. E. Zucker, T. L. Hendrickson, and C. A. Burrus, *Appl. Phys. Lett.*, **52**, 945 (1988).

CHAPTER
6

SEMICONDUCTOR
JUNCTION THEORY

6.1   INTRODUCTION

In previous chapters of this book we have examined the basic physical phenomena that control the electronic and optoelectronic properties of semiconductors. In the remainder of this book we will see how these special properties can be exploited to produce optoelectronic devices. Before going into the details of how different devices operate, it is important to pause and examine what devices are supposed to do in the arena of information processing.

As noted in the Introduction to this book, we live in the information age where the generation and manipulation of information is critical. Semiconductor devices play a key role in this generation and manipulation process. A pure semiconductor is almost never used as a device by itself. Semiconductor devices are produced when dopants, metals and insulators are incorporated in a precise manner on a semiconductor. *Junctions between differently doped semiconductors, between metals and semiconductors and between different semiconductors provide the building blocks of devices.*

In this chapter we will discuss the properties of several kinds of junctions. A most important junction is the *p-n* junction in which the nature of dopants is altered across a boundary to create a region that is *p*-type next to a region which is *n*-type. This junction has rectifying properties and can be used to produce strong nonlinear effects. Most optoelectronic devices are based on a *p-n* diode structure.

Another important class of junctions involves semiconductors and metals. Certain kinds of metal-semiconductor junctions called Schottky barriers can be used to provide a highly rectifying current-voltage relation that can be exploited for many device applications. Another class of junctions involves an insulator or a large gap semiconductor. In the case of Si-technology the insulator junction involves oxides or nitrides of Si. For GaAs technology one often uses AlGaAs.

The junctions we are about to discuss occasionally form important devices by themselves. This is especially true of optoelectronic devices where the *p-n* diode forms the basis of detectors, light emitting diodes, lasers and a variety of modulators. The topics discussed in this chapter have, most likely, been covered in detail by students using this book for an optoelectronics course. This chapter may then, in that case, be used simply for review.

6.2   THE UNBIASED P-N JUNCTION

$\mathcal{R}$ As noted in the previous section, the *p-n* junction is one of the most important junctions in solid-state electronics. The fabrication techniques used to form *p-* and *n*-type regions involve i) epitaxial procedures where the dopant species are simply switched at a particular instant in time; ii) ion-implantation in which the dopant ions are implanted at high energies into the semiconductor. The junction is obviously not as abrupt as in the case of epitaxial techniques; iii) diffusion of dopants into an oppositely doped semiconductor.

We will assume in our analysis that the *p-n* junction is abrupt, even though this is really only true for epitaxially grown junctions. Let us first discuss the properties of the junction in the absence of any external bias where there is no current flowing in the diode.

What happens when the *p-* and *n*-type materials are made to form a junction and there is no externally applied field? We know that in absence of any applied bias, there is no current in the system and the Fermi level is uniform throughout the structure.

This gives the schematic view of the junction shown in Fig. 6.1a. Three regions can be identified:

i) The *p*-type region at the far left where the material is neutral and the bands are flat. The density of acceptors exactly balances the density of holes;

ii) The *n*-type region in the far right where again the material is neutral and the density of immobile donors exactly balances the free electron density;

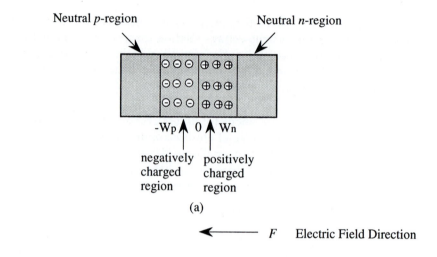

(a)

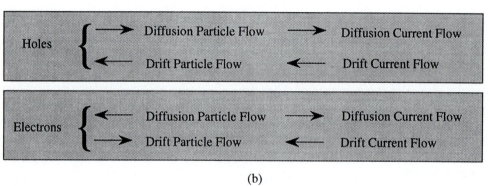

(b)

Figure 6.1: (a) An idealized model of the *p-n* junction without bias showing the neutral and the depletion areas. (b) A schematic showing various current and particle flow components in the *p-n* diode at equilibrium. For electrons, the current flow is in the direction opposite to that of the particle flow. Electrons that enter the depletion region from the *p*-side and holes that enter the depletion region from the *n*-side are swept away and are the source of the drift components.

iii) The depletion region where the bands are bent and a field exists which has swept out the mobile carriers leaving behind negatively charged acceptors in the *p*-region and positively charged donors in the *n*-region as shown in Fig. 6.1a.

In the depletion region, which extends a distance W_p in the *p*-region and a distance W_n in the *n*-region, an electric field exists. Any electrons or holes in the depletion region are swept away by this field. Thus a drift current exists which counterbalances the diffusion current which arises because of the difference in electron and hole densities across the junction.

In order to understand the diode properties, let us first identify all the current components flowing in the device. There is the electron drift current and electron diffusion current as well as the hole drift and hole diffusion current as shown in Fig. 6.1b. When there is no applied bias, these currents cancel each other *individually*. Let us consider these current components. The hole current density is

$$J_p(x) = e \left[\underbrace{\mu_p p(x) F(x)}_{\text{drift}} - \underbrace{D_p \frac{dp(x)}{dx}}_{\text{diffusion}} \right] = 0 \tag{6.1}$$

The ratio of μ_p and D_p is given by the Einstein relation

$$\frac{\mu_p}{D_p} = \frac{e}{k_B T} \tag{6.2}$$

We have for zero net current

$$\frac{\mu_p}{D_p} F(x) = \frac{1}{p(x)} \frac{dp(x)}{dx} \tag{6.3}$$

or in terms of the potential gradient $(F = -dV/dx)$, using the Einstein relation

$$\frac{-e}{k_B T} \frac{dV(x)}{dx} = \frac{1}{p(x)} \frac{dp(x)}{dx} \tag{6.4}$$

Let V_p and V_n denote the potentials in the neutral p-side and the neutral n-side, respectively, as shown in Fig. 6.2. Integrating from the left hand side to the right hand side of the p-n structure we get

$$\frac{-e}{k_B T} \int_{V_p}^{V_n} dV = \int_{p_p}^{p_n} \frac{dp}{p} \tag{6.5}$$

where p_p and p_n are the hole densities in the p-type and n-type neutral regions. We get, after integrating,

$$\frac{-e}{k_B T}(V_n - V_p) = \ell n \frac{p_n}{p_p} \tag{6.6}$$

Thus, the contact potential, or built-in potential, $V_{bi} = V_n - V_p$ is

$$\boxed{V_{bi} = \frac{k_B T}{e} \ell n \frac{p_p}{p_n}} \tag{6.7}$$

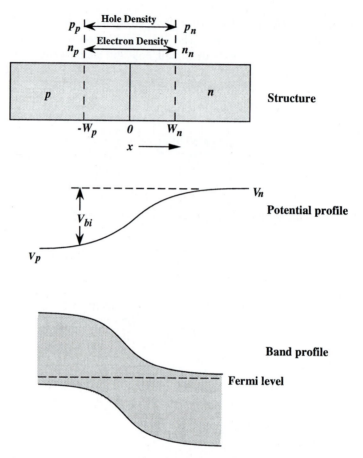

Figure 6.2: A schematic showing the p-n diode and the potential and band profiles. The voltage V_{bi} is the built-in potential at equilibrium. The expressions derived in the text can be extended to the cases where an external potential is added to V_{bi}.

If we had started with the electron drift and diffusion currents we would have obtained, by the same arguments,

$$V_{bi} = \frac{k_B T}{e}\ \ell n\ \frac{n_n}{n_p} \tag{6.8}$$

where n_n and n_p are the electron densities in the n-type and p-type regions. Remember that the law of mass action tells us that

$$n_n p_n = n_p p_p = n_i^2 \tag{6.9}$$

We can thus write the following equivalent expressions

$$\frac{p_p}{p_n} = e^{eV_{bi}/k_BT} = \frac{n_n}{n_p} \tag{6.10}$$

In this relation V_{bi} is the built-in voltage in the absence of any external bias. Under the approximations discussed later, a similar relation holds when an external bias V_{ext} is applied to alter V_{bi} to $V_{bi} + V_{ext}$, and will be used when we calculate the effect of external potentials on the current flow.

Let us now calculate the width of the depletion region for the diode under no applied bias. The calculation in the presence of a bias V_{ext} will follow the same approach and V_{bi} will simply be replaced by $V_{bi} + V_{ext}$. We note that in the depletion region, the mobile carrier density of electrons and holes is very small compared to the fixed background charge. This leads to the depletion approximation discussed earlier. With this approximation, there is a region of negative charge (due to acceptors) extending from the junction to the point W_p on the p-side and a region of positive charge (due to donors) extending the junction to W_n. The total negative and positive charge has the same magnitude, so that we have the equality

$$A\, W_p N_a = A\, W_n N_d \tag{6.11}$$

where A is the cross-section of the p-n structure, and N_a and N_d are the uniform doping densities for the acceptors and donors.

The Poisson equations for the structure are

$$\frac{d^2V(x)}{dx^2} = 0 \qquad -\infty < x < -W_p \tag{6.12}$$

$$\frac{d^2V(x)}{dx^2} = \frac{eN_a}{\epsilon} \qquad -W_p < x < 0 \tag{6.13}$$

$$\frac{d^2V(x)}{dx^2} = -\frac{eN_d}{\epsilon} \qquad 0 < x < W_n \tag{6.14}$$

$$\frac{d^2V(x)}{dx^2} = 0 \qquad W_n < x < \infty \tag{6.15}$$

The potential in the neutral p-side is

$$V(x) = V_p \qquad -\infty < x < -W_p \tag{6.16}$$

where V_p is defined as the potential in the neutral p-type region.

The electric field in the p-side of the depletion region is from the solution of the Poisson equation

$$F(x) = -\frac{eN_a x}{\epsilon} - \frac{eN_a W_p}{\epsilon} \qquad -W_p < x < 0 \qquad (6.17)$$

The electric field reaches a peak value at $x = 0$. The potential is given by integrating Eqn. 6.17,

$$V(x) = \frac{eN_a x^2}{2\epsilon} + \frac{eN_a W_p x}{\epsilon} + \frac{eN_a W_p^2}{2\epsilon} + V_p \qquad -W_p < x < 0 \qquad (6.18)$$

Carrying out a similar analysis for the n-side of the depletion region and n-side of the neutral region we get

$$\begin{aligned} V(x) &= V_n & W_n < x < \infty \\ F(x) &= 0 \end{aligned} \qquad (6.19)$$

where V_n is the potential at the neutral n-side.

$$F(x) = \frac{eN_d x}{\epsilon} - \frac{eN_d W_n}{\epsilon} \qquad 0 < x < W_n \qquad (6.20)$$

$$V(x) = -\frac{eN_d x^2}{2\epsilon} + \frac{eN_d W_n x}{\epsilon} - \frac{eN_d W_n^2}{2\epsilon} + V_n \qquad 0 < x < W_n \qquad (6.21)$$

From this discussion, the potential difference between points $-W_p$ and 0 is

$$V(0) - V(-W_p) = \frac{eN_a W_p^2}{2\epsilon} \qquad (6.22)$$

Similarly the potential difference between the points W_n and 0 is

$$V(W_n) - V(0) = \frac{eN_d W_n^2}{2\epsilon} \qquad (6.23)$$

Thus the built-in potential is

$$V(W_n) - V(-W_p) = V_{bi} = \frac{eN_d W_n^2}{2\epsilon} + \frac{eN_a W_p^2}{2\epsilon} \qquad (6.24)$$

Recalling that the charge neutrality gives us

$$N_d W_n = N_a W_p \qquad (6.25)$$

we get the following set of relationships:

$$W_p(V_{bi}) = \left\{ \frac{2\epsilon V_{bi}}{e} \left[\frac{N_d}{N_a(N_a + N_d)} \right] \right\}^{1/2} \tag{6.26}$$

$$W_n(V_{bi}) = \left\{ \frac{2\epsilon V_{bi}}{e} \left[\frac{N_a}{N_d(N_a + N_d)} \right] \right\}^{1/2} \tag{6.27}$$

$$W(V_{bi}) = W_p(V_{bi}) + W_n(V_{bi}) = \left(W_n^2(V_{bi}) + W_p^2(V_{bi}) + 2W_n(V_{bi})W_p(V_{bi}) \right)^{1/2}$$

$$W(V_{bi}) = \left[\frac{2\epsilon V_{bi}}{e} \left(\frac{N_a + N_d}{N_a N_d} \right) \right]^{1/2} \tag{6.28}$$

From these discussions, we can draw the following important conclusions about the diode:

i) The electric field in the depletion region peaks at the junction and decreases linearly towards the depletion region edges.

ii) The potential drop in the depletion region has a quadratic form.

We remind ourselves that this procedure can be extended to find the electric fields, potential and depletion widths for arbitrary values of V_p and V_n under quasi-equilibrium approximations. Thus we can directly use these equations when the diode is under external bias V, by simply replacing V_{bi} by $V_{bi} + V$. The applied bias can increase the total potential or decrease it as will be discussed later.

The results of the calculations carried out above are schematically shown in Fig. 6.3. Shown are the charge density and the electric field profiles. Notice that the electric field is nonuniform in the depletion region, peaking at the junction with a peak value (the sign of the field simply reflects the fact that in our study the field is pointing towards the negative x-axis)

$$F_m = -\frac{eN_d W_n}{\epsilon} = -\frac{eN_a W_p}{\epsilon} \tag{6.29}$$

Notice that the depletion in the p- and n-sides can be quite different. If $N_a \gg N_d$, the depletion width W_p is much smaller than W_n. Thus a very strong field exists over a very narrow region in the heavily doped side of the junction. *In such abrupt junction (p^+n or n^+p) the depletion region exists primarily on the lightly doped side.*

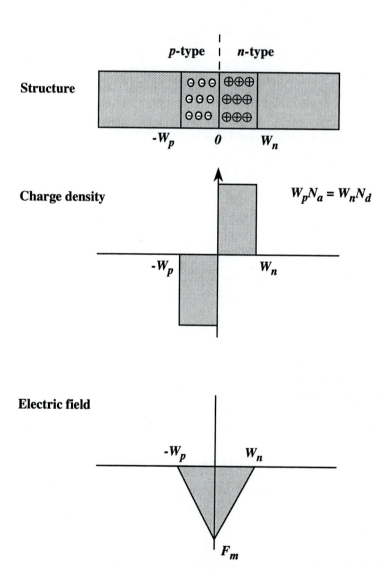

Figure 6.3: The p-n structure, with the charge and the electric field profile in the depletion region. Note that in the depletion approximation there is no charge or electric field outside the depletion region. The electric field peaks at the junction as shown.

In the depletion approximation we have assumed that the mobile carrier density is zero. However, this is not strictly true. Since we know the potential profile we can evaluate the mobile carrier density and the drift and diffusion currents. We will do this in the general case where an external bias is applied.

EXAMPLE 6.1 A silicon diode is fabricated by starting with an n-type ($N_d = 10^{16}$ cm^{-3}) substrate, into which indium is diffused to form as a p-type region doped at 10^{18} cm^{-3}. Assuming that an abrupt $p^+ - n$ junction is formed by the diffusion process, i) calculate the Fermi level positions in the p- and n-regions; ii) determine the contact potential in the diode; iii) calculate the depletion widths on the p- and n-side.

To find the positions of the Fermi levels, we can use either of the following equivalent relations. We assume complete ionization of the dopants. For the n-side we have

$$n_n = n_i \, \exp\left[\frac{-(E_{Fi} - E_{Fn})}{k_B T}\right]$$

or

$$n_n = N_c \, \exp\left[\frac{-(E_c - E_{Fn})}{k_B T}\right]$$

For the p-side we have

$$p_p = n_i \, \exp\left[\frac{(E_{Fi} - E_{Fp})}{k_B T}\right]$$

or

$$p_p = N_v \, \exp\left[\frac{(E_v - E_{Fp})}{k_B T}\right]$$

where $n_n = N_d; p_p = N_a$ and n_i, N_c, N_v are the intrinsic carrier concentration, conduction band effective density of states, and valence band effective density of states. Using the effective density of states relations, we have ($N_c = 2.8 \times 10^{19}$ cm^{-3}; $N_v = 1 \times 10^{19}$ cm^{-3} at 300 K)

$$
\begin{aligned}
E_{Fn} &= E_c + k_B T \, \ell n \, \frac{n_n}{N_c} \\
&= E_c - (0.026 \, eV) \times 7.937 \\
&= E_c - 0.206 \, eV \\
E_{Fp} &= E_v - k_B T \, \ell n \, \frac{p_p}{N_v} \\
&= E_v + (0.026 \, eV) \times 2.3 \\
&= E_v + 0.06 \, eV
\end{aligned}
$$

The built-in potential is given by

$$
\begin{aligned}
eV_{bi} &= E_g - 0.06 - 0.206 \, eV \\
&= 1.1 - 0.06 - 0.206 \\
&= 0.834 \, eV
\end{aligned}
$$

The reader should verify that the results obtained are equivalent to those obtained if we use the relations involving intrinsic carrier concentrations and Eqn. 6.8 for V_{bi}.

The depletion width on the p-side is given by

$$
\begin{aligned}
W_p(V_{bi}) &= \left\{ \frac{2\epsilon V_{bi}}{e} \left[\frac{N_d}{N_a(N_a + N_d)} \right] \right\}^{1/2} \\
&\cong \left\{ \frac{2 \times (11.9 \times 8.84 \times 10^{-12}\, F/m) \times 0.834(V)}{(1.6 \times 10^{-19} C)} \right. \\
&\qquad \left. \times \frac{10^{22}\, m^{-3}}{10^{24}\, m^{-3} \times (1.01 \times 10^{24}\, m^{-3})} \right\}^{1/2} \\
&= 3.2 \times 10^{-9}\, m = 32\ \text{\AA}
\end{aligned}
$$

The depletion width on the n-side is 100 times longer

$$
W_n(V_o) = 0.32\ \mu m
$$

This is an example of how abrupt the depletion region is on the heavily doped p-side. It is important to note again that even heavier doping can further reduce the depletion width and carriers can simply tunnel into the bands. This concept is used to make ohmic contacts, and is discussed further in this chapter

6.3   P-N JUNCTION UNDER BIAS

$\mathcal{R}$ Let us now consider the situation where an external potential is applied across the p and n regions. In the presence of the applied bias, the balance between the drift and diffusion currents will no longer exist and a net current flow will occur. In general, one needs a numerical treatment to understand the behavior of the p-n diode under bias. However, under quasi-equilibrium conditions, one can use the previous results for the biased diode as well.

In Fig. 6.4 we show the schematic profiles of the depletion region, potential profile, and the band profiles in equilibrium, forward bias, and reverse bias$p$-n diode in reverse bias. In forward bias V_f, the p- side is at a positive potential with respect to the n-side. In reverse bias case, the p-side is at a negative potential $-V_r$ with respect to the n-side. Remember that the way we plot the energy bands includes the negative electron charge so the energy bands have the opposite sign of the potential profile.

In the forward bias case, the potential difference between the n- and p-side is (V_f is taken as having a positive value)

$$
V_{Tot} = V_{bi} - V_f \tag{6.30}
$$

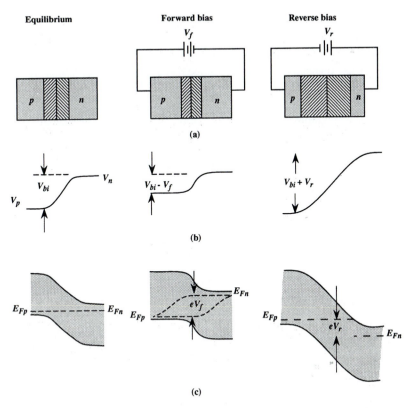

Figure 6.4: A schematic showing (a) the biasing of a p-n diode in the equilibrium, forward and reverse bias; (b) the voltage profile, and (c) the energy band profiles. In the forward bias, the potential across the junction decreases, while in reverse bias it increases. The quasi-Fermi levels are shown in the depletion region.

while for the reverse biased case it is (V_r is taken as having a positive value)

$$V_{Tot} = V_{bi} + V_r \tag{6.31}$$

Under the quasi-equilibrium approximations, the equations for electric field profile, potential profile, and depletion widths we had calculated in the previous section are directly applicable except that V_{bi} is replaced by V_{Tot}. Thus the depletion width and the peak electric field at the junction decrease under forward bias, while they increase under reverse bias, as can be seen from Eqns. 6.26 and 6.27 if V_{bi} is replaced by V_{Tot}.

EXAMPLE 6.2 Consider a $20\,\mu$m diameter p-n diode fabricated in silicon. The donor density is 10^{16} cm^{-3} and the acceptor density is 10^{18} cm^{-3}. Calculate the following in this diode at 300 K:

i) The depletion widths and the electric field profile under reverse biases of 0, 2, 5, and 10 V, and under a forward bias of 0.5 V.

ii) What are the charges in the depletion region for these biases?

Let us start with calculating V_{bi}, the built-in potential for the diode. This could be done as discussed in Example 6.1 .

$$V_{bi} = \frac{k_B T}{e} \ell n \frac{p_p}{p_n}$$

For our diode $p_p = 10^{18}$ cm^{-3}. The value of p_n is obtained by the law of mass action

$$p_n = \frac{n_i^2}{n_n} = \frac{(1.45 \times 10^{10} cm^{-3})^2}{10^{16} cm^{-3}} = 2.1 \times 10^4 cm^{-3}$$

Thus ($k_B T$ at 300 K = 0.026 eV)

$$V_{bi} = 0.026(V) \times 31.5 = 0.819 V$$

which is essentially the same as we got in Example 6.1 using a different approach.

The depletion widths are now (as in Example 6.1)

$$V_r = 0 \quad : \quad W_p(0.819 \ V) = 32 \ \text{\AA}$$
$$W_n(0.819 \ V) = 0.32 \ \mu m$$

$$V_r = 2 \ V \quad : \quad W_p(2.819 \ V) = 32 \ (\text{\AA}) \times \left(\frac{2.819}{0.819}\right)^{1/2} = 58.9 \ \text{\AA}$$
$$W_n(2.819 \ V) = 0.589 \ \mu m$$

$$V_r = 5 \ V \quad : \quad W_p(5.819 \ V) = 84.5 \ \text{\AA}$$
$$W_n(5.819 \ V) = 0.845 \ \mu m$$

$$V_r = 10 \ V \quad : \quad W_p(10.819 \ V) = 115.1 \ \text{\AA}$$
$$W_n(10.819 \ V) = 1.151 \ \mu m$$

$$V_f = 0.5 \ V \quad : \quad W_p(0.319 \ V) = 32 \ (\text{\AA}) \times \left(\frac{0.319}{0.819}\right)^{1/2} = 20.3 \ \text{\AA}$$
$$W_n(0.319 \ V) = 0.203 \ \mu m$$

The peak fields in the diode are given by

$$F_m = -\frac{e N_d W_n}{\epsilon}$$

$$F_m(V_r = 0 \ V) = -\frac{\left(1.6 \times 10^{-19} \ C\right) \left(10^{22} \ m^{-3}\right) \left(0.32 \times 10^{-6} \ m\right)}{\left(11.9 \times 8.84 \times 10^{-12} \ F/m\right)}$$

$$\begin{aligned}
&= -4.95 \times 10^6 \ V/m = -4.95 \times 10^4 \ V/cm \\
F_m(V_r = 2 \ V) &= -4.95 \times 10^4 \times \frac{W_n(2.819 \ V)}{W_n(0.819)} = -9.11 \times 10^4 \ V/cm \\
F_m(V_r = 5 \ V) &= -1.3 \times 10^5 \ V/cm \\
F_m(V_r = 10 \ V) &= -1.78 \times 10^5 \ V/cm \\
F_m(V_f = 0.5 \ V) &= -3.14 \times 10^4 \ V/cm
\end{aligned}$$

We can see that at a reverse bias of ~ 10 V, the peak field is beginning to approach the breakdown field for Si which is around 3×10^5 V/cm.

The charge in the n- or p-side depletion region is

$$Q = eN_dW_nA = eN_aW_pA$$

where A is the area of the diode.

$$\begin{aligned}
Q(V_r = 0) &= \left(1.6 \times 10^{-19} C\right) \times \left(10^{22} m^{-3}\right) \times \left(0.32 \times 10^{-6} m\right) \times \pi \left(10 \times 10^{-6} m\right)^2 \\
&= 1.61 \times 10^{-13} C
\end{aligned}$$

The charge can be obtained at other biases by simply using the various depletion widths.

6.3.1 Charge Injection and Current Flow

$\mathcal{R}$ We will now examine how the applied bias changes the various current components in the p-n diode. The presence of the bias increases or decreases the electric field in the depletion region. However, under moderate external bias, the *electric field in the depletion region is always higher than the field for carrier velocity saturation* $(F \gtrsim 10kVcm^{-1})$. *Thus the change in electric field does not alter the drift part of the electron or hole current in the depletion region. Regardless of the bias, electrons or holes that come into the depletion region are swept out and contribute to the same current independent of the field.* The situation is quite different for the diffusion current. Remember that the diffusion current depends upon the gradient of the carrier density. As the potential profile is greatly altered by the applied bias, the carrier profile changes accordingly, greatly affecting the diffusion current. Let us evaluate the mobile carrier densities across the depletion region. Recall that at no applied bias we have the relation (see Eqn. 6.7)

$$\frac{p_p}{p_n} = e^{eV_{bi}/k_BT} \tag{6.32}$$

In the presence of the applied bias, under the assumptions of quasi-equilibrium, the same mathematics used in the equilibrium case leads to the conditions

$$\frac{p(-W_p)}{p(W_n)} = e^{e(V_{bi}-V)/k_BT} \tag{6.33}$$

We assume that the injection of mobile carriers is small (low level injection) so that the majority carrier densities are essentially unchanged because of injection, i.e., $p(-W_p) = p_p$. Taking the ratio of Eqns. 6.32 and 6.33 we get (V has a positive value for forward bias and a negative value for reverse bias)

$$\frac{p(W_n)}{p_n} = e^{eV/k_BT} \tag{6.34}$$

where we have assumed that $p(-W_p) = p_p$. This equation suggests that the hole minority carrier density at the edge of the n-side depletion region can be increased dramatically if one applies a forward bias. Conversely, in reverse bias, this injection is greatly reduced. This is simply a consequence of the hole distribution being described by a Boltzmann distribution. Only those holes from the p-side that can overcome the potential barrier $V_{bi} - V$ can be injected over into the n-side, as shown in Fig. 6.5. *In forward bias, this barrier decreases so that more change can be injected across the barrier. In reverse bias, the barrier increases so that less charge can be injected.*

A similar consideration gives, for the electrons injected as a function of applied bias

$$\frac{n(-W_p)}{n_p} = e^{eV/k_BT} \tag{6.35}$$

The *excess carriers injected* across the depletion regions are

$$\Delta p_n = p(W_n) - p_n = p_n(e^{eV/k_BT} - 1) \tag{6.36}$$

$$\Delta n_p = n(-W_p) - n_p = n_p(e^{eV/k_BT} - 1) \tag{6.37}$$

From our discussions of Chapter 4 (Section 4.10.), we know that the excess minority carriers that are introduced will decay into the majority region due to recombination with the majority carriers. The decay is simply given by the appropriate diffusion lengths (L_p for holes; L_n for electrons). Thus the carrier densities of the minority carriers outside the depletion region are (see Section 4.10)

$$\begin{aligned} \delta p(x) &= \Delta p_n e^{(-(x-W_n)/L_p)} \\ &= p_n \left(e^{eV/k_BT} - 1 \right) e^{-(x-W_n)/L_p} \qquad x > W_n \end{aligned} \tag{6.38}$$

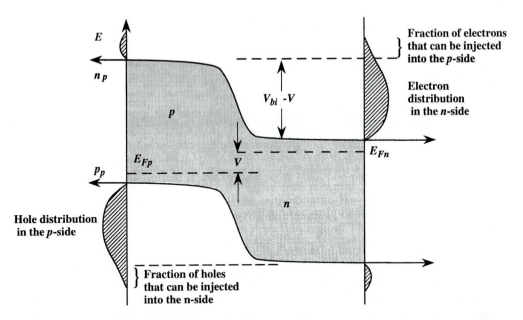

Figure 6.5: A schematic of the charge distribution in the n- and p-sides. The minority carrier injection (electrons from n-side to p-side or holes from p-side to n-side) is controlled by the applied bias as shown.

$$\delta n(x) \begin{array}{l} = \Delta n_p e^{(x+W_p)/L_n} \\ = n_p \left(e^{eV/k_BT} - 1\right) e^{(x+W_p)/L_n} \end{array} \quad \left\{ \begin{array}{l} x < -W_p \\ x \text{ is negative} \\ W_p \text{ is positive} \end{array} \right. \qquad (6.39)$$

We are now in a position to calculate the current in the diode as a function of the applied bias. We emphasize once again that we are assuming that the drift current is unaffected by the bias because of the assumption that the electric field in the depletion region is always larger than the field at which velocity saturates for the electrons and the holes. *Thus for the net current we only need to consider the excess electrons and holes injected as given by Eqns.6.38 and 6.39*

The diffusion current due to holes in the n-type material is (see Eqn. 4.23)

$$I_p(x) = -eAD_p \frac{d(\delta p(x))}{dx} = eA\frac{D_p}{L_p}(\delta p(x)) \qquad x > W_n \qquad (6.40)$$

where we have used Eqn. 6.38 to take the derivative of the excess hole density.

The hole current injected into the n-side is proportional to the excess hole density at a particular point. The total hole current injected into the n-side is given by the current at $x = W_n$ (after using the value of $\delta p(x = W_n)$ from Eqn. 6.38)

$$I_p(W_n) = e\, \frac{AD_p}{L_p}\, p_n \left(e^{eV/k_BT} - 1\right) \tag{6.41}$$

Similarly the total electron current injected into the p-side region is given by

$$I_n(-W_p) = \frac{eAD_n}{L_n}\, n_p \left(e^{eV/k_BT} - 1\right) \tag{6.42}$$

We assume initially that in the ideal diode there is no recombination of the electron and hole injected currents in the depletion region. Thus the total current can be simply obtained by adding the hole current injected across W_n and electron current injected across $-W_p$. The diode current is then

$$
\begin{aligned}
I(V) &= I_p(W_n) + I_n(-W_p) \\[2mm]
&= eA\left[\frac{D_p}{L_p}p_n + \frac{D_n}{L_n}n_p\right]\left(e^{eV/k_BT} - 1\right) \\[2mm]
I(V) &= I_o\left(e^{eV/k_BT} - 1\right) \tag{6.43}
\end{aligned}
$$

This equation, called the diode equation, gives us the current through a p-n junction under forward ($V > 0$) and reverse bias ($V < 0$). Under reverse bias, the current simply goes towards the value $-I_o$, where

$$\boxed{I_o = eA\left(\frac{D_p p_n}{L_p} + \frac{D_n n_p}{L_n}\right)} \tag{6.44}$$

Under forward bias the current increases exponentially with the applied forward bias. This strong asymmetry in the diode current is what makes the p-n diode attractive for many applications.

6.3.2   Minority and Majority Currents

$\mathcal{R}$ In the calculations for the diode current we have simply calculated the excess minority carrier diffusion current. This current was evaluated at its peak value at the edges of the depletion region. However, we can also see from Eqn. 6.38 that the diffusion current decreases rapidly in the majority region because of recombination. As the holes recombine with electrons in the n-region, an equal number of electrons are injected into the region. These electrons provide a drift current in the n-side to exactly balance the hole current that is lost through recombination. The drift part of the minority carrier is negligible because of the relatively low carrier density and very small electric field in the "neutral" region. Let us consider the hole diffusion current in the n-type region. This current is from Eqn. 6.40 using the value of $\delta p(x)$ from Eqn. 6.38

$$I_p(x) = e\ A\ \frac{D_p}{L_p}\ p_n\ e^{-(x-W_n/L_p)}\left(e^{eV/k_BT} - 1\right) ; x > W_n \qquad (6.45)$$

The total current is from Eqn. 6.43:

$$I = e\ A\left(\frac{D_p}{L_p}\ p_n + \frac{D_n}{L_n}\ n_p\right)\left(e^{eV/k_BT} - 1\right) \qquad (6.46)$$

Thus the electron current diode in the region is

$$
\begin{aligned}
I_n(x) \quad &= \quad I - I_p(x) \qquad\qquad\qquad x > W_n \\[2mm]
&= \quad eA\left[\frac{D_p}{L_p}\left(1 - e^{-(x-W_n)/L_p}\right)p_n \right.\\[2mm]
&\quad + \left.\frac{D_n}{L_n}n_p\right]\left(e^{eV/k_BT} - 1\right)
\end{aligned}
\qquad (6.47)
$$

As the hole current decreases from W_n into the n-side, the electron current increases correspondingly to maintain a constant current. A similar situation exists on the p- side region. As the electron injection current decays, the hole current compensates. A general picture shown schematically in Fig. 6.6 emerges from these discussions.

We see from the discussions of this section that the current flow through the simple p-n diode has some very interesting properties. We do not have the simple linear Ohm's law type behavior, but a strongly nonlinear and rectifying behavior. The current as shown in Fig. 6.7 saturates to a value I_o given by Eqn. 6.44 when a reverse bias is applied. Since this value is quite small, the diode is essentially nonconducting. On the other hand, when a positive bias is applied, the diode current increases exponentially and the diode becomes strongly conducting. The forward bias voltage at which the diode current becomes significant ($\sim$ mA) is called the cut-in voltage. This voltage is ~ 0.8V for Si diodes and ~ 1.2 V for GaAs diodes.

Most commercial p-n diodes for electronic applications are narrow diodes where the diode n and p side thicknesses are smaller than the diffusion length. In this case, the injected electron and hole minority charge decreases linearly from W_n to $W_{\ell n}$ and from $-W_p$ to $-W_{\ell p}$, respectively. Here, $W_{\ell n}$ and $W_{\ell p}$ are the diode n and p side widths.

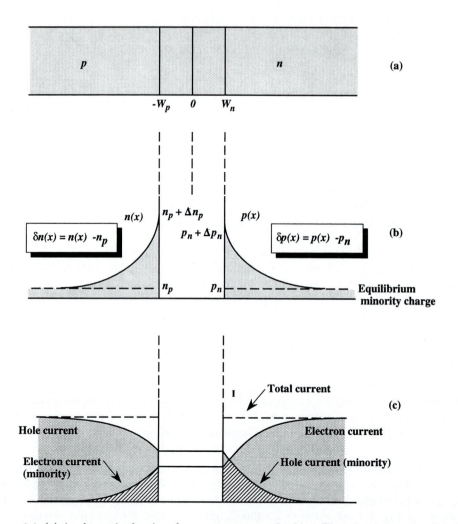

Figure 6.6: (a) A schematic showing the p-n structure under bias. The minority carrier distribution is shown in (b) while the form of the minority and majority current in the n-side is shown in (c). The minority carrier decays to essentially a zero value in a long diode. However, if the diode is narrow, the minority carrier continues to be significant up to the contact region, where it rapidly goes to zero. As the minority current (e.g., the hole current on the n-side) decays into the neutral region, the majority current (e.g., electron current on the n-side) increases so that the total current is uniform in the device.

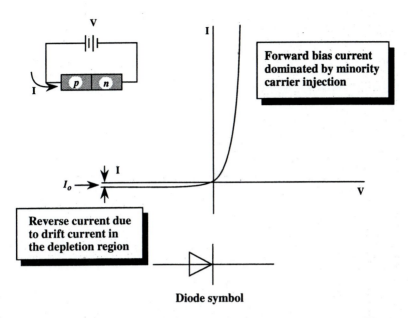

Figure 6.7: The highly nonlinear and rectifying I-V current of the *p-n* diode. The strong non-linear response makes the diode a very important device for a number of applications.

For the narrow diode, the prefactor of the diode current changes (i.e., the term L_n or L_p in the denominator is replaced by a smaller term $(W_{ln} - W_n)$ or $(|W_{lp} - W_p|)$). The prefactor becomes

$$I_o = eA \left(\frac{D_p p_n}{(W_{ln} - W_n)} + \frac{D_n n_p}{(|W_{lp} - W_p|)} \right) \qquad (6.48)$$

Thus the effect of the narrower diode is simply to change the value of the prefactor I_o. The diode still has the same rectifying properties. However, as the diode becomes narrower, the rectifying properties get worse since the reverse current (= I_o) starts to increase.

EXAMPLE 6.3 Consider an ideal diode model for a silicon *p-n* diode with $N_d = 10^{16}$ cm^{-3} and $N_a = 10^{18}$ cm^{-3}. The diode area is 10^{-3} cm^2.

The transport properties of the diode are given by the following values at 300 K:

$$n - \text{side} \begin{cases} \mu_p = 300 \ cm^2 \ V^{-1}s^{-1}; & \mu_n = 1300 \ cm^2 \ V^{-1}s^{-1} \\ D_p = 7.8 \ cm^2 \ s^{-1}; & D_n = 33 \ cm^2 \ s^{-1} \end{cases}$$

$$p-\text{side} \begin{cases} \mu_p = 100 \ cm^2 \ V^{-1}s^{-1}; & \mu_n = 280 \ cm^2 \ V^{-1}s^{-1} \\ D_p = 2.6 \ cm^2 \ s^{-1}; & D_n = 7.3 \ cm^2 \ s^{-1} \end{cases}$$

(Note that the mobility is a lot lower in the heavily doped p-side because of the increased ionized impurity scattering). Assume that $\tau_n = \tau_p = 10^{-6}s$. Calculate the diode current.

We need to calculate the minority carrier diffusion length to obtain the diode current. The hole diffusion length on the n-side is

$$L_p = \sqrt{D_p \tau_p} = \left(7.8 \ (cm^2 \ s^{-1}) \times 1.0 \times 10^{-6} \ s\right)^{1/2} = 2.79 \times 10^{-3} \ cm$$

The electron diffusion length on the p-side is

$$L_n = \sqrt{D_n \tau_n} = \left(7.3 \times 10^{-6}\right)^{1/2} = 2.7 \times 10^{-3} cm.$$

The prefactor I_o of the current is now

$$I_o = eA \left[\frac{D_p}{L_p} p_n + \frac{D_n}{L_n} n_p\right]$$

where

$$p_n = \frac{n_i^2}{n_n}; n_p = \frac{n_i^2}{p_p}$$

Using $n_i = 1.45 \times 10^{10} cm^{-3}$ and assuming that the dopants are fully ionized, we get

$$\begin{aligned} I_o &= \left(1.6 \times 10^{-19} C\right) \left(10^{-7} m^2\right) \left[\frac{7.8 \times 10^{-4} (m^2 s^{-1})}{2.79 \times 10^{-5} (m)} \times \left(2.1 \times 10^{10} m^{-3}\right)\right. \\ &\quad \left. + \frac{7.3 \times 10^{-4} (m^2 s^{-1})}{2.7 \times 10^{-5} (m)} \times 2.1 \times 10^{8} (m^{-3)}\right] \\ I_o &= 0.94 \times 10^{-14} A \end{aligned}$$

Note that the second term corresponding to the electron injection into the heavily doped p-side does not contribute much to the saturation current. The current is thus dominated by the hole current (the first term) component. Thus by proper doping one can ensure that the current is carried by either electron injection or hole injection. This is exploited in design of light emitting diodes as discussed in the next example.

The diode current is simply

$$I(V) = I_o \left[exp\left(\frac{eV}{k_B T}\right) - 1\right]$$

6.4 THE REAL DIODE: CONSEQUENCES OF DEFECTS

$\mathcal{R}$ In the calculations above we have assumed that the semiconductor is perfect, i.e., there are no defects and associated bandgap states which may lead to trapping, recombination, or generation terms. In Chapter 4 (Section 4.8) we had discussed the effects of bandgap states produced by defects. In our analysis of the diode ideal, we have assumed that the electrons and holes injected across the depletion region barrier, are not able to recombine with each other. Only when they enter the neutral regions are they able to recombine with the majority carriers. This recombination in the neutral region is described via the diffusion lengths L_n and L_p that appear in the expression for I_o.

In a real diode, a number of sources may lead to bandgap states. The states may arise if the material quality is not very pure so that there are chemical impurities present. The doping process itself can cause defects such as vacancies, interstitials, etc. Let us assume that the density of such deep level states is N_t. We will assume that the location of the deep level is at the center of the bandgap.

In Section 4.8 we had addressed this problem and shown how the Shockley, Hall, Read recombination comes about. The electron-hole recombination rate per unit volume is given by (see Eqn. 4.136; we denote τ_{nr} of Eqn. 4.135 by τ)

$$R_t = \frac{np}{\tau(n+p)} \tag{6.49}$$

where n and p are the electron-hole concentration, and the recombination time τ is given by (see Eqn. 4.133)

$$\tau = \frac{1}{N_t v_{th} \sigma} \tag{6.50}$$

where v_{th} is the thermal velocity of the electron (assumed same for the hole) and σ is the capture cross-section of the trap for the electron or hole. As electrons and holes enter the depletion region, one possible way they can cross the region without overcoming the potential barrier is to recombine with each other. This leads to an additional current flow mechanism. This current, called the generation-recombination current, must be added to the current calculated so far by us.

Note that at the edges of the depletion width we have from Eqns. 6.33 and 6.34, (using the relation $n_n p_n = n_i^2 = p_p n_p$)

$$n(-W_p)p_p(-W_p) = n_i^2 \, exp\left(\frac{eV}{k_B T}\right) = p(W_n)n_n(W_n) \tag{6.51}$$

We assume the np product remains constant in the depletion region as well.

This can be shown to be the case by detailed numerical calculations. We thus get (using $n \cong p$ for the maximum recombination rate) from Eqns. 6.49 and 6.51

$$R_t \cong \frac{n}{2\tau} \cong \frac{n_i}{2\tau}\, exp\left(\frac{eV}{2k_BT}\right) \tag{6.52}$$

The recombination current is now simply (current is equal to charge times volume times rate)

$$
\begin{aligned}
I_R &= eAWR_t = \frac{eAWn_i}{2\tau}\, exp\left(\frac{eV}{2k_BT}\right) \\
&= I^o_{GR}\, exp\left(\frac{eV}{2k_BT}\right)
\end{aligned}
\tag{6.53}
$$

where W is the depletion width. *At zero applied bias, a generation current of I_G balances out the recombination current.*

The generation-recombination current has an exponential dependence on the voltage as well, but the exponent is different. The generation-recombination current is

$$
\begin{aligned}
I_{GR} &= I_R - I_G = I_R - I_R(V = 0) \\
&= I^o_{GR}\left[exp\left(\frac{eV}{2k_BT}\right) - 1\right]
\end{aligned}
\tag{6.54}
$$

The total device current now becomes

$$I = I_o\left[exp\left(\frac{eV}{k_BT}\right) - 1\right] + I^o_{GR}\left[exp\left(\frac{eV}{2k_BT}\right) - 1\right]$$

or

$$I \cong I'_o\left[exp\left(\frac{eV}{mk_BT}\right) - 1\right] \tag{6.55}$$

The prefactor I^o_{GR} can be much larger than I_o for real devices. Thus at low applied voltages the diode current is often dominated by the second term. However, as the applied bias increases, the injection current starts to dominate. We thus have two regions in the forward I- V characteristics of the diode as shown in Fig. 6.8.

At low applied bias the plot of $\frac{eV}{k_BT}$ and $\log(I)$ has a slope of 1/2 which turns over to 1.0 at higher voltages. The parameter m of Eqn. 6.56 is called the *diode ideality factor*. If the diode is of high quality, m is close to unity, otherwise it approaches a value of 2.

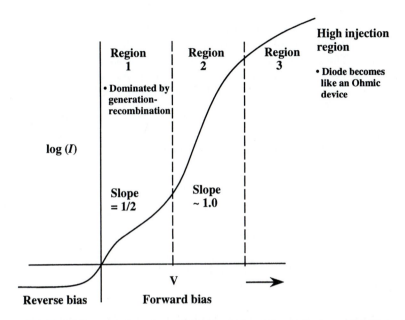

Figure 6.8: The I-V characteristics of a real diode. At low biases, the recombination effects are quite pronounced leading to a curve with slope 1/2. At higher biases the slope becomes closer to unity. At still higher biases the behavior becomes more ohmic as discussed in Section 6.5.1.

6.5 HIGH VOLTAGE EFFECTS IN DIODES

$\mathcal{R}$ In the derivations discussed so far for the diode properties, we have made several important assumptions. Some have been explicitly made while some are implicit in the equations we have used for current as a function of electric field. Many of these assumptions break down at large applied voltages.

6.5.1 Forward Bias: High Injection Region

One of the assumptions made by us was that the injection of minority carriers was at a fairly low level, so that essentially no voltage dropped across the bulk of the structure. All the voltage was assumed to be dropping over the depletion region. However, as the forward bias is increased, the injection level increases and eventually the injected minority carrier density becomes comparable to the majority carrier density. When this happens, an increasingly larger fraction of the external bias drops across the undepleted region. The diode current will then stop growing exponentially with the applied voltage, but will tend to saturate as shown as Region 3 in Fig. 6.8. The minority carriers that are injected move not only under diffusion effects but also under the electric field that is now present in the undepleted region. As the forward bias increases, the devices start to have more of an ohmic behavior, where the current-voltage relation is given by a simple linear expression. The current is now controlled by the resistance of the n- and p-type regions as well as the contact resistance. It must be noted, however, that at the high current densities involved, the device may heat and suffer burnout.

6.5.2   Reverse Bias: Impact Ionization

$\longrightarrow$ As the reverse bias is increased, the diode current may abruptly run away with the current being only limited by the external circuit (see Fig. 6.8). This phenomenon is called breakdown. An important cause for breakdown is impact ionization. After breakdown occurs at voltage V_{rb}, the current is given by

$$\boxed{I = \frac{|V_r - V_{rb}|}{R_L}} \tag{6.56}$$

where R_L is the resistance in the circuit and includes the effect of the diode resistance.

In Chapter 3 (Section 3.8) we had discussed the transport of carriers under very high electric fields and had discussed the impact ionization phenomena. Under very high electric fields, the electron (for example) acquires so much energy that it can scatter from an electron in the valence band knocking it into the conduction band. The final result is that instead of one initial electron to carry current, we have two electrons in the conduction band and one hole in the valence band. This results in current multiplication and the initial current I_o becomes

$$I_o' = M(V)I_o \tag{6.57}$$

where M is a factor that depends upon the impact ionization rate.

In Chapter 3 we had discussed the impact ionization coefficients α_{imp} and β_{imp} for the electrons and holes. These coefficients depend upon the material bandgap and the applied electric field. One can define a critical electric field F_{crit} for a given semiconductor at which the impact ionization rate becomes large. At F_{crit} α_{imp} or β_{imp} approaches $\sim 10^4$ cm^{-1} and values of F_{crit} for different materials are given in Appendix B. The value of 10^4 cm^{-1} is chosen so that an impact ionization occurs over a micron of distance which represents a typical device dimension of modern devices.

In the previous sections we have discussed the maximum electric field in a p-n junction as a function of applied bias. Once the applied bias becomes so large that $F_m \cong F_{crit}$, the impact ionization process starts to become dominant and the current shows a runaway behavior. If we consider a one-sided abrupt p^+n junction, the depletion width is essentially in the n-side, and the values of F_m and W are from Eqns.6.29 and 6.28,

$$F_m = \frac{eN_dW}{\epsilon} ; W \cong \left[\frac{2\epsilon V_r}{eN_d}\right]^{1/2} \tag{6.58}$$

where we have neglected the built-in field V_{bi} for the depletion width. The breakdown field V_{BD} is given by the reverse bias at which F_m becomes F_{crit}. This gives from Eqn. 6.58

$$\boxed{V_{BD} = \frac{\epsilon F_{crit}^2}{2eN_d}} \tag{6.59}$$

Note that as shown in Appendix B, F_{crit} has a dependence upon the doping density also.

In narrow bandgap materials, breakdown can also occur by Zener breaddown which involves band to band tunneling of electrons. The tunneling probability is given by

$$\boxed{T \approx exp\left(-\frac{4\sqrt{2m^*}E_g^{3/2}}{3e\hbar F}\right)} \tag{6.60}$$

where E_g is the bandgap of the semiconductor, m^* is the reduced mass of the electron-hole system, and F is the field in the depletion region.

EXAMPLE 6.4 A silicon p^+n diode has a doping of $N_a = 10^{19}$ cm^{-3}, $N_d = 10^{16}$ cm^{-3}. Calculate the 300 K breakdown voltage of this diode using the parameters given in Appendix B. If a diode with the same ϵ/N_d value were to be made from diamond, calculate the breakdown voltage.

The critical fields of silicon and diamond are (at a doping of 10^{16} cm^{-3}) $\sim 4 \times 10^5$ V/cm and 10^7 V/cm. The breakdown field is

$$V_{BD}(Si) = \frac{\epsilon(F_{crit})^2}{2eN_d} = \frac{(11.9)(8.85 \times 10^{-14} F/cm)(4 \times 10^5 V/cm)^2}{2(1.6 \times 10^{-19}C)(10^{16} cm^{-3})} = 51.7V$$

The breakdown for diamond is

$$V_{BD}(C) = 51.7 \times \left(\frac{10^7}{4 \times 10^5}\right)^2 = 32.3 \ kV!$$

One can see the tremendous potential of diamond for high power applications where the device must operate under high applied potentials.

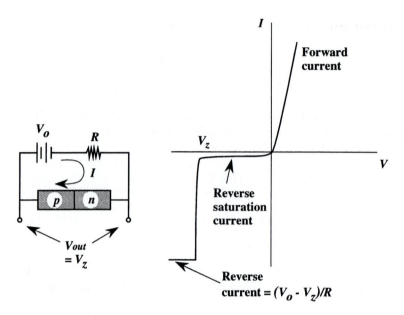

Figure 6.9: The use of the breakdown effect in the reverse biased *p-n* diode for a voltage clamping circuit. The current saturates to a value determined by the external circuit resistor, while the output voltage is clamped at the diode breakdown voltage. The circuit is thus very useful as a voltage regulator.

6.6   AC RESPONSE OF THE P-N DIODE

⟶   So far in this chapter we have discussed the dc characteristics of the *p-n* diode. It is important to note that most applications of diodes will involve transient or ac properties of the diode. The transient properties of the diode are usually not very appealing, especially for high speed applications. The *p-n* diode is a minority carrier device, i.e., it involves injection of electrons into a *p-* type region and holes into an *n*-side region. In the forward bias conditions where the diode is in a conducting state, the current flow is due to the minority charge injection. If this diode is to be switched, this excess charge must be removed. The device time response, therefore, depends upon how fast one can alter the minority charge that has been injected in the forward bias.

For the reverse biased case, where no minority charge injection occurs, the device speed can be quite high and is dominated by the device *RC* time constant. Let us examine the response of the *p-n* diode to large and small signals.

6.6.1   Small Signal Equivalent Circuit of a Diode

$\longrightarrow$ In order to analyze and design circuits using *p-n* diodes, it is essential to extract information on the diode small signal capacitance and resistance. The diode capacitance arises from two distinct regions of charges: i) The junction capacitance arises from the depletion region where there is a dipole of fixed positive and negative charge; and ii) The diffusion capacitance is due to the region outside the depletion region where minority carrier injection has introduced charges. Under reverse biased conditions, there are essentially no injected carriers and the junction capacitance dominates. The diffusion capacitance due to injected carriers dominates under forward bias conditions. The capacitance is in general defined by the relation

$$C = \left| \frac{dQ}{dV} \right| \tag{6.61}$$

We had calculated the junction depletion width earlier (see Eqn. 6.28 with V_{bi} replaced by the total bias $V_{bi} - V$)

$$W = \left[\frac{2\epsilon(V_{bi} - V)}{e} \left(\frac{N_a + N_d}{N_a N_d} \right) \right]^{1/2} \tag{6.62}$$

The depletion region charge is

$$|Q| = eA\, W_n N_d = eA\, W_p N_a \tag{6.63}$$

where we had shown earlier (see Eqns. 6.26 through 6.28)

$$W_n = \frac{N_a}{N_a + N_d} W; W_p = \frac{N_d}{N_a + N_d} W \tag{6.64}$$

Thus

$$|Q| = \frac{eA\, N_a N_d}{N_d + N_a} W = A \left[2e\epsilon(V_{bi} - V)\frac{N_d N_a}{N_d + N_a} \right]^{1/2} \tag{6.65}$$

The junction capacitance is then

$$
\begin{aligned}
C_j &= \left| \frac{dQ}{dV} \right| = \frac{A}{2} \left[\frac{2e\epsilon}{(V_{bi} - V)}\, \frac{N_a N_d}{N_a + N_d} \right]^{1/2} \\
&= \frac{A\epsilon}{W}
\end{aligned}
\tag{6.66}
$$

It is important to note that the capacitance is dependent upon the applied voltage.

For the forward biased diode, the injected charge density is quite large and can dominate the capacitance. The injected hole current; remember that charge is $I\tau_p$ and use $\tau_p D_p = L_p^2$; we also ignore 1 in the forward bias state)

$$Q_p = I\tau_p = eA \ L_p p_n \ e^{eV/k_BT} \tag{6.67}$$

The corresponding capacitance is

$$C_{diff} = \frac{dQ_p}{dV} = \frac{e^2}{k_BT} A \ L_p p_n \ e^{eV/k_BT} = \frac{e}{k_BT} I\tau_p \tag{6.68}$$

It is important to note here that the diffusion capacitance we calculated (in Eqn. 6.68) was based on the *total minority charge injected across the junction.* In a small signal response, *not all of the minority charge is modulated through the junction.* Some of the charge simply recombines in the neutral region. Thus the real diffusion capacitance of the small signal description is

$$\boxed{C_{diff} = K \frac{e}{k_BT} I\tau_p} \tag{6.69}$$

where K is a factor which is 1/2 for long base diodes and 2/3 for narrow base devices.

For small signal ac response, one can define the ac conductance of the diode as

$$\boxed{G_s = \frac{dI}{dV} = \frac{e}{k_BT} I(V)} \tag{6.70}$$

from the definition of the $I(V)$ function. At room temperature the conductance is (r_s is the diode resistance)

$$\boxed{G_s = \frac{1}{r_s} = \frac{I(mA)}{25.86} \Omega^{-1}} \tag{6.71}$$

Consider now a p-n diode which is forward biased at some voltage V_{dc} as shown in Fig. 6.10a. If an ac signal is now applied to the diode, the current output changes as shown schematically. We are interested in the ac impedance presented by the diode to the small signal excitation.

The equivalent circuit of the diode is shown in Fig. 6.10b and consists of the diode resistance r_s ($= G_s^{-1}$), the junction capacitance, and the diffusion capacitance.

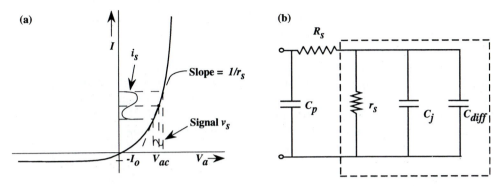

Figure 6.10: (a) A p-n diode is biased at a dc voltage V_{dc} and a small signal modulation is applied to it. (b) The equivalent circuit of a forward biased diode. At reverse bias the diffusion capacitance vanishes. The diode resistance is given by the differential slope of the I-V characteristics ($r_s = G_s^{-1}$). The resistance R_s is the series resistance of the neutral n- and p-regions, and the capacitance C_p is the capacitance associated with the diode packaging.

In the forward bias condition, the diffusion capacitance will dominate and we have the following relation between the current i_s and the applied voltage signal v_s:

$$i_s = G_s v_s + C_{diff} \frac{dv_s}{dt} \tag{6.72}$$

If we assume an input voltage which has a frequency ω ($v_s \sim v_s^o \, exp\,(j\omega t)$), we get

$$i_s = G_s v_s + j\omega C_{diff} v_s \tag{6.73}$$

and the admittance of the diode becomes

$$y = \frac{i_s}{v_s} = G_s + j\omega C_{diff} \tag{6.74}$$

Using a value of $K = 1/2$, we get the admittance

$$\boxed{y = \frac{eI}{k_B T} + \frac{j\omega e I \tau_p}{2 k_B T}} \tag{6.75}$$

In Fig. 6.10b we show the equivalent circuit of a packaged diode where we have the additional series resistance R_s associated with the diode n- and p-type neutral regions and a capacitance C_p associated with the diode packaging. As discussed, at forward bias the diffusion capacitance dominates, while at reverse bias the junction capacitance is dominant.

6.7   THE SCHOTTKY BARRIER DIODE

$\mathcal{R}$ A metal-semiconductor junction can, under appropriate conditions, be an active device capable of strongly nonlinear response. The resulting Schottky barrier diode and the Schottky barrier are widely used in extremely important applications, including the metal semiconductor photodetector discussed in Chapter 7. The Schottky diode has characteristics that are essentially similar to those of the *p-n* diode except that for many applications it has a much faster response.

6.7.1   Schottky Barrier Height

$\mathcal{R}$ The basic Schottky barrier is shown in Fig. 6.11a. The height of the barrier at the semiconductor-metal junction (Fig. 6.11b), defined as the difference between the semiconductor conduction band at the junction and the metal Fermi level, is called the Schottky barrier. While in an ideal case the barrier would be given by

$$e\phi_b = e\phi_m - e\phi_s + (E_c - E_{Fs}) = e\phi_m - e\chi_s \qquad (6.76)$$

in real metal-semiconductor juntions it is found that the Schottky barrier is independent of the metal species.

The electrons coming from the semiconductor into the metal face a barrier denoted by eV_{bi} as shown in Fig. 6.11b. The potential eV_{bi} is called the built-in potential of the junction and is given by

$$eV_{bi} = e\phi_m - e\phi_s \qquad (6.77)$$

The height of the potential barrier can be altered by applying an external bias as was the case in the *p-n* diode, and the junction can be used for rectification as we shall see below. One can have a *p*-type Schottky barrier using a metal-*p*-type semiconductor junction where we choose a metal so that the barrier exists for *p*-type transport. In this case, at equilibrium the electrons are injected from the metal to the semiconductor causing a negative charge on the semiconductor side. The bands are bent once again and a barrier is created for hole transport.

We have mentioned that it is found experimentally that the Schottky barrier height is *almost independent of the metal employed* as can be seen from Table 6.1. This can be understood qualitatively in terms of a model based upon nonideal surfaces. In this model the metal-semiconductor interface has a distribution of interface states which may arise from the presence of chemical defects (e.g., an oxide film) or broken bonds, etc. The defect region leads to a distribution of electronic levels in the bandgap at the interface. The distribution may be characterized by a neutral level ϕ_o having the

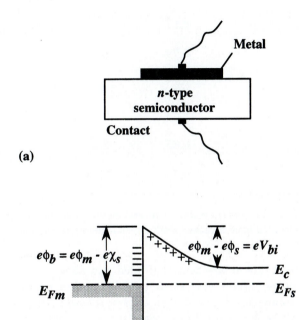

(a)

(b)

Metal-Semiconductor junction at equilibrium

Figure 6.11: (a) A schematic of a Schottky barrier junction. (b) The junction potential produced when the metal and semiconductor are brought together. Due to the built-in potential at the junction, a depletion region of width W is created.

Schottky Metal	n Si	p Si	nGaAs
Aluminum, Al	0.7	0.8	
Titanium, Ti	0.5	0.61	
Tungsten, W	0.67		
Gold, Au	0.79	0.25	0.9
Silver, Ag			0.88
Platinum, Pt			0.86
PtSi	0.85	0.2	
NiSi$_2$	0.7	0.45	

Table 6.1: Schottky barrier heights (in volts) for several metals on *n*- and *p*-type semiconductors. The barrier height is seen to have a rather weak dependence upon the metal used.

properties that states below it are neutral if filled and above it are neutral if empty. *If the density of bandgap states near ϕ_o is very large, then addition or depletion of electrons to the semiconductor does not alter the Fermi level position at the surface and the Fermi level is said to be pinned.* In this case, the Schottky barrier height is

$$e\phi_b = E_g - e\phi_o \qquad (6.78)$$

and is *almost independent of the metal used.* The model discussed above provides a qualitative understanding of the Schottky barrier heights. However, the detailed mechanism of the interface state formation and Fermi level pinning is quite complex.

6.7.2   Current Flow in a Schottky Barrier

$\mathcal{R}$ The current flow across a Schottky barrier can involve a number of different mechanisms. The most important and most desirable mechanism is that of thermionic emission in which electrons with energy greater than the barrier height $e(V_{bi} - V)$ can overcome the barrier and pass across the junction. Note that as the bias changes, the barrier to be overcome by electrons changes and the electron current injected is thus altered. This is shown schematically in Fig. 6.12.

In addition to thermionic emission, electrons can also tunnel through the barrier

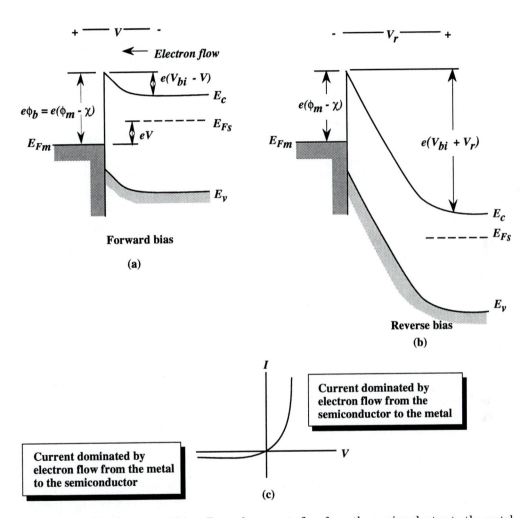

Figure 6.12: (a) The forward bias allows electrons to flow from the semiconductor to the metal side, increasing the current. (b) The reverse bias suppresses the electron flow from the semiconductor side while the flow from the metal side is unaffected. (c) The rectifying characteristics of the Schottky diode.

to generate current. This can be important if the semiconductor is heavily doped so that the depletion width is small. Other current mechanisms that are less important in high quality Schottky barriers may involve minority carrier injection, trap related recombination, etc. We will examine the dominant thermionic emission related current.

Thermionic Emission Current

If we assume that the tunneling current is negligible, the electrons that cross the metal-semiconductor junction must have energies greater than the barrier height at the junction. As an electron enters the semiconductor through a contact and travels towards the junction, its path in the neutral junction is simply determined by the drift-diffusion processes. At the junction it must pass over the barrier and only those electrons which have energies greater than the barrier will pass. The current is thus limited by the junction barrier and not the neutral region drift- diffusion in high quality semiconductors.

We assume that the electrons in the semiconductor region are distributed according to Boltzmann statistics. Thus the fraction of electrons with energy greater than the barrier of $(V_{bi} - V)$ is (V is positive for the forward biased diode and negative for the reverse biased case)

$$n_b = n_o \; exp \left[-e \frac{(V_{bi} - V)}{k_B T} \right] \qquad (6.79)$$

where n_o is the electron density in the neutral region. The density n_o is given in terms of the effective density of electrons N_c as

$$n_o = N_c \; exp \left[-\frac{(E_c - E_{Fs})}{k_B T} \right] \qquad (6.80)$$

Thus (note that the metal side barrier $e\phi_b = eV_{bi} + E_c - E_{Fs}$),

$$n_b = N_c \; exp \left[-\frac{(e\phi_b - eV)}{k_B T} \right] \qquad (6.81)$$

If the electrons are considered to be moving randomly, the average flux of electrons impinging on the metal-semiconductor barrier is $< v > n_b/4$, where $< v >$ is the average speed of the electrons. The corresponding current is then (A is the device area)

$$I_{sm} = \frac{eA < v >}{4} N_c \; exp \left[\frac{-e(\phi_b - V)}{k_B T} \right] \qquad (6.82)$$

When the applied bias V is zero, the current flow from the metal into the semiconductor I_{ms} must balance the current flow from the semiconductor to the metal. Thus

$$I_{ms} = I_{sm}(V = 0) = \frac{-eA < v >}{4} N_c \; exp - \left(\frac{e\phi_b}{k_B T} \right) \qquad (6.83)$$

When a potential V is applied, the *barrier seen by the electrons coming from the metal side is unchanged* and I_{ms} remains a constant ($= I_s$). Thus the net current at an applied bias V is, from Eqns. 6.82 and 6.83

$$I = I_{sm} - I_{ms} = I_s \left[exp\left(\frac{eV}{k_B T} \right) - 1 \right] \qquad (6.84)$$

For a Maxwell-Boltzmann distribution of electrons, the average velocity can be shown to be related to the temperature by the relation

$$<v> = \left(\frac{8 k_B T}{\pi m^*} \right)^{1/2} \qquad (6.85)$$

Substituting for the effective conduction band density N_c, we get for $I_s = I_{sm}(V = 0)$ from Eqn. 6.83

$$I_s = A \left(\frac{m^* e k_B^2}{2\pi^2 \hbar^3} \right) T^2 \, exp \left\{ -\left(\frac{e\phi_b}{k_B T} \right) \right\} \qquad (6.86)$$

The quantity in the parentheses is called the Richardson constant and is denoted by R^*. Its numerical value is given by

$$R^* = \frac{m^* e k_B^2}{2\pi^2 \hbar^3}$$

$$\cong 120 \left(\frac{m^*}{m_0} \right) A \, cm^{-2} K^{-2} \qquad (6.87)$$

A detailed formalism shows that the simple theory above gives too large a value of R^*. The more accurate value is ~ 60 A cm$^{-2}K^{-2}$. Also, one has to decide what value of effective mass to use in different semiconductors. It has been shown that the following values are appropriate for the conduction bands:

$$Si\,(<100> directions) \quad : \quad \frac{m^*}{m_0} = 2.05$$

$$Si\,(<111> direction) \quad : \quad \frac{m^*}{m_0} = 2.15$$

$$GaAs \quad : \quad \frac{m^*}{m_0} = 0.067$$

$$Ge\,(<100> direction) \quad : \quad \frac{m^*}{m_0} = 1.19$$

$$Ge\,(<111> direction) \quad : \quad \frac{m^*}{m_0} = 1.07$$

Approximate values for the effective Richardson constant taking these issues into account are: electrons in Si, $\sim 110\ A\ cm^{-2}K^{-2}$; electrons in GaAs, $8\ A\ cm^{-2}\ K^{-2}$; holes in Si, $\sim 32\ A\ cm^{-2}\ K^{-2}$; holes in GaAs $\sim 74\ A\ cm^{-2}\ K^{-2}$.

The saturation current in the Schottky barrier turns out to be much higher than that in a p-n diode with similar built-in voltage. This results in a turn on voltage for forward bias conducting state at a very low applied bias, but also results in a high reverse current. In Fig. 6.12, the band configurations in forward and reverse bias are shown along with the I-V characteristics. It is important to note that the electrons in the semiconductor side see a variable potential barrier as the applied bias changes, while the current from the metal side remains essentially unchanged.

EXAMPLE 6.5 A W-n-type Si Schottky barrier has a doping of 10^{16} cm^{-3} and an area of 10^{-3} cm^{2}.

(a) Calculate the 300 K diode current at a forward bias of 0.3 V.

(b) Consider an Si $p^{+} - n$ junction diode with the same area with doping of $N_a = 10^{19}$ cm^{-3} and $N_d = 10^{16}$ cm^{-3}, and $\tau_p = \tau_n = 10^{-6}$ s. At what forward bias will the p-n diode have the same current as the Schottky diode? $D_p = 10.5$ cm$^2/s$.

From Table 6.1 the Schottky barrier of W on Si is 0.67 V. Using an effective Richardson constant of 110 A cm$^{-2}K^{-1}$, we get for the reverse saturation current (see Eqn. 6.86)

$$I_s = (10^{-3}cm^2) \times (110A\ cm^{-2}K^{-2}) \times (300K)^2\ exp\left(\frac{-0.67(eV)}{0.026(eV)}\right)$$
$$= 6.37 \times 10^{-8}A$$

For a forward bias of 0.3 V, the current becomes (neglecting 1 in comparison to exp (0.3/0.026))

$$I = 6.37 \times 10^{-8}A\ exp(0.3/0.026)$$
$$= 6.53 \times 10^{-3}A$$

In the case of the p-n diode, we need to know the appropriate diffusion coefficients and lengths. The diffusion coefficient is 10.5 cm^2/s, and using a value of $\tau_p = 10^{-6}$s we get $L_p = 3.24 \times 10^{-3}$ cm. Using the results for the abrupt $p^{+} - n$ junction, we get for the saturation current ($p_n = 2.2 \times 10^4$ cm^{-3}) from Eqn. 6.48 (note that the saturation current is essentially due to hole injection into the n-side for a p^{+}-n diode)

$$I_o = (10^{-3}cm^2) \times (1.6 \times 10^{-19}C) \times \frac{(10.5cm^2/s^{-1})}{(3.24 \times 10^{-3}cm)} \times (2.25 \times 10^4 cm^{-3})$$
$$= 1.17 \times 10^{-14}A$$

This is an extremely small value of the current. At 0.3 V, the diode current becomes

$$I = I_s \; exp \; \left(\frac{eV}{k_B T} \right) = 1.2 \times 10^{-9} A$$

a value which is almost six orders of magnitude smaller than the value in the Schottky diode. For the p-n diode to have the same current that the Schottky diode has at 0.3 V, the voltage required is 0.71 V.

This example highlights the very important differences between Schottky and junction diodes. The Schottky diode turns on (i.e., the current is $\sim$1 mA) at 0.3 V while the $p - n$ diode turns on closer to 0.7 V.

6.7.3 Small Signal Circuit of a Schottky Diode

$\longrightarrow$ The small signal equivalent circuit of a Schottky diode is shown in Fig. 6.13. One has the parallel combination of the resistance

$$R_d = \frac{dV}{dI} \tag{6.88}$$

and the differential capacitance of the depletion region. The depletion capacitance has the form which we had calculated earlier for the p-n diode

$$C_d = A \left[\frac{eN_d \epsilon}{2(V_{bi} - V)} \right]^{1/2} \tag{6.89}$$

These circuit elements are in series with the series resistance R_s (which includes the contact resistance and the resistance of the neutral doped region of the semiconductor) and the parasitic inductance. Finally, one has to include the device geometry capacitance

$$C_{geom} = \frac{\epsilon A}{L} \tag{6.90}$$

where L is the device length.

There is a very important difference between the equivalent circuit of the Schottky diode and the p-n diode discussed in this Chapter. This has to do with the *absence of the diffusion capacitance which dominates the forward bias capacitance of a p-n diode.* This allows a very fast response of the Schottky diode which is exploited in a number of applications, as we shall see at the end of this chapter.

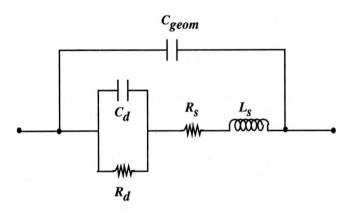

Figure 6.13: Equivalent circuit of a Schottky diode. The Schottky diode being a majority carrier device does not have the diffusion capacitance that hampers p-n diode performance in the forward bias mode.

EXAMPLE 6.6 Consider the Schottky and p-n diodes of Example 6.5. Compare the ac performance of the diodes when the devices are forward biased at 10 mA.

The applied bias for the Schottky diode corresponding to a 10 mA forward current is given by

$$V = \frac{k_B T}{e} \ell n \frac{I}{I_s} = 0.026(V) \times 11.7$$
$$= 0.31 \; V$$

We need to find the built-in voltage V_{bi} which is equal to the Schottky barrier height minus the value of $\frac{E_c - E_{Fs}}{e}$. The Fermi level position in the neutral semiconductor (E_{Fs}) with respect to the conduction band is given by

$$E_c - E_{Fs} \cong \frac{k_B T}{e} \ell n \left(\frac{N_c}{N_d} \right)$$
$$\cong 0.026(eV) \times \ell n \left\{ \frac{2.8 \times 10^{19} (cm^{-3})}{10^{16} (cm^{-3})} \right\}$$
$$= 0.2 \; eV$$

The built-in potential is

$$V_{bi} = \phi_b - \frac{1}{e}(E_c - E_F) = 0.66 - 0.2$$
$$= 0.46V$$

The diode capacitance is now

$$C_d = 10^{-3}(cm^2) \sqrt{\frac{(1.6 \times 10^{-19}C) \times (8.84 \times 10^{-14} \times 11.9 F/cm) \times (10^{16} cm^{-3})}{2 \times (0.16(V))}}$$

$$= 72.5 pF$$

The small signal resistance of the Schottky (and p-n) diode is given by the derivation of the voltage with respect to current

$$R = \frac{k_B T}{e} \frac{1}{I} = \frac{0.026(V)}{10^{-2}(A)} = 2.6\Omega$$

In the p-n diode, the junction capacitance and the small signal resistance will be the same as those in the Schottky diode. However, we now have to consider the diffusion capacitance.

The diffusion capacitance is given by (see Eqn. 6.69)

$$C_{diff} = \frac{e}{k_B T} I \tau_p = \frac{10^{-2}(A) \times 10^{-6}(s)}{(0.026V)} = 3.8 \times 10^{-7} F$$

The diffusion capacitance completely dominates the pn diode response. The RC time constant for the Schottky diode is 1.88×10^{-10} s, while that of the p-n diode is 9.9×10^{-7} s. Thus the pn diode is almost 1000 times slower.

6.8 HETEROSTRUCTURE JUNCTIONS

$\mathcal{R}$ We have discussed in chapter 2 the effects of heterostructures on the elecronic properties of a structure. Heterostructure based devices have made their presence felt in a number of technologies. The area of optoelectronics has benefitted perhaps the most from heterostructures.

In Fig.6.14 we show some of the motivations for the use of heterostructures based devices. We have already dicussed the important bandstructure related issues in heterostructures in Chapter 2, Section 2.9. There we discussed the importance of band discontinuities in determining the properties of the heterostructure.

An important benefit in heterojuntion diodes is the improvement in the injection efficiency. This injection efficiency is important for light emitting devices as well as for bipolar transistors. In Appendix D we discuss this important issue and the advantages of heterojunctions.

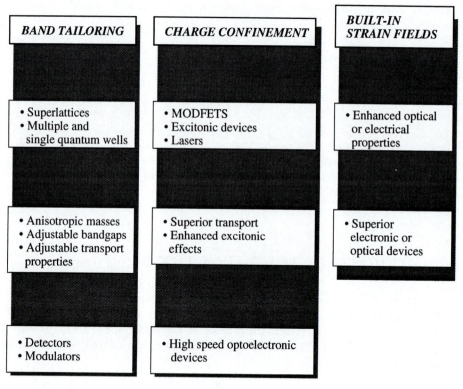

Figure 6.14: Semiconductor heterostructure properties and the devices that benefit from them.

6.9   CHAPTER SUMMARY

$\mathcal{R}$ In this chapter we have addressed one of the most important semiconductor structures—the p-n diode. We have observed how the use of the special properties of semiconductors—viz. large change in carrier concentrations controllable by external fields together with the basic transport, recombination phenomena—lead to a highly nonlinear device response. We have also examined the Schottky barrier diode and the ohmic contact which form important components of the optoelectronic devices. The summary tables (Tables 6.3-6.6) highlight the key findings of this chapter.

6.10   PROBLEMS

Section 6.2
6.1 An abrupt GaAs p-n diode has $N_a = 10^{17}$ cm^{-3} and $N_d = 10^{15}$ cm^{-3}.

a) Calculate the Fermi level positions at 300 K in the p and n regions.

b) Draw the equilibrium band diagram and determine the contact potential V_{bi}.

6.2 Consider the sample discussed in Problem 6.1. The diode has a diameter of 50μm. Calculate the depletion widths in the n and p regions. Also calculate the charge in the depletion regions and plot the electric field profile in the diode.
6.3   An abrupt silicon p-n diode at 300 K has a doping of $N_a = 10^{18}$ cm^{-3}, $N_d = 10^{15}$ cm^{-3}. Calculate the built-in potential and the depletion widths in the n and p regions.
6.4   A Ge p-n diode has $N_a = 5 \times 10^{17}$ cm^{-3} and $N_d = 10^{17}$ cm^{-3}. Calculate the built-in voltage at 300 K. At what temperature does the built in voltage decrease by 1%?

Section 6.3
6.5 The diode of Problem 6.1 is subjected to bias values of: i) $V_f = 0.1$ V; ii) $V_f = 0.5$ V; iii) $V_r = 1.0$V; iv) $V_r = 5.0$ V. Calculate the depletion widths and the maximum field F_m under these biases.
6.6 Consider a p^+n Si diode with $N_a = 10^{18}$ cm^{-3} and $N_d = 10^{16}$ cm^{-3}. The hole diffusion coefficient in the n-side is 10 cm^2/s and $\tau_p = 10^{-7}$ s. The device area is 10^{-4} cm^2. Calculate the reverse saturation current and the forward current at a forward bias of 0.4 V at 300 K.
6.7   Consider a p^+n silicon diode with area 10^{-4} cm^2. The doping is given by $N_a = 10^{18}$ cm^{-3} and $N_d = 10^{17}$ cm^{-3}. Plot the 300 K values of the electron and hole currents I_n and I_p at a forward bias of 0.4 V. Assume $\tau_n = \tau_p = 1\mu$s and neglect recombination

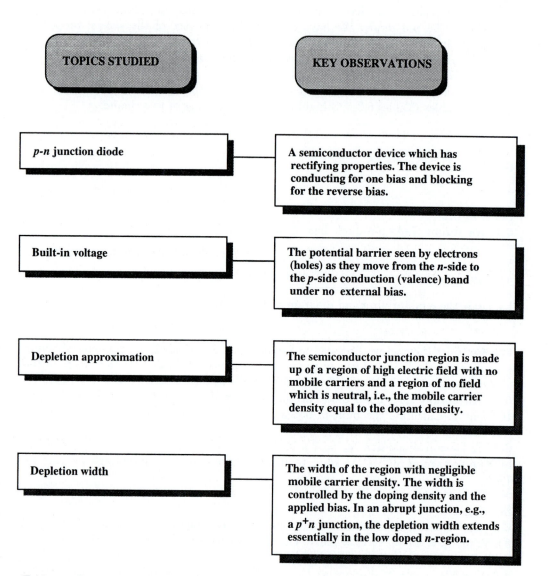

Table 6.2: Summary table.

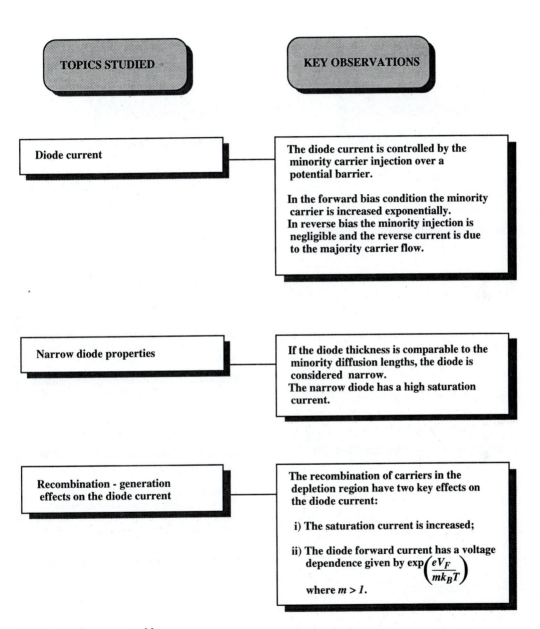

TOPICS STUDIED

KEY OBSERVATIONS

Diode current

The diode current is controlled by the minority carrier injection over a potential barrier.

In the forward bias condition the minority carrier is increased exponentially.
In reverse bias the minority injection is negligible and the reverse current is due to the majority carrier flow.

Narrow diode properties

If the diode thickness is comparable to the minority diffusion lengths, the diode is considered narrow.
The narrow diode has a high saturation current.

Recombination - generation effects on the diode current

The recombination of carriers in the depletion region have two key effects on the diode current:

i) The saturation current is increased;

ii) The diode forward current has a voltage dependence given by $\exp\left(\dfrac{eV_F}{mk_BT}\right)$

where $m > 1$.

Table 6.3: Summary table.

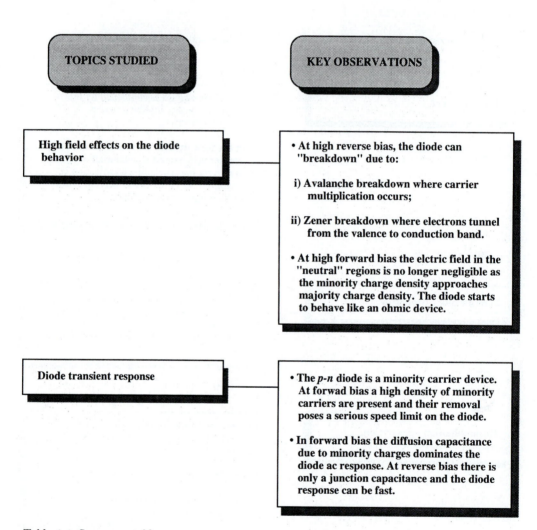

TOPICS STUDIED	KEY OBSERVATIONS
High field effects on the diode behavior	• At high reverse bias, the diode can "breakdown" due to: i) Avalanche breakdown where carrier multiplication occurs; ii) Zener breakdown where electrons tunnel from the valence to conduction band. • At high forward bias the elctric field in the "neutral" regions is no longer negligible as the minority charge density approaches majority charge density. The diode starts to behave like an ohmic device.
Diode transient response	• The *p-n* diode is a minority carrier device. At forwad bias a high density of minority carriers are present and their removal poses a serious speed limit on the diode. • In forward bias the diffusion capacitance due to minority charges dominates the diode ac response. At reverse bias there is only a junction capacitance and the diode response can be fast.

Table 6.4: Summary table.

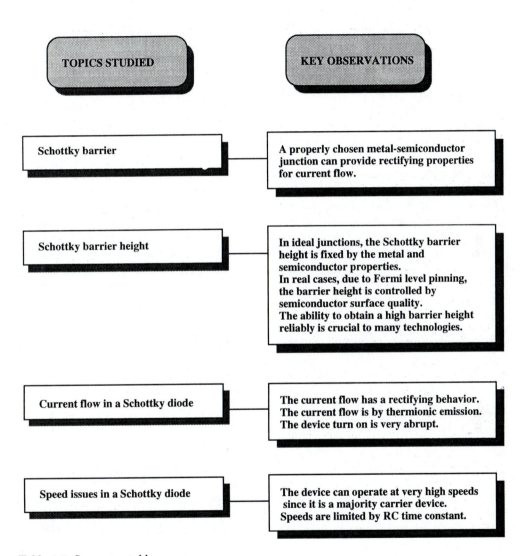

Table 6.5: Summary table.

effects. $D_n = 20$ cm^2/s; and $D_p = 10$ cm^2/s.

6.8 A GaAs LED has a doping profile of $N_a = 10^{17}$ cm^{-3}, $N_d = 10^{18}$ cm^{-3} at 300 K. The minority carrier time is $\tau_n = 10^{-8}$ s; $\tau_p = 5 \times 10^{-9}$ s. The electron diffusion coefficient is 100 cm^2 s^{-1} while that of the holes is 20 cm^2 s^{-1}. Calculate the ratio of the electron injected current to the total current.

6.9 The diode of Example 6.8 has an area of 1mm^2 and is operated at a forward bias of 0.5 V. Assume that 50% of the minority carriers injected recombine with the majority charge to produce photons. Calculate the rate of the photon generation in the n- and p-side of the diode.

6.10 Consider a GaAs p-n diode with a doping profile of $N_a = 10^{16}$ cm^{-3}, $N_d = 10^{17}$ cm^{-3} at 300 K. The minority carrier lifetimes are $\tau_n = 10^{-7}$ s; $\tau_p = 10^{-8}$ s. The electron and hole diffusion coefficients are 150 cm^2/s and 24cm^2/s, respectively. Calculate and plot the minority carrier current density in the neutral n and p regions at a forward bias of 1.0 V.

6.11 Consider a p-n diode made from InAs at 300 K. The doping is $N_a = 10^{16}$ cm$^{-3} = N_d$. Calculate the saturation current density if the electron and hole density of states masses are 0.02 m_0 and 0.4 m_0, respectively. Compare this value with that of a silicon p-n diode doped at the same levels. The diffusion coefficients are $D_n = 800$ cm^2/s; $D_p = 30$ cm^2/s. The carrier lifetimes are $\tau_n = \tau_p = 10^{-8}$s for InAs. For the silicon diode use the values $D_n = 30$ cm^2/s; $D_p = 10$ cm^2/s; $\tau_n = \tau_p = 10^{-7}$s.

6.12 Consider a p-n diode in which the doping is linearly graded. The doping is given by

$$N_d - N_a = Gx$$

so that the doping is p-type at $x < 0$ and n-type at $x > 0$. Show that the electric field profile is given by

$$F(x) = \frac{e}{2\epsilon}G\left[x^2 - \left(\frac{W}{2}\right)^2\right]$$

where W is the depletion width given by

$$W = \left[\frac{12\epsilon\,(V_{bi} - V)}{eG}\right]^{1/3}$$

6.13 A silicon diode is being used as a thermometer by operating it at a fixed forward bias current. The voltage is then a measure of the temperature. At 300 K, the diode voltage is found to be 0.6 V. How much will the voltage change if the temperature changes by 1 K?

Section 6.4

6.14 Consider a GaAs p-n diode with $N_a = 10^{17}$ cm^{-3}, $N_d = 10^{17}$ cm^{-3}. The diode area is 10^{-3} cm^2 and the minority carrier mobilities are (at 300 K) $\mu_n = 3000$ cm^2/V-s; $\mu_p = 200$ cm^2/V$-$s. The electron-hole recombination times are 10^{-8}s($\tau_p = \tau_n = \tau$). Calculate the diode current at a reverse bias of 5 V. Plot the diode forward bias current including generation recombination current between 0.1 V and 1.0 V.

6.15 A long base GaAs abrupt *p-n* junction diode has an area of 10^{-3} cm^2, $N_a = 10^{18}$ cm^{-3}, $N_d = 10^{17}$ cm^{-3}, $\tau_p = \tau_n = 10^{-8}$ s, $D_p = 6$ cm^2 s^{-1} and $D_n = 100$ cm^2 s^{-1}. Calculate the 300 K diode current at a forward bias of 0.3 V and a reverse bias of 5 V. The electron-hole recombination time in the depletion regions is 10^{-7}s.

Section 6.5

6.16 The critical field for breakdown of silicon is 4×10^5 V/cm. Calculate the *n*-side doping of an abrupt p^+n diode that allows one to have a breakdown voltage of 30 V.

6.17 Consider an abrupt p^+n GaAs diode at 300 K with a doping of $N_d = 10^{16}$ cm^{-3}. Calculate the breakdown voltage. Repeat the calculation for a similarly doped p^+n diode made from diamond. Use Appendix B for the data you may need.

6.18 What is the width of the potential barrier seen by electrons during band to band tunneling in an applied field of 5×10^5 V/cm in GaAs, Si and In$_{0.53}$Ga$_{0.47}$As($E_g = 0.8$ V)?

6.19 The tunneling probability in a triangular potential barrier was discussed in Chapter 1. If the electron effective mass is 0.5 m_0, and the semiconductor bandgap is 1.0 eV, at what applied field does the tunneling probability become 10^{-10}?

6.20 Consider an *Si* $p-n$ diode with $N_a = 10^{18}$ cm^{-3}; $N_d = 10^{18}$ cm^{-3}. Assume that the diode will break down by Zener tunneling if the peak field reaches 10^6 V/cm. Calculate the reverse bias at which the diode will break down.

Section 6.6

6.21 A $p^+ - n$ silicon diode has an area of 10^{-2} cm^2. The measured junction capacitance is given by (at 300 K)

$$\frac{1}{C^2} = 5 \times 10^8 (2.5 - 4V)$$

where C is in units of μF and V is in volts. Calculate the built-in voltage and the depletion width at zero bias. What are the dopant concentrations of the diode?

6.22 In a long base n^+p diode, the slope of the C_{diff} versus I_F plot is $1.6 \times 10^{-5} F/A$. Calculate the electron lifetime, the stored charge, and the value of the diffusion capacitance at $I_F = 1$ mA.

6.23 Consider a Si p^+n diode with a long base. The diode is forward biased (at 300 K) at a current of 2 mA. The hole lifetime in the *n*-region is 10^{-7} s. Assume that the depletion capacitance is negligible and calculate the diode impedance at the frequency of 100 KHz, 100 MHz, and 500 MHz.

Section 6.7

6.24 Assume the ideal Schottky barrier model with no interface states for an *n*-type Si with $N_d = 10^{16}$ cm^{-3}. The metal work function is 4.5 eV and the Si electron affinity is 4 eV. Calculate the Schottky barrier height, built-in voltage, and depletion width at no external bias. Assume $T = 300$ K.

6.25 A GaAs sample has a uniform *n*-type doping of 10^{16} cm^{-3} and a uniform density of surface states given by 5×10^{12} cm^{-2} eV^{-1}. The position of the neutral level at the surface is 0.5 eV above the valence band. Calculate how much the position of the Fermi level at the surface will alter away from ϕ_o if the depletion width in the semiconductor is a) 0.2 μm; b) 0.5 μm. Calculate the surface voltages that produce these depletion widths. Remember that the depletion region charge is equal to the surface state charge.

6.26 The capacitance of a Pt-n-type GaAs Schottky diode is given by

$$\frac{1}{\left(C(\mu F)\right)^2} = 1.0 \times 10^5 - 2.0 \times 10^5 V$$

The diode area is 0.1 cm^2. Calculate the built-in voltage V_{bi}, the barrier height, and the doping concentration.

6.27 Calculate the mean thermal speed of electrons in Si and GaAs at 77 K and 300 K. $m_{Si}^* = 0.3 \ m_0$; $m_{GaAs}^* = 0.067 \ m_0$.

6.28 Calculate the saturation current density in an Au Schottky diode made from n-type GaAs at 300 K. Use the Schottky barrier height values given in Table 6.2.

6.29 Consider an Au n-type GaAs Schottky diode with 50 μm diameter. Plot the current voltage characteristics for the diode between a reverse voltage of 2 V and a forward voltage of 0.5 V.

6.30 A Schottky barrier is formed between Al and n-type silicon with a doping of 10^{16} cm^{-3}. Calculate the theoretical barrier if there are no surface states. Compare this with the actual barrier height. Use the data in the text.

6.31 Calculate and plot the I-V characteristics of a Schottky barrier diode between W and n-type Si doped at 5×10^{16} cm^{-3} at 300 K. The junction area is 1 mm^2. Plot the results from a forward current of 0 to 100 mA.

6.11   REFERENCES

- **General**

    - M. S. Tyagi, *Introduction to Semiconductor Materials and Devices*, John Wiley and Sons, New York (1991).

    - B. G. Streetman, *Solid State Electronic Devices*, Prentice-Hall, Englewood Cliffs, NJ (1980).

    - E. S. Yang, *Fundamentals of Semiconductor Devices*, McGraw-Hill, New York (1978).

- **Diode Breakdown**

    - M. H. Lee and S. M. Sze, "Orientation Dependence of Breakdown Voltage in GaAs," *Solid State Electronics 23*, 1007 (1980).

    - S. M. Sze, *Physics of Semiconductor Devices*, John Wiley and Sons, New York (1981).

    - S. M. Sze and G. Gibbons, *Applied Physics Letters*, 8, 112 (1986).

- **Temporal Response of Diodes**

- R. H. Kingston, "Switching Time in Junction Diodes and Junction Transistors," *Proc. IRE*, 42, 829 (1954).

- M. S. Tyagi, *Introduction to Semiconductor Material and Devices*, John Wiley and Sons, New York (1991).

CHAPTER
7

OPTOELECTRONIC
DETECTORS

7.1 INTRODUCTION

One of the most important roles of information processing devices is to detect information. In electronics, the presence or absence of information is detected by devices such as field effect transistors, bipolar transistors, diodes, etc. These electronic "detectors" have properties such as high gain, tunability, low noise operation, etc., which have given electronics such a powerful role in information processing systems. However, these devices are not suitable for detecting optical frequency ($\sim 10^{14}$ Hz) signals. To detect optical signals, the light has to be converted into an electrical signal (current or voltage) which can then be amplified by one of the electronic devices mentioned above. In this chapter we will discuss the optoelectronic detectors which exploit the interaction of photons with electrons in a semiconductor to detect light.

While a great deal of progress has been made in the area of optical detectors, compared to microwave detectors, the field is relatively immature. Optoelectronic system designers still do not have wavelength or frequency selective detectors, or tunable detectors. This puts a tremendous handicap on the systems designer, as we will discuss in Chapter 13. The fact that, in spite of these handicaps, optoelectronics systems can compete with electronic systems is a testament to the tremendous potential of this technology.

In this chapter we will discuss the various optical detectors that are available. In the next chapter we will discuss the important noise considerations that limit detector performance as well as the various detection schemes that are used.

7.2 OPTICAL ABSORPTION IN A SEMICONDUCTOR

$\longrightarrow$ In order for a semiconductor device to be useful as a detector, some property of the device should be affected by radiation. The most commonly used property is the conversion of light into electron-hole pairs which can be detected in a properly chosen electric circuit.

When light impinges on a semiconductor, it can scatter an electron in the valence band into the conduction band. This process, called the absorption of a photon, was discussed in Chapter 4. In order to take the electron from the fully occupied valence band to the empty conduction band, the photon energy must be at least equal to the bandgap of the semiconductors. We will summarize some of the results discussed in Chapter 4 (Sections 4.4 and 4.8), which the reader should review to understand this and the next chapter.

The photon absorption process is strongest when the photon can directly cause an electron in the valence band to go into the conduction band. Since the photon momentum is extremely small on the scale of the electron momentum, the conservation of momentum requires that the electron-hole transitions are *vertical in k-space*, as shown in Fig. 7.1a . Such transitions are only possible near the bandedge for direct bandgap semiconductors. For such semiconductors one can write the absorption coefficient as

$$\alpha(\hbar\omega) = \frac{2\pi e^2 \hbar}{3 n_r c m_0^2 \epsilon_0} \frac{|p_{cv}|^2}{\hbar\omega} \frac{\sqrt{2}(m_r^*)^{3/2}(\hbar\omega - E_g)^{1/2}}{\pi^2 \hbar^3} \tag{7.1}$$

where m_r^* is the reduced e-h mass, n_r is the refractive index, $\hbar\omega$ the photon energy, E_g the band gap and p_{cv} is a momentum matrix element which allows the transition to take place. For direct gap semiconductors, when the various values for the constants are plugged into Eqn. 7.1, the absorption coefficient turns out to be $\hbar\omega$, E_g in eV (see Example 4.5)

$$\boxed{\alpha(\hbar\omega) \cong 4 \times 10^6 \left(\frac{m_r^*}{m_0}\right)^{3/2} \frac{(\hbar\omega - E_g)^{1/2}}{\hbar\omega} \ cm^{-1}} \tag{7.2}$$

When a semiconductor does not have a direct bandgap, vertical k transitions are not possible and the electrons can absorb a photon only if a phonon (or lattice vibration) participates in the process as shown in Fig. 7.1b. Such processes are not as strong as the ones which do not involve a phonon. They lead to an absorption coefficient which has the form

$$\alpha_{\text{indirect}} = (K_0 + K_1(T))(\hbar\omega - E_g)^2 \tag{7.3}$$

where K_0 is a constant and $K_1(T)$ is a temperature dependent factor. As temperature increases, $K_1(T)$ increases and the absorption coefficient increases. This is because the temperature allows more lattice vibrations to be present in the material. The prefactors

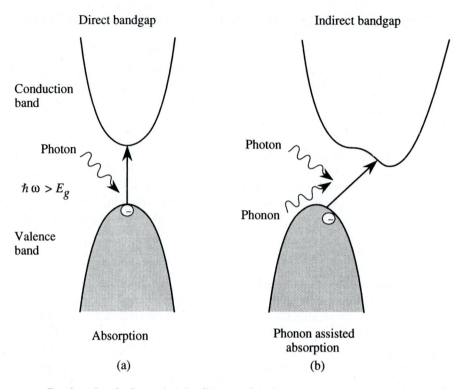

Figure 7.1: Band to band absorption in direct and indirect semiconductors. (a) An electron in the valence band "absorbs" a photon and moves into the conduction band. Momentum conservation ensures that only vertical transitions are allowed. (b) In indirect semiconductors a phonon or lattice vibration must participate to take an electron from the top of the valence band to the bottom of the conduction band. This is a comparatively weak process.

K_0 and K_1 are such that the absorption coefficient for indirect gap materials are typically a factor of 100 smaller as compared to the direct gap case for the same value of photon energy above bandgap ($\hbar\omega - E_g$).

As can be seen from Eqn. 7.1, the absorption coefficient is zero above a cutoff wavelength given by λ_c, where

$$\boxed{\lambda_c = \frac{hc}{E_g} = \frac{1.24}{E_g(eV)}(\mu m)} \qquad (7.4)$$

where E_g is the semiconductor bandgap. We note, however, that as discussed in Chapter 5, the electron-hole interaction produces excitons which modify the optical spectra near the bandedge. Thus, optical absorption can occur a few meV below the bandgap. In Fig. 7.2 we show the bandgap and cutoff wavelengths for several semiconductors along with the relative response of the human eye. In Fig. 7.3 we show the absorption coefficients for several different semiconductors. Materials like GaAs, InP, InGaAs, etc., have strong

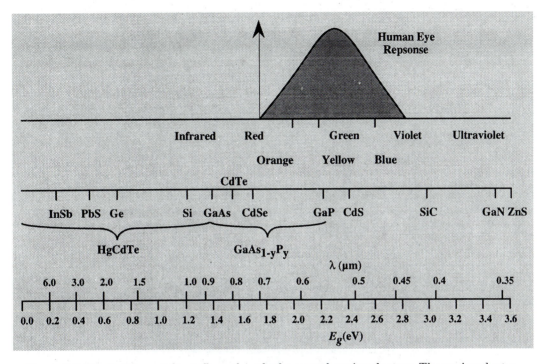

Figure 7.2: The bandgap and cutoff wavelengths for several semiconductors. The semiconductor bandgaps range from 0 (for $Hg_{0.84}Cd_{0.15}Te$) to well above 3 eV, providing versatile detection systems.

optical absorption at the bandedges because the optical absorption can occur without a phonon participation. On the other hand, Si and Ge have an indirect bandgap, and the absorption strength is weak near the bandedge. However, this does not mean that these materials cannot be used as detectors (unfortunately they cannot be used as lasers, as we will see in the next chapter). For detection of an optical signal, the light should be absorbed. If L is the length of the sample, the fraction of incident light absorbed in the sample is

$$1 - exp\left(-\alpha L\right) \tag{7.5}$$

Thus, for strong absorption, we must have

$$L > \frac{1}{\alpha(\hbar\omega)} \tag{7.6}$$

Thus, if Si is to absorb at GaAs laser emission ($\hbar\omega \sim 1.45$ eV), one needs a material thickness of 10-20 μm. On the other hand, a Ge detector would require an interaction length of only ~ 1 μm, even though Ge is an indirect gap material.

The electron-hole pair generation by "band to band" transition, i.e., an electron

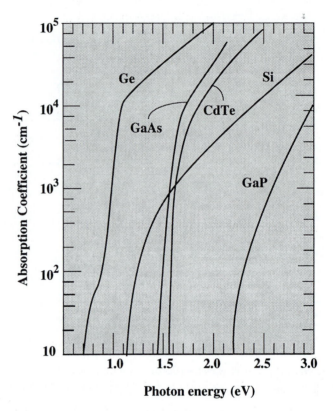

Figure 7.3: Absorption coefficients in different semiconductors. Notice the slow rise in absorption coefficients in indirect gap semiconductors (Si, Ge). (After M. Shur, *Physics of Semiconductor Devices*, Prentice-Hall, New Jersey (1990).)

being transferred from a valence band to a conduction band, is not the only way of detecting photons. In "extrinsic" detectors, a semiconductor is doped with a particular impurity which creates electron states in the bandgap as shown schematically in Fig. 7.4. Extrinsic detectors form an important class of detectors especially for detection for long wavelength radiation. Intrinsic band to band detectors need a very narrow bandgap for long wavelength radiation detection, and it is difficult to fabricate high quality devices from such materials. On the other hand, in extrinsic detectors, the radiation energy can be much smaller than the bandgap. In fact, extrinsic detectors can operate at wavelengths up to 120 μm at low temperatures using certain impurities in Ge or Si. In Table 7.1 we show the ionization energies and the cutoff wavelengths of a number of important extrinsic impurities. The absorption coefficient for the extrinsic absorption is, however, quite small (~ 10 cm^{-1}) so that a thick sample is needed.

Once the absorption coefficient for a semiconductor is known, one needs to know the rate at which electron-hole pairs will be generated. To calculate the rate of

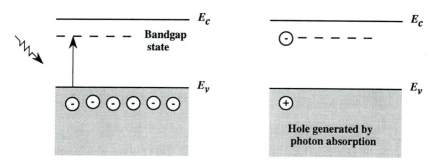

Figure 7.4: In an extrinsic detector, a deep level is introduced in the bandgap, and photons can cause transitions to these levels. The photon energy can be smaller than the bandgap in this case. The figure on the left shows a valence band full of electrons. Once a photon is absorbed, one of the valence band electrons is transferred to an impurity level as shown on the right.

SEMICONDUCTOR	IMPURITY	IONIZATION ENEREGY	WAVELENGTH CUTOFF for DETECTION (μm)
Ge	Au	0.15	8.3
	Cu	0.041	30
	Zn	0.0333	38
	B	0.0104	120
Si	In	0.155	8
	Bi	0.0706	18
	B	0.044	28

Table 7.1: Ionization energies and cutoff wavelengths for impurity doped germanium and silicon. (After R. J. Keyes, *Optical and Infrared Detectors*, Springer-Verlag, Berlin (1980).)

e-h pair generation, consider an optical beam with intensity $P_{op}(0)$ impinging upon a semiconductor per unit area. The intensity at a point x is given by (intensity has units of W/cm^2)

$$P_{op}(x) = P_{op}(0) \, exp\,(-\alpha x) \qquad (7.7)$$

The energy absorbed per second per unit area in a thickness region of thickness dx, between points x and $x + dx$ is (dx is very small)

$$
\begin{aligned}
P_{op}(x + dx) - P_{op}(x) &= P_{op}(0)\,[exp\,(-\alpha(x + dx)) - exp(-\alpha x)] \\
&= P_{op}(0)\,[exp\,(-\alpha x)]\,\alpha dx
\end{aligned}
\qquad (7.8)
$$

If this absorbed energy produces *e-h* pairs of energy $\hbar\omega$, the rate of the carrier generation is G_L (rate per unit volume)

$$\boxed{G_L = \frac{\alpha P_{op}(x)}{\hbar\omega} = \alpha J_{ph}(x)} \qquad (7.9)$$

where J_{ph} is the photon flux density impinging at point x (flux has units of cm^{-2} s^{-1}).

When light impinges upon a semiconductor and generates *e-h* pairs, the detector performance depends upon collecting these carriers and thus changing the conductivity of the material or generating a voltage signal. In the absence of an electric field or a concentration gradient, the *e-h* pairs will recombine with each other and not generate a detectable signal. An important property of the detector is described by its responsivity which gives the current produced by a certain optical power. The responsivity R_{ph} is defined by

$$\boxed{R_{ph} = \frac{I_L/A}{P_{op}} = \frac{J_L}{P_{op}}} \qquad (7.10)$$

where I_L is the photocurrent produced in a device of area A and J_L is the photocurrent density. The quantum efficiency of the detector is defined by

$$
\begin{aligned}
\eta_Q &= \frac{J_L/e}{P_{op}/\hbar\omega} \\
&= R_{ph}\,\frac{\hbar\omega}{e}
\end{aligned}
\qquad (7.11)
$$

The quantum efficiency essentially tells us how many carriers are collected for each photon impinging on the detector. Improving η_Q is one of the important design considerations for the detector and will be discussed later.

The responsivity of a detector has a strong dependence upon the wavelength of the impinging photons. If the wavelength is above the cutoff wavelength, the photons will not be absorbed and no photocurrent will be generated. When the wavelength is smaller than λ_c, the photon energy will be larger than the bandgap energy and the

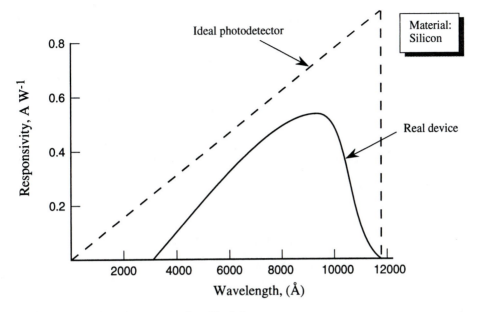

Figure 7.5: The responsivity curve of an ideal detector.

difference will be released as heat. Thus, even though the photon energy increases above the bandgap, it still produces the same number of e-h pairs. Thus the responsivity starts to decrease as shown in Fig.7.5

To collect the electron-hole pairs generated by light one needs an electric field. This can be generated by either simply applying a bias across an undoped semiconductor or by using a p-n diode. The former choice leads to the photoconductive detector in which the e-h pairs change the conductivity of the semiconductor. The p-n (or p-i-n) diode is widely used as a detector and exploits the built-in electric fields present at the junction together with an applied reverse bias to collect electrons and holes. The p-n diode can be used in a variety of modes, depending upon the applied bias and load configurations.

In addition to the diodes, transistors can also be used to detect optical signals. Phototransistors are widely used in optoelectronic technology. Phototransistors offer high gain due to the transistor gain. In the next few sections we will discuss the various modes of operation.

EXAMPLE 7.1 The momentum matrix element p_{cv} has a value of 23 eV for GaAs. Calculate the absorption coefficient for GaAs for an incident optical beam with energy of 1.7 eV. Assume that $E_g = 1.43$ eV.

The absorption coefficient for GaAs was calculated in Example 4.6 in Chapter 4. There we found that for GaAs the coefficient is

$$\alpha(\hbar\omega) = 5.6 \times 10^4 \frac{(\hbar\omega - E_g)^{1/2}}{\hbar\omega} \ cm^{-1}$$

where $\hbar\omega$ and E_g are expressed in eV. For our case we get

$$\alpha(\hbar\omega = 1.7eV) = 4.21 \times 10^4 \frac{(0.27)}{1.7} = 6.7 \times 10^3 cm^{-1}$$

EXAMPLE 7.2 A Ge detector is to be used for an optical communication system using a GaAs laser with emission energy of 1.43 eV. Calculate the depth of the detector needed to be able to absorb 90% of the optical signal entering the detector.

From Fig. 7.3 we see that at $\hbar\omega = 1.43$ eV, $\alpha \cong 2.5 \times 10^4$ cm^{-1}. The length of material required to absorb 90% of the light is given by

$$L = -\frac{1}{\alpha}\ell n \ (1 - 0.9) = \frac{2.3}{(2.5 \times 10^4 cm^{-1})} = 0.92 \ \mu m$$

Thus a rather thin region of Ge can absorb a large fraction of light emitted from a GaAs laser. Of course, if the light was emitted by an $In_{0.53}Ga_{0.47}As$ laser ($\hbar\omega \sim 0.8$ eV), Ge would not be as suitable a material.

EXAMPLE 7.3 An optical intensity of 10 W/cm^2 at a wavelength of 0.75 μm is incident on a GaAs detector. Calculate the rate at which electron-hole pairs will be produced at this intensity at 300 K. If the *e-h* recombination time is 10^{-9} s, calculate the excess carrier density.

From Fig. 7.3, the absorption coefficient of GaAs at this wavelength is $\sim 7 \times 10^3$ cm^{-1}. The *e-h* generation rate is (0.75 μm wavelength is equivalent to a photon of 1.65 eV)

$$G_L = \frac{\alpha P_{op}}{\hbar\omega} = \frac{(7 \times 10^3 \ cm^{-1})(10 \ W cm^{-2})}{(1.65 \times 1.6 \times 10^{-19} \ J)} = 2.65 \times 10^{23} \ cm^{-3} \ s^{-1}$$

The excess carrier density is

$$\begin{aligned} \delta n = \delta p = G_L \tau \ &= \ (2.65 \times 10^{23} \ cm^{-3} \ s^{-1})(10^{-9} \ s) \\ &= \ 2.65 \times 10^{14} \ cm^{-3} \end{aligned}$$

7.3 MATERIALS FOR OPTICAL DETECTORS

$\mathcal{R}$ Essentially all optical detectors based upon semiconductors depend upon converting an optical signal to an electrical signal. This involves the creation of electron-hole pairs. Since essentially all semiconductors have some electronic response to an optical signal provided that the wavelength of the photons is properly chosen, there is a wide selection of materials to choose from. Even indirect bandgap materials which cannot be used for the reverse process of efficient light emission can be excellent detectors. In the following sections we will discuss some important detectors and some of the material issues involved. In this section we will discuss some broader driving forces in the choice of semiconductor detector materials.

Substrate Availability
Semiconductor technology presently uses only a very few materials for the substrates on which active devices are grown. These are Si, GaAs, Ge and InP. Other substrates are available but are either available in very small sizes (a few millimeters), or have a high defect density, or are simply too expensive ($> \$100.00$ per cm^2). As a result, the choice of semiconductor materials for detectors is limited to those which have a good lattice matching with the aforementioned substrates. In some cases it is possible to grow a lattice mismatched epitaxial layer and cause dislocations to be trapped near the interfacial region between the epilayer and the substrate. However, this is not an optimum situation since reliability of devices based on such an approach is not very good.

Semiconductor alloys are widely used for detectors and one can find compositions which lattice match to some substrate.

Long Distance Communication Applications
For long distance communications, the photons with wavelength of 1.55 μm or 1.3 μm are used since the transmission losses in an optical fiber are very low at these wavelengths. A detector is thus needed to respond to these energies. Clearly GaAs cannot respond at these photon energies because its cutoff wavelength is ~ 0.8 μ m.

Among compound semiconductors, the alloy systems of InGaAs, InGaAsP, GaAlSb, HgCdTe can all be tailored to respond at these energies as shown in Fig. 7.6. The materials with potential for emission at 1.55 μm and 1.3 μm (i.e., E_g corresponds to these wavelengths) are identified. The detector bandgap must be somewhat lower than the photon energy so that the absorption coefficient is significant. The most widely used detector material is $In_{0.53}Ga_{0.47}As$ lattice matched to InP for long haul communication. This material has a matured technology, partly because it is also used for high frequency modulation doped field effect transistors.

In addition to compound semiconductors, Ge is also used for long haul communication. This detector is used as an avalanche photodetector to improve the gain of the device.

Local Area Networks
In the local area networks (LANs), where the optical signal has to propagate about a

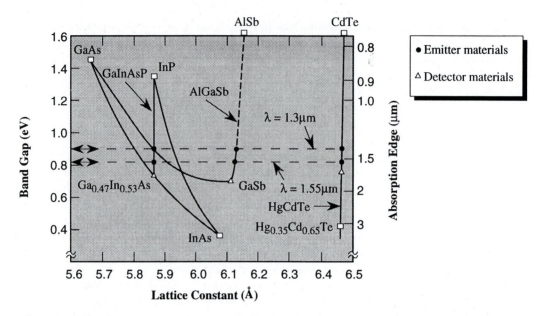

Figure 7.6: Bandgap versus lattice constant for some important compound semiconductors that can be exploited for long distance communications. (Based on *Semiconductors and Semimetals*, Vol. 22, ed. W. T. Tsang, Academic Press, New York (1985).)

kilometer, GaAs based emitters can be used. These devices emit at ~ 0.8 μm and are much cheaper than the devices emitting at 1.55 μm and 1.3 μm. The compound semiconductor detectors used for long distance communication can be used for LANs as well, but Si forms a good detector material. Silicon avalanche photodiodes are used widely for LAN applications.

Long Wavelength Detection

An important application of detectors is in the area of thermal imaging for night vision on medical diagnostics. The area of interest ranges from wavelengths going all the way to $20\mu m$. Detectors for such applications must be either based on very narrow bandgap materials, extrinsic defect levels or heterostructure concepts. Among narrow bandgap materials important choices are HgCdTe alloys, PbTe, PbSe, InSb. Extrinsic detectors based on Si and Ge implanted with impurities can also be used.

An important area of applications of quantum wells is in the area of intersub-band detectors. GaAs/AlGaAs technology can be used for such detectors.

High Speed Detectors

An important advance in high speed detector response has been the recent discovery of "low temperature GaAs." This material is grown at very low substrate temperature where a large number of defects are incorporated into the material. These defects decrease the *e-h* recombination time to ~ 1 ps in contrast to about 1 ns for high quality GaAs. The very short response time leads to high speed optical detection systems.

In Table 7.2 we present a brief summary of the different material systems and their key properties that can be exploited for detectors.

7.4 PHOTOCURRENT IN A P-N DIODE

$\longrightarrow$ When light impinges upon a semiconductor to create electron-hole pairs, some of the carriers are collected at the contact and lead to the photocurrent. Let us consider a long *p-n* diode in which excess carriers are generated uniformly at a rate G_L. Fig. 7.7 shows a *p-n* diode with a depletion region of width W. The electron-hole pairs generated in the depletion region are swept rapidly by the electric field existing in the region. Thus the electrons are swept into the *n*-region while the holes are swept into the *p*-region. The photocurrent arising from the photons absorbed in the depletion region is thus

$$I_{L1} = A \cdot e \int_0^{x'} G_L \cdot dx = A \cdot e G_L W \tag{7.12}$$

where A is the diode area and we have assumed a uniform generation rate in the diode. *Since the electrons and holes contributing to I_{L1} move under high electric fields, the response is very fast, and this component of the current is called the prompt photocurrent.*

In addition to the carriers generated in the depletion region, *e-h* pairs are generated in the neutral *n*- and *p*-regions of the diode. On physical grounds, we may expect that holes generated within a distance L_p (the diffusion length) of the depletion region edge ($x = 0$ of Fig. 7.7) will be able to enter the depletion region from where the electric field will sweep them into the *p*-side. Similarly, electrons generated within a distance L_n of the $x' = 0$ side of the depletion region will also be collected and contribute to the current. Thus the photocurrent should come from all carriers generated in a region $(W + L_n + L_p)$. A quantitative analysis reaches the same conclusion as shown below.

We will use the diode theory and the approximations used in Chapter 6 to obtain the photocurrent. We start with the continuity equation assuming that *e-h* pairs are generated uniformly at a rate G_L. The steady state continuity equation for holes in the *n*-region is (see Eqn. 4.164 and add the generation term G_L)

$$D_p \frac{\partial^2 \delta p_n}{\partial x^2} - \frac{\delta p_n}{\tau_p} + G_L = 0 \tag{7.13}$$

where D_p and τ_p are the hole (minority carrier) diffusion coefficient and recombination time. The excess carrier density is $\delta p_n = p(x) - p_n$. We use the boundary conditions

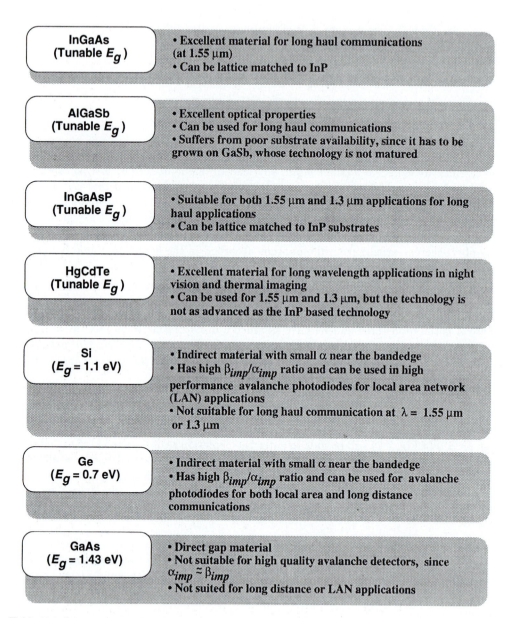

Table 7.2: Important semiconductor systems for detectors.

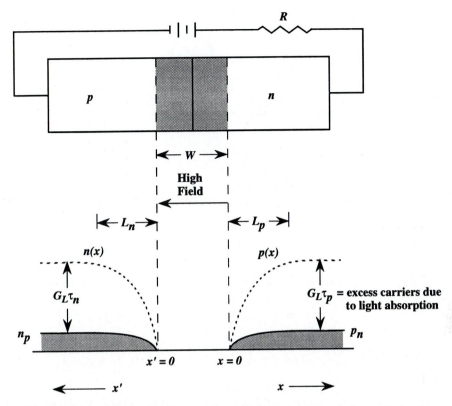

Figure 7.7: A schematic of a *p-n* diode and the minority carrier concentration in absence and presence of light. The minority charge goes to zero at the depletion region edge due to the high field which sweeps the charge away. The equilibrium minority charge is p_n and n_p in the *n*- and *p*-sides, respectively.

(the voltage V across the diode is positive for forward bias and negative for reverse bias)

$$\delta p(x \longrightarrow \infty) = G_L \tau_p \tag{7.14}$$

$$\delta p(x = 0) = p_n \left[exp \frac{eV}{k_B T} - 1 \right] \tag{7.15}$$

where $x = 0$ is the depletion region boundary as shown in Fig. 7.7. We assume that there is no recombination in the depletion width, i.e., $W < L_p = (D_p \tau_p)^{1/2}$, and the *n*- region is larger than L_p. To obtain the solution of the continuity equation with the generation rate, we note that the solution to Eqn. 7.13 is given by the sum of the solution to the homogeneous equation and the particular equation. The homogeneous equation results for $G_L = 0$ and has the form

$$\frac{d^2 \delta p_n'}{dx^2} - \frac{\delta p_n'}{L_p^2} = 0$$

with a solution for the long diode (see Eqn. 4.168) as

$$\delta p_n' = A \ exp \ \left(-\frac{x}{L_p}\right)$$

The particular equation has the form

$$\frac{\delta p_n''}{L_n^2} = \frac{G_L}{D_p}$$

or $\delta p_n'' = G_L \tau_p$. The resulting solution is thus

$$\delta p_n = A \ exp \ \left(-\frac{x}{L_p}\right) + G_L \tau_p$$

Using the boundary condition given by Eqn. 7.15 at $x = 0$, we obtain the value for the constant A. This gives the final solution for the excess holes in the neutral n-region

$$\delta p(x) = \left[p_n \left\{ exp \ \left(\frac{eV}{k_B T}\right) - 1\right\} - G_L \tau_p\right] \ exp \ \left(\frac{-x}{L_p}\right) + G_L \tau_p \qquad (7.16)$$

If the diode is operated in the short circuit mode so that the diode voltage is zero, or in a reverse bias mode, we can assume that $\delta p(x = 0)$ is essentially zero. This gives us

$$\delta p(x) = G_L \tau_p \left[1 - exp \ \left(\frac{-x}{L_p}\right)\right] \qquad (7.17)$$

so that the hole current, due to carriers absorbed in the neutral n-region, is

$$I_{pL} = A e D_p \left.\frac{d\delta p}{dx}\right|_{x=0} = e G_L L_p A \qquad (7.18)$$

The electron current can be similarly calculated, so that the total current, due to carriers in the neutral region and the depletion region, is

$$\boxed{I_L = I_{nL} + I_{pL} + I_{L1} = e G_L (L_p + L_n + W) A} \qquad (7.19)$$

It is important to note that the photocurrent contribution from the neutral regions has a slower time response since carriers are collected by diffusion under almost no electric fields. In case the widths of the neutral regions d_p and d_n of the diode are smaller than L_p and L_n, and if ohmic boundary conditions are used ($\delta p(d_n) = \delta n(d_p)$)

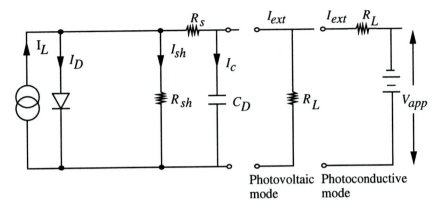

Figure 7.8: The equivalent circuit of a photodiode. The device can be represented by a photocurrent source I_L feeding into a diode. The device's internal characteristics are represented by a shunt resistor R_{sh} and a capacitor C_D. R_s is the series resistance of the diode. In the photovoltaic mode (used for solar cells and other devices) the diode is connected to a high resistance R_L, while in the photoconductive mode (used for detectors) the diode is connected to a load R_L and a power supply.

$= 0$), we can assume that half the carriers generated in the neutral regions contribute to the photocurrent. The current is then

$$I_L = eG_L \left(W + \frac{d_n}{2} + \frac{d_p}{2} \right) A \qquad (7.20)$$

It must be noted that the *e-h* pair generation is not uniform with depth but decreases with penetration depth. Thus G_L has to be replaced by an average generation rate for an accurate description. *It is also important to note that the photocurrent flows in the direction of the reverse current of the diode.*

The total current in the diode connected to the external load, as shown in Fig. 7.8, is given by the light generated current and the diode current in the absence of light. In general, if the voltage across the diode is V, the total current is (note that the photocurrent flows in the opposite direction to the forward bias diode current)

$$I = I_L + I_0 \left[1 - exp \left\{ \frac{e(V + R_s I)}{mk_B T} \right\} \right] \qquad (7.21)$$

where R_s is the diode series resistance, m the ideality factor, and V is the voltage across the diode. As shown in Fig. 7.8, the photodiode can be used in one of two configurations. In the photovoltaic mode, used for solar cells, there is no external bias applied. The photocurrent passes through an external load to generate power. In the photoconductive mode, used for detectors, the diode is reverse biased and the photocurrent is collected.

7.4.1   Application to a Solar Cell

An important use of the p-n diode is to convert optical energy to electrical energy as in a solar cell. The solar cell operates without an external power supply and relies on the optical power to generate current and voltage. To calculate the important parameters of a solar cell consider the case where the diode is used in the open circuit mode so that the current I is zero. This gives, for Eqn. 7.21

$$I = 0 = I_L - I_0 \left[exp \left(\frac{eV_{oc}}{mk_BT} \right) - 1 \right] \tag{7.22}$$

where V_{oc} is the voltage across the diode and is known as the open circuit voltage. We get for this voltage

$$\boxed{V_{oc} = \frac{mk_BT}{e} \ell n \left(1 + \frac{I_L}{I_0} \right)} \tag{7.23}$$

At high optical intensities the open circuit voltage can approach the semiconductor bandgap. In the case of Si solar cells for solar illumination (without atmospheric absorption), the value of V_{oc} is roughly 0.7 eV.

A second limiting case in the solar cell is the one where the output is short circuited, i.e., $R = 0$ and $V = 0$. The short circuit current is then

$$I = I_{sc} = I_L \tag{7.24}$$

A plot of the diode current in the solar cell as a function of the diode voltage then provides the curve shown in Fig. 7.9. In general, the electrical power delivered to the load is given by

$$P = I \times V = I_L V - I_0 \left[exp \left(\frac{eV}{k_BT} \right) - 1 \right] V \tag{7.25}$$

The maximum power is delivered at a voltage and current value of V_m and I_m as shown in Fig. 7.9.

The conversion efficiency of a solar cell is defined as the rate of the output electrical power to the input optical power. When the solar cell is operating under maximum power conditions, the conversion efficiency is

$$\eta_{conv} = \frac{P_m}{P_{in}} \times 100(\text{percent}) = \frac{I_m V_m}{P_{in}} \times 100(\text{percent}) \tag{7.26}$$

Another useful parameter in defining solar cell parameters is the fill factor F_f, defined as

$$F_f = \frac{I_m V_m}{I_{sc} V_{oc}} \tag{7.27}$$

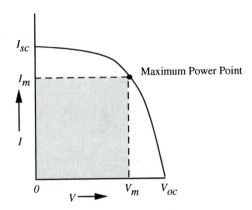

Figure 7.9: The relationship between the current and voltage delivered by a solar cell. The open circuit voltage is V_{oc} and the short circuit current is I_{sc}. The maximum power is delivered at the point shown.

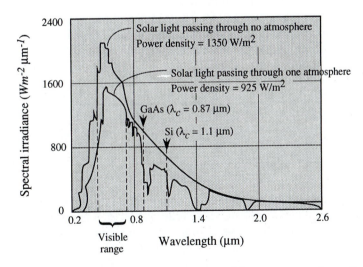

Figure 7.10: The spectral irradiance of the solar energy. The spectra are shown for no absorption in the atmosphere and for the sea level spectra. Also shown are the cutoff wavelengths for GaAs and Si. (Based on M. Shur, *Physics of Semiconductor Devices*, Prentice-Hall, New Jersey (1990).)

In most solar cells the fill factor is ~ 0.7.

In the solar cell conversion efficiency, it is important to note that photons which have an energy $\hbar\omega$ smaller than the semiconductor bandgap will not produce any electron- hole pairs. *Also, photons with energy greater than the bandgap will produce electrons and holes with the same energy (E_g) regardless of how large ($\hbar\omega - E_g$) is.* The excess energy $\hbar\omega - E_g$ is simply dissipated as heat. Thus the solar cell efficiency depends quite critically on how the semiconductor bandgap matches with the solar energy spectra. In Fig. 7.10 we show the solar energy spectra. Also shown are the cutoff wavelengths for silicon and GaAs. GaAs solar cells are better matched to the solar spectra and provide greater efficiencies. However, the technology is more expensive when compared to Si technology. Thus GaAs solar cells are used for space applications while silicon (or amorphous silicon) solar cells are used for applications where cost is a key factor.

7.4.2   Amorphous Silicon Solar Cells

$\mathcal{R}$ As we have noted in Chapters 1 and 2, most of the properties of semiconductors arise from the crystalline nature of the material. This includes the concept of bandgap and bandedges. The atoms in a crystalline material are arranged in a perfect lattice and as a result it is quite expensive to fabricate single crystal devices. In contrast amorphous materials have an arrangement of atoms which is not perfect, but on a local scale the atoms still have the same number of nearest and second neighbor atoms as in a perfect crystal.

Amorphous silicon films are deposited by chemical vapor deposition techniques at fairly low temperatures ($\sim 600°C$). In the deposition process a good deal of hydrogen is incorporated in the film which appears to benefit the film quality by tying up broken bonds in the Si lattice. Amorphous silicon films can be deposited on almost any substrate which makes them very inexpensive and versatile.

The electronic properties of amorphous silicon (a-Si) are not as bad as one would naively expect. The lack of long range order produces a material with poor transport properties, but the presence of short range order has important consequences. The density of states of a typical a - Si film are shown in Fig. 7.11. Unlike its crystalline counterpart, there is a high density of allowed states in the bandgap region. However, these states are strongly "localized," i.e., their extent in space is quite small. As one moves out of the bandgap region, the states became "extended" or "free" like the Bloch functions of the perfect crystal. When the nature of the electronic states change from localized to extended, the corresponding mobility changes from very low (≤ 1.0 cm^2 V^{-1} s^{-1}) to reasonably high values (~ 10 cm^2 V^{-1} s^{-1}). The point in energy where this transition occurs is called the mobility edges. The mobility edges define an effective bandgap for

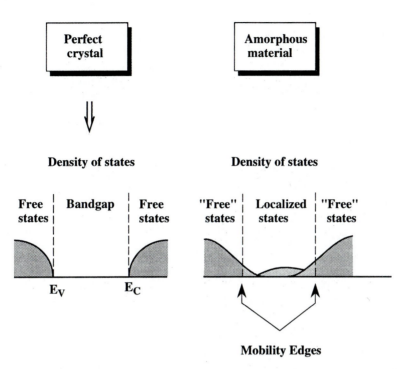

Figure 7.11: A schematic of the effect of disorder in an amorphous material on the density of states. A comparison is shown for the perfect crystal and an amorphous material. Mobility edges represent effective bandedges in the amorphous material.

the a-Si film and in most films the effective gap is ~ 1.6 eV.

An important positive aspect of a-Si is that the vertical k selection rule which produces low optical absorption in crystalline (indirect) Si no longer applies. This is because the k-selection rule is applicable strictly in perfect crystals where the electronic states have the plane wave ($\sim e^{ik \cdot r}$) form. As a result a-Si can have very high absorptions coefficient as shown in Fig. 4.12 and one needs a very thin film ($\sim \mu$m) to absorb the solar energy.

The ease of fabricating a-Si films makes them very attractive for an alternative energy source. The a-Si films can be made on tapes which can be rolled up. Even sheets with area running into square meters have been demonstrated. The technology is also pollution free and relies on a never ending natural source.

Before ending the section on solar cells, it is important to note that heterojunctions can also be used for solar cells. The use of more than one bandgap can improve the conversion efficiency. The technology is, of course, expensive and is used only in space applications where efficiency is of utmost importance.

EXAMPLE 7.4 Consider a long Si p-n junction that is reverse biased with a reverse bias voltage of 2 V. The diode has the following parameters (all at 300 K):

Diode area,	A =	10^4 μm^2
p-side doping,	N_a =	2×10^{16} cm^{-3}
n-side doping,	N_d =	10^{16} cm^{-3}
Electron diffusion coefficient	D_n =	20 cm^2/s
Hole diffusion coefficient,	D_p =	12 cm^2/s
Electron minority carrier lifetime,	τ_n =	10^{-8} s
Hole minority carrier lifetime,	τ_p =	10^{-8} s
Electron-hole pair generation rate by light,	G_L =	10^{22} cm^{-3} s^{-1}

Calculate the photocurrent. The electron length diffusion length is

$$L_n = \sqrt{D_n \tau_n} = \left[(20)(10^{-8})\right]^{1/2} = 4.5 \ \mu m$$

The hole diffusion length is

$$L_p = \sqrt{D_p \tau_p} = \left[(12)(10^{-8})\right]^{1/2} = 3.46 \ \mu m$$

To calculate the depletion width, we need to find the built-in voltage,

$$V_{bi} = \frac{k_B T}{e} \ \ell n \left(\frac{N_a N_d}{n_i^2}\right) = 0.026 \ \ell n \left(\frac{(2 \times 10^{16})(10^{16})}{(1.5 \times 10^{10})^2}\right) = 0.715 \ V$$

The depletion width is now

$$\begin{aligned} W & = \left\{\frac{2\epsilon_s}{e} \left(\frac{N_a + N_d}{N_a N_d}\right) (V_{bi} + V_R)\right\}^{1/2} \\ & = \left\{\frac{2(11.9)(8.85 \times 10^{-14})}{(1.6 \times 10^{-19})} \left(\frac{(2 \times 10^{16} + 10^{16})}{(2 \times 10^{16})(10^{16})}\right) (2.715)\right\}^{1/2} \\ & = 0.73 \ \mu m \end{aligned}$$

We see in this case that L_n and L_p are larger than W. The prompt photocurrent is thus a small part of the total photocurrent. The photocurrent is now

$$\begin{aligned} I_L & = eAG_L(W + L_n + L_p) \\ & = (1.6 \times 10^{-19} \ C)(10^4 \times 10^{-8} \ cm^2)(10^{22} \ cm^{-3} \ s^{-1})(8.69 \times 10^{-4} \ cm) \\ & = 0.137 \ mA \end{aligned}$$

The photocurrent is much larger than the reverse saturation current I_0 and its direction is the same as the reverse current.

EXAMPLE 7.5 Consider an Si solar cell at 300 K with the following parameters:

Area,	A	$=$	$1.0\ cm^2$
Acceptor doping,	N_a	$=$	$5 \times 10^{17}\ cm^{-3}$
Donor doping,	N_d	$=$	$10^{16}\ cm^{-3}$
Electron diffusion coefficient,	D_n	$=$	$20\ cm^2/s$
Hole diffusion coefficient,	D_p	$=$	$10\ cm^2/s$
Electron recombination time,	τ_n	$=$	$3 \times 10^{-7}\ s$
Hole recombination time,	τ_p	$=$	$10^{-7}\ s$
Photocurrent,	I_L	$=$	$25\ mA$

Calculate the open circuit voltage of the solar cell.

To find the open circuit voltage, we need to calculate the saturation current I_0, which is given by

$$I_0 = A \left[\frac{eD_n n_p}{L_n} + \frac{eD_p P_n}{L_p} \right] = A e n_i^2 \left[\frac{D_n}{L_n N_a} + \frac{D_p}{L_p N_d} \right]$$

Also,

$$L_n = \sqrt{D_n \tau_n} = \left[(20)(3 \times 10^{-7}) \right]^{1/2} = 24.5\ \mu m$$

$$L_p = \sqrt{D_p \tau_p} = \left[(10)(10^{-7}) \right]^{1/2} = 10.0\ \mu m$$

Thus,

$$I_0 = (1)(1.6 \times 10^{-19})(1.5 \times 10^{10})^2 \left[\frac{20}{(24.5 \times 10^{-4})(5 \times 10^{17})} + \frac{10}{(10 \times 10^{-4})(10^{16})} \right]$$

$$= 3.66 \times 10^{-11}\ A$$

The open circuit voltage is now

$$V_{oc} = \frac{k_B T}{e} \ell n \left(1 + \frac{I_L}{I_0} \right) = (0.026)\ell n \left(1 + \frac{25 \times 10^{-3}}{3.66 \times 10^{-11}} \right) = 0.53\ V$$

EXAMPLE 7.6 A single solar cell of area 1 cm^2 has a photocurrent of $I_L = 25$ mA and a diode saturation current of 3.66×10^{-11} A at 300 K. a) Calculate the open circuit voltage and short circuit current of the solar cell; b) calculate the power extracted from each cell if the fill factor is 0.8; c) if a solar power system requires a power of 10 W at a voltage level of 10 V, calculate the number of solar cells needed in series and the number of rows in parallel for such a solar cell array. (This diode has the same features as the diode considered in Example 7.5.)

The open circuit voltage was calculated in Example 7.5 and is 0.53 V.

The short circuit current is simply $I_L = 25$ mA.

The power per solar cell is

$$P = 0.8 I_{sc} V_{oc} = 0.8(25 \times 10^{-3})(0.53) = 1.06\ mW$$

The number of solar cells needed in series to produce an output voltage of 10 V is (each cell produces approximately $V_M \sim (F_f)^{1/2} \sim 0.9 V_{oc}$)

$$N(series) = \frac{10}{0.9 \times 0.53} \sim 24 \text{ cells}$$

The number of rows needed to produce a power of 10 W is now ($I_m \sim 0.9 I_{sc}$)

$$N(parallel) = \frac{10W}{10V(25 \times 10^{-3} \times 0.9A)} = 45 \text{ rows}$$

Thus the system needs a total of 1080 solar cells to meet the specifications.

7.5   THE PHOTOCONDUCTIVE DETECTOR

$\longrightarrow$ The photoconductive detector is the simplest of the detectors and consists of a simple region of semiconductor across which a bias is applied as shown in Fig. 7.12a. When light with a proper wavelength impinges upon the semiconductor, e-h pairs are created which are then collected by the electric field. The change in current is detected by a circuit of the form shown in Fig. 7.12b. *An important benefit of the photoconductive detector is the gain in the device, i.e., one can collect more than one electron (or hole) for each photon impinging.* Let us examine the operation of the photoconductive detector and the gain mechanism.

When light impinges on the i-region, e-h pairs are generated which change the material conductivity. The electric field in the device causes the electrons and holes to move in opposite directions, leading to current. *The carriers are present in the system until they either recombine or are collected at the contacts.* Consider the case where we have an n-i-n structure with an e-h recombination rate R_{eh} which is equal to the photogeneration rate

$$R_{eh} = \delta n / \tau_p = G_L \qquad (7.28)$$

where τ_p is the effective recombination time for the excess carriers. Let us assume that we have a lightly doped n-type device where the electrons dominate the conductivity. In the absence of the light signal, the conductivity is (n_0 and p_0 are the electron and hole densities in dark)

$$\sigma_o = e \left(\mu_n n_0 + \mu_p p_0 \right) \qquad (7.29)$$

If the optical signal generates an excess carrier density of $\delta n = \delta p$, the conductivity becomes

$$\sigma = e \left[\mu_n (n_0 + \delta n) + \mu_p (p_0 + \delta p) \right] \qquad (7.30)$$

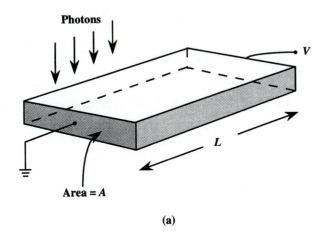

(a)

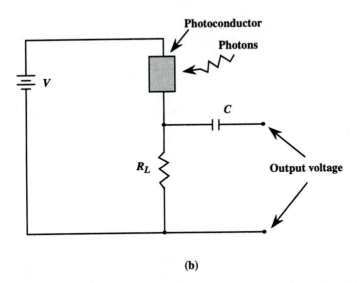

(b)

Figure 7.12: (a) Geometry of a photoconductor of length L and area A. (b) A typical bias circuit for a photodetector. Light causes a change in the resistance of the photoconductor. A blocking capacitor may be used if only the ac signal is to be detected.

The excess carrier density is given by (see Eqn. 7.28)

$$\delta p = \delta n = G_L \tau_p \qquad (7.31)$$

where G_L is the e-h pair generation rate. The change in the conductivity of the material due to the optical signal is called the photoconductivity and is given by Eqn. 7.29 and 7.30 as

$$\Delta\sigma = e\delta p(\mu_n + \mu_p) \qquad (7.32)$$

In the presence of an electric field F, the current density is given by

$$J = (J_d + J_L) = (\sigma_0 + \Delta\sigma)F \qquad (7.33)$$

where J_d is the dark current density of the detector. The photocurrent is thus

$$
\begin{aligned}
I_L = J_L \cdot A &= e\delta p(\mu_n + \mu_p)AF \\
&= eG_L\tau_p(\mu_n + \mu_p)AF
\end{aligned}
\qquad (7.34)
$$

One must keep in mind that $\mu_n F$ and $\mu_p F$ represent the electron and hole velocities. In high electric fields, $\mu_n F$ and $\mu_p F$ simply are the saturation velocities independent of the field. Let us define the transit time of the electrons in the device by

$$t_{tr} = \frac{L}{\mu_n F} \qquad (7.35)$$

The photocurrent now becomes, after expressing $\mu_n F$ in terms of t_{tr} and L using Eqn. 10.34,

$$I_L = eG_L \left(\frac{\tau_p}{t_{tr}}\right)\left(1 + \frac{\mu_p}{\mu_n}\right)AL \qquad (7.36)$$

This is the photocurrent generated in the circuit. We may define the primary photocurrent as

$$I_{Lp} = eG_L AL \qquad (7.37)$$

This would be the photocurrent if each e-h pair simply contributed one charge at the contact, i.e., if there was no gain in the device. The gain of the photoconductive detector is now

$$\boxed{G_{ph} = \frac{I_L}{I_{Lp}} = \frac{\tau_p}{t_{tr}}\left(1 + \frac{\mu_p}{\mu_n}\right)} \qquad (7.38)$$

The gain in the device arises because the electron goes around the circuit several times before it can recombine with a photogenerated hole. Each time the electron goes through the circuit it contributes to the current as shown schematically in Fig. 7.13.

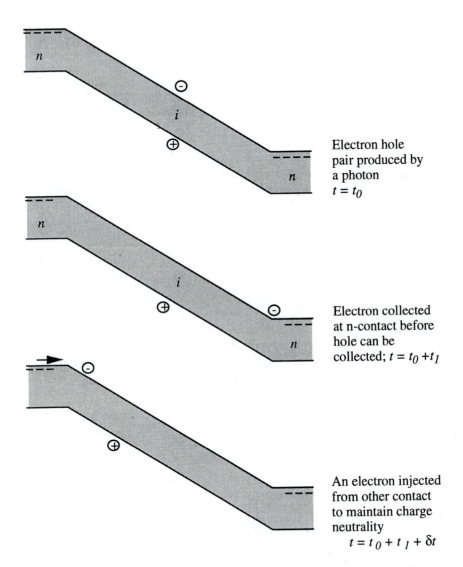

Electron hole
pair produced by
a photon
$t = t_0$

Electron collected
at n-contact before
hole can be
collected; $t = t_0 + t_1$

An electron injected
from other contact
to maintain charge
neutrality
$t = t_0 + t_1 + \delta t$

Figure 7.13: A schematic picture of how gain is produced in an n-i-n detector due to several "round trips" an electron can make before the hole recombines either at the contact or with an electron in the semiconductor i-region.

If τ_p is large and t_{tr} is small, a very large gain can be produced. In Si based devices where τ_p can be very large, gains of 1000 or more can be obtained. However, the improved gain comes at the expense of speed, since the speed is controlled by τ_p and not t_{tr}. Thus the gain bandwidth product is essentially constant.

While the photoconductive n-i-n detector can produce a large gain, it suffers from the presence of a large dark current noise in the detector. As can be seen from Eqn. 7.29, even in dark the device can have a high conductivity and thus have a large dark current. In contrast, a reverse biased p-n or p-i-n diode has a very low dark current, thus allowing a large signal to noise ratio.

EXAMPLE 7.7 Consider a GaAs photoconductor (n-type) with a length of 25 μm and an area of 10^{-6} cm^2. The minority carrier lifetime is 10^{-7} s. A voltage of 5 V is applied across the detector. Calculate the gain of the device using a constant mobility model with $\mu_n = 8000$ cm^2/V-s and $\mu_p = 1000$ cm^2/V-s. Also calculate the gain if an appropriate velocity-field relation is used. The electron transit time is (for a constant mobility model)

$$t_{tr} = \frac{L}{\mu_n F} = \frac{L^2}{\mu_n V} = \frac{(25 \times 10^{-4} \ cm)^2}{(8000 \ cm^2 V^{-1} s^{-1})(5 \ V)} = 1.56 \times 10^{-10} \ s$$

The gain is

$$G_{ph} = \frac{\tau_p}{t_{tr}} \left(1 + \frac{\mu_p}{\mu_n}\right) = \frac{10^{-7} \ s}{1.56 \times 10^{-10} \ s} \left(1 + \frac{1000}{8000}\right) = 641.2$$

If we assume a proper velocity-field relation, $v(e) \cong 1.5 \times 10^7$ cm/s; $v(h) \cong 2 \times 10^6$ cm/s at the applied field of 2 kV/cm, the transit time is

$$t_{tr} = \frac{L}{v(e)} = \frac{25 \times 10^{-4}}{1.5 \times 10^7} = 1.67 \times 10^{-10} \ s$$

The gain is

$$G_{ph} = \frac{10^{-7}}{1.67 \times 10^{-10}} \left(1 + \frac{2 \times 10^6}{1.5 \times 10^7}\right) = 678.6$$

The error produced is minimal at the low fields of the detector. If the applied field was higher, the error in using the constant mobility model would become larger.

7.6   THE P-I-N PHOTODETECTOR

$\longrightarrow$ An important mode of operation of the p-n (or p-i-n) diode under illumination is when the diode is under reverse bias conditions. The reverse bias is, however, not so strong that there are breakdown effects as in the avalanche photodiode to be discussed in the next section.

A schematic of the band profile of a p-i-n detector is shown in Fig. 7.14. Since the device is in reverse bias, the diode current in dark is I_0 and is independent of the

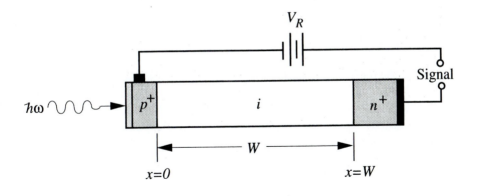

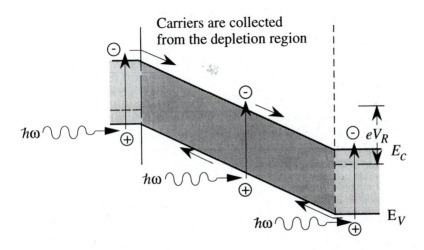

Figure 7.14: A cross-section and energy band profile of a *p-i-n* detector structure. Carriers generated in the depletion region are collected and contribute to the current. If the intrinsic region is thick, the photocurrent is dominated by carriers collected from the depletion region since the carriers generated in the neutral regions contribute a smaller fraction of photocurrent. Since the photocurrent is dominated by the prompt photocurrent, the device response is fast.

applied bias. The photocurrent I_L is essentially due to the carriers generated in the depletion region (*i*-region) that are collected. The diode is reverse biased so that the entire *i*-region is depleted and has a strong electric field. The device response is fast since the photocurrent is primarily due to the prompt photocurrent discussed in Section 7.4.1. The maximum current that can be collected is (we assume that the intrinsic region is larger than the diffusion lengths of the electrons or holes)

$$I_L = eA \int_0^W G_L(x)dx \qquad (7.39)$$

where W is the depletion width. In this expression we will account for the fact that as photons penetrate a material, their intensity decreases through absorption. The generation rate at a point x is given from Eqn. 7.9 by

$$G_L(x) = \alpha J_{ph}(0) \; exp(-\alpha x) \qquad (7.40)$$

where $J_{ph}(0)$ is the photon flux (number per cm^2 per second) at $x = 0$. The photocurrent is then

$$I_L = eA J_{ph}(0) \left[1 - exp\left(-\alpha W\right)\right] \qquad (7.41)$$

If R is the reflectivity of the surface (i.e., the fraction of photons that actually go into the device is $1 - R$), the photocurrent is

$$I_L = eA J_{ph}(0)(I - R) \left[1 - exp(-\alpha W)\right]$$

One measure of the detector efficiency is the ratio of the photocurrent density to the incident flux,

$$\eta_{det} = \frac{I_L}{A J_{ph}(0)} = (I - R) \left[1 - exp(-\alpha W)\right] \qquad (7.42)$$

For high efficiency one must have a small R (by placing anti-reflective coatings) and a long W. However, if W is too long the electron transit time that controls the device speed becomes too large, reducing the device speed. High speed devices have W of about a micron or less and can operate at speeds in excess of 10 GHz.

7.6.1   Material Choice and Frequency Response of a P-I-N Detector

The foremost issue in the detector design is to work with a material which has a good absorption coefficient for the frequencies to be detected. For communication applications where GaAs/AlGaAs ($\hbar\omega \sim 1.45$ eV) sources are used (usually for local area networks), Si detectors are adequate unless high speeds are required. The Si detectors must have an absorption length of $> 10 \; \mu$m. For longer wavelength applications, Ge detectors are used. An important wavelength is the 1.55 μm used for long haul communications since the fiber propagation loss is a minimum. For high speed applications one requires direct gap detectors so that the active absorption length can be brought down to a micron or less. Thus InGaAs detectors are now widely used for long haul communication applications (see Chapter 13).

In the case of night vision applications, materials like Hg$_x$Cd$_{1-x}$Te, InAs, and InSb, which have very narrow bandgaps, are used. One problem with the narrow gap materials is a high dark current (I_0 of the diode). To reduce I_0, the devices are cooled often to liquid He temperature. The various materials available and their bandgaps are shown in Fig. 7.2.

Once a material choice is made, the important issues in device design are:

i) *Minimizing surface reflection:* This is usually done by applying anti-reflective coatings which can reduce the reflection losses from as high as 40% to 2-3%;

ii) *Maximizing the absorption in the depletion region:* For high detector efficiency one must have as large absorption as possible in the depletion width as seen from Eqn. 7.42. However, for speed it is not always a good option to increase the depletion width. Often metal mirrors can be used to increase the optical interaction length of the device by causing the optical beam to take more than one path through the device ;

iii) *Minimizing carrier recombination:* For improved efficiency one also wants as little recombination as possible in the depletion region. This requires the use of high purity material so that there is no trap related recombination;

iv) *Minimize transit time:* For the purpose of high speed, the transit times must be minimized, which means the depletion region should be as short as possible.

In addition to the above issues, the device time response is controlled by the circuit issues. The equivalent circuit of the diode is given by Fig. 7.7. We assume that the diode output is fed into an amplifier. The diode capacitance C_D is, for the reverse bias case, (see Eqn. 6.67)

$$C_D = \frac{\epsilon A}{W} \tag{7.43}$$

where W is the device depletion width. The diode has a series resistance R_s and conductance G_D. For high frequency response the capacitance and resistance are each to be minimized, which usually means reducing the area A since if W is increased too much the device is limited by transit time effects.

If the device capacitance and resistance are optimized, the transit time limits the device response. The effect of the transit time is to prevent the photocurrent from following the optical power modulation at high frequencies. The transit time is controlled by the width of the depletion region and the saturation velocity.

Consider an input optical flux of the form

$$\Phi(t) = \Phi_o + \phi_1 e^{i\omega t} \tag{7.44}$$

To study the temporal response of the detector to this input optical signal we start with the current continuity equations for the electrons and holes,

$$
\begin{aligned}
\frac{\partial p}{\partial t} &= -\frac{\delta p}{\tau} + G - \frac{1}{e}\nabla \cdot J_p \\
\frac{\partial n}{\partial t} &= -\frac{\delta n}{\tau} + G + \frac{1}{e}\nabla \cdot J_n
\end{aligned}
\tag{7.45}
$$

We now assume that in the depletion region of the reverse biased *p-i-n* diode, the current flow is dominated by drift process and the carriers move with saturation velocities. Thus we have ($v_s(h)$, $v_s(e)$ are the hole and electron saturation velocities),

$$
\begin{aligned}
J_p &= e v_s(h) p \\
J_n &= e v_s(e) n
\end{aligned}
\tag{7.46}
$$

The total current is the sum of the conduction current and the displacement current

$$
J = J_p + J_n + \epsilon_s \frac{\partial F}{\partial t}
\tag{7.47}
$$

We now carry out a small signal analysis for this problem by assuming that the carrier generation, carrier densities, and current all have dc and ac components that follow the optical signal, i.e.,

$$
\begin{aligned}
G(x,t) &= G_0(x) + G_1(x) \, exp \, (i\omega t) \\
p(x,t) &= p_0(x) + p_1(x) \, exp \, (i\omega t) \\
n(x,t) &= n_0(x) + n_1(x) \, exp \, (i\omega t)
\end{aligned}
\tag{7.48}
$$

Using these equations in the current continuity equations give us

$$
\begin{aligned}
\frac{\partial J_{p1}}{\partial x} - i\omega \frac{J_{p1}}{v_s(h)} &= -eG_1 \\
\frac{\partial J_{n1}}{\partial x} + i\omega \frac{J_{n1}}{v_s(e)} &= eG_1
\end{aligned}
\tag{7.49}
$$

Note that the generation rate is $G_1 = \alpha\phi_1 exp(-\alpha x)$. These equations are subject to the boundary conditions

$$
\begin{aligned}
J_{n1}(0) &= 0 \\
J_{h1}(W) &= 0
\end{aligned}
$$

The solution to the differential equations then becomes

$$
J_{p1}(x) = -\alpha e \phi_1 \left[\frac{e^{-\alpha x} - e^{-\alpha W} e^{\frac{i\omega(W-x)}{v_s(h)}}}{(\alpha - i\omega/v_s(h))} \right]
$$

$$
J_{n1}(x) = -\alpha e \phi_1 \left[\frac{e^{\frac{i\omega x}{v_s(e)}} - e^{-\alpha x}}{\alpha - i\omega/v_s(h)} \right]
$$

The total current in the device is now

$$
J(\omega) = \frac{1}{W} \int_0^W \left[(J_{n1}(x) + J_{p1}(x)) + \epsilon \frac{\partial E}{\partial t} \right] dx
\tag{7.50}
$$

Defining the electron and hole transit times as

$$t_{tr}^e = \frac{W}{v_s(e)}$$

$$t_{tr}^h = \frac{W}{v_s(h)}$$

we get for the time dependent currently (with time dependence $e^{i\omega t}$),

$$
\begin{aligned}
J(\omega) &= e\phi_1\alpha W \left[\frac{e^{-\alpha W} - 1}{\alpha W(\alpha W - i\omega t_{tr}^h)} + \frac{e^{-\alpha W}(e^{i\omega t_{tr}^h} - 1)}{+i\omega t_{tr}^h(\alpha W - i\omega t_{tr}^h)} \right] \\
&+ e\phi_1\alpha W \left[\frac{1 - e^{i\omega t_{tr}^e}}{i\omega t_{tr}^e(\alpha W + i\omega t_{tr}^e)} + \frac{1 - e^{-\alpha W}}{\alpha W(\alpha W + i\omega t_{tr}^e)} \right] \\
&+ \frac{i\omega\epsilon V}{W}
\end{aligned}
\tag{7.51}
$$

If we examine the short circuit current (by equating $V = 0$), we see that the magnitude of the current decreases as ωt_{tr}^e or ωt_{tr}^h becomes larger than unity. If we assume that the transit time effects are dominated by hole velocity, the magnitude drops by $\sqrt{2}$ at roughly $\omega t_{tr}^h = 2.4$.

One may want to improve the device response by decreasing W and thus decreasing t_{tr}. However, as can be seen from Eqn. 7.51, in this case the current response also decreases. A reasonable compromise between high frequency response and high quantum efficiency is to have W between $1/\alpha$ and $2/\alpha$. If $W = 1/\alpha$, the cutoff frequency of the devices is approximately given by

$$f_{3dB} \cong \frac{2.4}{2\pi t_{tr}} \simeq \frac{0.4 v_s}{W} \simeq 0.4\alpha v_s \tag{7.52}$$

where v_s is the slower of the electron or hole saturation velocities.

From this expression and the quantum efficiency results, there is a clear trade-off between quantum efficiency, input wavelength (which determines α) and speed. In Fig. 7.15, we show some results for Si, Ge, and InGaAs detectors which illustrate this trade-off. Note that Si detectors cannot be used for the 1.55 μm radiation important for long distance fiber optic communication.

EXAMPLE 7.8 Consider a silicon p-i-n photodiode with an intrinsic region of width 10 μm. Light from a GaAs laser at energy $\hbar\omega = 1.43$ eV impinges upon the diode. The optical power is 1 W/cm^2. Calculate the photocurrent density in the detector.

The photon flux incident on the detector is

$$\Phi_0 = \frac{P_{op}}{\hbar\omega} = \frac{1\ W/cm^{-2}}{1.43(1.6 \times 10^{-19}\ J)} = 4.37 \times 10^{18}\ cm^{-2}\ s^{-1}$$

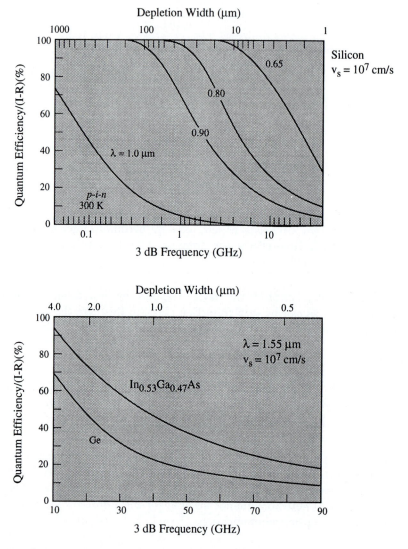

Figure 7.15: a) The 3dB frequency and quantum efficiency tradeoffs in a silicon detector. Results are shown for several wavelengths. b) The high frequency-quantum efficiency tradeoffs in $In_{0.53}Ga_{0.47}As$ and Ge detectors for 1.55 μm radiation.

The absorption coefficient for Si at GaAs wavelength (i.e., photons with energy 1.43 eV) is ~ 700 cm^{-1}. The photocurrent density is (assuming no reflection losses)

$$
\begin{aligned}
J_L &= e\Phi_o \{1 - exp - (\alpha W)\} \\
&= (1.6 \times 10^{-19})(4.37 \times 10^{18}) \left\{ 1 - exp\left(-700 \times 10^{-3}\right) \right\} \\
&= 0.352 \ A/cm^2
\end{aligned}
$$

One can see from this example that Si detectors are capable of producing acceptable response to GaAs photons. Since Si technology is so advanced, one uses Si detectors for GaAs lasers.

7.7 THE AVALANCHE PHOTODETECTOR

$\longrightarrow$ In addition to the *p-i-n* detector discussed in the previous section, an important class of detectors uses the impact ionization or avalanche process to obtain very high gain devices. While in the *p-i-n* detector the gain of the detector can, at most, be unity; in the avalanche photodetector (APD), very large gains can be achieved.

In Chapter 3, Section 3.7.1, we had discussed the basis for the avalanche process in which a high energy electron (hole) creates an electron-hole pair. Usually this process, occurring at high electric fields, limits the high power operation of electronic devices, but in APDs it is exploited to multiply carriers generated by a photon. The electron and holes generated by photon absorption can be used to cause avalanching, as shown in Fig. 7.16. The output signal is thus enhanced.

It must be noted that the avalanche process requires the initial electron to have an energy somewhat greater than the bandgap energy as discussed in Chapter 3. The impact ionization coefficients for the electrons and holes are denoted by α_{imp} and β_{imp}. In Appendix B we show the value of α_{imp} and β_{imp} for some important semiconductors.

Because of carrier multiplication, the APD has a very high gain and is thus used widely for optical communication systems. However, since the multiplication process is random, the device is quite noisy. The noise level depends upon the carrier multiplication factor and the α_{imp}/β_{imp} ratio. A number of material systems including Ge, Si, and many III-V compound semiconductors have been used in designing photodetectors. In the next section we will discuss some of the design issues for APDs. The noise issues will be discussed in the next chapter.

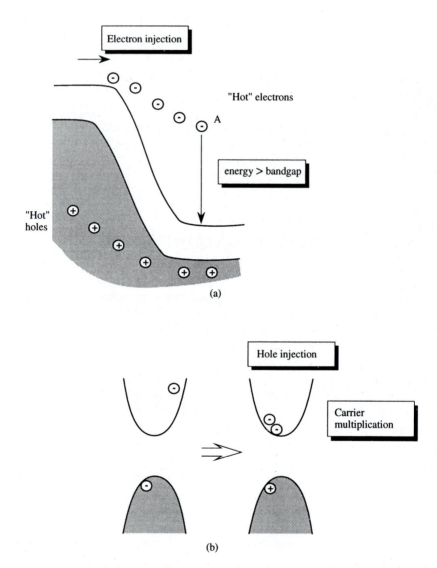

Figure 7.16: (a) A schematic of a reverse biased *p-n* junction. The electron *A* has an energy greater than the bandgap of the semiconductor. (b) A "hot" electron in the conduction band interacts with an electron in the valence band to generate two electrons and a hole, as shown.

7.7.1 APD Design Issues

As in the p-i-n detector, the first design issue for an APD is to have a depletion region which is thick enough to allow absorption of the optical signal. This region has to be $\sim 1/\alpha_{imp}(\hbar\omega)$ and can range from a micron for direct gap semiconductors to several tens of microns for indirect gap materials. The absorbing region and the avalanche region are kept distinct in general (especially if the absorbing region is larger than a micron) because of the difficulty of maintaining a constant high electric field over a long region. Typically electric fields of $\gtrsim 10^5$ V/cm are needed for the avalanche process. If the field is nonuniform local charge oscillations can develop and the device output becomes difficult to control and predict.

An important structure for APDs is the "reach through" structure shown in Fig. 7.17. The structure has an n^+-p-π-p^+ configuration. Photon absorption occurs in the undoped π region with thickness W_{abs}, while the avalanche region has a thin extent in the n^+-p junction with width W_{aval}. One can design the structure so that the field in the absorbing region is high enough that all the carriers move with saturation velocity ($v_s(e)$ or $v_s(h)$). Also, either electrons or holes can be chosen to be injected into the avalanche region. The avalanche process should be initiated by the carrier with the high impact ionization coefficient (electrons in our discussion) to optimize the device response.

In the APD, the current gain is very much dependent upon not only the bias applied, but also thermal fluctuations. Thus heat sinking is crucial in such devices. Also, guard rings are introduced to minimize the electric fields around the p-n junction edges of the device. The guard ring involves an n-dopant which produces a π region in the p part of the device and thus there is a lower field at the $n^+\pi$ region of the guard ring and breakdown is avoided at the edges.

Germanium APDs suffer from a number of inherent problems which limit the useful gain of the devices to $\sim$ 10-20. The causes for these problems are high dark current due to the larger thermal generation of e-h pairs. It is also difficult to obtain high quality substrates for Ge.

Use of heterojunction APDs has been increasing, and III-V compound semiconductors have produced some of the highest performance devices. In case of direct bandgap semiconductors, one does not need the long absorbing region and devices can be built with thin absorbing and avalanching regions (which are the same physical region) as shown in Fig. 7.18 for the $In_{0.53}Ga_{0.47}As$ APD.

Carrier Multiplication

The APD takes advantage of the carrier multiplication that occurs as electrons and holes move through a semiconductor at a very high electric field. At such high electric fields,

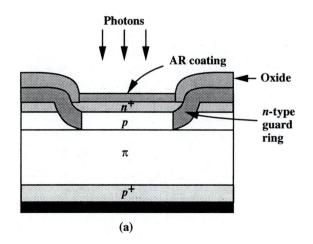

(a)

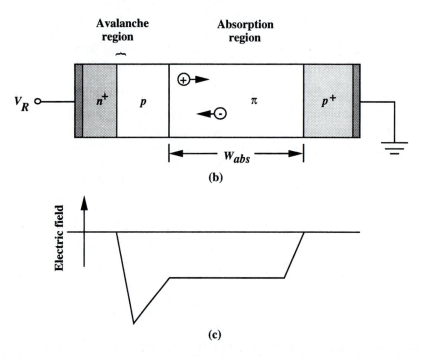

Figure 7.17: (a) A schematic of the reach through APD. (b) The cross-section of the APD showing the regions for absorption and avalanching. In the structure shown, the electrons are responsible for starting the multiplication process. (c) The electric field profile in the APD structure. The strong field at the n^+p junction causes the avalanche process.

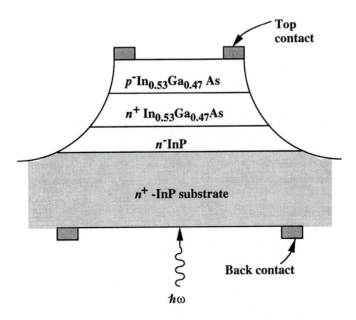

Figure 7.18: A schematic cross-section of a back illuminated InGaAs/InP avalanche photodiode. Due to the short absorption region, in the direct gap semiconductors, one can have the same region for absorption and avalanche processes.

the carriers are moving with saturation velocity and the current is simply proportional to the number of electrons and holes. The electron current has a spatial increment given by

$$\delta i_e = i_e \alpha_{imp} dx + i_h \beta_{imp} dx \tag{7.53}$$

where the first term is the contribution to excess electrons due to the electron current over a distance dx and the second term is due to the excess electrons due to the hole current. We thus have

$$\frac{di_e}{dx} = i_e \alpha_{imp} + i_h \beta_{imp} \tag{7.54}$$

A similar equation exists for the hole current derivative. Even though both the electron and hole current change with spatial position, the total current is constant over the device

$$I = i_e(x) + i_h(x) = \text{constant} \tag{7.55}$$

Let us consider a case where we have an avalanche region extending from $x = 0$ to $x = W$ and at $x = W$, only electrons are injected. Thus we have

$$i_h(W) = 0; \quad i_e(W) = I \tag{7.56}$$

Expressing $i_h(x)$ in terms of $i_e(x)$ and I by using Eqn. 7.55, we get from Eqn. 7.54

$$\frac{di_e(x)}{dx} - (\alpha_{imp} - \beta_{imp})i_e(x) = \beta_{imp} I \tag{7.57}$$

This equation is of the form

$$\frac{dy}{dx} + Py = Q \tag{7.58}$$

and has a standard solution given by

$$i_e(x) = \frac{i_e(0) + \int_0^x \beta_{imp} I \ exp \ \left\{ - \int_0^x (\alpha_{imp} - \beta_{imp}) dx' \right\} dx}{exp \ \left\{ - \int_0^x (\alpha_{imp} - \beta_{imp}) dx' \right\}} \tag{7.59}$$

The multiplication factor for the device can be defined by the relation

$$M_e = \frac{I}{i_e(0)} = \frac{i_e(W)}{i_e(0)} \tag{7.60}$$

i.e., the ratio of the electron current injected at $x = 0$ and the total current in the device. We get

$$M_e = \frac{i_e(0) + i_e(W) \int_0^W \beta_{imp} \ exp \ \left\{ - \int_0^x (\alpha_{imp} - \beta_{imp}) dx' \right\} dx}{i_e(0) \ exp \ \left\{ - \int_0^W (\alpha_{imp} - \beta_{imp}) dx \right\}} \tag{7.61}$$

Using standard solutions for the integral

$$exp \left\{ - \int_0^W (\alpha_{imp} - \beta_{imp}) dx \right\} = 1 - \int_0^W (\alpha_{imp} - \beta_{imp}) \ exp \left\{ - \int_0^x (\alpha_{imp} - \beta_{imp}) dx' \right\} dx \tag{7.62}$$

we get

$$M_e = \frac{1}{1 - \int_o^W \alpha_{imp} \ exp\{ - \int_o^x (\alpha_{imp} - \beta_{imp}) dx' \} dx} \tag{7.63}$$

The condition for breakdown is when $M_e \longrightarrow \infty$, i.e., when

$$\int_o^W \alpha_{imp} \ exp\{ - \int_o^x (\alpha_{imp} - \beta_{imp}) dx' \} dx = 1 \tag{7.64}$$

If the avalanche process is taking under a uniform electric field, the values of $\alpha_{imp}, \beta_{imp}$ have no spatial dependence and we have

$$M_e = \frac{1}{1 - \alpha_{imp} \int_o^W exp\{ -(\alpha_{imp} - \beta_{imp}) x \} dx}$$

$$= \frac{1}{1 - \frac{\alpha_{imp}}{\alpha_{imp} - \beta_{imp}} [1 - exp\{ -(\alpha_{imp} - \beta_{imp}) W \}]} \tag{7.65}$$

If α_{imp} and β_{imp} are the same, we simply get

$$M_e \longrightarrow \frac{1}{1 - \alpha_{imp}W} \tag{7.66}$$

The values of M_e (or M_h) that can be obtained in real devices are limited by other processes that may occur at high fields, especially in narrow bandgap semiconductors. For example, tunneling current due to band to band tunneling limits M to ~ 10 in detection using $In_{0.53}Ga_{0.47}As$ material.

In an experimental setup, two factors related to circuit parameters limit the multiplication level reached. One is the series resistance R_s between the junction and the diode terminals. The second factor comes from the fact that once multiplication starts, the device temperature increases and this reduces α_{imp} and β_{imp} and thus limits $M_e(M_h)$. In a diode with a breakdown voltage V_B, the experimentally observed multiplication factor can be fitted to the following relation:

$$M = \frac{1}{\left[1 - \left(\frac{V - IR}{V_B}\right)\right]^n} \tag{7.67}$$

where n is a parameter depending upon the device design, R is an effective resistance which includes the series resistance R_S and any thermal effects, V is the applied bias.

7.7.2 APD Bandwidth

A key attraction of the APD's is the high gain that can be achieved in the device. Thus the device is suitable for detection of very low photon intensities. However, a price has to be paid in terms of the device bandwidth and noise. In an APD having the general configuration of Fig. 7.17, the device response is limited by three important times:

i) the transit time across the absorbing region

$$t_{tr}(e) = \frac{W_{abs}}{v_s(e)} \tag{7.68}$$

ii) the time required for the avalanche process to develop, t_A;

iii) the transit time for the holes generated during the avalanche process to transmit through the absorbing region back to the p-region.

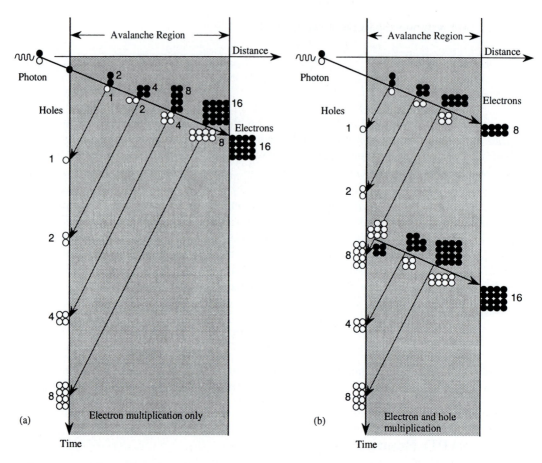

Figure 7.19: The avalanche build-up process shown as a function of time. (a) Only electrons are causing the carrier multiplication. (b) Both electrons and holes are causing the multiplication. (After J. Gowar, *Optical Communication Systems*, Prentice-Hall, Englewood Cliffs, New Jersey (1989).)

$$t_{tr}(h) = \frac{W_{abs}}{v_s(h)} \qquad (7.69)$$

The avalanche delay time t_A depends upon the value of α_{imp}/β_{imp}. If $\alpha_{imp} \gg \beta_{imp}$, just a single pass of the electrons across W_{aval} generates the entire avalanche process as shown in Fig. 7.19. However, for a general α_{imp}/β_{imp} ratio, the delay time is

$$t_A = \frac{M \beta_{imp} W_{aval}}{\alpha_{imp} v_s(e)} \qquad (7.70)$$

The overall device response time is then (defining $\beta_{imp}/\alpha_{imp} = k$)

$$\tau \cong \frac{W_{abs} + Mk\ W_{aval}}{v_s(e)} + \frac{W_{abs} + W_{aval}}{v_s(h)} \tag{7.71}$$

It is important to note that for large multiplication values, the produce $\frac{M}{\tau}$ is a constant. The gain-bandwidth product thus remains a constant for a device with large multiplication. This outcome is similar to that observed in the case of photoconductors as well. Thus, if one wants a very high detectivity for very weak optical signals, one has to sacrifice the device speed. Another important conclusion from the Eqn. 7.71 is that gain-bandwidth product can be optimized by choosing a material system with $\alpha_{imp} \gg \beta_{imp}$.

EXAMPLE 7.9 Consider a typical avalanche photodiode with the following parameters:

Incident optical power,	$P_{op} \cdot A$	$=$	$50\ mW$
Efficiency,	η_{det}	$=$	90%
Optical frequency,	ν	$=$	$4.5 \times 10^{14}\ Hz$
Breakdown voltage,	V_B	$=$	$35\ V$
Diode voltage,	V	$=$	$34\ V$
Dark current,	I_0	$=$	$10\ nA$
Parameter n' for the multiplication		$=$	2

Assume that the series resistance is negligible. Calculate the a) multiplication factor; b) photon flux; c) photocurrent.

a) The multiplication factor from Eqn. 7.67 is

$$M = \left[1 - \left(\frac{34}{35}\right)^2\right]^{-1} = 16.67$$

b) The photon flux is

$$I_{ph} = \frac{P_{op}A}{h\nu} = \frac{(50 \times 10^{-3}\ W)}{(6.625 \times 10^{-34}\ Js)(4.5 \times 10^{14}\ Hz)}$$
$$= 1.68 \times 10^{17}\ s^{-1}$$

c) The unmultiplied photocurrent is

$$I_L = e\eta_{det}I_{ph} = (1.6 \times 10^{-19}\ C)(1.51 \times 10^{17}\ s^{-1})$$
$$= 24.16\ mA$$

The multiplied photocurrent is

$$M \cdot I_L = (24.16\ mA)(16.67) = 0.4\ A$$

7.8   THE PHOTOTRANSISTOR

$\longrightarrow$ While the APD discussed above provides very high gain detection, it is an inherently high noise device due to the random nature of the carrier multiplication process. Another device which can produce gain and function as a detector is the bipolar transistor. The phototransistor, the name for the bipolar device used for optical detection, provides high gain due to the transistor action. The device is also a low noise device when compared to an APD.

The bipolar transistor is essentially a device with two coupled p-n diodes, as shown in Fig. 7.20a. The n-p-n transistor is discussed in Appendix D, where we present the underlying operation principles of current gain and transistor action. In this subsection, we will use the results derived in Appendix D to understand how the phototransistor works.

When a bipolar device is biased in the forward active mode (i.e., the emitter-base junction (EBJ) is forward biased and the base-collector junction (BCJ) is reverse biased), the band profile has the form shown in Fig. 7.20b. Normally, in a bipolar transistor, the injection of a small base current causes a small change in the forward bias across the EBJ, causing a large injection of current in the forward biased junction. If the base width is small and the base is of high quality, essentially all of this current is collected in the collector. The current gain, defined as the ratio of the collector and base current, can be quite high (see Example 7.10).

In the case of the phototransistor, the base current is not provided by an external supply (often there is no base contact on the phototransistor), but via an optical signal. Light shining on the device creates electron-hole pairs. These pairs are generated throughout the device, although in an HBT, the emitter can be made of a layer bandgap material and hence, designed to be transparent.

The phototransistor doping levels are designed so that the EBJ depletion width is quite small, while the BCJ depletion width is large, so that the optical signal is absorbed primarily in the BCJ depletion region. This also requires a small width for the base. The photogenerated holes in the BCJ depletion region provide the base current, as shown in Fig. 7.20b. The fraction of the photon current absorbed in the BCJ depletion region is

$$\frac{I_{ph}(BCJ)}{I_{ph}(0)} = \eta \ exp \ (-\alpha W_{bn})[1 - \ exp \ (-\alpha W_{BCJ})] \qquad (7.72)$$

where η is the quantum efficiency of the material ($\eta \sim 1$ for a high quality material);

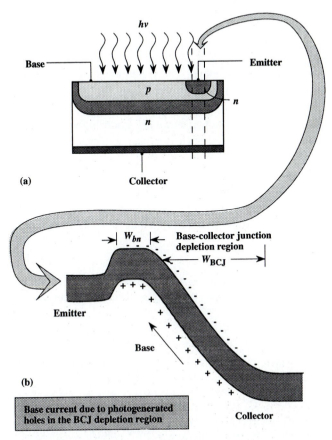

Figure 7.20: (a) A schematic of the phototransistor. (b) Band diagram of the phototransistor which is in the open base mode. Holes generated in the reverse biased base-collector junction region provide a base current signal which causes the electrons to be injected from the emitter.

the second term is the loss suffered by absorption in the neutral base; and the third term (in the brackets) is the absorption in the depletion region of width W_{BCJ}.

If the phototransistor is designed so that W_{bn} is so small that $exp\,(-\alpha W_{bn}) \sim 1$ and the value of W_{BCJ} is large,

$$I_{ph}(BCJ) \approx \eta I_{ph}(0) \tag{7.73}$$

In this case, essentially all of the optical signal provides a base current, since the high field in the BCJ depletion region injects the holes generated in that region into the base. The optical gain of the device is then $(I_B = eI_{ph}(BCJ))$

$$\Gamma_G = \frac{I_c}{eI_{ph}(0)} = \frac{\eta I_c}{I_B} = \eta\beta \tag{7.74}$$

where β is the current gain of the device. The dependence of β on the device parameters is discussed in Appendix D (see also Example 7.10).

The phototransistor does not have a very good high frequency response due to the very large capacitance associated with the base-collector junction. However, it finds important uses due to its low noise and high gain.

Phototransistors can be designed using heterojunction bipolar transistors. As discussed in Appendix D, heterojunctions allow one to achieve very high emitter efficiencies and yet achieve low base resistance. Thus, both gain speed and device speed are improved.

EXAMPLE 7.10 This example will examine the dependence of current gain on the minority carrier recombination in the base. Consider a silicon *npn* transistor with the following parameters at 300 K:

Emitter doping,	N_{de}	=	5×10^{17} cm^{-3}
Base doping,	N_{ab}	=	10^{17} cm^{-3}
Base width,	W_{bn}	=	1.0 μm
Diffusion coefficient,	$D_b = D_e$	=	20 cm^2/s

Calculate the current gain for the two cases: i) minority carrier lifetime for the electrons and holes $= 10^{-6}$ s; ii) minority carrier lifetime $= 10^{-8}$ s. Such a reduction in lifetime can be obtained in silicon by introducing defects (see Appendix D for the mathematics used in this Example).

In the first case, the diffusion lengths are

$$L_e = \sqrt{D_e \tau_e} = (20 \times 10^{-6})^{1/2} = 44.7 \ \mu m$$
$$L_b = 44.7 \ \mu m$$

Also we have

$$p_{eo} = \frac{n_i^2}{N_{de}} = \frac{(1.5 \times 10^{10})^2}{5 \times 10^{17}} = 4.5 \times 10^2 \ cm^{-3}$$

$$n_{bo} = \frac{n_i^2}{N_{ab}} = \frac{(1.5 \times 10^{10})^2}{10^{17}} = 2.25 \times 10^3 \ cm^{-3}$$

The current gain α is

$$\alpha = \left[1 - \frac{p_{eo}D_e W_{bn}}{n_{bo}D_b L_e}\right]\left[1 - \frac{W_{bn}^2}{2L_b^2}\right]$$

$$= \left[1 - \frac{(4.5 \times 10^2)(20)(1.0 \times 10^{-4})}{(2.25 \times 10^3)(20)(44.7 \times 10^{-4})}\right]\left[1 - \frac{(1.0 \times 10^{-4})^2}{2(44.7 \times 10^{-4})^2}\right]$$

$$= (1 - 4.47 \times 10^{-3})(1 - 2.5 \times 10^{-4}) = (0.9955)(0.9998)$$

$$= 0.9953$$

The current gain is dominated in this case by the emitter efficiency since the base transport factor (second term) is close to unity. The current gain β is

$$\beta = \frac{\alpha}{1 - \alpha} = 210.7$$

In the second case, we have

$$L_e = \sqrt{D_e \tau_e} = (20 \times 10^{-8})^{1/2} = 4.47 \; \mu m$$
$$L_b = 4.47 \; \mu m$$

The current gain α is

$$\alpha = \left[1 - \frac{(4.5 \times 10^2)(20)(1.0 \times 10^{-4})}{(2.25 \times 10^3)(20)(4.47 \times 10^{-4})}\right]\left[1 - \frac{(1 \times 10^{-4})^2}{2(4.47 \times 10^{-4})^2}\right]$$

$$= \left[1 - 4.47 \times 10^{-3}\right]\left[1 - 2.5 \times 10^{-2}\right] = (0.9955)(0.974)$$

$$= 0.9696$$

In this case the base transport factor is dominating the current gain reduction. The gain β is

$$\beta = 3.19$$

One sees a big loss in β as the minority carrier time decreases.

7.9 METAL-SEMICONDUCTOR DETECTORS

$\longrightarrow$ An extremely important class of photodetectors involves the use of a Schottky barrier produced between a metal and a lightly doped semiconductor. We have discussed the workings of the Schottky barrier in Chapter 6. As discussed in that chapter, we had seen that a key advantage of the Schottky barrier device is that being a majority carrier device, it does not suffer from speed delays arising from minority carrier lifetime issues.

Schottky barrier based devices involve two kinds of configuration. In Fig. 7.21a we show a device which is a simple mesa structure with an n^+ layer on a semi-insulating substrate. The active absorbing layer is lightly doped ($N_d \sim 10^{15}$ cm^{-3}) and a thin semitransparent metal layer is deposited on it. The metal film is thick enough to allow the Schottky barrier formation (~300-400 Å) but thin enough to allow light to pass through. For high performance the metal film is coated with dielectric anti-reflection coatings and the device area is kept as small as 10^{-5} cm^2 (~50 μm diameter mesa diodes).

The band profile of the Schottky barrier diode is shown in Fig. 7.21b along. Also shown are the Schottky barrier height $e\phi_{bn}$ and the potential drop across the barrier. When light impinges upon the diode, the diode can respond in two important regimes:

i) $\hbar\omega > e\phi_{bn}$: In this case, electrons can be excited in the metal barrier to overcome the Schottky barrier height. As a result, a photocurrent will flow in the device. This current will add to the dark current in the reverse bias diode.

ii) $\hbar\omega > E_g$: In this case, e-h pairs will be created in the semiconductor. As in the case of the photodiode, the carriers generated in the depletion region will be swept out to produce photocurrent.

In high speed devices the depletion region is less than a micron so that device speeds can be extremely high. With proper design Schottky barrier diodes can operate up to 150 GHz.

A second class of metal semiconductor detectors is the metal-semiconductor-metal (MSM) detector in which two Schottky barriers are placed in a planar geometry close to each other. In actual design the approach used is the interdigitated scheme shown in Fig. 7.22a. The spacing between the fingers is ~1-5 μm so that when a bias is applied between the contacts, the region between the fingers can be completely depleted.

As seen in Fig. 7.22b, when a bias is applied across the fingers, one junction becomes reverse biased, while the other one becomes forward biased. However, since the semiconductor is depleted, the current in the forward biased junction is not the usual high electron forward bias current. Instead, the dark current in the forward biased junction is due to the hole current injected from the metal over the barrier $e\phi_{bp}$ as shown in Fig. 7.22b. As a result, under a strong applied bias, the dark current of the device is equal to the reverse saturation currents from electrons and holes. As discussed in Chapter 6, we have for the dark current density

$$J = A_n^* T^2 e^{-e\phi_{bn}/k_B T} + A_p^* T^2 e^{-\phi_{bp}/k_B T}$$

where A_n^* and A_p^* are the electron and hole effective Richardson constants. The dark current density is usually higher than that achievable in p-i-n diodes for reasons discussed in Chapter 6. However, sufficiently low dark current can be achieved for most applications.

The MSM detectors are found to have internal gain, often at even low applied biases where impact ionization cannot occur. This suggests the possibility of photoconductive gain enhanced by traps which may capture and re-emit either electrons or holes. MSM diodes have been fabricated in both GaAs and InGaAs systems. Thus these devices can be applicable in both local area networks and long haul communication systems. It is also important to point out that MSM detectors are very attractive for OEIC applications.

7.10 QUANTUM WELL INTERSUBBAND DETECTOR

↜→ An important application of detectors is in the area of the detection of long wavelength radiation (λ ranging from 5-20μm). If a direct band to band transition is to be used for such detectors, the bandgap of the materal has to be very small. An important material system in which the bandgap can be tailored from 0 to 1.5 eV is the HgCdTe alloy (see Figs. 7.2 and 7.6). The system is widely used for thermal imaging, night vision applications, etc. However, the small bandgap HgCdTe is a very "soft" material which is very difficult to process. Thus the device yield is rather poor. The quantum well intersubband detector offers the advantages of long wavelength detection using established technologies such as the GaAs technology.

In Fig. 7.23, we show a quantum well which is doped so that the ground state has a certain electron density and the excited state is unoccupied. As discussed in Chapter 4, Section 4.6, when a photon with energy equal to the intersubband separation impinges upon the quantum well, the light is absorbed and the ground state electron is scattered into the excited state.

For the absorption process to produce an electrical signal, one must have the following conditions satisfied:

i) the ground state electrons should not produce a current. If this is not satisfied, there will be a high dark current in the detector. The electrons in the ground state carry current by thermionic emission over the band discontinuity. At low temperatures this process can be suppressed.

ii) It should be possible to extract the excited state electrons from the quantum well so that a signal can be produced. The excited electron state should, therefore, be designed to be near the top of the quantum well barrier, so that the excited electrons can be extracted with ease by an applied electric field.

In Chapter 4, Section 4.6, we had seen that absorption coefficients for the intersubband transitions can approach 10^4 cm^{-1}. Thus a series of multiquantum wells with an effective width of ~ 1.0 μm can be used for efficient detection of light. Also, it is possible to tailor the intersubband separation by either the well width variation or

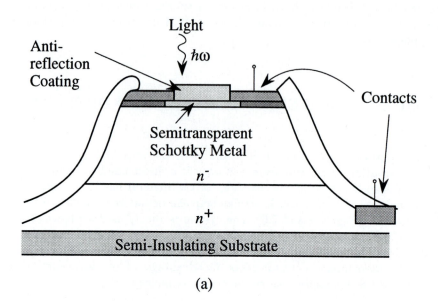

(a)

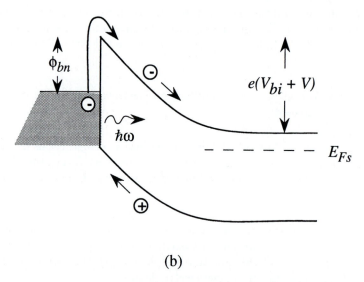

(b)

Figure 7.21: (a) A schematic of the Schottky barrier detector. (b) The band profile of the detector. V_{bi} is the built-in voltage and V is the applied bias.

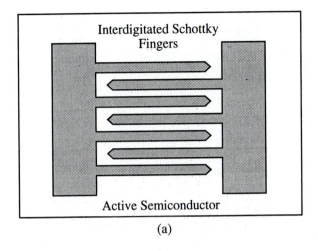

(a)

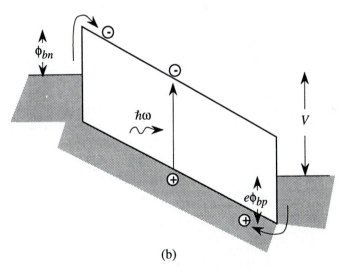

(b)

Figure 7.22: (a) A schematic of the MSM detector using interdigitated Schottky fingers. (b) Band profile of the MSM photodiode under an applied bias.

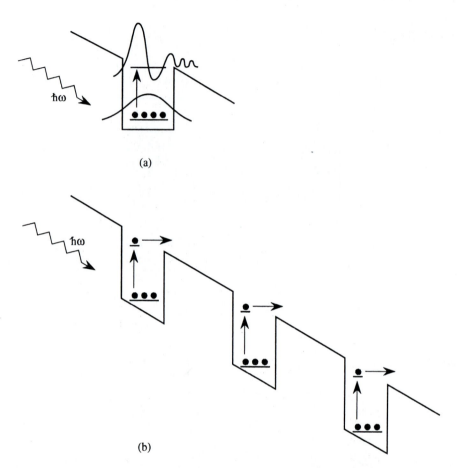

Figure 7.23: (a) A schematic of the electron wavefunction in the ground state and the excited state. The photon causes transitions to the excited state from where the electrons are collected as shown in (b).

by the barrier height variation. Thus an extremely versatile device can be fabricated.

We had discussed in Chapter 4 that for the conduction band quantum well in which the central cell symmetry of the states is pure s-type, the light is absorbed only if it is incident z-polarized. Here z is the quantum well growth direction. Thus, for vertical incidence, there is no absorption. This is an important drawback for such detectors. One way to overcome this problem is to use etched mirrors on the surface to reflect vertically incident light so that it has a z-polarization. Another way to avoid this is to use the intersubband transitions in the valence band where, due to the mixed nature of the HH and LH states, the z-polarization rule is not valid. However, the poor hole transport properties and the difficulty in reducing the dark current reduce the detector performance. In Fig. 7.24, we show a comparison of the band to band and intersubband

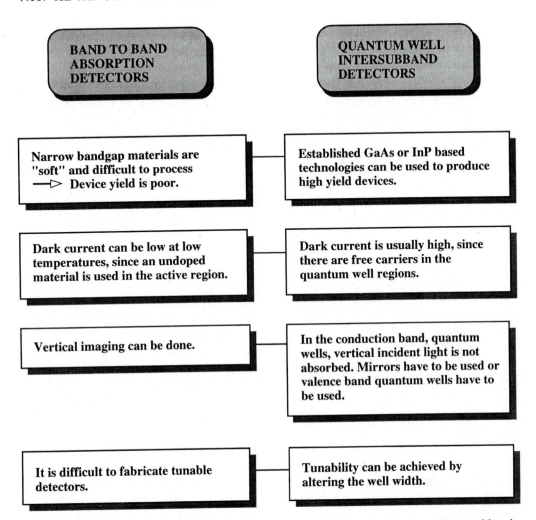

Figure 7.24: A comparison of band to band absorption detectors and quantum well intersubband detectors for long wavelength detectors

detectors for long wavelength detection.

7.11 ADVANCED DETECTORS

$\mathcal{R}$ Important issues in detectors are: i) *tunability*; ii) *speed*; and iii) *integration*. The issue of tunability is an important one and involves mainly the use of different material systems with bandgap that fits a given range of photon energies. A key challenge in this context is the detection of long wavelength photons ($\lambda \sim 10$ - 14 μm). This range is important for night vision applications, thermal imaging (for medical applications), and vision through fog. Two approaches are being pursued for long wavelength detection. These involve the use of narrow gap materials such as HgCdTe alloy or InAsSb alloy or multiquantum well structures. In narrow gap alloys the difficulty arises due to the inherent "softness" of the material which makes it easier to produce defects and makes processing difficult. Nevertheless, focal plane arrays are now made from HgCdTe alloys and their heterostructures.

The intra-subband levels in a quantum well can be tailored to any arbitrary separation by adjusting the quantum well geometry. If the lower level is filled with electrons, photons can be detected by exciting electrons to upper subbands from which they can be collected.

The speed of detectors is essentially controlled by the RC time constant and the transit time for the carriers. The device speed design issues are thus similar to those for electronic devices. The devices are made as small as possible and for high speed materials such as InGaAs, it is possible to have bandwidths approaching 150 GHz with current technology.

An important class of detectors for high speed and integration is the Schottky metal-semiconductor detector. This detector is easy to fabricate and has an extremely high speed. Schottky detectors have been shown to have 3dB bandwidths approaching 160 GHz.

The advent of heteroepitaxy has allowed one to make a number of advances in detector technology. We have already discussed the inter-subband detector for long wavelength application. A number of other devices has been proposed and demonstrated using the concepts of heterostructures often coupled with the ability to abruptly alter the doping in epitaxial growth. We will give a brief review of some such devices.

APDs Based on Quantum Wells

In our discussions on the APD performance issues, we noted that the ratio of α_{imp} and β_{imp} is quite important. In the next chapter, when we discuss noise in APDs, we will see that for low noise applications one must have $\alpha_{imp} >> \beta_{imp}$ or $\beta_{imp} >> \alpha_{imp}$. For a given material system, the relation between α_{imp} and β_{imp} depends upon the details of the bandstructure and the relative scattering rates of electrons and holes. It is, therefore, not possible to alter the α_{imp}/β_{imp} ratio for a given semiconductor. However, in quantum well structures, the ratio can be considerably altered.

A number of experimental studies have confirmed the potential of quantum wells to alter α_{imp}/β_{imp}. The theoretical understanding of the processes involved is not quite complete although several effects seem to be important. As shown in Fig. 7.25, when electrons and holes travel across a heterostructure, the transport process of the

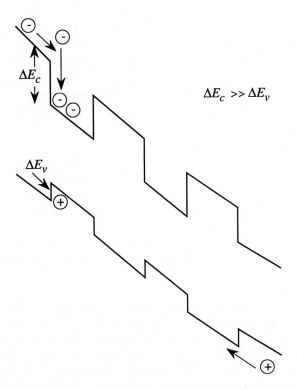

ΔE_c

$\Delta E_c \gg \Delta E_v$

ΔE_v

Figure 7.25: The use of quantum well structures to alter α_{imp}/β_{imp}. In the figure, electron impact ionization is enhanced due to the large gain in the energy of electrons as they enter the well region. Asymmetry in ΔE_c and ΔE_V as well as difference in electron, hole transport can alter α_{imp}/β_{imp}.

individual carriers can be considerably altered. Several causes may lead to the change in α_{imp}/β_{imp}. These are:

i) A strong asymmetry in $\Delta E_c/\Delta E_V$: If ΔE_c is much larger than ΔE_V as shown in Fig. 7.25, as electrons go across the barrier they gain an energy ΔE_c when they enter the well region. The holes, on the other hand, only gain an energy ΔE_c. As a result, α_{imp} can be enhanced since a larger fraction of electrons have enough energy to impact ionize.

ii) Suppression of hole transport by trapping in wells: The value of α_{imp}/β_{imp} can be altered if the energy gain of one of the carriers (say the hole) can be suppressed. This could occur if the holes with heavier masses get trapped in the valence band quantum well during their transport. This would reduce β_{imp} and thus alter α_{imp}/β_{imp}.

iii) Carrier carries scattering related "multiplication:" In this approach, applicable at low electric fields, one assumes that a certain fraction of electrons and holes are trapped in the quantum wells. "Hot" electrons or holes can knock these trapped carriers out and as a result enhance the current. This process does not involve band to band multiplication. It involves allowing trapped carriers to become mobile.

The use of quantum well structures in APDs has resulted in some of the lowest noise photodetectors. Both GaAs and Inp based technologies have been used.

Modulated Barrier Photodiodes

In our discussion of the Schottky barrier photodiode we have seen how a built-in potential barrier can be exploited for photodetection. However, the Schottky barrier height is a material parameter which cannot be altered. A device called modulated barrier diode or camel diode (because of its band profile) offers the versatility of a variable barrier height. The device structure is shown in Fig. 7.26 and consists of a $n^+p^+n^-$ structure. The p^+ region is kept thin enough that it is completely depleted at zero bias as shown in Fig. 7.26. As shown in Fig. 7.26, the barrier to carrier injection, say from the n^+ side, can be varied by an external bias.

The modulated barrier photodiode can be made from a heterostructure where the top region can be a transparent wide bandgap material so that the optical absorption occurs in the p^+n^- depletion region. When photons are absorbed in the device, electrons and holes are produced as shown in Fig. 7.26b. The electrons are swept away by the field while the holes are trapped in the triangular potential well. The hole lifetime is then determined by recombination time with electrons and thermionic emission over the barrier. Due to the large and tailorable hole lifetime, a high gain can be obtained in the device.

Exciton Based Detectors

In Chapter 5, we have discussed the excitonic phenomenon in semiconductors. The excitons produce sharp resonances in the absorption spectra just below the bandedges. As discussed in Chapter 5, these resonances are particularly strong in quantum well structures. Since the width of these resonances is ~ 2 meV, it is possible to exploit them for a variety of detection schemes. In Chapter 12, we will discuss how the absorption and photocurrent can be exploited for a variety of optoelectronic devices. The sharpness of the excitonic resonances allows one to use them for wavelength selective detection, an area of great interest in optical communications. However, as discussed in Chapter 5 and 12, a number of technology challenges remain for excitonic devices.

7.12   CHAPTER SUMMARY

$\mathcal{R}$ In this chapter we have examined optoelectronic devices which can convert an optical signal to an electrical signal. Most of these devices depend upon band to band

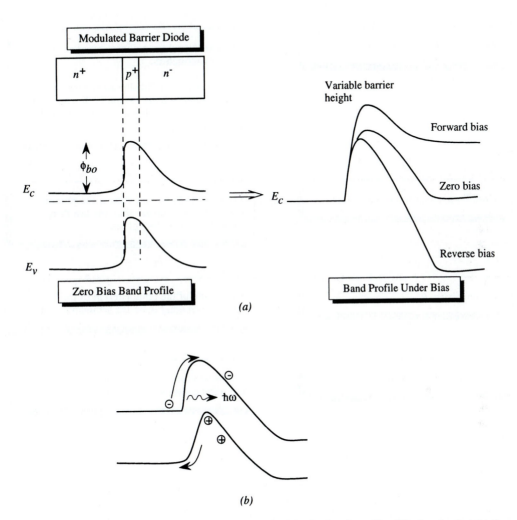

Figure 7.26: (a) A modulated barrier diode along with the band profile. The barrier height for carrier injection from the n^+ side can be tailored by varying the applied bias as shown. (b) Electron and hole generation when light impinges on the diode.

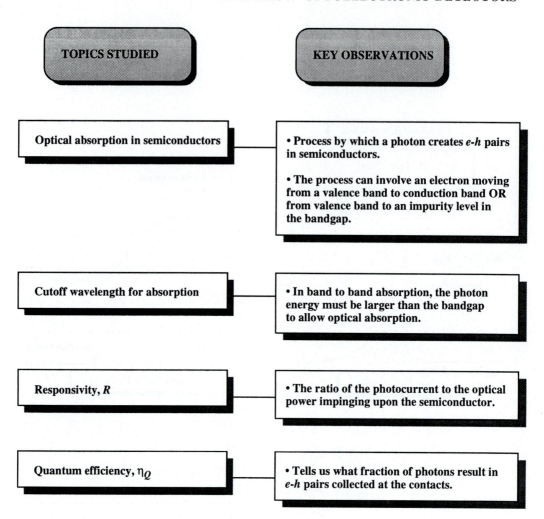

Table 7.3: Summary table.

transitions in which electrons are transferred from the valence band to the conduction band by the photons. The key findings of this chapter are summarized by the chapter summary tables (Tables 7.3-7.5).

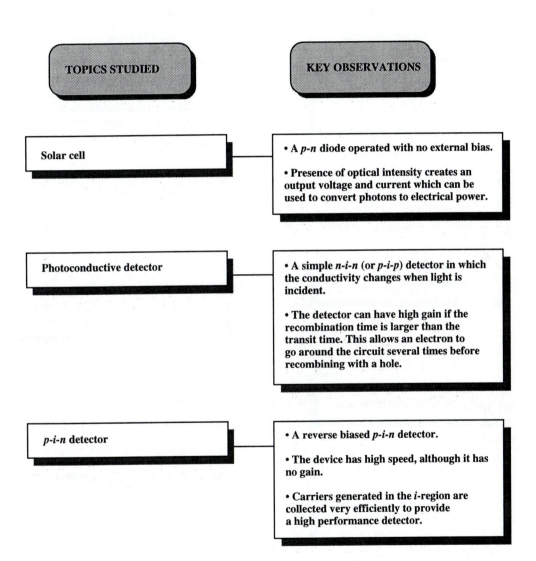

TOPICS STUDIED

KEY OBSERVATIONS

Solar cell

- A *p-n* diode operated with no external bias.

- Presence of optical intensity creates an output voltage and current which can be used to convert photons to electrical power.

Photoconductive detector

- A simple *n-i-n* (or *p-i-p*) detector in which the conductivity changes when light is incident.

- The detector can have high gain if the recombination time is larger than the transit time. This allows an electron to go around the circuit several times before recombining with a hole.

p-i-n detector

- A reverse biased *p-i-n* detector.

- The device has high speed, although it has no gain.

- Carriers generated in the *i*-region are collected very efficiently to provide a high performance detector.

Table 7.4: Summary table.

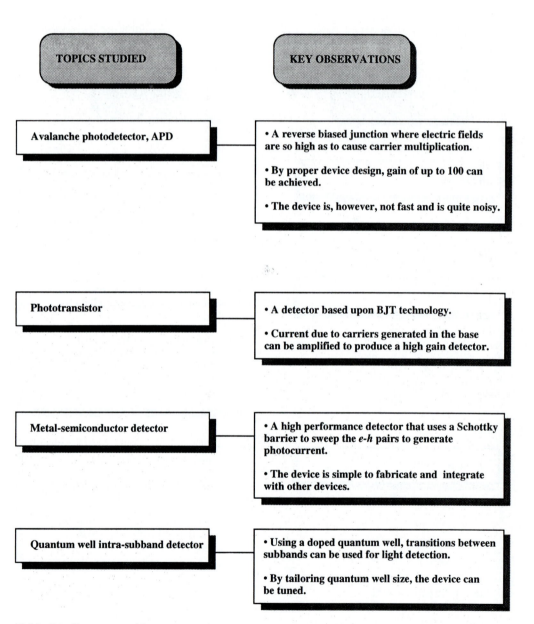

Table 7.5: Summary table.

7.13   PROBLEMS

Section 7.2

7.1 The bandgap of the $Hg_{1-x}Cd_x$Te alloy is given by the expression

$$E_g(x) = -0.3 + 1.9x \quad (eV)$$

Calculate the composition of an alloy which gives a cutoff wavelength of a) 10 μm; b) 5.0 μm.

7.2 Calculate the cutoff wavelength for a GaAs detector. If the cutoff wavelength is to be decreased to 0.7 μm, how much AlAs must be added to a GaAs? Assume that the bandgap of $Ga_{1-x}Al_x$As is given by

$$E_g(x) = 1.43 + 1.25x \quad (eV)$$

as long as $x \leq 0.4$.

7.3 Calculate the absorption coefficient for GaAs for photons with energy 1.8 eV. Calculate the fraction of this light absorbed in a GaAs sample of thickness of 0.5 μm.

7.4 An optical power density of 1 W/cm^2 is incident on a GaAs sample. The photon energy is 2.0 eV and there is no reflection from the surface. Calculate the excess electron-hole carrier densities at the surface and 0.5 μm from the surface. The e-h recombination time is 10^{-8} s.

7.5 Assume that all the photons in an optical beam produce an electron-hole pair in a Ge detector. If all the carriers are collected calculate the responsivity for photon energies of a) 0.7 eV; b) 1.0 eV; c) 2.0 eV.

Section 7.4

7.6 Consider a long Si p-n junction with a reverse bias of 1 V at 300 K. The diode has the following parameters:

Diode area,	A	$= 1\ cm^2$
p-side doping,	N_a	$= 3 \times 10^{17}\ cm^{-3}$
n-side doping,	N_d	$= 10^{17}\ cm^{-3}$
Electron diffusion coefficient,	D_n	$= 12\ cm^2/s$
Hole diffusion coefficient,	D_p	$= 8\ cm^2/s$
Electron minority carrier lifetime,	τ_n	$= 10^{-7}\ s$
Hole minority carrier lifetime,	τ_p	$= 10^{-7}\ s$
Optical absorption coefficient,	α	$= 10^3\ cm^{-1}$
Optical power density,	P_{op}	$= 10\ W/cm^2$
Photon energy,	$\hbar\omega$	$= 1.7\ eV$

Calculate the photocurrent in the diode.

7.7 Consider a long Si p-n junction solar cell with an area of 4 cm^2 at 300 K. The solar cell has the following parameters:

n-type doping,	N_d	$=$	$10^{18}\ cm^{-3}$
p-type doping,	N_a	$=$	$3 \times 10^{17}\ cm^{-3}$
Electron diffusion coefficient,	D_n	$=$	$15\ cm^2/s$
Hole diffusion coefficient,	D_p	$=$	$7.5\ cm^2/s$
Electron minority carrier lifetime,	τ_n	$=$	$10^{-7}\ s$
Hole minority carrier lifetime,	τ_p	$=$	$10^{-7}\ s$
Photocurrent,	I_L	$=$	$1.0\ A$
Diode ideality factor,	m	$=$	1.25

Calculate the open circuit voltage of the diode. If the fill factor is 0.75, calculate the maximum power output.

7.8 Consider the solar cell of problem 7.7. A solar system is to be developed from such cells to deliver a power of 15 W at a voltage level of 5 V. Calculate the total number of solar cells needed.

Section 7.5

7.9 Consider a GaAs photoconductor in which carriers are generated at a rate of $G_L = 10^{20}$ cm^{-3} s^{-1}. The device area is (10 μm ×10 μm) and the length is 10 μm. The material parameters are:

Background doping,	N_d	$=$	$10^{16}\ cm^{-3};\ N_a = 0$
Applied voltage,	V_{app}	$=$	$1.0\ V$
Electron velocity at the applied field,	v_e	$=$	$6 \times 10^6\ cm/s$
Hole velocity at the applied field,	v_h	$=$	$10^6\ cm/s$
Electron lifetime,	τ_n	$=$	$10^{-7}\ s$
Hole lifetime,	τ_p	$=$	$10^{-8}\ s$

Calculate the excess carrier concentration, the steady state photocurrent, and the photoconductive gain.

7.10 Consider a silicon photoconductor at 300 K with the following parameters:

Background doping,	N_d	$=$	$10^{15}\ cm^{-3}$
Electron mobility,	μ_n	$=$	$1200\ cm^2/V - s$
Hole mobility,	μ_p	$=$	$400\ cm^2/V - s$
Electron lifetime,	τ_n	$=$	$10^{-6}\ s$
Hole lifetime,	τ_p	$=$	$5 \times 10^{-7}\ s$
Detector area,	A	$=$	$10^{-4}\ cm^2$
Detector length,	L	$=$	$100\ \mu m$

A bias of 5 V is applied to the detector. Calculate the dark current. If light falls on the detector to produce a generation rate of 10^{21} cm^{-3} s^{-1}, calculate the excess concentration, the photoconductivity, and device gain.

7.11 Consider the detector of Problem 10.10. Calculate the gain-bandwidth product of the photoconductive detector.

Section 7.6

7.12 Consider a long silicon p-n photodiode at 300 K on which light from a GaAs laser ($\hbar\omega = 1.43$ eV) is impinging. The optical power density is 10^{-2} W/cm^2. The diode has the following parameters:

Device area,	A	$= 10^{-6}\ cm^2$
n-type doping,	N_d	$= 5 \times 10^{16}\ cm^{-3}$
p-type doping,	N_a	$= 10^{17}\ cm^{-3}$
Electron diffusion coefficient,	D_n	$= 20\ cm^2/s$
Hole diffusion coefficient,	D_p	$= 12\ cm^2/s$
Electron minority carrier lifetime,	τ_n	$= 1.5 \times 10^{-7}\ s$
Hole minority carrier lifetime,	τ_p	$= 10^{-7}\ s$
Absorption coefficient,	α	$= 700\ cm^{-1}$

A reverse bias voltage of 5 V is applied to the diode. Assume that the carrier generation rate is uniform. Calculate the prompt photocurrent and the total photocurrent in the detector.

7.13 Consider a GaAs p-i-n detector with an intrinsic layer width of $1.0\mu m$. Optical power density (photon energy 1.6 eV) of 0.1 W/cm^2 impinges upon the detector. The absorption coefficient for the active region is 10^4 cm^{-1}. Calculate the prompt photocurrent of the device. The device area is 10^{-4} cm^2.

7.14 Consider a silicon p-i-n photodetector in which the i layer is 10.0 μm thick. Calculate the maximum quantum efficiency of this detector if only light absorbed in the undoped region contributes to the photocurrent. The absorption coefficient is 10^3 cm^{-1}. Also calculate the minimum thickness of the i-region needed to ensure a quantum efficiency of 0.8. There are no reflection losses.

Section 7.7

7.15 Using the data for α_{imp} and β_{imp} in the text, calculate the ratio $k = \beta_{imp}/\alpha_{imp}$ for Si, Ge, and GaAs as a function of electric field.

7.16 An avalanche photodetector has an avalanche region of 5 μm and the electric field is such that $\alpha_{imp} = \beta_{imp} = 10^4$ cm^{-1}. Calculate the multiplication factor of the device.

Section 7.8

7.17 Consider a Si n-p-n transistor with the following parameters:

$$D_b = 20\ cm^2\ s^{-1} \qquad D_e = 10\ cm^2\ s^{-1}$$
$$N_{de} = 5 \times 10^{18}\ cm^{-3} \qquad N_{ab} = 5 \times 10^{16}\ cm^{-3}$$
$$N_{dc} = 5 \times 10^{14}\ cm^{-3} \qquad W_b = 1.0\ \mu m$$
$$\tau_B = \tau_E = 10^{-7}\ s \qquad n_i^2 = 2.25 \times 10^{20}\ cm^{-3}$$
$$A = 10^{-2}\ cm^2$$

Calculate the collector current in the active mode with an applied emitter base bias of 0.5 V. What is the collector current when the base current is now increased by 20%?

7.18 A Si npn transistor at 300 K has an area of 1 mm^2, base width of 1.0 μm, dopings of $N_{de} = 10^{18}$ cm^{-3}, $N_{ab} = 10^{17}$ cm^{-3}, $N_{dc} = 10^{16}$ cm^{-3}. The minority carrier lifetimes are $\tau_E = 10^{-7} = \tau_B$; $\tau_C = 10^{-6}$ s. Calculate the collector current in the active mode for a) $V_{BE} = 0.5$ V, b) $I_E = 2.5$ mA, and c) $I_B = 5$ μA. The base diffusion coefficient is $D_b = 20$ cm^{2}s^{-1}.

7.19 Consider the transistor of Problem 7.17 above. Calculate the ratio of photons absorbed in the neutral base and in the depletion region of the base collector junction when the BCJ is reverse biased at 3 volts. Assume that $\alpha = 5 \times 10^3$ cm^{-1}. Calculate the optical gain in the forward active mode. Assume that the quantum efficiency is unity. Calculate the gain for light incident through the emitter.

7.20 Discuss the conflicts between designing a high speed bipolar transistor and a high gain phototransistor.

Section 7.9

7.21 Design a GaAs/Al$_{0.3}$Ga$_{0.7}$As quantum well intersubband detector to detect 14 μm radiation using the conduction subbands. Assume that 60% of the bandgap discontinuity is in the conduction band for the GaAs/AlGaAs system.

7.22 Discuss the polarization selection rules for conduction band inter-subband detectors. Discuss the reasons why the rules are relaxed in valence subband detectors.

7.14 REFERENCES

- **General**

  - Excellent discussions on detectors appear in *Semiconductors and Semimetals*, Vol. 22 part D, ed. W. T. Tsang, Academic Press, Orlando (1985).

  - P. K. Bhattacharya, *Semiconductor Optoelectronic Devices*, Prentice-Hall, Englewood Cliffs, NJ (1994).

  - J. Gowar, *Optical Communication Systems*, Prentice-Hall, Englewood Cliffs, NJ (1989).

  - M. Ito and O. Wada, *IEEE J. Quantum Electronics*, **QE-22**, 1073 (1986).

  - J. I. Pankove, *Optical Processes in Semiconductors*, Dover Publications, New York (1977).

  - J. Singh, *Physics of Semiconductors and Their Heterostructures*, McGraw-Hill, New York (1993).

  - J. Wilson and J. F. B. Hawkes, *Optoelectronics: An Introduction*, Prentice-Hall, Englewood Cliffs, NJ (1983).

CHAPTER
8

NOISE AND THE PHOTORECEIVER

8.1 INTRODUCTION

In the previous chapter we have discussed important semiconductor detectors that are used in modern optoelectronic systems. We have also discussed the electrical response of these detectors to an optical input. An important issue that we have not discussed is the noise in the detection process. Noise is present in every signal and in every detection system and represents an undesirable component of the signal that interferes with a successful interpretation of the received signal.

Noise has many sources. For example, in a lecture hall if some students are gossiping, the noise is produced on top of the professor's lecture. On the other hand, if the professor has a bad cold, the lecture may contain noise produced by wheezing and coughing. In optoelectronic systems, noise is a serious problem that limits the performance of detectors. The noise is produced in the optical source, the optoelectronic detector, and the amplifier circuit. We will examine these noise sources in this chapter.

The performance of a detection system is closely tied to the coding scheme used to transmit information. The information to be sent on an optical wave may involve coding the information in the intensity, amplitude, frequency, or phase of the wave. The detection system must then be able to decode this signal by using an appropriate circuit. At present, the dominant coding scheme involves intensity modulation (IM) with direct detection. This coding has a relatively high noise level, compared to coding schemes that involve frequency, phase, or amplitude modulation. However, since the state of the art in semiconductor optoelectronic systems is not so developed, and amplitude, frequency, and phase modulation schemes are not fully exploited. However, as devices evolve, these schemes will be used to fully exploit the potential of optical communication. We start this chapter with a brief review of the coding schemes that can be used in sending and detecting information.

8.2   MODULATION AND DETECTION SCHEMES

$\mathcal{R}$ Optoelectronic detectors are used for a variety of applications, one of the most important being in the area of communications. The information to be transmitted is coded into a carrier beam (an optical beam for optical communication) which is then transmitted over an appropriate medium. The detector system is responsible for decoding the information sent. The design and performance of the detection system is intimately tied to the coding scheme used. At present, limitations placed by the performance of semiconductor optoelectronic sources and detectors do not allow one to use the full range of coding schemes available. Nevertheless, we will examine these schemes briefly since eventually, as devices advance, the system designer will be offered these choices.

The optical beam that carries the information is characterized by its amplitude, frequency, or wavelength and phase as well as its intensity (which is determined by the amplitude). All of these parameters can be modulated to code information provided that devices exist to code and decode the parameters.

8.2.1   Amplitude Modulation

$\mathcal{R}$ Amplitude modulation (AM) is an important modulation scheme widely used to send information using microwaves. The field of the carrier wave is represented by

$$F_c = F_{co} sin\omega_c t \tag{8.1}$$

and that of the modulation signal by

$$F_m = F_{mo} sin\omega_m t \tag{8.2}$$

In amplitude modulation, the maximum amplitude F_{co} of the carrier wave is made proportional to the instantaneous modulating voltage $F_{mo} sin\omega_m t$. The modulation index is defined as

$$m = \frac{F_{mo}}{F_{co}} \tag{8.3}$$

The amplitude of the amplitude modulated carrier becomes (see Fig. 8.1)

$$
\begin{aligned}
A &= F_{co} + F_m = F_{co} + F_{mo} sin\omega_m t \\
&= F_{co} + mF_{co} sin\omega_m t \\
&= F_{co}(1 + msin\omega_m t)
\end{aligned}
\tag{8.4}
$$

The field associated with the modulated carrier wave is now

$$
\begin{aligned}
F &= Asin\omega_c t = F_{co}(1 + msin\omega_m t)sin\omega_c t \\
&= F_{co}sin\omega_c t + \frac{mF_{co}}{2}cos(\omega_c - \omega_m)t - \frac{mF_{co}}{2}cos(\omega_c + \omega_m)t
\end{aligned}
\tag{8.5}
$$

The amplitude modulated carrier contains three terms : i) the unmodulated carrier term (see Fig. 8.1); ii) an upper side band (USB) with frequency $\omega_c + \omega_m$; iii) a lower side band (LSB) with frequency $\omega_c - \omega_m$.

It must be kept in mind that the optical carrier frequencies are in the range of $10^{14} - 10^{15}$ Hz while the modulating frequencies are (limited by the electronics and the optical transmitter (lasers)) $\sim 10^{10}$ Hz. In order to be able to decode the information being sent, it should be possible to have a detection system that can isolate the USB or the LSB. The theory behind the detection will be discussed in Section 8.6, but is is important to note that while in the microwave domain it is possible to do such detection, in the optical regime severe challenges still remain.

8.2.2 Frequency Modulation

$\mathcal{R}$ Another important modulation scheme used widely in transmission of signals is the frequency modulation (FM) approach, shown in Fig. 8.2. As the name implies, the carrier signal frequency is modulated in this approach. Once again the unmodulated carrier wave can be written as

$$
F_c = F_{co} sin(\omega_c t + \phi)
\tag{8.6}
$$

where we have included a phase term for completeness. If the modulating signal is (we choose a cosine term for simplicity)

$$
F_m = F_{mo} cos\omega_m t
\tag{8.7}
$$

the *frequency of the modulated signal is*

$$
f = f_c(1 + k\, F_{mo} cos\omega_m t)
\tag{8.8}
$$

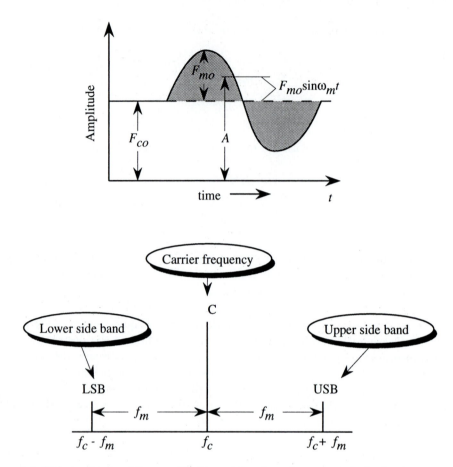

Figure 8.1: The amplitude of the amplitude modulated carrier wave and the frequency components of the carrier.

where k is the proportionality constant which is dependent upon the details of how the modulation is done. The extremes in the frequency of the modulated carrier occur at value

$$f = f_c(1 \pm k \ F_{mo}) \tag{8.9}$$

The modulated carrier can be represented by

$$F = F_{co}sin[\mathcal{F}(\omega_c, \omega_m)] = F_{co}sin\theta \tag{8.10}$$

where the function $\mathcal{F}$ is yet to be determined. To determine the function $\mathcal{F}$ or the angle θ at any time t, we must integrate ω with respect to time

$$\theta = \int \omega dt = \int \omega_c(1 + k \ F_{mo}cos\omega_m t)dt$$

$$= \omega_c \int (1 + k \ F_{mo} cos\omega_m t) dt$$

$$= \omega_c t + \frac{k \ F_{mo}\omega_c sin\omega_m t}{\omega_m}$$

$$= \omega_c t + \frac{k \ F_{mo} f_c sin\omega_m t}{f_m} \tag{8.11}$$

Defining

$$\delta = k \ F_{mo} f_c \tag{8.12}$$

we get

$$\theta = \omega_c t + \frac{\delta}{f_m} sin\omega_m t. \tag{8.13}$$

We now get, for the modulated carrier signal

$$F = F_{co} sin \left(\omega_c t + \frac{\delta}{f_m} sin\omega_m t \right)$$

$$= F_{co} sin \left(\omega_c t + m_f sin\omega_m t \right) \tag{8.14}$$

where m_f is the modulation index. The FM signal has a number of frequencies present unlike the AM signal which had the USB and LSB present. The particular form of the FM signal allows one to write the signal in terms of a Bessel function expansion of the form

$$\begin{aligned} F = \ & F_{co} \left\{ J_o(m_f) sin\omega_c t \right. \\ + \ & J_1(m_f) \left[sin(\omega_c + \omega_m)t - sin(\omega_c - \omega_m)t \right] \\ + \ & J_2(m_f) \left[sin(\omega_c + 2\omega_m)t + sin(\omega_c - 2\omega_m)t \right] \\ + \ & J_3(m_f) \left[sin(\omega_c + 3\omega_m)t - sin(\omega_c - 3\omega_m)t \right] \\ + \ & \left. \ldots \right\} \end{aligned} \tag{8.15}$$

As a result, the signal has not only the carrier frequency ω_c but a number of other frequency terms. The functions J_m fall off as m increases, so that only the first five or six terms are important. Nevertheless, it is clear that the FM signal requires a larger spectral width for transmission. The higher bandwidth needed is, however, compensated by the lower noise that is possible in FM detection systems.

Closely related to the FM scheme is the phase modulation scheme where the modulating signal modulates the phase of a carrier signal.

In the discussion above, we have discussed the modulation by an analog signal. These days, increasingly, digital coding schemes are being used. In this case, amplitude,

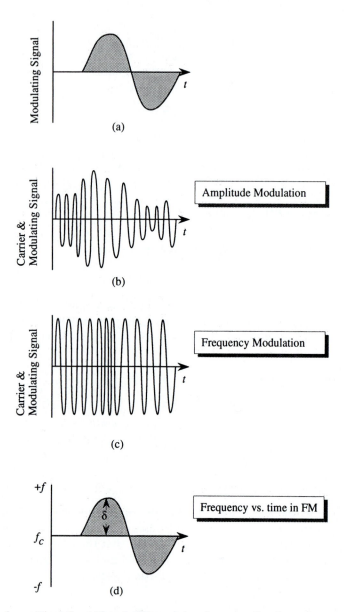

Figure 8.2: A schematic of the AM and FM modulated carrier signal.

frequency, or phase is not varied continuously, but can be one of two values corresponding to 0 and 1 bits. The modulation schemes are then called amplitude, frequency, or phase shift keying, i.e., ASK, FSK, and PSK, respectively.

In order to use the approaches described so far, the optical transmitter should be extremely stable in its amplitude, frequency, and phase. We will discuss these requirements and the limitations of modern lasers, which are used as transmitters, in Chapters 10 and 11. In addition to the source, the detection system also needs to be of extremely high performance. We will discuss, in Section 8.6, the detection approaches used for these "coherent" detection schemes. The term "coherent" detection is used because of the need to have frequency (phase) stable sources and detection systems. These schemes are extremely promising because of the very low noise that is introduced in these schemes, and because of the possibility of super wideband frequency division multiplexing. These issues will be discussed in Chapter 13. However, at present, no commercial system uses these coherent schemes because of the extremely high cost of the system, mainly due to the immature state of optoelectronic devices, as far as coherence is concerned. In fact, the present state of coherent systems for optical communication has been compared to the pre-1930 state of radio communication.

8.2.3 Intensity Modulation

We will discuss in Section 8.6 the theory behind detection of AM or FM signals. As noted earlier, the devices needed to carry out such detection are not perfected at present and as a result the most common approach used for transmitting signals is by intensity modulation (IM). The intensity modulation is the simplest modulation scheme in which the optical power of an optical source is modulated to send a signal. The signal is then directly detected by the detection. Such a modulation is compatible with detectors that have been discussed in the previous chapter. However, two important disadvantages exist in the intensity modulation scheme. The noise levels produced are quite high and the full bandwidth of the optical system cannot be used. In Chapter 13 we will discuss the bandwidth issues. In this chapter we will focus on the noise produced during the detection of intensity modulated transmission and its direct detection.

8.3 NOISE AND DETECTION LIMITS

In the previous chapter, we have discussed a number of important optical detectors. To assess the quality of the photodetectors, important figure of merit parameters have been developed. An important parameter refers to the weakest (dc) source of radiation that can be detected.

To understand the detection limits we will discuss the noise of the detectors.

8.3.1 Shot Noise

$\longrightarrow$ The electrons (or photons) which make up the device current optical signal are discrete particles carrying discrete charge. There are fluctuations in the number of the particles impinging upon a detector or passing through an electrode due to the statistical fluctuations present in any ensemble. Thus, if one sits at an electrode and counts the number of electrons coming in a certain time interval Δt, the number will vary as shown schematically in Fig. 8.3. *The shorter the time interval of interrogation, the larger the variation.*

We assume that the variation in the number of electrons (the treatment also applies to photons) coming in during a time interval Δt is given by a Poisson distribution. According to this statistical distribution, if $a\Delta t$ is the *average number of particles passing in a time interval* Δt, the probability of getting N particles in the interval Δt is (for N much larger than one)

$$P(N, \Delta t) = \frac{1}{\sqrt{2\pi(a\Delta t)}} exp\left(-\frac{(N - a\Delta t)^2}{2a\Delta t}\right) = \frac{1}{\sqrt{2\pi\overline{N}}} exp\left(-\frac{\Delta N^2}{2\overline{N}}\right) \qquad (8.16)$$

where $\overline{N}$ is the average value ($= a\Delta t$) and ΔN is the fluctuation from the average value. This probability function is shown in Fig. 8.3b. This function maximizes when

$$N = \overline{N} = a\Delta t \qquad (8.17)$$

as expected. The rms deviation of the Poisson distribution, i.e., the noise, is found, from statistics, to be

$$\sqrt{\overline{(\Delta N)^2}} = \sqrt{\overline{(N - \overline{N})^2}} = \sqrt{\overline{N}} \qquad (8.18)$$

It is important to point out that this noise, called the shot noise, would occur in the photon stream that is impinging upon a detector, or the current flowing in the detector resulting from the e-h pair generation since both events involve discrete particles. As a result, the output current has a noise shown schematically in Fig. 8.3c.

The average signal in the device is given by $\overline{N}$ ($= a\Delta t$). An important device parameter is the ratio between the signal generated and the random noise. The signal to noise ratio (SNR) for the shot noise limited detector becomes (from Eqn. 8.18)

$$\boxed{SNR = \frac{\overline{N}}{\sqrt{\overline{N}}} = \sqrt{\overline{N}} = \sqrt{a\Delta t}} \qquad (8.19)$$

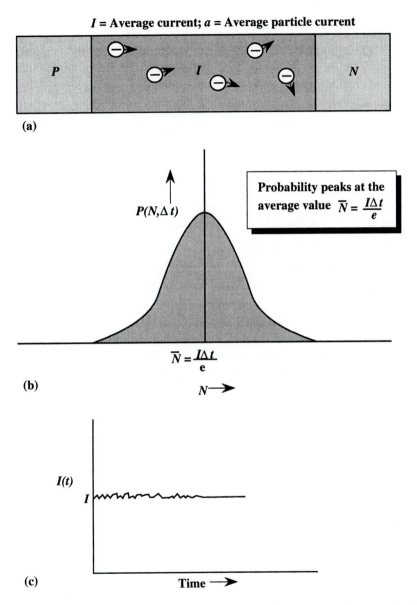

Figure 8.3: (a) The electrons in a semiconductor are moving randomly with a certain distribution function. (b) The probability of finding N electrons crossing an area A in time interval Δt. The average particle current is a so that the mean value of the particle number is $\overline{N} = a\Delta t$. (c) Schematic of the current flow in a device. The statistical variations result in noise in the current.

Note that if the current through the device is I, the quantity a, which is the particle current, is given by

$$a = \frac{I}{e} \qquad (8.20)$$

We see from Eqn. 8.19 that the SNR increases with the decrease in the observation time interval Δt. Denoting the bandwidth of the device by f, we have

$$f \cong \frac{1}{2\Delta t} \qquad (8.21)$$

$$SNR = \sqrt{\frac{a}{2f}} \qquad (8.22)$$

As can be expected on physical grounds, the SNR gets worse as the device is operated at higher frequencies. The rms noise in the current of the detector is called the shot noise current (I_{sh}) and is (from Eqns. 8.18, 8.19 and 8.20)

$$\boxed{I_{sh} = e \, \frac{\sqrt{(\Delta N)^2}}{\Delta t} = \frac{e\sqrt{a\Delta t}}{\Delta t} = \sqrt{2eIf}} \qquad (8.23)$$

In addition to the shot noise, there are many other sources of noise encountered by the detection system. Often the signal produced by a detector in response to an optical signal is amplified by a receiver amplifier. The receiver amplifier produces its own noise. Another source of noise is due to the background thermal noise associated with the blackbody radiation. At a given temperature T, according to the Planck's theory of blackbody radiation, radiation of all different wavelengths exist according to the distribution function

$$N(\lambda) = \frac{2c}{\lambda^4 \left[exp \, \frac{\hbar\omega}{k_B T} - 1 \right]} \qquad \text{number/s/(area} \times \text{unit wavelength)} \qquad (8.24)$$

The photons with energy $\hbar\omega > E_g$ will create noise by creating *e-h* pairs. *However, if $k_B T$ is small compared to the bandgap, the noise level due to the thermal background is negligible. Thus at most optical frequencies, the thermal noise is not significant.* However, the thermal noise plays an important role in the receiver amplifier resistance noise, as will be discussed later.

An important figure of merit for a detector is the minimum detectable signal that would produce the same rms output as that generated by the noise. This is given by the noise equivalent power (NEP). Let us consider the case of the detector that is limited by the shot noise. The optical power density P_{op} needed to produce a photocurrent I_L is (see Eqn. 7.10)

$$P_{op} \cdot A = \frac{I_L \, \hbar\omega}{\eta_Q e} \qquad (8.25)$$

For the noise equivalent power we equate the noise current I_{sh} to I_L. This gives, from Eqn. 8.23,

$$I_L = I_{sh} = [2e(I_L + I_0)f]^{1/2} \tag{8.26}$$

where I_0 is the dark current of the detector. *If $I_0 \ll I_L$, we get*

$$I_L = 2ef \tag{8.27}$$

The optical power required at bandwidth f is now from Eqn. 8.25

$$\boxed{P_{op} \cdot A = NEP = \frac{2\hbar\omega f}{\eta_Q}} \tag{8.28}$$

The NEP of the ideal quantum detector is given by the above equation with the quantum efficiency $\eta_Q = 1$.

If $I_0 \gg I_L$ we have, from Eqn. 8.26,

$$I_L \sim (2eI_0 f)^{1/2} \tag{8.29}$$

and

$$\boxed{P_{op} \cdot A = NEP = \frac{(2eI_0 f)^{1/2}\hbar\omega}{\eta_Q e}} \tag{8.30}$$

The detectivity of the detector is defined by

$$\boxed{D = \frac{1}{NEP}} \tag{8.31}$$

The detectivity or NEP depends upon the area of the detector as well as the bandwidth of the detector. A quantity called the specific detectivity, D^*, is defined which accounts for the variable bandwidth and detector area:

$$\boxed{D^* = \frac{(Af)^{1/2}}{NEP}} \tag{8.32}$$

In the choice of a detector, *once the bandwidth requirements are met, one chooses a detector with the highest D^* values.* In Example 8.3 below, we discuss typical values of the parameters discussed above.

In the discussion above we have not included the effects of the random carrier multiplication processes that occur in an avalanche device. In an APD, the absorbed photocarriers are injected into an avalanche region and not only is the injected photocurrent multiplied, the noise is also amplified. We will discuss this issue in the next subsection.

EXAMPLE 8.1 A flux of 8×10^6 particles/s impinges on a detector. Calculate the maximum bandwidth at which the SNR for the device is unity for a shot noise limited case.

From Eqn. 8.19

$$SNR \quad = \quad 1 = \sqrt{\frac{8 \times 10^6 \ s^{-1}}{2f \ s^{-1}}}$$

$$\text{or} \quad f = 4 \times 10^6 \ Hz$$

EXAMPLE 8.2 A shot noise limited detector is to operate at 1 GHz. Calculate the current level needed to ensure an SNR of 100 (or 40 dB).

The particle current is given by Eqn. 8.19 and 8.21 as

$$a \quad = \quad (SNR)^2 (2f)$$
$$= \quad (100)^2 (2 \times 10^9 s^{-1}) = 2 \times 10^{13} \text{particles/s}$$

The electric current is

$$I = ea = (1.6 \times 10^{-19} C)(2 \times 10^{13} s^{-1}) = 3.2 \times 10^{-6} \ A$$

EXAMPLE 8.3 A detector has the following parameters:

Detector area,	A	$=$	$1 \ cm^2$
Detection wavelength,	λ	$=$	$1.0 \ \mu m$
Quantum efficiency,	η_Q	$=$	20%
Bandwidth,	f	$=$	$1 \ Hz$
Noise current,	I_{sh}	$=$	$10 \ pA$

Calculate the a) responsivity, b) noise equivalent power, c) detectivity, and d) specific detectivity of the device.

a) The responsivity is given by

$$R = \frac{\eta_Q e}{\hbar \omega} = \frac{(0.2)(1.6 \times 10^{-19} \ C)}{1.242 \times 1.6 \times 10^{-19} \ J} = 0.16 \ A/W$$

b) The noise equivalent power is from Eqns. 8.26 and 8.27 (equating $I_L = I_{sh}$) as

$$NEP = \frac{I_{sh} \hbar \omega}{\eta_Q e} \quad = \quad \frac{(10 \times 10^{-12} \ A)(1.242 \times 1.6 \times 10^{-19} \ J)}{(0.2)(1.6 \times 10^{-19} \ C)}$$
$$= \quad 62 \ pW$$

c) The detectivity is

$$D = \frac{1}{NEP} = 1.61 \times 10^{10} \ W^{-1}$$

d) The specific detectivity is

$$
\begin{aligned}
D^* &= D(Af)^{1/2} = 1.61 \times 10^{10}(1)^{1/2} \\
&= 1.61 \times 10^{10} \ cm - Hz^{1/2} \ W
\end{aligned}
$$

8.3.2 Noise in Avalanche Photodetectors

↝ In the previous subsection we have discussed the shot noise that arises from the discrete nature of photons or electrons that lead to the photocurrent. In an APD, the carrier multiplication factor amplifies the shot noise further so that the device noise increases as the multiplication factor increases. The factor by which the noise increases is called the noise factor. A treatment has been given by McIntyre which tells us (for a set of approximations) what the noise factor is for a given multiplication and impact ionization coefficients α_{imp} and β_{imp}. According to this treatment, when the multiplication is initiated by electrons, the noise factor is

$$F_e = M_e \left[1 - (1 - k)(M_e - 1)^2 / M_e^2 \right] \tag{8.33}$$

and if the initiation is by holes, the factor is

$$F_h = M_h \left[1 + \frac{(1 - k)}{k} \frac{(M_h - 1)^2}{M_h^2} \right] \tag{8.34}$$

where $k = \beta_{imp}/\alpha_{imp}$. These equations tell us that if the electron ionization coefficient α_{imp} is much larger than β_{imp}, the electrons should be used to cause the carrier multiplication.

Let us consider a device in which carriers are photogenerated in the absorbing region as shown in Fig. 8.4 and electrons are injected into the avalanche region at $x = 0$. The hole current at $x = W$ is zero since there are no holes injected from that side. Let us consider an electron-hole pair generated at a point x. The electron moves towards the right (towards $x = W$) and the hole moves towards $x = 0$. In this transit they create further carriers and the multiplication for this pair is

$$M(x) = 1 + \int_0^x \beta_{imp} M(x')dx' + \int_x^W \alpha_{imp} M(x')dx' \tag{8.35}$$

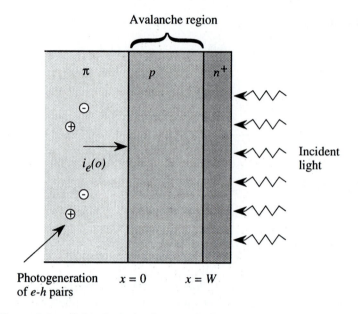

Figure 8.4: The model used for the noise factor calculation assumes that electrons injected at $x = 0$ initiate the multiplication. At $x = W$, there are no holes so that the hole current is zero here.

where the first integral gives the hole contribution and the second integral gives the electron contribution. If we differentiate Eqn. 8.35, we get

$$\frac{dM(x)}{dx} = -(\alpha_{imp} - \beta_{imp})M(x) \tag{8.36}$$

which has the solution

$$M(x) = M(0)exp\left\{-\int_o^x (\alpha_{imp} - \beta_{imp})dx'\right\} \tag{8.37}$$

The term $M(0)$ refers to the carriers generated at x=0, and in this case, since only the electrons are the source of multiplication, $M(0)$ is defined as M_e, the electron multiplication factor.

If we substitute $x = W$ in Eqn. 8.35 we get

$$\begin{aligned} M(W) &= 1 + \int_o^W \beta_{imp} M(x')dx' \\ &= 1 + \frac{k}{(1-k)} \int_o^W (\alpha_{imp} - \beta_{imp})M(x')dx' \end{aligned} \tag{8.38}$$

Substituting for the integrand from Eqn. 8.36, we get

$$
\begin{aligned}
M(W) &= 1 - \frac{k}{1-k} \int_o^W dM(x') \\
&= 1 + \frac{k}{1-k} [M(0) - M(W)]
\end{aligned}
$$

or

$$
M(W) - 1 = k \{M(0) - 1\} \tag{8.39}
$$

At this point it is also useful to exploit the same approach to evaluate an integral that will be used later for the noise calculation. Thus we have

$$
\begin{aligned}
2 \int_o^W \beta_{imp} M^2(x) dx &= \frac{2k}{(1-k)} \int_o^W (\alpha_{imp} - \beta_{imp}) M^2(x) dx \\
&= \frac{-2k}{(1-k)} \int_o^W M(x) dM \\
&= \frac{-k}{1-k} \int_o^W d(M^2(x)) \\
&= \frac{k}{1-k} \{M^2(0) - M^2(W)\} \tag{8.40}
\end{aligned}
$$

Let us now consider the basis of the noise theory. We assume that the carriers generated at a point x are multiplied by the factor M and the carriers are generated randomly (i.e., they satisfy the Poisson distribution) and carry the full shot noise.

An analysis for the shot noise was developed in Section 8.31 and it was found that the noise in the current was given by

$$
(I_{sh})^2 = 2 \, e \, If \tag{8.41}
$$

where f is the bandwidth of the detector. The average current is I and there is no multiplication. It is useful to describe the noise per unit bandwidth (shot noise spectral density I_{sh}^*)

$$
(I_{sh}^*)^2 = 2 \, e \, I \tag{8.42}
$$

The electron current generated in an element dx is given by

$$
di_e(x) = \{\alpha_{imp} i_e(x) + \beta_{imp} i_h(x)\} \, dx \tag{8.43}
$$

and associated with this current is a shot noise spectral density $2edi_e(x)$. Due to the multiplication process, both the current $di_e(x)$ and the rms current noise $\{2edi_e(x)\}^{1/2}$

are multiplied by $M(x)$. The total mean square noise spectral density (I_M^*) is a sum of the noise originating from the total avalanche region

$$(I_M^*)^2 = \int_o^W M^2(x) 2e \; di_e(x) \qquad (8.44)$$

Integrating by parts, we have

$$
\begin{aligned}
\int_o^W M^2(x) di_e(x) &= M^2(x) i_e(x) \big|_o^W - \int_o^W i_e(x) M(x) \frac{dM(x)}{dx} dx \\
&= M^2(W) i_e(W) - M^2(0) i_e(o) \\
&\quad + 2 \int_o^W (\alpha_{imp} - \beta_{imp}) i_e(x) M^2(x) dx
\end{aligned}
\qquad (8.45)
$$

where we have used Eqn. 8.36 for dM/dx to simplify the integral.

To evaluate the integral on the right side of Eqn. 8.45, we note that the total current is

$$I = i_e(x) + i_h(x) = M(0) i_e(0) = i_e(W) \qquad (8.46)$$

since at $x = W$, there is no hole current. Also,

$$\frac{di_e(x)}{dx} = \alpha_{imp} i_e(x) + \beta_{imp} i_h(x) \qquad (8.47)$$

so that we have, after elimination $i_h(x)$ from Eqns. 8.46 and 8.47,

$$(\alpha_{imp} - \beta_{imp}) i_e(x) = \frac{di_e(x)}{dx} - \beta_{imp} I \qquad (8.48)$$

Substituting this in the integrand on the right hand side of Eqn. 8.45, we get (for Eqn. 8.45)

$$
\begin{aligned}
\int_o^W M^2(x) di_e(x) &= M^2(W) i_e(W) - M^2(0) i_e(0) \\
&\quad + 2 \int_o^W M^2(x) di_e(x) - 2I \int_o^W \beta_{imp} M^2(x) dx
\end{aligned}
\qquad (8.49)
$$

This gives

$$\int_o^W M^2(x) di_e(x) = M^2(0) i_e(0) - M^2(W) i_e(W) + 2I \int_o^W \beta_{imp} M^2(x) dx \qquad (8.50)$$

Using Eqn. 8.46 for the first term and Eqn. 8.40 for the integral on the right hand side, we get

$$\int_0^W M^2(x) di_e(x) = \{M(0) - M^2(W)\} I + \frac{Ik}{(1-k)} \{M^2(0) - M^2(W)\} \qquad (8.51)$$

Finally, substituting in Eqn. 8.44, we get for the noise spectral density

$$(I_M^*)^2 = 2eI \left[\{M(0) - M^2(W)\} + \frac{k}{(1-k)} \{M^2(0) - M^2(W)\} \right]$$

$$= 2eI \left\{ M(0) + \frac{kM^2(0)}{(1-k)} - \frac{M^2(W)}{(1-k)} \right\} \tag{8.52}$$

The total mean square noise is obtained by further adding the multiplied mean square noise produced from the initial photocurrent. This term is

$$M^2(0)(I_{sh}^*)^2 = M^2(0) \cdot 2ei_e(0) = 2_e IM(0) \tag{8.53}$$

after using Eqn. 8.46. The noise factor is the ratio of the total noise to the multiplied shot noise, i.e.,

$$F_e = 1 + \frac{(I_M^*)^2}{2eIM(0)}$$

$$= 1 + 1 + \frac{\{kM^2(0) - M^2(W)\}}{(1-k)} \tag{8.54}$$

We now use Eqn. 8.39 to express $M(W)$ in terms of $M(0) = M_e$. This finally gives

$$F_e = 2 + \{kM_e^2 - 2kM_e - (1-k)\} / M_e$$

or

$$\boxed{F_e = M_e \left\{ 1 - \frac{(1-k)(M_e-1)^2}{M_e^2} \right\}} \tag{8.55}$$

If holes were to initiate the injection, one would get the results given by Eqn. 8.34. If M_e is large, one gets the noise factor to approach kM_e. Thus, if $\alpha_{imp} >> \beta_{imp}(k << 1)$, the noise is suppressed. Normally we cannot control the ratio of α_{imp} and β_{imp}, but as discussed in Chapter 7, the use of quantum well structures can alter this ratio and lead to very low noise APDs.

8.3.3 Noise in Resistors and Electronic Devices

$\longrightarrow$ In the previous section we discussed the effect of the finite number of photons or electron-hole pairs generated during detection on the detector noise. The fluctuations from the mean was responsible for the shot noise. There are other fluctuations present when current flows through a resistor or an electronic device such as an FET or a bipolar device. Since resistors and transistors are invariably involved in an optoelectronic detector (receiver), one must examine these additional sources of noise.

We have noted earlier in Chapter 2 on our discussion on the distribution function of electrons in a material that at equilibrium the carrier distribution is given by the Fermi Dirac distribution function which can be approximated by the Maxwell-Boltzmann distribution. The distribution function is characterized by the lattice temperature which tells us the "spread" in the carrier energies and momenta. As shown in Fig. 8.5a, the carriers do not have a single momentum k, but have a distribution of momenta. When an electric field F is applied, and the lattice temperature is T, the electrons gain energy from the electric field and their distribution function is approximately described by a temperature known as electron temperature T_e. In fact, the distribution function $f_e(T, F)$ can be schematically written as

$$f_e(T, F) = f^o(T_e) + g(T, F) \qquad (8.56)$$

where $f^o(T_e)$ is the equilibrium distribution function at temperature T_e and $g(T, F)$ is a deviation due to the electric field and represents the fact that on an average the electrons have a drift along the electric field. A schematic view of this is shown in Fig. 8.5b.

The random velocities that the electrons have in a Maxwell-Boltzmann distribution function are an important source of noise in the device current or voltage. Consider a resistor having a resistance R. If a voltage measurement is done on the resistor in absence of any bias, the voltage will be zero. However, if an extremely sensitive voltmeter is used, one will see a random fluctuation in the voltage, as shown in Fig. 8.6. This fluctuation occurs since the electrons have a certain velocity distribution so that even though the average velocity in the absence of a field is zero, electrons are moving randomly creating small potential fluctuations. These potential fluctuations are superimposed upon any signal that is produced by actually applying an electric field.

To calculate the thermal noise due to the finite extent of the distribution function, note that the mean kinetic energy of the electrons in the material is approximately $k_B T$. The power associated with the random fluctuations is

$$P_n = k_B T \Delta f \qquad (8.57)$$

where Δf is the bandwidth over which the noise is measured. If we now consider a resistor as a noise generator, the noise power is

$$P_n = \frac{V^2}{R} = \frac{(V_n/2)^2}{R} = \frac{V_n^2}{R} \qquad (8.58)$$

where V_n is the noise voltage. Equating the two noise powers we get

$$\boxed{V_n = \sqrt{4k_B T \Delta f R}} \qquad (8.59)$$

This gives the rms voltage across a resistor. The noise is white noise, i.e., it does not depend upon the frequency at which it is measured. The thermal noise is also called the Johnson noise.

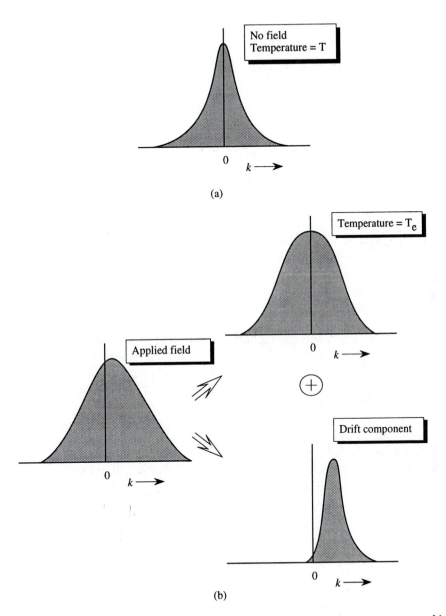

Figure 8.5: (a) The equilibrium distribution characterized by a carrier temperature which is the same as the lattice temperature. (b) In the presence of an electric field, the electrons get heated and drift in the field direction. The distribution function can be solved into a hot carrier random distribution and a drift distribution as shown.

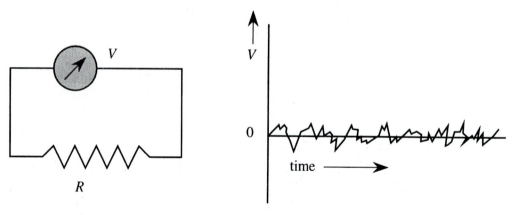

Figure 8.6: The voltage measurement versus time across a resistor.

If a current consists of several resistances connected in series or in parallel, one simply finds the equivalent resistance which is then used in the noise voltage expression.

In addition to the noise in resistors (or inductors), active devices such as transistors also contribute noise. The source of noise in these devices may be either due to particle number fluctuations (shot noise) or due to carrier velocity distributions. In a field effect transistor, the gate leakage current leads to a shot noise term and one has an additional noise in the channel due to the distribution in the electron velocities in the device channel. In the bipolar device, noise is produced due to the shot noise fluctuations in the base current, and the thermal noise from the collector current.

The channel noise is usually treated as a thermal noise, but one has to remember that one should use the carrier temperature T_e discussed above in the noise expression. In the next section we will discuss the noise and sensitivity issues in the receiver-amplifier and will discuss the important contributions to noise coming from the detector, the load resistors and the amplifier in a reception system.

8.3.4   Flicker Noise

$\longrightarrow$ Another source of device noise is the flicker or the 1/f noise which has the special behavior that its spectral distribution is inversely proportional to the frequency. This noise can usually be ignored at high frequencies except in some amplifiers where upconversion can create noise at high frequencies.

The flicker noise is due to the random trapping and recombination effects involving bandgap states in device structures. As the frequency increases, the slow response of the trap states reduces the noise.

The magnitude of the 1/f noise drops below the thermal noise floor at a corner frequency which is typically between 100 kHz and 100 MHz depending on the device quality. In the next section we will examine the mathematical representation of the flicker noise.

EXAMPLE 8.4 Consider an amplifier operating over a frequency of 100 to 120 MHz. The amplifier has a 10KΩ input resistor. Calculate the rms noise voltage feeding into the amplifier at 300 K.

The rms noise voltage is

$$
\begin{aligned}
V_n &= \sqrt{4k_B T \Delta f R} \\
&= \left(4 \times 1.38 \times 10^{-23} (JK^{-1}) \times 300(K) \times (120 - 100) \times 10^6 \ (s^{-1}) \times 10^4 (\Omega)\right)^{1/2} \\
&= 5.76 \times 10^{-5} \ V
\end{aligned}
$$

8.4 THE RECEIVER AMPLIFIER

$\longrightarrow$ The photodetectors discussed in Chapter 7 are rarely used independently in an optical information processing system. Usually the optical signal coming in and the resultant photocurrent is quite weak and must be amplified before it can be used for further processing. In detectors such as p-i-n diodes, there is no gain in the device which makes it essential to have an amplification circuit along with the detector. For high sensitivity, internal gain can be provided by a phototransistor, an n-i-n photoconductive detector or an avalanche photodiode. In Fig. 8.7 we show a comparison of the advantages and disadvantages of the various detectors.

The p-i-n detector and the APD have emerged as the most important detectors for high speed high sensitivity applications. The p-i-n is based on direct gap materials to provide a short absorption region and high speed. The APD's can be based on Si, Ge and compound semiconductors such as InGaAs. In an APD one has to maintain very stable voltage and temperature values which makes the system costly and somewhat unreliable especially if it is placed in a region that is difficult to access. For example, p-i-n diodes and not APDs are used in undersea regenerators for long distance optical communications.

A general receiver amplifier circuit is shown in Fig. 8.8. The amplification is provided by a transistor which is either a field effect transistor or a bipolar transistor. The choice of the transistor is a very important issue in the receiver design. One obviously wants a device with high gain and low noise. An additional constraint on the

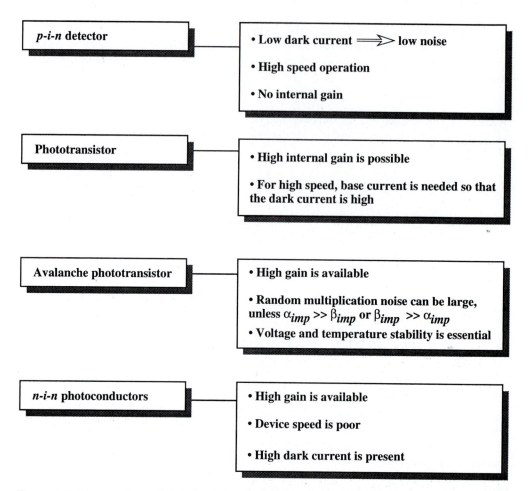

Figure 8.7: A comparison of the advantages and disadvantages of various detectors. The *p-i-n* and APD devices are chosen for high speed applications. An amplifier is used in the receiver circuit.

choice of the transistor may come for integrated photoreceivers where the detector and amplifier are on the same chip. We will consider the constraints imposed by integration in Section 8.7.

The FET is usually used in the receiver amplifier because of the very high frequencies up to which the FET can operate and the low noise level that can be achieved. Advances in FET technology have come from the use of the compound semiconductors and from the use of heterostructure based devices such as the MODFET. We will identify some typical noise figures for these devices as we discuss the receiver noise. An HBT can also be used in the photoreceiver, and recent advances in HBT technology are being exploited for integrated HBT-detector receivers.

Let us start with a p-i-n detector which is part of the front end of a photoreceiver. The APD detector based photoreceiver can be studied by a simple extension of the p-i-n study. We have in Fig. 8.8 the block diagram of the photoreceiver. We will focus only on the preamplifier. The equalizer is used to reshape the input pulse which is distorted during the transmission and detection. The post amplifier further amplifies the pulse and the filter sets the bandwidth of the receiver. Since the noise depends upon the bandwidth, the filter also limits the noise.

The equivalent circuit of a high-impedance FET front end with the detector is shown in Fig. 8.8. The load resistor R_L is used to bias the photodiode and the Johnson noise from thermal fluctuations is given by a noise current

$$\overline{i_R^2} = \frac{4k_B T B}{R_T} \tag{8.60}$$

where R_T is the total resistance that is the sum of R_L and the amplifier resistance and B is the bandwidth. The expression above is modified somewhat because of the specific details of the shape of the optical pulse. For example, for a rectangular pulse the noise current is modified as

$$\overline{i_R^2} = \frac{4k_B T I_2 B}{R_T} \tag{8.61}$$

where I_2 is called Personick integral with a value ~ 0.55. For other shaped pulses, the Personick integral has a slightly different value.

In addition to the Johnson noise, we have the shot noise due to the currents in the input circuit. These currents are

$$I_T(t) = I_g + I_D + I_s(t) \tag{8.62}$$

where I_g is the sum of the gate leakage current and the shunt currents, I_D is the diode dark current ($= I_o$ for the reverse bias diode), and $I_s(t)$ is the current due to the optical signal. In high performance systems where the optical signal is extremely weak we can ignore $I_s(t)$ in the shot noise calculations. Thus we have for the shot noise

$$\overline{i_T^2} = 2e(I_g + I_D) I_2 B \tag{8.63}$$

Let us now address the channel noise of the FET. This noise also has a source similar to the Johnson noise, i.e., it arises from the distribution function of the electrons as they move from the source to the drain of the FET. However, as discussed in the previous sections, the carriers are not described by a lattice temperature but by a carrier temperature T_e which is greater than the lattice temperature. The FET noise can be written in terms of the noise voltage per unit bandwidth

$$\overline{V_{FET}^2} = \frac{4k_B T E}{g_m} \tag{8.64}$$

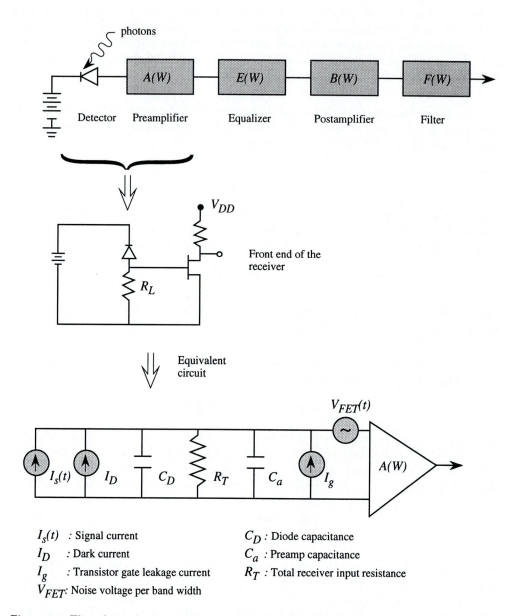

Figure 8.8: The schematic of a receiver amplifier. The front end of the amplifier is shown along with the equivalent circuit and the circuit parameters.

where E is the excess-noise factor due to the electron temperature effects, and g_m is the channel transconductance. We wish to rewrite the voltage noise in terms of a current noise. This can be done by defining the input admittance

$$Y_{in} = \frac{1}{R_L} + i\omega C_T \qquad (8.65)$$

where C_T is the total capacitance of the front end. Referring to Fig. 8.8,

$$C_T = C_D + C_a = C_T + C_{gs} + C_{gd} + C_s \qquad (8.66)$$

where C_D is the diode capacitance, C_{gs} and C_{gd} are the gate source and gate drain capacitances of the transistor and C_s is the stray capacitance due to packaging and wire bonds. Using this input admittance, the channel current noise becomes for the bandwidth B,

$$\overline{i_c^2} = \frac{4k_B T E}{g_m} \left[\frac{I_2 B}{R_L^2} + (2\pi C_T)^2 I_3 B^3 \right] \qquad (8.67)$$

where I_3 is another Personick integral with value around 0.1. Note that the noise current in the channel has a B^3 dependence due to the front end capacitance. It is thus extremely important to reduce the capacitance for high bandwidth applications.

We finally consider the 1/f noise of the FET which has the form

$$\overline{i_f^2} = \frac{4k_B T E}{g_m} (2\pi C_T)^2 f_c I_c B^2 \qquad (8.68)$$

where f_c is the 1/f noise carrier frequency and I_c has been calculated to be ~ 0.12.

We now have the total noise current of the amplifier

$$\overline{i_{preamp}^2} = \overline{i_R^2} + \overline{i_T^2} + \overline{i_c^2} + \overline{i_f^2} \qquad (8.69)$$

In addition, we can add to this noise the noise of the postamplifier and other circuits, to obtain the total amplifier current noise. Denoting the postamplifier noise by $\overline{i_p^2}$, we have

$$\overline{i_{amp}^2} = \overline{i_{preamp}^2} + \overline{i_p^2} \qquad (8.70)$$

The receiver discussed above is the high impedance (HZ) receiver. Another important receiver is the transimpedance (TZ) amplifier in which a feedback capacitor C_F and resistor R_F is used. The noise of the TZ amplifier is calculated as above except that R_L is replaced by R_F. The amplifiers cutoff frequency is limited by $R_L C_T$ (for the HZ) and $R_F C_F$ (for the TZ). Since C_F can be made smaller than C_T, the TZ amplifier has a better frequency response. However, since R_F is smaller than R_L, the TZ amplifier has a higher noise.

Let us now consider the receiver noise for the case where the detector is an avalanche photodiode.

8.4.1 APD Receiver Noise

$\longrightarrow$ We have noted in the previous sections that the multiplication process of the APD generates noise because of the random nature of the multiplication process. The multiplication related noise is described by the excess noise factor F_e discussed earlier and the multiplication M. The shot noise can now be written as in analogy with the p-i-n case

$$\overline{i^2_{APD}} = 2e \left(I_{Du} + M^2 F_e I_{DM} \right) I_2 B \tag{8.71}$$

where I_{Du} = unmultiplied dark current due to the detector, and FET gate leakage current.

I_{DM} = the dark current of the APD which is multiplied.

The first term on the right hand side of Eqn. 8.71 simply gives the p-i-n noise current discussed earlier and was denoted by $\overline{i^2_T} = (\overline{i^2_{p-i-n}})$. Thus we have

$$\overline{i^2_{APD}} = \overline{i^2_{p-i-n}} + 2eM^2 F_e I_{DM} I_2 B \tag{8.72}$$

The remaining contributions to the noise in the amplifier are the same as those discussed for the p-i-n photoreceiver.

EXAMPLE 8.5 Consider a photoreceiver with the following device parameters:

Load resistance(adjusted so that $R_L C_T \cong \pi/B$) $R_L = 10^6 - 10^{-3} \Omega$

FET leakage current $I_g = 10 \ nA$
p − i − ndark current $I_D = 70 \ nA$

FET transconductance $g_m = 50 \text{millisiemanns}$

FET excess noise factor $E = 2.0$

Total preamp capacitance $C_T = 1.0 \ pF$

p − i − ndiode capacitance $C_D = 0.5 \ pF$

FET 1/f noise corner frequency $f_c = 25 \ MHz$

Calculate the total noise current of the photoreceiver as a function of the bandwidth.

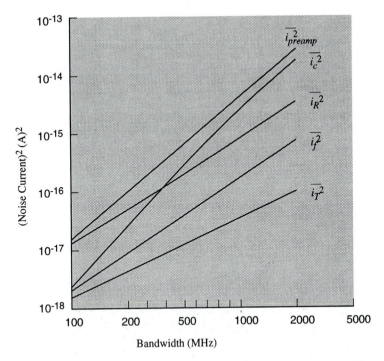

Figure 8.9: The noise current arising from various sources in the receiver amplifier of Example 8.5.

Note that the receiver resistance is adjusted to ensure that the device can operate at the desired bit rate. This makes the noise $\overline{i_R^2}$ proportional to B^2 and the channel noise proportional to B^3. The individual noise contributions are shown in Fig. 8.9. It is clear that at high bit rates, the FET noise arising from the channel dominates all noise sources. It is thus critical to obtain a high performance FET for a photoreceiver.

8.5   DIGITAL RECEIVER SENSITIVITY

$\longrightarrow$ We will now consider the effect of the receiver noise on the overall performance of a digital communication system. Let us assume that the signal current noise has a Gaussian form. In Fig. 8.10a we show the signal produced in a digital receiver when the input optical pulse corresponds to the 0 or 1 state. The signal has a spread due to the noise and the width of the Gaussian pulse is given by the standard deviation σ. As shown in Fig. 8.10a, the overlapping region of the signals at 0 and 1 state represents the probability of error. Assuming that the decision level D to distinguish between 0 and 1 is set symmetrically between S_0 and S_1, the bit error rate is

$$BER = \frac{1}{\sqrt{2\pi}} \int_Q^\infty exp\left(-\frac{x^2}{2}\right) dx = \frac{1}{2} erfc\left(\frac{Q}{\sqrt{2}}\right) \qquad (8.73)$$

where

$$Q = \frac{D - S_o}{\sigma}$$

or

$$Q = \frac{|D - S_1|}{\sigma} \qquad (8.74)$$

The higher the receiver sensitivity, the larger will $|D - S_o|$ or $|D - S_1|$ be, and smaller will the BER be. In Fig. 8.10b we show the BER as a function of the parameter Q. For a BER of 10^{-9}, a guideline value for optical communication, $Q = 6$.

Let us consider a *p-i-n* photoreceiver where the rms noise current defines the signal width σ

$$\sigma = \left(\overline{i_{amp}^2}\right)^{1/2} \qquad (8.75)$$

If the detected optical power in the state 0 and 1 is p_0 and p_1, we have from Eqns. 8.74 and 8.75,

$$\begin{aligned} D - S_1 &= D - \frac{e\lambda p_1}{hc} = -Q\left(\overline{i_{amp}^2}\right)^{1/2} \\ D - S_0 &= D - \frac{e\lambda p_0}{hc} = Q\left(\overline{i_{amp}^2}\right)^{1/2} \end{aligned} \qquad (8.76)$$

where we assume that the noise level in the receiver is not affected much by the optical signal and is determined primarily by the detector dark current and the transistor noise.

The difference between the optical power level in the two states is

$$p_1 - p_0 = \frac{2Q\, hc\, \left(\overline{i_{amp}^2}\right)^{1/2}}{e\lambda} \qquad (8.77)$$

and

$$D = \frac{1}{2} \frac{e\lambda(p_1 + p_0)}{hc} \qquad (8.78)$$

The optimum decision point for the detector is midway between the two power levels. The mean power received by the receiver is $\bar{p}$ and is given by

$$\eta\bar{p} = p_1 f + p_0(1 - f) \qquad (8.79)$$

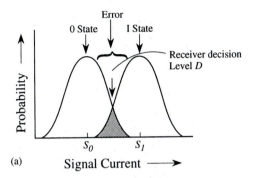

(a)

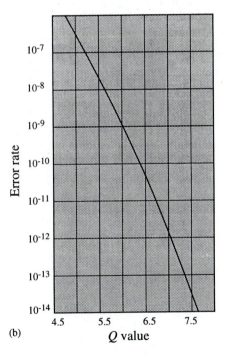

(b)

Figure 8.10: (a) Signal currents corresponding to a 0 and 1 state of the input pulse. (b) Dependence of the BER on the parameter Q which represents the spacing between S_0 and S_i of figure (a).

where η is the efficiency of power conversion for the amplifier and includes both the detector quantum efficiency and the optical coupling efficiency. The probability of 1's and 0's in the signal is given by f and $1 - f$ and for most cases we can assume $f = 1/2$. Thus

$$\eta\bar{p} = \frac{1}{2}(p_1 + p_0) \qquad (8.80)$$

Defining the extinction ratio r in the off/on ratio of the optical power

$$r = \frac{p_0}{p_1} \qquad (8.81)$$

we get

$$\boxed{\eta\bar{p} = \frac{(1+r)Q \; hc \; \left(\overline{i_{amp}^2}\right)^{1/2}}{(1-r)e\lambda}} \qquad (8.82)$$

This is the minimum detectable power in the p-i-n photoreceiver and defines the sensitivity of the receiver. The BER requirements are reflected by the presence in this equation of the parameter Q. If the optical signals are coming so that p_0 is zero, $r = 0$ and we have

$$\boxed{\eta\bar{p} = \frac{Q \; hc}{e\lambda} \; \left(\overline{i_{amp}^2}\right)^{1/2}} \qquad (8.83)$$

The minimum detectable power needed to reach a level of BER decreases with increasing λ, as long as λ is below the cutoff wavelength. Usually the sensitivity of the receiver is defined in dBm as discussed in Example 8.6 below.

EXAMPLE 8.6 Consider the photoreceiver discussed in Example 8.5. Calculate the sensitivity of the receiver in a communication system assuming that the ratio between optical power for the 0 and 1 state is 0. The required BER is 10^{-9}, and the wavelength is 1.3 μm.

The sensitivity is given by

$$\eta\bar{p}_{pin} = \frac{Q \; hc}{e\lambda} \; \left(\overline{i_{amp}^2}\right)^{1/2} \qquad (8.84)$$

or

$$\eta\bar{p}_{pin}(dBm) = 10log\left(\frac{\eta\bar{p}_{pin}}{10^{-3}(W)}\right) \qquad (8.85)$$

For $\lambda = 1.3$ μm with $Q = 6$ for a BER of 10^{-9}, we have

$$\frac{Qhc}{e\lambda} = 5.7V \qquad (8.86)$$

Using the values of $\overline{i_{amp}^2}$ calculated in Example 8.5, (assuming that the front end receiver dominates the noise) we get the receiver sensitivity. For example, at 2 GHz the detector sensitivity is -30 dBm while at 500 MHz it is -38.6 dBm.

8.5.1 APD Receiver Sensitivity

$\longrightarrow$ We have noted earlier than in an APD, the multiplication process adds to the noise of the detector. However, the signal also gets multiplied and, therefore, an optimization point occurs where the receiver sensitivity is maximum. We can extend the case of the p-i-n receiver to the APD receiver by writing in analogy to the previous case

$$
\begin{aligned}
\overline{i_1^2} &= \text{noise for the state 1} \\
&= \overline{i_a^2} + 2e\left(\frac{e\lambda}{hc}\right)M^2 F(M)I_1 B p_1
\end{aligned}
\tag{8.87}
$$

where the first term comes from the unmultiplied current and the second term comes from the excess noise of the multiplied current. I_1 is a Personick integral with a value of ~ 0.5 for rectangular pulses. Similarly for the zero state

$$
\begin{aligned}
\overline{i_o^2} &= \text{noise for the state 0} \\
&= \overline{i_a^2} + 2e\left(\frac{e\lambda}{hc}\right)M^2 F(M)I_1 B p_0
\end{aligned}
\tag{8.88}
$$

For simplicity, we assume that the dark current generated by the APD is zero and is only coming from the receiver. This gives, as for the the p-i-n receiver,

$$
D - \frac{e\lambda M p_1}{hc} = Q\left[\overline{i_a^2} + 2e\left(\frac{e\lambda}{hc}\right)M^2 F(M)I_1 B p_1\right]^{1/2}
\tag{8.89}
$$

$$
D - \frac{e\lambda M p_0}{hc} = Q\left[\overline{i_a^2} + 2e\left(\frac{e\lambda}{hc}\right)M^2 F(M)I_1 B p_0\right]^{1/2}
\tag{8.90}
$$

As in the case of the p-i-n receiver, when we solve for the receiver sensitivity, we get

$$
\begin{aligned}
\eta \bar{p}_{APD} = {}& \frac{Qhc}{e\lambda}\left(\frac{1+r}{1-r}\right)\left\{eQF(M)I_1 B\left(\frac{1+r}{1-r}\right)\right. \\
&\left. + \left[(2eQF(M)I_1 B)^2 \frac{r}{(1-r)^2} + \frac{<i_a^2>}{M^2}\right]^{1/2}\right\}
\end{aligned}
\tag{8.91}
$$

If the optical source has a value $r = 0$, i.e., no photons in the off state, we get

$$
\eta \bar{p}_{APD} = \frac{Qhc}{e\lambda} \left[\frac{\left(\overline{i_a^2} \right)^{1/2}}{M} + eQF(M)I_1 B \right]
$$

$$
= \frac{\eta \bar{p}_{pin}}{M} + \frac{hc}{\lambda} Q^2 F(M) I_1 B \tag{8.92}
$$

We have already discussed the excess noise factor $F(M)$ which depends upon the ratio of α_{imp} to β_{imp}. The APD multiplication factor has an important effect on the state receiver sensitivity. We see from Eqn. 8.92 that the first term in the sensitivity decreases as M increases, while the second term containing $F(M)$ increases as M increases. The sensitivity reaches an optimum value at a particular value of $M = M_{optimum}$ for a given value of $\overline{i_a^2}$ and the ratio of α_{imp} to β_{imp}.

In Fig. 8.11 we show the dependence of $\eta \bar{p}_{APD}$ on M for several values of α_{imp}/β_{imp} ratio. Other parameters used for the results are also given in Fig. 8.11. The sensitivity in dBm is defined by

$$
\eta \bar{p}_{APD}(dBm) = 10 log \left(\frac{\eta \bar{p}_{APD}}{1 \times 10^{-3} watt} \right) \tag{8.93}
$$

The extreme importance of the ionization coefficient ratio k is apparent from the results in Fig. 8.11. The very small value of k in silicon shows has attractive silicon is for low noise APD receivers. *However, keep in mind that Si will not respond to the 1.3 μm system considered in Fig. 8.11.*

If the dark current I_{DM} of the APD is not zero, the receiver sensitivity can be shown to be

$$
\eta \bar{p}_{APD} = \left[\left(\frac{\eta \bar{p}_{pin}}{M} \right)^2 + 2e \left(\frac{hc}{e\lambda} \right)^2 Q^2 F(M) I_{DM} I_2 B \right]^{1/2}
$$

$$
+ \left(\frac{hc}{\lambda} \right) Q^2 F(M) I_1 B \tag{8.94}
$$

where the second term in the bracket arises from the multiplication of the APD dark current.

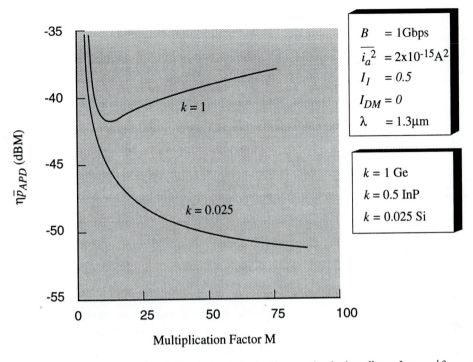

Figure 8.11: The importance of the impact ionization ratio k (smaller of α_{imp}/β_{imp} or β_{imp}/α_{imp}) and carrier multiplication of an APD photoreceiver sensitivity.

8.6 HETERODYNE AND HOMODYNE DETECTION OF SIGNALS

$\mathcal{R}$ As discussed in Section 8.2, in modern communication systems, information to be transmitted is coded onto an electromagnetic wave by modulating some feature of the wave. The feature to be modulated can, in principle, be the amplitude of the wave, the frequency of the wave, the phase of the wave, or the intensity of the wave. We have discussed the intensity modulation direct detection (IMDD) approach in which the optical source is simply used as a noisy source of light. The coherence of the system is of no consequence in such systems. At present, all commercial optical systems rely on the IMDD approach which is reliable and relatively inexpensive. However, as seen in the previous sections, the receiver adds a considerable amount of noise to the optical signal.

The coherent detection systems which rely on the frequency or phase coherence of the optical signal can reach extremely low noise levels and can, if optimally designed, add no additional noise to the noise already present in the incoming signal. However, these systems are extremely demanding in terms of device needs for both sources and receivers. At present, the state of the art in optoelectronic devices is such that these systems can only be used in laboratories or in controlled field tests. We will consider, in this section, how one decodes a signal coded in frequency or phase by the use of a local oscillator.

Consider a carrier wave with electric field given by

$$F_c(t) = A_c cos(\omega_c t + \phi_c) \tag{8.95}$$

where A_c, ω_c and ϕ_c represent the amplitude, frequency and phase of the wave. If the signal impinges on a detector directly, the detector output will provide an averaged irradiance if the detector is slow compared to $1/\omega_c$. The output will be

$$
\begin{aligned}
O_d &= R\ A_c^2 < cos^2(\omega_c t + \phi) > \\
&= \frac{R\ A_c^2}{2}
\end{aligned}
\tag{8.96}
$$

where R is the responsivity of the detector. Thus a detector that is slow will only respond to A_c^2, and it will not be possible to use frequency or phase modulation, if direct detection is used. However, it is possible to overcome this drawback by using the concept of heterodyne or homodyne detection.

In the scheme used for heterodyne detection, the incoming signal is mixed with the signal of a local oscillator as shown in Fig. 8.12. The local oscillator generates a signal of the form

$$F_{LO}(t) = A_{LO} cos(\omega_{LO} + \phi_{LO}) \tag{8.97}$$

where ω_{LO} and ω_c are chosen to be quite close. The detector output is now

$$
\begin{aligned}
O_d &= R\left\langle [F_c(t) + F_{LO}(t)]^2 \right\rangle \\
&= R\left\langle F_c^2(t) + F_{LO}^2(t) + 2F_c(t) \cdot F_{LO}(t) \right\rangle \\
&= R\left[\frac{A_c^2}{2} + \frac{A_{LO}^2}{2} + 2A_c A_{LO} \left\langle cos(\omega_c t + \phi_c) \cdot cos(\omega_{LO} t + \phi_{LO}) \right\rangle \right]
\end{aligned}
\tag{8.98}
$$

The average of the cosine terms is given by

$$
\begin{aligned}
\langle cos(\omega_c t + \phi_c) \cdot cos(\omega_{LO} t + \phi_{LO}) \rangle &= \frac{1}{2} \langle cos\left[(\omega_c + \omega_{co})t + (\phi_c + \phi_{LO})\right] \\
&+ cos\left[(\omega_c - \omega_{LO})t + (\phi_c - \phi_{LO})\right] \rangle
\end{aligned}
\tag{8.99}
$$

In the heterodyne detection scheme, ω_c and ω_{LO} are chosen to be very close to each other so that the detector can follow the temporal variation of the term $cos[(\omega_c - \omega_{LO})t +$

$(\phi_c - \phi_{LO})$]. This term is of main interest in the detection scheme. Using an approach shown in Fig. 8.12 where a bandpass filter centered around $(\omega_c - \omega_{LO})$ only allows this term to emerge, one gets the final output of the system

$$O_d = R \; A_c \; A_{LO} \; cos \left[(\omega_c - \omega_{LO})t + (\phi_c - \phi_{LO}) \right] \qquad (8.100)$$

The output depends upon the amplitude, frequency and phase of the carrier signal and thus can be used to decode any kind of modulation.

In a homodyne detection, the local oscillator frequency is matched exactly to the carrier frequency. The mixed signal output now becomes

$$\begin{aligned} O_d \;\; = \;\; & R \left[\frac{A_c^2}{2} + \frac{A_{LO}^2}{2} \right. \\ & + \;\; A_c A_{LO} \;\; cos(\phi_c - \phi_{LO}) + A_c A_{LO} \left(\; cos(2\omega_c t + \phi_c + \phi_{LO}) \right) \right] \quad (8.101) \end{aligned}$$

By inserting a low pass filter after the detector, the term

$$O_d = R \; A_c \; A_{LO} \; cos(\phi_c - \phi_{LO}) \qquad (8.102)$$

can be obtained so that both amplitude and phase modulation can be carried out in homodyne detection.

Apart from the advantage of being able to use a wide variety of modulation schemes, coherent detection (i.e., heterodyne and homodyne detection) has advantages of having better signal to noise properties over direct detection. At present, however, it is difficult to carry out coherent detection in optical communications. This is because of the difficulty in maintaining lasers at a stable frequency or phase. Thus one uses direct detection in optical communications. In microwave communications, however, where highly stable oscillators are available this is not a problem. The significance of these issues will be discussed in Chapter 13.

8.7   ADVANCED PHOTORECEIVERS: OEICs

$\mathcal{R}$ The tremendous advantages of integrating devices on the same chip in electronics naturally suggest that the same be done with electronic and optoelectronic devices. The OEICs have, however, proven to be a difficult challenge. Integrating transistors and resistors and capacitors for electronic ICs was not so difficult because of the compatibility of the fabrication process. However, the integration of the laser with its driver (a FET or a bipolar transistor) or a photodetector with an amplifier is proving to be quite

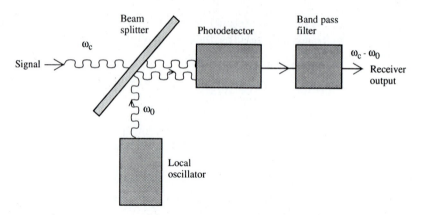

Figure 8.12: A schematic of the heterodyne detection. A beam splitter (mixer) mixes the incoming modulated signal with a coherent signal generated by a local oscillator. The output signal in the receiver can then detect amplitude, frequency or phase modulation.

a challenge because of the inherent incompatibilities in these devices. For example, the laser requires a *p-i-n* structure which is quite different from the structure of any transistor. This incompatibility requires that if the system is to remain planar, deep etching and regrowth must be carried out as shown in Fig. 8.13.

The area of regrowth is still in its infancy. Also, it is clear that when a material is etched, certain defects are left on the surface so that, after regrowth, bandgap states are created which can trap electrons. A great deal of effort in OEIC technology is being devoted to develop less damaging etches and better regrowth techniques.

In OEICs, if regrowth is to be completely avoided, one can integrate devices on different levels (i.e., on different heights on the wafer). This, however, is not an optimum approach, although it does provide working circuits. A number of schemes have been reported and are being pursued to advance OEIC technology. Nearly all kinds of different electronic and optical devices have been integrated, and recent results have indicated performance levels approaching those in hybrid technologies. Of course, it is expected that as progress continues, the OEICs will achieve better performance and certainly better reliability than hybrid circuits. Advances in OEIC technology are essential to realize the full promise of optoelectronics.

8.8 CHAPTER SUMMARY

$\mathcal{R}$ In this chapter we have examined the noise present in detectors and in electronic devices used for amplification. Important issues in photoreceiver sensitivity have also been discussed. The key findings of this chapter are summarized by the chapter summary

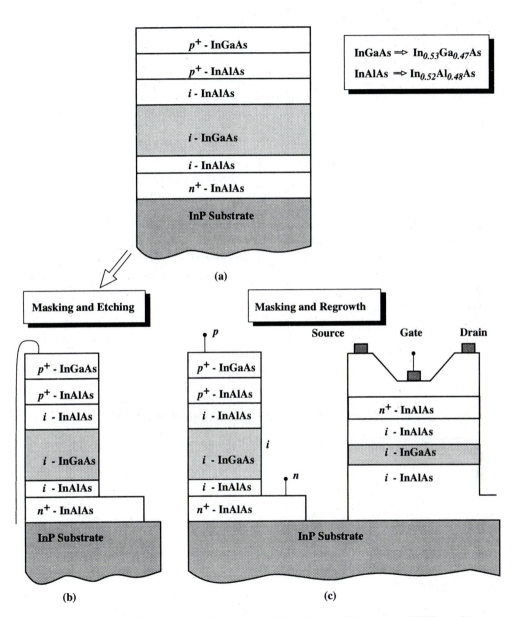

Figure 8.13: One of the many approaches that can be taken to fabricate an OEIC, in this case a p-i-n-FET detector. (a) A layer is grown epitaxially with a structure corresponding to a p-i-n detector. (b) The layer is masked and a section is etched. (c) Regrowth is carried out to grow a MODFET in the etched section.

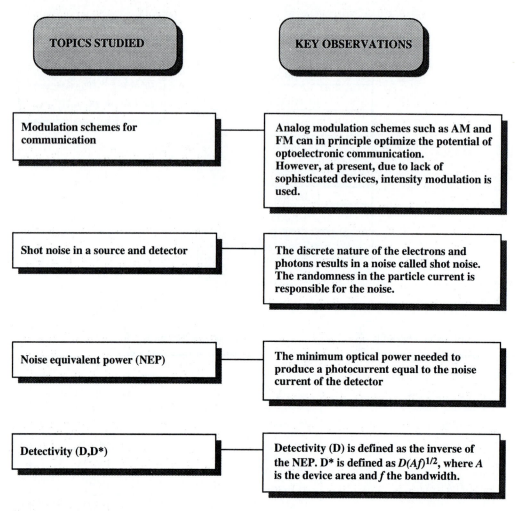

TOPICS STUDIED	KEY OBSERVATIONS
Modulation schemes for communication	Analog modulation schemes such as AM and FM can in principle optimize the potential of optoelectronic communication. However, at present, due to lack of sophisticated devices, intensity modulation is used.
Shot noise in a source and detector	The discrete nature of the electrons and photons results in a noise called shot noise. The randomness in the particle current is responsible for the noise.
Noise equivalent power (NEP)	The minimum optical power needed to produce a photocurrent equal to the noise current of the detector
Detectivity (D,D*)	Detectivity (D) is defined as the inverse of the NEP. D* is defined as $D(Af)^{1/2}$, where A is the device area and f the bandwidth.

Table 8.1: Summary table.

tables (Tables 8.1-8.3).

8.9   PROBLEMS

Section 8.3

8.1 A detector receives digital data from an optical source. On an average, the bit 1 has 50 photons in it and the bit 0 has 25 photons. Assuming that the photons obey a Poisson distribution, estimate the bit error rate arising from the random noise.

8.2 A *p-i-n* detector which is limited by shot noise is to be operated at 10 GHz.

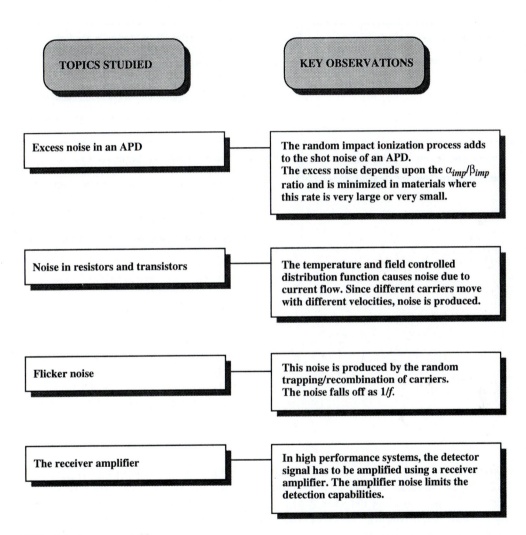

TOPICS STUDIED	KEY OBSERVATIONS
Excess noise in an APD	The random impact ionization process adds to the shot noise of an APD. The excess noise depends upon the α_{imp}/β_{imp} ratio and is minimized in materials where this rate is very large or very small.
Noise in resistors and transistors	The temperature and field controlled distribution function causes noise due to current flow. Since different carriers move with different velocities, noise is produced.
Flicker noise	This noise is produced by the random trapping/recombination of carriers. The noise falls off as $1/f$.
The receiver amplifier	In high performance systems, the detector signal has to be amplified using a receiver amplifier. The amplifier noise limits the detection capabilities.

Table 8.2: Summary table.

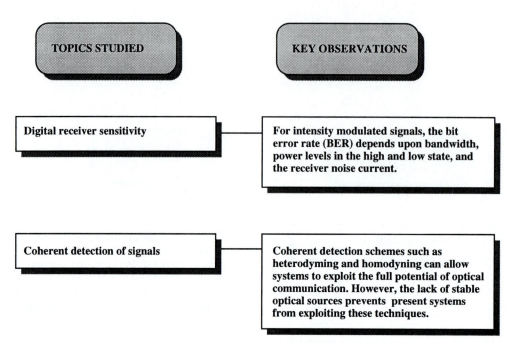

Table 8.3: Summary table.

Calculate the photocurrent needed to ensure a signal to noise ratio of 60 dB.

8.3 The D^* value for an HgCdTe detector for 10 μm radiation is found to be 5×10^{10} cm $\text{Hz}^{1/2}W^{-1}$ at 77 K. The detector is to be used in an imaging array which is to operate at a bandwidth of 100 Hz. The area of the detector is 10^{-4} cm^2. Calculate the minimum power that can be detected.

8.4 A detector material is to be selected for a night vision imaging application. The detector area is to be 10^{-4} cm^2 and it should be able to detect radiation with power levels one millionth of the daytime power levels. The imaging system is to operate at a bandwidth of 100 Hz. The daytime power levels are 0.1 $W/$cm^2. Calculate the D^* of the detector needed.

8.5 A transimpedance amplifier is to be used in a *p-i-n* photoreceiver. The system is described by the following parameters:

Feedback resistance	$R_F = 10^5 - 10^3 \Omega$
FET leakage current	$I_g = 5 \ nA$
$p-i-n$ dark current	$I_D = 50 \ nA$
FET transconductance	$2m = 60$ millisiemanns
FET excess noise factor	$E = 2.0$
Preamp capacitance	$C_T = 1.5 \ pF$
Diode capacitance	$C_D = 0.2 \ pF$
FET 1/f corner frequency	$f_c = 20 \ MHz$

Calculate the various noise contributions for the photoreceiver at 100 Mbps and 1 Gbps. Assume that $R_F C_T = \pi/B$ is maintained so that the receiver is not RC limited.

Section 8.3

8.6 Consider a p-i-n photoreceiver with a total noise current given by $\overline{i_a^2} = 2 \times 10^{-15} \ A^2$ at $B = 1$ Gbps. The receiver is used in a communication system with a BER $= 10^{-6}$ and $\lambda = 1.3 \ \mu$m. The system is such that in the extinction ratio is $r = 0$. Calculate the receiver sensitivity. If an APD were to be used with zero dark current, calculate the optimum sensitivity that can be achieved. What is the optimum multiplication factor? Assume that $k= 0.5$ and $B = 1$ Gbps.

8.7 In the previous problem, calculate the sensitivity of the APD if the APD dark current is 10 nA.

8.8 In most optical sources used for optical digital communication, the source emits photons in both ON and OFF state so that the extinction ratio $r = p_o/p_1 > 0$. Calculate the sensitivity of a p-i-n receiver in which $\overline{i_a^2} = 2 \times 10^{-15} A^2$, BER $= 10^{-9}, \lambda = 1.3 \ \mu$m as a function of r. Calculate the results as r goes from 0 to 1.

8.9 An APD photoreceiver is used in a 1.3 μm communication system operating at 1 Gbps. The k value for the APD material is 0.5 and the dark current in the APD is zero. The noise in the amplifier is $\overline{i_a^2} = 2 \times 10^{-15} A^2$. Calculate the optimum receiver sensitivity at a BER $= 10^{-9}$. How does the sensitivity change if due to a temperature change the APD dark current changes to 100 nA?

8.10 Consider the APD photoreceiver of the previous problem. Calculate the optimum multiplication factor. If a voltage instability changes the biasing so that the optimum multiplication factor drops to half its value, calculate the change in the receiver sensitivity. The dark current is zero.

8.10   REFERENCES

- **General**

  - An excellent discussion of photoreceiver design is given in articles by T. Kaneda, by S. R. Forrest, and by J. C. Campbell in *Semiconductors and Semimetals*, Volume 22, part D, Academic Press, New York (1985).

  - J. Gowar, *Optical Communication Systems*, Prentice-Hall, Englewood Cliffs, NJ (1989).

  - R. J. McIntyre, "Multiplication Noise in Uniform Avalanche Diodes," *IEEE Transactions on Electron Devices*, ED-13, 164 (1966).

  - The integrals I_1, I_2 discussed in the receiver noise are discussed by S. D. Personick, *Bell System Technical Journal*, 52, 843 (1973).

CHAPTER 9

THE LIGHT EMITTING DIODE

9.1 INTRODUCTION

$\mathcal{R}$ Optical signal generation in the optical regime is essential for a variety of passive and active information processing functions. Optical signals are used for communication where the superior overall performance of optical fibers as a carrier of information is rapidly replacing metallic cables. Optical signals are needed for display devices to project information. Optical beams are required for memory systems based on optical reading. Both incoherent and coherent light emitting devices are needed for a variety of applications in communication and display. Light emitting devices are also being considered for applications involving active decision making roles. As a result switches, and logic devices are being proposed and demonstrated using light emitting devices.

The light emitting diode (LED) is one of the simplest optoelectronic devices which has found important applications as a display device as well as an optical signal generator for optical communication. Compared to the laser diode, the fabrication of the LED is very simple since it does not require any special optical cavity for its operation. However, one pays the price in terms of low optical output, broad and incoherent spectra, and slow device response. In this chapter we will explore the physics behind the operation of the LED . The laser diode will be explored in the next two chapters. These two devices are providing a major thrust in the advancement of optoelectronic information systems. The reader is advised to review our discussions in Chapter 4 on optical processes in semiconductors (Sections 4.4 and 4.7).

9.2 MATERIAL SYSTEMS FOR THE LED

$\mathcal{R}$ The simplicity of the light emitting diode (LED) makes it a very attractive device for display and communication applications. It gives way to the laser diode in applications where modulation speeds above ~ 5 GHz are needed or where spectrally pure optical output is needed. The spectral width of the optical output of an LED is of the order of $k_B T$ which translates into a wavelength spread of 300-400 Å at room temperature. Although this is a large value, the LED produces a single color to the human eye. Thus the LEDs can be used very effectively in color displays. An important recent application of LEDs is in providing tail lights in automobiles, an application that could make LEDs very important commercial devices.

The basic LED is a *p-n* junction which is forward biased to inject electrons and holes into the *p*- and *n*-sides respectively. The injected minority charge recombines with the majority charge in the depletion region or the neutral region. *In direct band semiconductors, this recombination leads to light emission since radiative recombination dominates in high quality materials. In indirect gap materials, the light emission efficiency is quite poor and most of the recombination paths are non-radiative which generate heat rather than light.* We will now examine the important issues which govern the material systems that are used for LEDs.

Light emitting devices are one class of devices that have given impetus to the compound semiconductor industry. Since Si is an indirect gap material, and the radiative recombination is very poor, this material, which dominates all other areas of electronics, finds itself handicapped when it comes to light emission.

In Appendix B, Fig. B.1, we show the bandgaps of a variety of compound semiconductors along with their lattice constants. The direct gap regions are denoted by the solid line while the indirect gap materials are denoted by the dashed line. As can be seen, a wide range of combinations is available to the device designer.

We will briefly outline some of the important considerations in choosing a semiconductor for LEDs or laser diodes. The reader should revisit the discussion on materials used for detectors (section 7.3).

Emission Energy: The light emitted from the device is very close to the semiconductor bandgap since the injected electrons and holes are described by quasi-Fermi distribution functions. The desire to have a particular emission energy may arise from a number of motivations. In Fig. 7.2 we show the response of the human eye to radiation of different wavelengths. Also shown are the bandgaps of some semiconductors. If a color display is to be produced that is to be seen by people, one has to choose an appropriate semiconductor. Very often one has to choose an alloy since there is a greater flexibility in the bandgap range available. In Fig. 9.1 we show the loss characteristics of an optical fiber. As can be seen, the loss is lowest at 1.55 μm and 1.3 μm. If optical communication sources are desired, one must choose materials which can emit at these wavelengths. This is especially true if the communication is long haul, i.e., over hundreds or even thousands of kilometers. Materials like GaAs which emit at 0.8 μm can still be used for local area networks (LANs) which involve communicating within a building or local areas. The importance of the light energy in communication applications is examined in greater detail in Chapter 13.

Substrate Availability: Almost all optoelectronic light sources depend upon epitaxial crystal growth techniques where a thin active layer (a few microns) is grown on a substrate (which is ~ 200 μm). The availability of a high quality substrate is extremely important in epitaxial technology. If a substrate that lattice matches to the active device layer is not available, the device layer may have dislocations and other defects in it. These can seriously hurt the device performance. The important substrates that are available for light emitting technology are GaAs and InP. A few semiconductors and their alloys can match with these substrates. The lattice constant of an alloy is the weighted mean of the lattice constants of the individual components, i.e., the lattice constant of the alloy $A_x B_{1-x}$ is

$$a_{all} = x a_A + (1 - x) a_B \tag{9.1}$$

where a_A and a_B are the lattice constants of A and B. Semiconductors that cannot lattice match with GaAs or InP have an uphill battle for technological success. The crystal grower must learn the difficult task to grow the semiconductor on a mismatched substrate without allowing dislocation to propagate into the active region (see discussions on strained heteroepitaxy in Chapter 1).

Important semiconductor materials exploited in optoelectronics are the alloy $Ga_x Al_{1-x} As$ which is lattice matched very well to GaAs substrates; $In_{0.53}Ga_{0.47}As$ and $In_{0.52}Al_{0.48}As$ which are lattice matched to InP; InGaAsP which is a quaternary material whose composition can be tailored to match with InP and can emit at 1.55 μm; and GaAsP which has a wide range of bandgaps available. Recently there has been a considerable interest in large bandgap materials such as ZnSe, ZnS, SiC, and GaN to produce devices that emit blue or green light. The motivation is for superior display technology and for high density optical memory applications (a shorter wavelength allows reading of smaller features). Reliable SiC LEDs are now available in the commercial market, although only a few suppliers can meet the technology challenge.

It is important to keep in mind that alloys like GaAlAs and GaAsP become indirect at certain compositions as shown in Fig. 9.2. For efficient light emission, one needs to work in the direct gap region. However, with a suitable impurity, one can obtain light emission in an indirect bandgap material. In Fig. 9.3 we show some important material systems used in light emitting devices, along with some of their special properties.

As we saw in Chapter 4, the momentum conservation causes strong radiative transitions to occur only in direct gap semiconductors. *Some indirect gap materials can,*

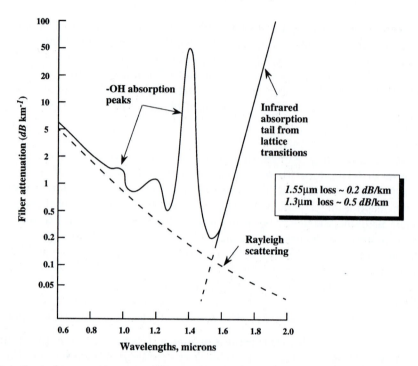

Figure 9.1: Optical attenuation vs. wavelength for an optical fiber. Primary loss mechanisms are identified as absorption and scattering. See Chapter 13 for details on how the loss mechanisms affect device requirements.

however, have a reasonable radiative efficiency if they are doped with certain impurities. The impurities create levels in the bandgap and photon absorption is allowed by moving electrons into these levels. The absorption and emission rates are, however, smaller than those for direct gap semiconductors. The GaAsP alloy system is one semiconductor system in which impurity levels have been widely used to produce LEDs. However, since the light emission efficiency is poor, it has not been possible to use these impurity levels to produce laser diodes. The need for higher radiative efficiencies for laser diodes will become clear when we discuss their physics in Section 9.3.

9.3 OPERATION OF THE LED

$\longrightarrow$ The LED is a forward biased *p-n* diode in which electrons and holes are injected into a region where they recombine. In general, the electron-hole recombination process can occur by radiative and nonradiative channels. Under the condition of minority carrier recombination or high injection recombination, as shown in Chapter 4 (Sections 4.7 and 4.8), one can define a lifetime for carrier recombination. If τ_r and τ_{nr} are the radiative

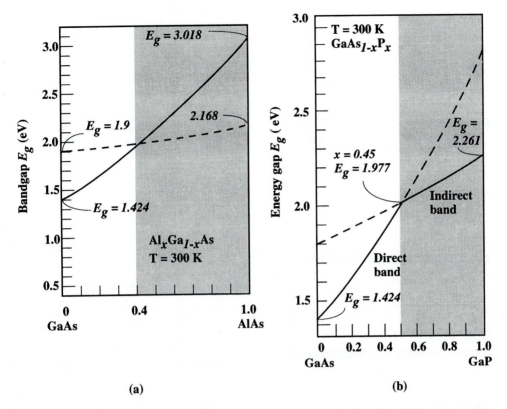

Figure 9.2: Bandgap of (a) $Al_xGa_{1-x}As$ and (b) $GaAs_{1-x}P_x$ as a function of alloy composition. Note that the bandgap changes from direct to indirect as shown. (After H. C. Casey and M. B. Panish, *Heterostructure Lasers*, Academic Press, New York (1978).)

and non-radiative lifetimes, the total recombination time is (for, say, an electron)

$$\frac{1}{\tau_n} = \frac{1}{\tau_r} + \frac{1}{\tau_{nr}} \tag{9.2}$$

The internal quantum efficiency for the radiative processes is then defined as

$$\eta_{Qr} = \frac{\frac{1}{\tau_r}}{\frac{1}{\tau_r} + \frac{1}{\tau_{nr}}} = \frac{1}{1 + \frac{\tau_r}{\tau_{nr}}} \tag{9.3}$$

In high quality direct gap semiconductors, the internal efficiency is usually close to unity. In indirect materials the efficiency is of the order of 10^{-2} to 10^{-3}.

Before starting the discussion on light emission, let us remind ourselves of some important definitions and symbols used in this chapter:

I_{ph} : photon current = number of photons passing a cross-section per second.

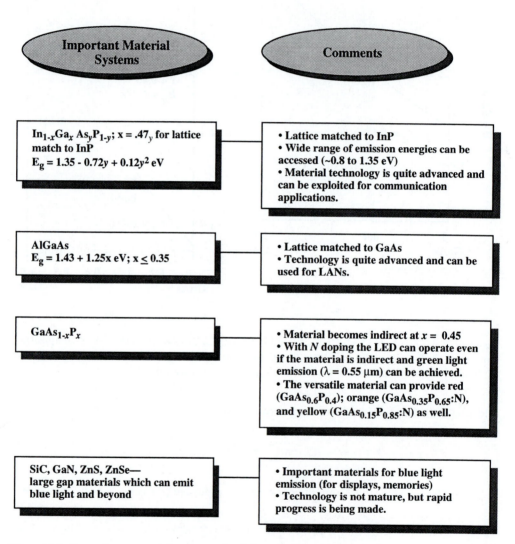

Figure 9.3: Important material systems for light emission. LEDs can be made from indirect materials with an appropriate impurity, but the emission efficiency is low.

J_{ph} : photon current density $\quad = \quad$ number of photons passing a unit area per second.

P_{op} : optical power intensity $\quad = \quad$ energy carried by photons per second per area.

9.3.1 Carrier Injection and Spontaneous Emission

$\longrightarrow$ The LED is essentially a forward biased *p-n* diode as shown in Fig. 9.4. Electrons and holes are injected as minority carriers across the diode junction and they recombine either by radiative recombination or non-radiative recombination. The diode must be designed so that the radiative recombination can be made as strong as possible.

The theory of the *p-n* diode was discussed in detail in Chapter 6. In the forward bias conditions the electrons are injected from the *n*-side to the *p*-side while holes are injected from the *p*-side to the *n*-side. As noted in Chapter 6, the forward bias current is dominated by the minority charge diffusion current across the junction. The diffusion current, in general, consists of three components: i) minority carrier electron diffusion current, ii) minority carrier hole diffusion current; and iii) trap assisted recombination current in the depletion region of width W. These current densities have the following forms respectively (see Eqns. 6.41, 6.42, and 6.43 of Chapter 6):

$$J_n = \frac{eD_n n_p}{L_n}\left[exp\left(\frac{eV}{k_BT}\right) - 1\right] \tag{9.4}$$

$$J_p = \frac{eD_p p_n}{L_p}\left[exp\left(\frac{eV}{k_BT}\right) - 1\right] \tag{9.5}$$

$$J_{GR} = \frac{en_iW}{2\tau}\left[exp\left(\frac{eV}{2k_BT}\right) - 1\right] \tag{9.6}$$

where τ is the recombination time in the depletion region and depends upon the trap density. *The LED is designed so that the photons are emitted close to the top layer and not in the buried layer as shown in Fig. 9.4. The reason for this choice is that photons emitted deep in the device have a high probability of being reabsorbed. Thus one prefers to have only one kind of carrier injection for the diode current.* Usually the top layer of the LED is *p*-type, and for photons to be emitted in this layer one must require the diode current to be dominated by the electron current (i.e., $J_n \gg J_p$). The ratio of the electron current density to the total diode current density is called the injection efficiency γ_{inj}. Thus we have

$$\gamma_{inj} = \frac{J_n}{J_n + J_p + J_{GR}} \tag{9.7}$$

If the diode is pn^+, $n_p \gg p_n$, and, as can be seen from Eqns. 9.4 and 9.5, J_n becomes much larger than J_p. If, in addition, the material is high quality so that the recombination current is small, the injection efficiency approaches unity.

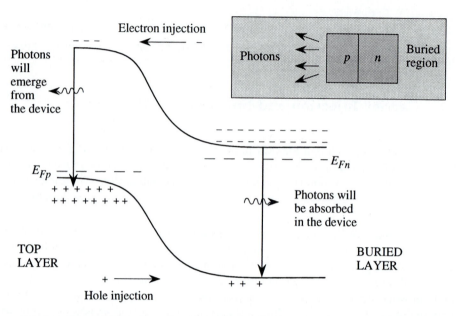

Figure 9.4: In a forward biased p-n junction, electrons and holes are injected as shown. In the figure, the holes injected into the buried n region will generate photons which will not emerge from the surface of the LED. The electrons injected will generate photons which are near the surface and have a high probability to emerge.

Once the minority charge (electrons) is injected into the doped neutral region (p-type), the electrons and holes will recombine to produce photons. They may also recombine non-radiatively via defects or via phonons. The radiative recombination process was discussed in Chapter 4, and we will briefly review it for the direct bandgap semiconductors.

As discussed in Chapter 4, the radiative process is "vertical," i.e., the k-value of the electron and that of the hole are the same in the conduction and valence bands, respectively. From Fig. 9.5 we see that the photon energy and the electron and hole energies are related by

$$\hbar\omega - E_g = \frac{\hbar^2 k^2}{2}\left[\frac{1}{m_e^*} + \frac{1}{m_h^*}\right] = \frac{\hbar^2 k^2}{2m_r^*} \tag{9.8}$$

where m_r^* is the reduced mass for the e-h system. The electron and hole energies are related to the photon energy by the relations

$$E^e = E_c + \frac{\hbar^2 k^2}{2m_e^*} = E_c + \frac{m_r^*}{m_e^*}(\hbar\omega - E_g) \tag{9.9}$$

$$E^h = E_v - \frac{\hbar^2 k^2}{2m_h^*} = E_v - \frac{m_r^*}{m_h^*}(\hbar\omega - E_g) \tag{9.10}$$

If an electron is available in a state k, and a hole is also available in the state k (i.e., the Fermi functions for the electrons and holes satisfy $f^e(k) = f^h(k) = 1$), the radiative recombination rate is given by (see Eqns. 4.83 and 4.84 and Example 4.6)

$$W_{em} = \frac{1}{\tau_o} = \frac{e^2 n_r \hbar \omega}{3\pi \epsilon_o m_0^2 c^3 \hbar^2} |p_{cv}|^2 \tag{9.11}$$

where n_r is the refractive index of the semiconductor, m_0 the free electron mass, and p_{cv} the momentum matrix elements between the conduction and valence bands. It turns out that p_{cv} does not vary too much between semiconductors and has a value given by the equation

$$\frac{2p_{cv}^2}{m_0} \simeq 22 \ eV \tag{9.12}$$

Thus the emission rate turns out to be (see Examples 4.6 or 9.1)

$$\boxed{W_{em} \sim 1.14 \times 10^9 \hbar\omega(eV) \qquad s^{-1}} \tag{9.13}$$

and the recombination time becomes ($\hbar\omega$ is expressed in electron volts)

$$\boxed{\tau_o = \frac{0.88}{\hbar\omega(eV)} \qquad ns} \tag{9.14}$$

The recombination time discussed above is the shortest possible spontaneous emission time since we have assumed that the electron has a unit probability of finding a hole with the same k-value.

When carriers are injected into the semiconductors (see Section 4.7) the occupation probabilities for the electron and hole states are given by the appropriate quasi-Fermi levels. *The e-h recombination process is determined by the spontaneous emission which implies that the photon density of the emission is quite low so that stimulated emission is not significant.* The emitted photons leave the device volume so that the photon density never becomes high in the e-h recombination region. In a laser diode the situation is different, as we shall see later. The photon emission rate is given by integrating the emission rate W_{em} over all the electron-hole pairs after introducing the appropriate Fermi functions. There are several important limits of the spontaneous rate which were discussed in Chapter 4. The reader is advised to review the discussions in Section 4.7. We will summarize them here:

i) *In the case where the electron and hole densities n and p are small (non-degenerate case), the Fermi functions have a Boltzmann form ($exp(-E/k_BT)$). The* recombination rate then becomes (see Eqn. 4.111)

$$R_{spon} = \frac{1}{2\tau_o} \left(\frac{2\pi\hbar^2 m_r^*}{k_B T m_e^* m_h^*} \right)^{3/2} n \, p \tag{9.15}$$

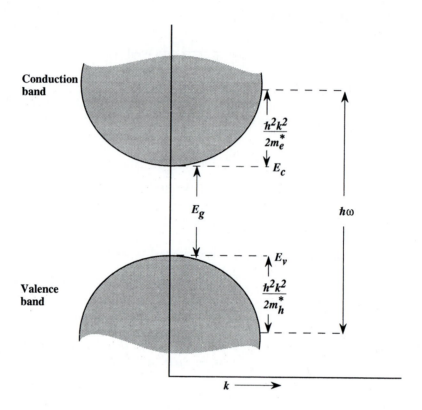

Figure 9.5: A schematic of the *E-k* diagram for the conduction and valence bands. Optical transitions are vertical; i.e., *k*-vector of the electron in the valence band and in the conduction band is the same.

The rate of photon emission depends upon the product of the electron and hole densities. If we were to define the lifetime of a single electron injected into a lightly doped ($p = N_a \leq 10^{17} \text{cm}^{-3}$) p-type region with hole density p, it would be given from Eqn. 9.15 by

$$\frac{R_{spon}}{n} = \frac{1}{\tau_r} = \frac{1}{2\tau_o} \left(\frac{2\pi\hbar^2 m_r^*}{k_B T m_e^* m_h^*} \right)^{3/2} p \qquad (9.16)$$

The time τ_r in this regime is very long (hundreds of nanoseconds), as shown in Fig. 9.6 and becomes smaller as p increases.

ii) *In the case where electrons are injected into a heavily doped p-region (or holes are injected into a heavily doped n-region), the function $f^h(f^e)$ can be assumed to be unity.* The spontaneous emission rate is then (see Eqn. 4.108)

$$R_{spon} \sim \frac{1}{\tau_o} \left(\frac{m_r^*}{m_h^*} \right)^{3/2} n \tag{9.17}$$

for electron concentration n injected into a heavily doped p-type region and

$$R_{spon} \sim \frac{1}{\tau_o} \left(\frac{m_r^*}{m_e^*} \right)^{3/2} p \tag{9.18}$$

for hole injection into a heavily doped n-type region.

The minority carrier lifetimes (i.e., n/R_{spon}) play a very important role not only in LEDs, but also in diodes and bipolar devices. In this regime the lifetime of a single electron (hole) is independent of the holes (electrons) present since there is always a unity probability that the electron (hole) will find a hole (electron). The lifetime is now essentially τ_o, as shown in Fig. 9.6.

iii) *Another important regime is that of high injection where $n = p$ is so high that one can assume $f^e = f^h = 1$ in the integral for the spontaneous emission rate.* The spontaneous emission rate is (see Eqn. 4.109)

$$\boxed{R_{spon} \sim \frac{n}{\tau_o} \sim \frac{p}{\tau_o}} \tag{9.19}$$

and the radiative lifetime ($n/R_{spon} = p/R_{spon}$) is τ_o.

iv) *A regime that is quite important for laser operation is one where sufficient electrons and holes are injected into the semiconductor to cause "inversion."* As will be discussed later, this occurs if $f^e + f^h \geq 1$. If we make the approximation $f^e \sim f^h = 1/2$, for all the electrons and holes at inversion, we get the relation (see Eqn. 4.115)

$$R_{spon} \sim \frac{n}{4\tau_o} \tag{9.20}$$

or the radiative lifetime at inversion is

$$\boxed{\tau_r \sim \frac{\tau_o}{4}} \tag{9.21}$$

This value is a reasonable choice to calculate for the spontaneous emission rate in lasers near threshold.

The radiative recombination depends upon the radiative lifetime τ_r and the non-radiative lifetime τ_{nr}. To improve the efficiency of photon emission one needs a value

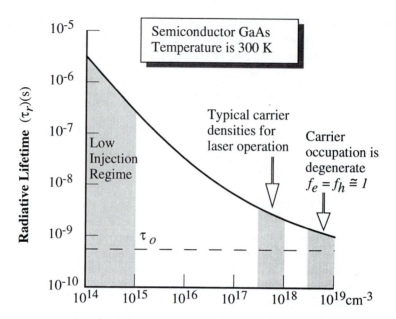

N_d (for holes injected into an n-type semiconductor)

$n = p$ (for excess electron hole pairs injected into a region)

Figure 9.6: Radiative lifetimes of electrons or holes in a direct gap semiconductor as a function of doping or excess charge. The figure gives the lifetimes of a minority charge (a hole) injected into an n-type material. The figure also gives the lifetime behavior of electron-hole recombination when excess electrons and holes are injected into a material as a function of excess carrier concentration.

of τ_r as small as possible and τ_{nr} as large as possible. To increase τ_{nr} one must reduce the material defect density. This includes improving surface and interface qualities.

From our discussion τ_r can be reduced by increasing the p-type doping in the region where the injected electrons recombine with holes. However, this reduces the injection efficiency γ_{inj} as can be seen from Eqns. 9.4 and 9.7. The total internal quantum efficiency is

$$\eta_{int} = \gamma_{inj}\eta_{Qr} \qquad (9.22)$$

Thus to maximize the value of η_{int}, one needs to optimize the p-side doping so that it is not so low that η_{Qr} is poor, but not so high that γ_{inj} is low.

EXAMPLE 9.1 In most direct gap semiconductors, the momentum matrix element p_{cv} is given by

$$\frac{2p_{cv}^2}{m_0} \cong 22eV \text{ or } \frac{p_{cv}^2}{m_0} = 11eV$$

Calculate the electron-hole recombination time τ_o for GaAs. Remember that τ_o is defined as the recombination time for an electron in state k to recombine with a hole in state k, i.e., $f^e = 1$, $f^h = 1$.

The spontaneous emission rate is given by

$$W_{em} = \frac{1}{\tau_o} = \frac{4}{3} \frac{e^2}{4\pi\epsilon_o} \frac{n_r \hbar\omega |p_{cv}|^2}{m_o^2 c^3 \hbar^2}$$

Using $\hbar\omega = 1.5$ eV, $n_r = 3.5$, and $|p_{cv}|^2$ as given, we get

$$\frac{1}{\tau_o} = \frac{4}{3} \frac{\left(1.6 \times 10^{-19}C\right)^2 \times (3.5) \times \left(1.5 \times 1.6 \times 10^{-19} J\right)}{(4 \times 3.1416 \times 8.85 \times 10^{-12} F/m) \times (9.1 \times 10^{-31} kg)}$$

$$\times \frac{\left(11.0 \times 1.6 \times 10^{-19} \ J\right)}{(3 \times 10^8 \ m/s)^3 \times (1.05 \times 10^{-34} \ Js)^2}$$

$$= 1.67 \times 10^9 \ s^{-1}$$

or

$$\tau_o = 0.6 \ ns$$

EXAMPLE 9.2 Calculate the e-h recombination time when an excess electron and hole density of 10^{15} cm^{-3} is injected into a GaAs sample at room temperature.

Since 10^{15} cm^{-3} or 10^{21} m^{-3} is a very low level of injection, the recombination time is given by Eqn. 9.16 as

$$\frac{1}{\tau_r} = \frac{1}{2\tau_o} \left(\frac{2\pi\hbar^2 m_r^*}{k_B T m_e^* m_h^*} \right)^{3/2} p$$

$$= \frac{1}{2\tau_o} \left(\frac{2\pi\hbar^2}{k_B T m_e^* + m_h^*} \right)^{3/2} p$$

Using $\tau_o = 0.6$ ns and $k_B T = 0.026$ eV, we get for $m_e^* = 0.067 m_0$, $m_h^* = 0.45 m_0$,

$$\frac{1}{\tau_r} = \frac{10^{21} m^{-3}}{2 \times (0.6 \times 10^{-9} \ s)} \left[\frac{2 \times 3.1416 \times (1.05 \times 10^{-34} \ Js)^2}{(0.026 \times 1.6 \times 10^{-19} J) \times (0.517 \times 9.1 \times 10^{-31} \ kg)} \right]^{3/2}$$

$$\tau_r = 5.7 \times 10^{-6} \ s \cong 9.5 \times 10^3 \tau_o$$

We see from this example, that at low injection levels, the carrier lifetime can be very long. Physically, this occurs because at such a low injection level, the electron has a very small probability of finding a hole to recombine with.

EXAMPLE 9.3 In two n^+p GaAs LEDs, $n^+ >> p$ so that the electron injection efficiency is 100% for both diodes. If the non-radiative recombination time is 10^{-7}s, calculate the 300 K internal radiative efficiency for the diodes when the doping in the p-region for the two diodes is 10^{16} cm^{-3} and 5×10^{17} cm^{-3}.

When the p-type doping is 10^{16} cm^{-3}, the hole density is low and the e-h recombination time for the injected electrons is given by Eqn. 9.16 as

$$\frac{1}{\tau_r} = \frac{1}{2\tau_o} \left(\frac{2\pi\hbar^2 m_r^*}{k_B T m_e^* m_h^*} \right)^{3/2} p$$

From the previous example, we can see that for p equal to 10^{16} cm^{-3}, we have (in the previous example the value of p was ten times smaller)

$$\tau_r = 5.7 \times 10^{-7} s$$

In the case where the p doping is high, the recombination time is given by the high density limit (see Eqn. 9.17) as

$$\frac{1}{\tau_r} = \frac{R_{spon}}{n} = \frac{1}{\tau_o} \left(\frac{m_r^*}{m_h^*} \right)^{3/2}$$

$$\tau_r = \frac{\tau_o}{0.05} \sim 20\tau_o \sim 12 \ ns$$

For the low doping case, the internal quantum efficiency for the diode is

$$\eta_{Qr} = \frac{1}{1 + \frac{\tau_r}{t_{nr}}} = \frac{1}{1 + (5.7)} = 0.15$$

For the heavier doped p-region diode, we have

$$\eta_{Qr} = \frac{1}{1 + \frac{10^{-7}}{20 \times 10^{-9}}} = 0.83$$

Thus, there is an increase in the internal efficiency as the p doping is increased. Of course, as discussed in the text, this increase cannot continue with p doping since eventually the injection efficiency will decrease.

EXAMPLE 9.4 Consider a GaAs p-n diode with the following parameters at 300 K:

Electron diffusion coefficient,	D_n =	$30 \ cm^2/V - s$
Hole diffusion coefficient,	D_p =	$15 \ cm^2/V - s$
p-side doping,	N_a =	$5 \times 10^{16} \ cm^{-3}$
n-side doping,	N_d =	$5 \times 10^{17} \ cm^{-3}$
Electron minority carrier lifetime,	τ_n =	$10^{-8} \ s$
Hole minority carrier lifetime,	τ_p =	$10^{-7} \ s$

Calculate the injection efficiency of the LED assuming no recombination due to traps.

The intrinsic carrier concentration in GaAs at 300 K is 2×10^6 cm^{-3}. This gives

$$n_p = \frac{n_i^2}{N_a} = \frac{(2 \times 10^6)^2}{5 \times 10^{16}} = 8 \times 10^{-5} \ cm^{-3}$$

$$p_n = \frac{n_i^2}{N_d} = \frac{(2 \times 10^6)^2}{5 \times 10^{17}} = 8 \times 10^{-6} \ cm^{-3}$$

The diffusion lengths are

$$L_n = \sqrt{D_n \tau_n} = \left[(30)(10^{-8})\right]^{1/2} = 5.47 \ \mu m$$

$$L_p = \sqrt{D_p \tau_p} = \left[(15)(10^{-7})\right]^{1/2} = 12.25 \ \mu m$$

The injection efficiency is now (assuming no recombination via traps)

$$\gamma_{inj} = \frac{\frac{eD_n n_{po}}{L_n}}{\frac{eD_n n_{po}}{L_n} + \frac{eD_p p_{no}}{L_p}} = 0.98 \tag{9.23}$$

EXAMPLE 9.5 Consider the p-n^+ diode of the previous example. The diode is forward biased with a forward bias potential of 1 V. If the radiative recombination efficiency $\eta_{Qr} = 0.5$, calculate the photon flux and optical power generated by the LED. The diode area is 1mm^2.

The electron current injected into the p-region will be responsible for the photon generation. This current is

$$I_n = \frac{AeD_n n_{po}}{L_n} \left[exp\left(\frac{eV}{k_B T}\right) - 1\right]$$

$$= \frac{(10^{-2} \ cm^2)(1.6 \times 10^{-19} \ C)(30 \ cm^2/s)(8 \times 10^{-5} \ cm^{-3})}{5.47 \times 10^{-4} \ cm} \left[exp\left(\frac{1}{0.026}\right) - 1\right]$$

$$= 0.35 \ mA$$

The photons generated per second are

$$I_{ph} = \frac{I_n}{e} \cdot \eta_{Qr} = \frac{(0.35 \times 10^{-3} \ A)(0.5)}{1.6 \times 10^{-19} \ C}$$

$$= 1.09 \times 10^{15} \ s^{-1}$$

Each photon has an energy of 1.41 eV (= bandgap of GaAs). The optical power is thus

$$Power = (1.09 \times 10^{15} \ s^{-1})(1.41)(1.6 \times 10^{-19} \ J)$$

$$= 0.25 \ mW$$

It must be kept in mind that not all of this light will be useful due to reabsorption or reflection from the GaAs-air interface. The next subsection discusses the overall efficiency of the diode.

9.4   EXTERNAL QUANTUM EFFICIENCY

$\longrightarrow$ In the previous section we have seen how photons are created in an LED structure. For these photons to emerge from the device, great care must be taken in the design of the LED. There are three main loss mechanisms for the emitted photons: i) the emitted photons can be reabsorbed in the semiconductor by creating an electron-hole pair; ii) a certain fraction of photons will be reflected back at the semiconductor-air interface; and iii) some photons impinge upon the surface with angles greater than the critical angle thus suffering total internal reflection.

To minimize the absorption of the photons, it is essential that the photons be emitted near the surface so that a good fraction of the photons do not have to travel long distances to the surface. This criterion was considered in our discussion of the injection efficiency γ_{inj} in the previous subsection. Note that for direct gap materials, a photon can only travel a micron or so before getting absorbed. *It must be noted, however, that the active emission volume cannot be placed too close to the surface, otherwise non-radiative recombination processes mediated by surface defects will reduce the device efficiency.*

Photons that are able to make it to the semiconductor-air surface have to suffer reflection from the surface, as shown in Fig. 9.7a. Those that are reflected are lost. If n_{r2} is the refractive index of the semiconductor and n_{r1} the index of air, the reflection coefficient is (for vertical incident light),

$$R = \left(\frac{n_{r2} - n_{r1}}{n_{r2} + n_{r1}} \right)^2 \tag{9.24}$$

This loss is called the Fresnel loss. For a GaAs LED, if we choose $n_{r2} = 3.66$, $n_{r1} = 1.0$, we get a loss of 0.33, i.e., 33% of the photons cannot get through. To avoid this excessive loss, usually the device is encapsulated in a dielectric dome, as shown in Fig. 9.7b. The dielectric has a refractive index of ~ 1.6 and this allows a greater fraction of photons to emerge.

Finally, one has the loss of photons due to total internal reflection. If light impinges at a surface from a region of high refractive index ($n_{r2} > n_{r1}$), it is totally reflected back if the angle of incidence is greater than a critical angle θ_C where

$$\theta_C = sin^{-1} \left(\frac{n_{r1}}{n_{r2}} \right) \tag{9.25}$$

For the GaAs-air surface, the critical angle is 15.9°. Once again, use of the dome encapsulation suppresses this loss.

In addition to the three loss mechanisms discussed above, for many applica-

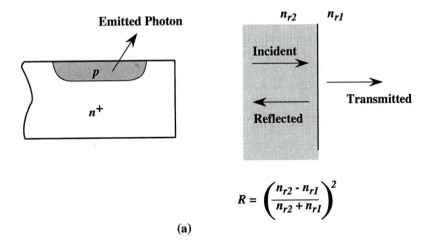

(a)

$$R = \left(\frac{n_{r2} - n_{r1}}{n_{r2} + n_{r1}}\right)^2$$

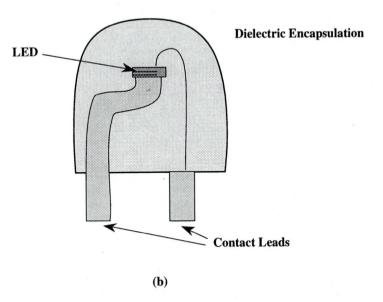

(b)

Figure 9.7: (a) The LED structure along with a schematic of the reflection and transmission of light at the semiconductor surface. (b) The dielectric encapsulation used to improve the transmission of photons generated. The presence of the dielectric reduces the reflection losses from the GaAs-air surface.

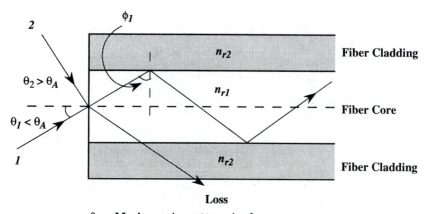

Loss

θ_A = **Maximum Acceptance Angle**

$\phi = \phi_C$ = **Critical Angle for Total Internal Reflection**

Figure 9.8: An optical fiber is made up of a core region of index n_{r1} and an outer cladding region of index n_{r2} $(n_{r1} > n_{r2})$. If the light is incident at an angle $\theta_1 < \theta_A$ so that the light suffers total internal reflection in the fiber, the light is coupled successfully into the fiber. Optical waves coming at an angle greater than θ_A are not able to propagate in the fiber.

tions, the photons have to be coupled into special devices. For example, for optical communications, the light must be coupled into an optical fiber. If light is to be coupled into an optical fiber it must enter it on an angle so that it suffers total internal reflection, as shown in Fig. 9.8. If n_{r1} and n_{r2} are the refractive indices of the core and the cladding region, the maximum angle of acceptance, θ_A, is given by (the result is derived in Chapter 13)

$$\theta_A = sin^{-1} \left(n_{r1}^2 - n_{r2}^2 \right)^{1/2} = sin^{-1}(A_n) \qquad (9.26)$$

where A_n is called the numerical aperture of the fiber.

The photons emerging from the LED have an angular distribution between $\theta = 0$ and $\theta = \pi/2$. Let us assume that the distribution has a form

$$I_{ph}(\theta) = I_0 cos\theta \qquad (9.27)$$

so that there is higher probability of photons emerging normally where the photons travel the least distance through the semiconductor and thus suffer the least losses. A source which has the cosine distribution given by Eqn. 9.27 is called a Lambertian source. The fraction of the light coupled into the fiber is then

$$\boxed{\eta_{fiber} = \frac{\int_0^{\theta_A} I_{ph}(\theta)sin\theta d\theta}{\int_0^{\pi/2} I_{ph}(\theta)sin\theta d\theta} = sin^2\theta_A} \qquad (9.28)$$

This value for η_{fiber} is quite small ($\sim 10\%$) so that a great deal of design has to focus on achieving better coupling efficiency, especially if the LEDs are to be used in fiber optical communication systems.

EXAMPLE 9.6 Calculate the fraction of electrons that are reflected when light is incident normally from GaAs at an interface with air. Also calculate the incident angle for total internal reflection. How would the results change if light was emitted into glass with an index of 1.5?

The refractive index of GaAs is 3.66 while that of air is 1.0. The reflection coefficient for normal incidence is

$$R = \frac{(n_{r2} - n_{r1})^2}{(n_{r2} + n_{r1})^2} = \left(\frac{2.66}{4.66}\right)^2 = 0.33$$

Thus 33% of the photons are reflected back and do not escape to the outside of the device. The angle of total internal reflection is

$$\theta_c = sin^{-1}\left(\frac{n_{r1}}{n_{r2}}\right) = sin^{-1}\left(\frac{1.0}{3.66}\right) = 15.9°$$

If n_{r1} was to change to 1.5 we get

$$R = 0.18; \theta_c = 24.2°$$

Thus a much higher fraction of photons would escape to the outside.

EXAMPLE 9.7 Consider an LED in which the photon output obeys a cosine law for its intensity. The light is to be coupled to an optical fiber which has a refractive index of 1.5 for the core and 1.40 for the outer cladding layer. Calculate the maximum angle of acceptance and the coupling efficiency for the fiber.

The maximum angle of acceptance is

$$\theta_A = sin^{-1}\left(n_{r1}^2 - n_{r2}^2\right)^{1/2} = 14.1°$$

The coupling efficiency is

$$\eta_{fiber} = sin^2\theta_A = 0.06$$

Thus only 6% of the optical output is able to couple into the optical fiber.

9.5 ADVANCED LED STRUCTURES

$\longrightarrow$ As we have discussed in the previous sections, important issues for LED technology are internal and external quantum efficiency, spectral purity of the light output, and temporal response. Although it is difficult to improve spectral purity and temporal response without using laser diodes, a number of advances have been made in the LED technology.

9.5.1   Heterojunction LED

If the LED is made from a single semiconductor, there are a number of problems that reduce the device efficiency. An important problem is that in a homojunction LED (i.e., a device based on a single semiconductor), the photon emission volume must be close to the surface so that the emitted photons are not reabsorbed. Since near the surface the semiconductor quality is usually not very good due to the presence of defect states, this causes a great deal of surface state mediated non-radiative recombination. In addition to this problem, the electrons injected from the n^+ side into the p-region can diffuse over long distances before recombining with holes. Thus the effective volume from which photons are emerging is quite large.

The heterojunction LED resolves these problems by injecting charge from a larger bandgap material in a narrow gap active region. Fig. 9.9 gives a schematic of the LED. Electrons and holes are injected from the wide gap n- and p-regions into the narrow gap active region. The electrons cannot enter the wide gap p-region below the active region and thus do not suffer from poor surface conditions. *The photons emitted are also not absorbed in the top or bottom region since the photon energy is smaller than the bandgap of the n- or p-region.*

The heterojunction LEDs are made by epitaxial processes, and the active region is kept to $\sim$0.1-0.2 μm. The materials commonly used are GaAs/AlGaAs grown on GaAs and InGaAsP/InP and InGaAs/InGaAsP grown on InP substrates.

9.5.2   Edge Emitting LED

An important issue in optical communication is the efficiency with which the light emitted by an LED couples into an optical fiber. We had discussed this coupling efficiency and seen that one needs a highly collimated beam for efficient coupling. The heterostructure technology is exploited to fabricate the edge emitting LED shown schematically in Fig. 9.10. As we shall see later, the device looks almost like a laser diode. The difference is that in the laser diode exceptional care is taken to produce a high quality optical cavity to ensure optical feedback.

An important ingredient of the edge emitting LED is the wide gap cladding layers which confine not only the electrons and holes to the active layer, but also cause

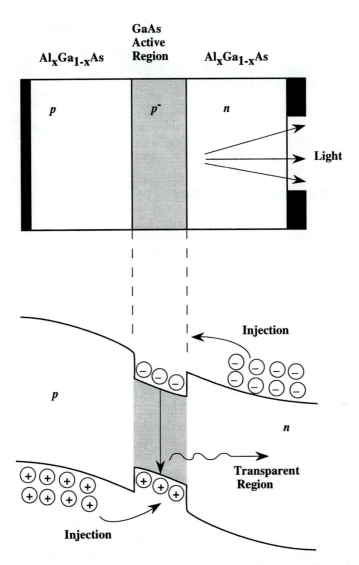

Figure 9.9: A heterojunction LED uses a narrow gap semiconductor for the active region. The photons that are emitted are not reabsorbed in the top or bottom layers, which are transparent to the emitted radiation.

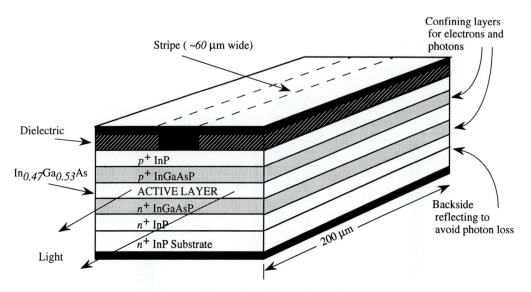

Figure 9.10: A schematic of an edge emitting LED. The active region is $In_{0.47}Ga_{0.53}As$ ($E_g = 0.8$ eV) surrounded by confining layers fabricated from InGaAsP ($E_g \sim 1.0$ eV). The confining layers cause the light to be coupled out through the edge of the device.

the emitted photons to travel along the LED axis and emerge from the edge of the device. This optical cavity will be discussed in more detail when we discuss the laser diode.

Due to the superior collimation of the edge emitting LED ($\sim 30°$ width perpendicular to the layer and $\sim 120°$ parallel to the layer) the coupling efficiency to a fiber is greatly improved.

9.5.3 Surface Emitting LED

An important class of LEDs is the surface emitting LED first realized by Burrus and Dawson in 1970. A schematic of this LED is shown in Fig. 9.11. An optical fiber is butt coupled to the LED by etching a hole in the LED and attaching the fiber by epoxy resin. The LED itself is a heterostructure LED with a thin active region of low bandgap surrounded by wide gap regions.

The photons emitted are directly coupled to the optical fiber. In various advanced structures a microlens is placed on the LED to improve the coupling efficiency.

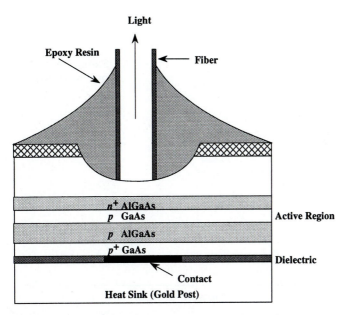

Figure 9.11: A schematic of a surface emitting LED. A heterostructive LED has an optical fiber butt coupled to it with an epoxy resin. The light generated in the active p-GaAs is fed into the fiber directly. In advanced designs, a microlens is used to improve the coupling.

9.6 LED PERFORMANCE ISSUES

$\longrightarrow$ The LED depends upon the spontaneous emission process to provide light from the injected electrons and holes. As a result there are simplifications in the fabrication and design of the LED, but one has to pay a price in performance when an LED is compared to a laser diode. This comparison will be clear after our discussions on the laser diode. For the LED the important performance issues are represented by the light-current characteristics, the spectral purity of the light output, the time response of the LED to external electrical signals and the temperature dependence of the output. The importance of the spectral purity (i.e., spread in wavelengths of the output light) is quite critical for high performance optical communication systems, as will be examined in Chapter 13.

9.6.1 Light-Current Characteristics

When a current I is passing through the forward bias diode, a certain fraction of the current is converted to light. If η_{Tot} represents the total efficiency of this conversion,

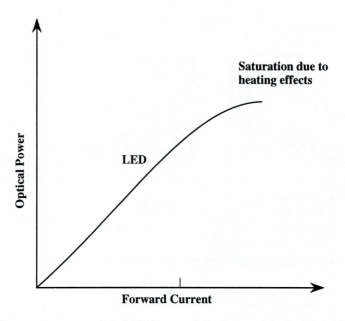

Figure 9.12: The output power of an LED is essentially linear with the injected currents.

the photon current that emerges from the diode is

$$
\begin{aligned}
I_{ph} &= \text{Number of photons per second} \\
&= \eta_{Tot} \cdot \frac{I}{e}
\end{aligned}
\tag{9.29}
$$

In general, η_{Tot} depends upon the injected current since the carrier radiative lifetime τ_r depends upon carrier injection level. However, in an LED this dependence is quite weak so that the $I_{ph} - I$ characteristics are essentially linear as shown in Fig. 9.12. At very high injection the light output starts to saturate as the device heats and the radiative recombination efficiency decreases.

In surface emitting LEDs, it is often found that the light output decreases sublinearly with current, an effect that cannot be simply explained by heating effects. It appears that at high drive current the photon density in the LED becomes large enough that stimulated emissions start to occur. This emission is in the plane of the LED so that the photons, emitted perpendicular to the surface, decrease. Such LEDs are called superluminescent LEDs and behave similar to the laser diode discussed in the next chapter.

9.6.2 Spectral Purity of LEDs

The spectral purity or the linewidth of the emitted radiation is an important characteristic of optical devices. The importance of the spectral purity of the emitted light depends upon applications. If the LED is to be used in a display device, the spectral purity is not an issue. However, in optical communication applications the spectral purity is a critical issue. Light pulses of different wavelengths travels through an optical fiber at different speeds. Thus a signal gets distorted if the optical beam has a large wavelength spread. This will be examined in detail in Chapter 13. As can be seen from the discussion in Chapter 4, Section 4.7, the emission spectrum is essentially determined by the product $(\hbar\omega - E_g)^{1/2} f^e(E^e)f^h(E^h)$. This is the convoluted product of the electron and hole occupation probabilities. At low injection, this width is of the order of k_BT. At high injection the width is (n is the total charge density)

$$\Delta E \sim \frac{n}{N_c}\, k_BT \qquad (9.30)$$

where N_c is the effective bandedge density of states. A typical emission spectrum for an LED is shown in Fig. 9.13. The linewidth (full width at half maxima) is seen to be of the order of 20 nm (200 Å) at room temperature. This is obviously a broad spectrum. However, for many applications this width is adequate. *In fact, LEDs are widely used in optical communication as long as the signal does not have to be sent over long distances.* However, for long distance communication, the LED output is not adequate and a laser diode must be used.

9.6.3 LED Temporal Response

An important use of light emission devices is the conversion of an electrical signal to an optical signal. A typical circuit utilizing the LED for generating optical signals is shown in Fig. 9.14a. The information could be transmitted as an analog or digital signal.

The LED is basically a forward biased *p-n* diode in which minority charge is injected into an active recombination region. To be able to modulate the output of the device, it must be possible to modulate the injected carriers. Thus apart from external parasitics, a key issue in the device speed is the time taken to extract the charge. This time is controlled by the carrier recombination time. Let us consider a simple model of the response function of the LED.

In Fig. 9.14b we show a typical LED structure considered as a 1-dimensional system. Carriers are injected from the *n*- to the *p*-side. The continuity equation for the carriers is

$$\frac{\partial n(x)}{\partial t} = -\frac{n(x)}{\tau} + D_n \frac{\partial^2}{\partial x^2} \qquad (9.31)$$

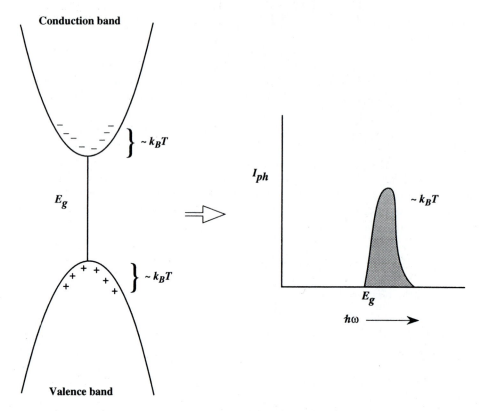

Figure 9.13: In an LED, the electrons and holes are distributed over an energy width of $\sim k_B T$. Since all the e-h pairs contribute to the optical output, the LED output is quite broad with a width roughly equal to $k_B T$. The shape of the output depends upon the carrier occupation function and the density of states function.

where the first term is due to carrier recombination (including non-radiative processes) and the second term is the diffusive component of the particle current flow. The electric field in a forward biased junction are small and so we ignore the drift current. Consider a small signal bias applied to the diode resulting in a carrier density

$$n(x,t) = n_o(x) + n_1(x)exp(i\omega t) \tag{9.32}$$

Substituting this in Eqn. 9.31, we get, comparing the dc and ac components

$$D_n \frac{\partial^2 n_o}{\partial x^2} \quad - \quad \frac{n_o}{\tau} = 0 \tag{9.33}$$

$$D_n \frac{\partial^2 n_1}{\partial x^2} \quad - \quad \frac{n_1(1+i\omega\tau)}{\tau} = 0 \tag{9.34}$$

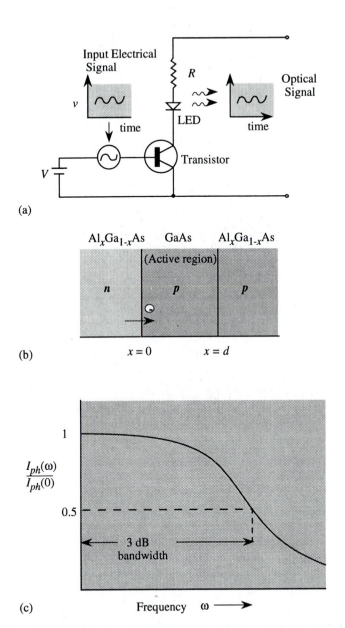

Figure 9.14: (a) A schematic of a circuit for modulation of the output of an LED. (b) Geometry of the LED used to study the intrinsic temporal response to an ac signal. (c) The response of the LED falls with frequency as shown. The 3dB bandwidth is indicated.

Defining the lengths

$$L_n = (D_n \tau)^{1/2} \tag{9.35}$$

$$L_n(\omega) = \left[\frac{D_n \tau}{1 + i\omega\tau} \right]^{1/2} \tag{9.36}$$

we get

$$\frac{\partial^2 n_1(x)}{\partial x^2} = \frac{n_1(x)}{L_n^2(\omega)} \tag{9.37}$$

We can define the temporal response of the LED by the response

$$r(\omega) = \frac{e J_{ph1}(\omega)}{J_1(\omega)} \tag{9.38}$$

i.e., the ratio of the ac part of the photon current density to the ac part of the electron current density. Let us assume that only electrons are responsible for the current flow. Then we have

$$n_1(x = 0) = n_1(0) \tag{9.39}$$

We also assume that all the carriers recombine by the time they reach the edge of the active region d; i.e., $L_n \ll d$

$$n_1(d) = 0 \tag{9.40}$$

This gives the solution of the Eqn. 9.37

$$n_1(x) = n_1(0) exp\left(\frac{-x}{L_n(\omega)} \right) \tag{9.41}$$

and the photon current density becomes

$$J_{ph}(\omega) = \frac{1}{\tau} \int_o^d n_1(x) dx \tag{9.42}$$

$$= \frac{n_1(0) L_n(\omega)}{\tau} \tag{9.43}$$

Also,

$$J_1(\omega) = e D_n \frac{\partial n_1(x)}{\partial x} \tag{9.44}$$

$$= -e \frac{D_n n_1(0)}{L_n(\omega)} \tag{9.45}$$

This gives for the temporal response function

$$r(\omega) = \frac{|L_n(\omega)|^2}{\tau D_n} = \frac{1}{(1 + \omega^2 \tau^2)^{1/2}} \tag{9.46}$$

This expression shows the importance of the recombination time τ in the bandwidth limits that can be reached in LEDs. The modulated bandwidth f_c is defined at the frequency where the power is one-half that of the zero frequency value

$$f_c = \frac{\omega_c}{2\pi} = \frac{1}{2\pi\tau} \tag{9.47}$$

and

$$\frac{1}{\tau} = \frac{1}{\tau_r} + \frac{1}{\tau_{nr}} \tag{9.48}$$

For high quality devices $\tau \sim \tau_r$. The LED response as a function of frequency is shown in Fig. 9.14c.

In Fig. 9.6 we had shown the dependence of the radiative lifetime on the carrier density or the doping of the active region. As the LED drive current increases, the recombination time decreases and the modulation bandwidth thus decreases. This dependence is, indeed, seen experimentally as shown in Fig. 9.15. Note that the carrier density in the active region is proportional to J/d, where J is the current density and d is the active region thickness.

The modulation bandwidth can also be increased by increasing the doping of the active region. The higher p-doping allows the injected electrons to recombine with holes in a shorter time.

The ultimate bandwidth is controlled by the time τ_o for an e-h recombination (with $f_e = f_h = 1$) which, as discussed earlier, has a value of ~ 0.5 ns. Thus, the LED cutoff frequency can approach a gigahertz.

The limits arising from the radiative recombination is a key difference between LEDs and laser diodes. The laser diodes operate not under conditions of spontaneous emission, but stimulated emission. As we shall see later, the stimulated emission depends upon the photon density present and can result in recombination times approaching 10 ps.

EXAMPLE 9.8 An AlGaAs(n) - GaAs(i) - AlGaAs(p) LED is operated at 300 K at current density levels of 100 A/cm^2, 500 A/cm^2 and 1000 A/cm^2. The width of the active region is 1.0 μm. Calculate the 3 dB cutoff frequency for the two operating points. Assume that the temporal response is limited by the e-h recombination time and the efficiency is unity. Use the information given in Fig. 9.6.

The current density in the LED is (for unity efficiency)

$$J = \frac{end}{\tau_r}$$

or

$$\frac{n}{\tau_r} = \frac{J}{ed} = \frac{p}{\tau_r}$$

For a current density of 100 A/cm^2 we have

$$\frac{n}{\tau_r} = \frac{(100 \ A/cm^2)}{(1.6 \times 10^{-19} \ C)(1.0 \times 10^{-4} cm)} = 6.25 \times 10^{24} \ cm^{-3} \ s^{-1}$$

From Fig. 9.6 we see that this value occurs for the combination

$$n \cong 6 \times 10^{16} \ cm^{-3}; \tau_r \sim 10^{-8} \ s$$

Thus the cutoff frequency is

$$f_c = \frac{1}{2\pi\tau_r} = 15.9 \ MHz$$

When the drive current is 500 A/cm^{-2}, we get

$$\frac{n}{\tau_r} = 3.38 \times 10^{25} \ cm^{-3} \ s^{-1}$$

Once again from Fig. 9.6, we see that this occurs when

$$n \cong 3 \times 10^{17} \ cm^{-3}; \tau_r \cong 8 \times 10^{-9} \ s$$

The cutoff frequency is now

$$f_c = 20 \ MHz$$

When the current density is 1000 A/cm^2, we get

$$f_c \cong 35 \ MHz$$

EXAMPLE 9.9 Consider the diode of the previous example. Calculate the cutoff frequencies at the same injection levels if the diode active region width is 0.1 μm.

For the 0.1 μm case we have

$$J = 100 \ A/cm^2$$
$$\frac{n}{\tau_r} = 6.25 \times 10^{25} \ cm^{-3} \ s^{-1}$$

Using Fig. 9.6 we now get

$$f_c = 35 \ MHz$$

$$J = 500 \ A/cm^2$$
$$\frac{n}{\tau_r} = 3.125 \times 10^{26} \ cm^{-3} \ s^{-1}$$

so that, from Fig. 9.6,

$$\tau_r = 3 \times 10^{-9} \ s$$

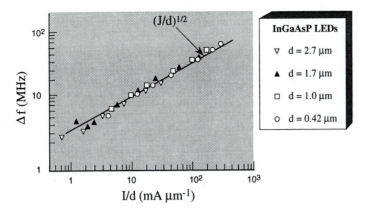

Figure 9.15: Modulation bandwidth as a function of normalized drive current density for In-GaAsP LEDs with different active region thicknesses. (After O. Wada, Y. Yamakoshi, M. Abe, Y. Yishitoni, and T. Sakwai, *IEEE J. Quantum Electronics*, QE-17, 174 (1981).)

and

$$f_c = 53 MHz$$

$$J = 1000 A/cm^2$$
$$\frac{n}{\tau_r} = 6.25 \times 10^{27} \ cm^{-3} \ s^{-1}$$
$$\tau_r \cong 2 \times 10^{-9} \ s$$

and

$$f_c = 80 \ MHz$$

We see that the diode with the narrower width has a better time response for the same current density.

9.6.4   Temperature Dependence of LED Emission

In the previous subsection on light output as a function of current, we had noted that the output power saturates at high current due to heating effects. The temperature of the diode influences the *e-h* recombination efficiency through two phenomenon: i) Leakage of injected carriers into the contact regions at high temperatures; ii) Auger processes that contribute to non-radiative recombination.

In the LED which is a forward biased *p-n* diode, carriers are injected from the doped sides into an active region where they recombine. Ideally, the carriers should thermalize in the active region as shown schematically in Fig. 9.16a and recombine to emit photons. However, as temperature increases, the distribution of the injected charge is wider as shown in Fig. 9.16b, so that an increasing fraction of the charge can leak across the active region. The leakage current will not contribute to the photon emission process and thus reduce the optical power.

As an LED is pumped harder by increasing the drive current, the device heats, and as a result, the leakage current increases. The leakage current depends upon the device design. For example, if the active region is wide, the leakage may be small. The leakage current can be quite serious in most devices and contribute as much as 20-30% of the total current. One can suppress the LED heating at high drive current by using a pulsed source of current.

The Auger process is another important mechanism for non-radiative recombination. We have discussed the physics behind the Auger process in Chapter 4. The Auger process involves three carriers in the initial state which could be 2 electrons and a hole or one electron and two holes. The end product after the Auger recombination is one hot electron or hole with no emission of a photon. The recombination rate is strongly dependent on carrier density ($\propto n^3$), the bandgap of the material operating temperature and details of the bandstructure. The net effect is that for narrow gap based LEDs ($E_g < 1.0\ eV$), the Auger process can be an important source of non-radiative recombination. The process also has a strong temperature dependence.

The net effect of the carrier leakage and Auger processes produces an optical output from a LED which has a form

$$I_{ph} = I_{ph}(0)exp\ \left(-\frac{T}{T_1}\right) \tag{9.49}$$

where T_1 is a temperature that depends upon the material bandgap and the LED design parameters. The temperature T_1 should be as large as possible to ensure temperature independence of LEDs. For 1.3 μm LEDs based on InGaAsP, the value of T_1 is in the range of 180-200 K while for GaAs based LEDs it is 300-350 K.

In addition to the optical power dependence upon temperature, it is important to note that the bandgap of all semiconductors decreases with temperature. As a result, the peak of the emission spectra of the LED will shift to higher wavelengths as temperature increases. This shift is an approximately 3.5 ÅK^{-1} for GaAs LEDs and 6 ÅK^{-1} for InGaAsP LEDs.

9.7   OUTPUT POWER AND MODULATION BANDWIDTH

$\longrightarrow$ In addition to high modulation speed, for many applications one needs high output power. These demands include applications in communications and display systems. The

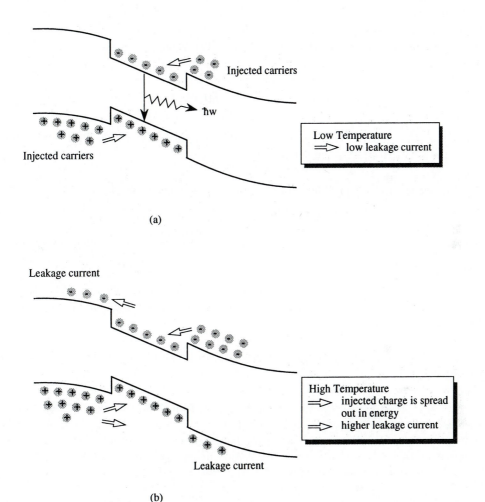

Figure 9.16: (a) A schematic of a charge injection in a LED at low temperature. All the injected carriers recombine in the active region. (b) At higher temperatures, due to the energy spread ($\sim k_B T$) of the injected carriers, a greater fraction of charge can leak, reducing the radiative efficiency.

use of LEDs in automobile brake lights are an example where brightness is of importance. As high power LEDs become available, other applications will emerge.

The two demands of power and bandwidth usually conflict with each other. We have discussed the dependence of the LED bandwidth on the radiative recombination time. For high optical power one needs a device with a long active region as can be seen from the following considerations. The photon current is

$$
\begin{aligned}
I_{ph} &= \eta \frac{I}{e} \\
&= \eta \frac{ndA}{\tau_r}
\end{aligned}
\tag{9.50}
$$

where d is the width of the active region, η is the efficiency of the LED, A the diode area, and n is the injected carrier density.

We see from Eqn. 9.50 that the optical power generated by the LED is

$$
\begin{aligned}
\text{Optical Power} &= I_{ph}\hbar\omega \\
&= \eta \frac{\hbar\omega\ n\ d\ A}{\tau_r}
\end{aligned}
\tag{9.51}
$$

Note that as discussed in Section 9.4, care has to be taken in LED design to ensure a high external quantum efficiency so that the generated optical power is not lost.

To increase the optical power of the device, one has the following design considerations:

i) Injected carrier density: It would appear from Eqn. 9.51 that by increasing n, the output power could increase indefinitely. However, this is not possible because of the problems of heating, Auger recombination and high leakage current as discussed in the previous section. Thus the maximum injection density is limited to about $10^{18}\ cm^{-3}$. At such injection densities, τ_r is about 1 to 2 nanoseconds.

ii) Device area: Increasing the device area can obviously increase the output power. Thus, for applications where output power is the only consideration, the device area can be very large. However, one cannot increase the device area indefinitely due to constraints placed by device fabrication technology. As the device area increases, the probability that the device will have defects also increases. This can cause catastrophic failure of the device. Also, as the device area is increased, the capacitance of the device increases and thus the device modulation speed suffers.

iii) Active gain thickness: The output power can be increased if the active region thickness d is increased. This causes a limitation on the bandwidth of the LED since

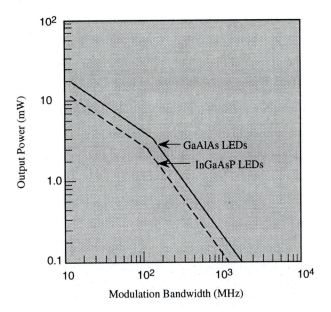

Figure 9.17: Output power versus 3-dB bandwidth in LEDs (After R. H. Saul, T. P. Lee, and C. A. Burrus, *Semiconductors and Semimetals*, Volume 22, part C, Academic Press, New York (1985).)

transit time effects dominate the device speed. The transit time is controlled by the diffusion process and is given by

$$t_{tr} \cong \frac{d^2}{2D}$$

where D is the diffusion coefficient of the slower carrier (usually the holes).

From the above discussion, it is clear that there is a tradeoff between high output power and modulation bandwidth of LEDs. LEDs have been shown to achieve an output power of 15 mW with a bandwidth of 17 MHz. On the other hand, bandwidths of about 2.0 GHz have been achieved at a power of 0.2 mW. In Fig. 9.17, we show some results on the output power and available bandwidth.

9.8 LED RELIABILITY ISSUES

$\mathcal{R}$ LEDs are important devices for display and communication technologies. It is important that the light output of the device remain constant in time if the system is to operate reliably. LEDs failures can be classified into three key categories, as shown in Fig. 9.18. The first category is called in "infant failure," where devices that have suffered serious processing related damage fail during the initial "burn in" period. The "burn in" involves operating the device at a high power level for up to 100 hours. Devices that survive this burn in usually have a large mean time to failure (MTTF). Advanced fabrication and processing techniques including starting from a high quality substrate with low defect density can greatly reduce the infant failures.

A small fraction of devices can fail shortly after the burn in due to extreme combinations of the random defects that are present in every device. This combination of "bad luck" results in what is called freak failure.

Most of the devices survive the first few hundred hours of operation and then go on to have a MTTF of up to 10^6 hours for GaAs based devices and up to 10^9 hours of InP based 1.3 μm devices. These are adequate for system applications. The MTTF is defined differently depending upon the system demands. In communication systems the optical power coupled into the fiber determines to a large extent (along with the optical wavelength) the repeater spacing (see Chapter 13). As a result MTTF is defined in terms of the time taken for the power output to decrease to a certain fraction of the original value. In a system with enough tolerance MTTF may be defined as the time taken for the power to drop by 50% (-3 dB). However, in systems designed with high repeater spacings the MTTF may be defined as the time taken for a 20% (-1 dB) power drop.

Failure in LEDs is gradual in contrast to laser diodes where the failure is catastrophic. Since the light output of the LED is linear with current, the failure mechanism involves an increase in the non-radiative recombination due to various kinds of defects. In the early stages of LED and LD development, an important defect mechanism involved migration of dislocations into the active device region. These defects which contribute strong non-radiative centers are called dark line defects (they appear as dark lines in electroluminescence). These defects create a catastrophic failure in LDs, but in LEDs the effect is a gradual decrease in light output. The dark line defect formation process involves a climb of substrate defects into the active region. This is found to occur in GaAs based devices but not in the InP based 1.3 μm devices. As a result, InP based devices have a longer MTTF.

Testing and controlling reliability of LEDs (and lasers) remains an important challenge. Researchers are constantly looking for testing techniques that allow elimination of devices that have a poor MTTF without long time testing. Also, the influence of substrate and processing (including metals used in contact formation) is still being actively studied.

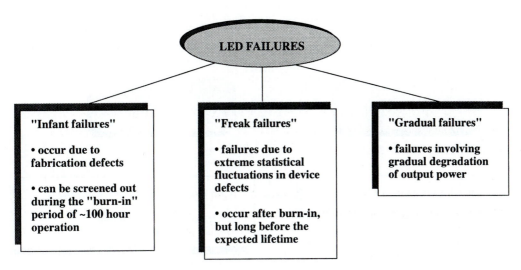

Figure 9.18: LED failures can be classified into three categories as shown in this chart.

9.9   SUMMARY

The LED is an important optical source which finds uses in many applications, including those in display systems and optical communication. The main advantages of the LED are the simplicity of the fabrication process and the easy incorporation of the device (which is a simple *p-n* diode) in most circuitry. The key drawbacks of the LED are the broad spectrum of the emitted light and the difficulty in pushing the modulation bandwidths above a few gigahertz.

The summary tables (Tables 9.1-9.3) highlight the important areas discussed in this chapter.

9.10   PROBLEMS

Section 9.2

9.1 Identify a composition of $In_xGa_{1-x}As_yP_{1-y}$ which has a bandgap corresponding to the photon wavelength of 1.55 μm and which has a lattice constant equal to that of InP. Assume that the lattice constants and bandgaps scale linearly with the composition of the alloy. Use the bandgap and lattice constant values given in the text.

9.2 Identify the semiconductors (direct and indirect) that can be used for blue light emission.

9.3 Consider the alloys $Hg_xCd_{1-x}Te$, $In_xGa_{1-x}As$ and $GaAs_xSb_{1-x}$. All of these alloys can be exploited for LEDs emitting near the minima of the optical fiber attenuation.

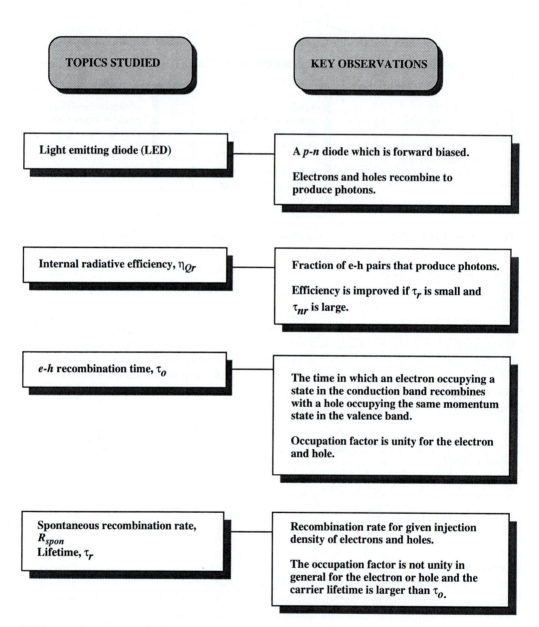

Table 9.1: Summary table.

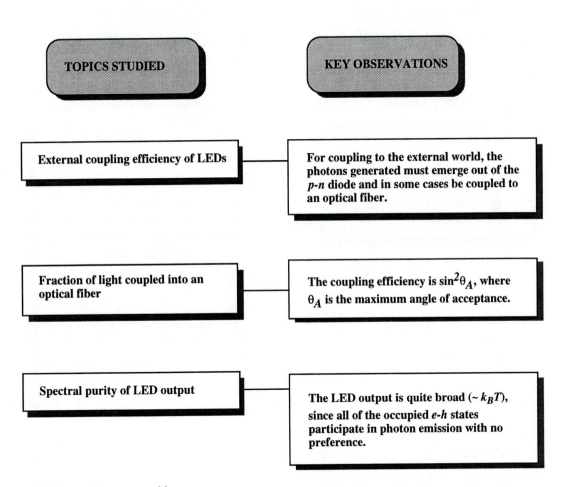

Table 9.2: Summary table.

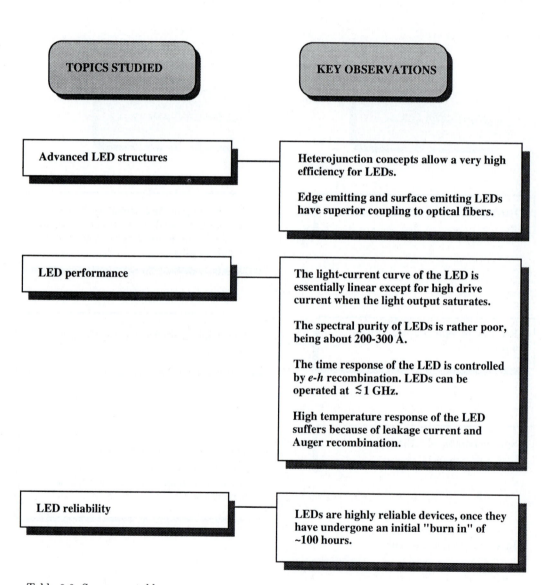

Table 9.3: Summary table.

Calculate the change in the emission energy for a 1% change in the alloy composition for each of these systems.

9.4 An important emerging semiconductor for LEDs is the material SiC. Calculate the emission wavelength for LEDs made from SiC. Discuss the improvement possible in the memory density if SiC were to be used, instead of GaAs, in reading optical discs.

Section 9.3

9.5 Calculate the e-h radiative recombination time τ_o (i.e., for $f^e = 1 = f^h$) for carriers in $Hg_{1-x}Cd_xTe$ $(E_g(eV) = -0.3 + 1.9x)$ for x between 0.5 and 1.0. The momentum matrix element is given by

$$\frac{2p_{cv}^2}{m_0} = 22 \ eV$$

Assume that the refractive index is 3.7 and is independent of composition.

9.6 Consider an $In_{0.53}Ga_{0.47}As$ sample at 300 K in which excess electrons and hole are injected. The excess density is 10^{15} cm^{-3}. Calculate the rate at which photons are generated in the system. The bandgap is 0.8 eV and carrier masses are $m_e^* = 0.042m_0$; $m_h^* = 0.4m_0$. Also calculate the photon emission rate if the same density is injected at 77 K. (Use the low injection approximation.) Assume that the refractive index and the momentum matrix element is the same as in GaAs (given in the text).

9.7 Determine the error in the position of the Fermi levels E_{Fn} and E_{Fp} calculated from the Boltzmann and Joyce-Dixon approximations for GaAs when the free carrier density $n = p$ ranges from 10^{15} cm^{-3} to 10^{18} cm^{-3}. Calculate the results at 77 K and 300 K.

9.8 Consider a GaAs p-n^+ junction LED with the following parameters at 300 K:

Electron diffusion coefficient,	D_n	=	$25 \ cm^2/s$
Hole diffusion coefficient,	D_p	=	$12 \ cm^2/s$
n-side doping,	N_d	=	$5 \times 10^{17} \ cm^{-3}$
p-side doping,	N_a	=	$10^{16} \ cm^{-3}$
Electron minority carrier lifetime,	τ_n	=	$10 \ ns$
Hole minority carrier lifetime,	τ_p	=	$10 \ ns$

Calculate the injection efficiency of the LED assuming no trap related recombination.

9.9 The diode in Problem 9.8 is to be used to generate an optical power of 1 mW. The diode area is 1 mm^2 and the external radiative efficiency is 20%. Calculate the forward bias voltage required.

9.10 Consider the GaAs LED of Problem 9.8. The LED has to be used in a communication system. The binary data bits 0 and 1 are to be coded so that the optical pulse output is 1 nW and 50 μW. If the external efficiency factor is 10%, calculate the forward bias voltages required to send the 0's and 1's. The LED area is 1 mm^2.

Section 9.4

9.11 Light from a GaAs LED is to be coupled to an external transmission system. A dome of a dielectric is placed over the LED. Calculate the refractive index of the dome if the reflection coefficient for the normally incident photons from the GaAs into the dome is to be 10%.

9.12 The light from a GaAs LED is coupled into an optical fiber which has refractive indices of 1.51 and 1.47 for the core and the cladding layers. Calculate the maximum angle of acceptance for the fiber. The LED has a Lambertian (cosine) output. Calculate the coupling efficiency for the diode.

Section 9.5

9.13 Discuss why in a homojunction LED it is important to ensure that the electron injection current is dominant, but in a heterojunction LED this is not important.

9.14 Discuss the reasons why the edge emitting LED has a better coupling efficiency to an optical fiber than a surface emitting LED.

Section 9.6

9.15 Consider an AlGaAs/GaAs heterojunction LED. The injection densities for electrons and holes are equal and are both 10^{17} cm^{-3} in the active GaAs region. Calculate the position of the emission peak energy if E_g(GaAs) is 1.43 eV. Calculate the shift in the peak position if the injection density is increased to 10^{18} cm^{-3}. The temperature is 300 K.

9.16 In the previous problem, calculate the spectral width at the two injection densities at 300 K and 77 K.

9.17 Discuss whether a GaAs LED or an Si LED will have a broader spectral width if the light output power is the same.

9.18 A heterojunction LED based on GaAs is biased at 100 A/cm^2 current density at 300 K. The active layer of the LED is 0.5 μm. Calculate the cutoff frequency of the diode. If the temperature of operation changes to 400 K, calculate the change in the cutoff frequency. Assume that the cutoff frequency is limited by the radiative lifetime. Use the information provided by Fig. 9.6.

9.19 Consider two heterojunction LEDs based on GaAs biased at 100 A/cm^2. In one case, the active region is undoped while, in the other, the region is p-type doped at 10^{18} cm^{-3}. The active layer thickness is 1.0 μm. Compare the radiative lifetime limited modulation bandwidth of the two LEDs at 300 K.

9.20 A GaAs LED is to be used in a local area network. The fiber system used puts a restriction on the light emitter that the peak wavelength should not shift by more than ± 50 Å. Calculate the level of temperature control needed for this device.

Section 9.7

9.21 A GaAs LED is to be designed with an output power of 5.0 mW. The maximum device area that can be allowed is 100 μm^2. Estimate the thickness of the active region needed. The efficiency of the device is 20%. The maximum injection density is 10^{18} cm^{-3}.

9.22 A 0.1 mm² SiC LED has a total overall optical efficiency of 2%. If the maximum current density is limited to 50 kA/cm², calculate the output power of this device. How does this value compare to the optical power of a typical flashlight you may have at home?

9.11   REFERENCES

- **General**

  - H. Kressel and J. K. Butler, *Semiconductor Lasers and Heterojunction LEDs*, Academic Press, New York (1977).

  - R. Baets, "Heterostructures in III-V Optoelectronic Devices," *Solid State Electronics*, 30, 1175 (1987).

  - Articles in *Semiconductors and Semimetals*, ed. W. T. Tsang, Volume 22, part C, Academic Press, New York (1985).

  - W. T.Tsang, "High Speed Photonic Devices," *High Speed Semiconductor Devices*, ed. S. M. Sze, Wiley-Interscience, New York (1990).

CHAPTER
10

LASER DIODE: STATIC PROPERTIES

10.1   INTRODUCTION

The LED discussed in the previous chapter is an important optical source which finds uses in many applications, including those in display systems and optical communication. The LED is not, however, a device of choice for many high performance applications. The main advantages of the LED are the simplicity of the fabrication process and the easy incorporation of the device (which is a simple *p-n* diode) in most circuitry. The key drawbacks of the LED are the broad spectrum of the emitted light and the difficulty in pushing the modulation bandwidths above a gigahertz. We have already discussed the sources of these limitations. In Fig. 10.1 we show the important limitations of the LED. The laser diode is able to overcome these limitations by exploiting special properties of optical cavities and stimulated emission. As a result the semiconductor laser diode provides an extremely sharp emission line with linewidth up to two orders of magnitude narrower than that of an LED. The modulation bandwidth of the laser diode approaches 50 GHz and can be even higher in principle. Also, because of its superior spatial coherence, the laser beam does not spread as much as beams from other sources and can thus be focused to give a very high intensity.

What are the basic differences that allow the laser diode to have such a performance edge over the LED? This is the subject of this chapter. In this chapter we will focus on the static properties of lasers while in the next chapter we will address the dynamic properties of the laser diode.

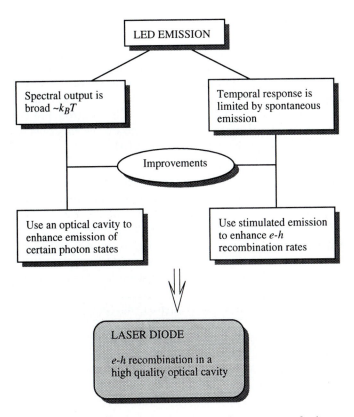

Figure 10.1: A schematic description of how the LED performance can be improved by exploiting an optical cavity.

10.2 SPONTANEOUS AND STIMULATED EMISSION

$\longrightarrow$ The semiconductor laser diode operates as a forward bias p-n junction just as the LED studied in the previous sections. *However, while the structure appears similar to the LED as far as the electron and holes are concerned, it is quite different from the point of view of the photons.*

As in the case of the LED, electrons and holes are injected into an active region by forward biasing the laser diode. At low injection, these electrons and holes recombine radiatively via the spontaneous emission process to emit photons. However, the laser structure is so designed that at higher injections the emission process occurs by stimulated emission. In chapter 4 we have discussed the difference between spontaneous and stimulated emission. The stimulated emission process provides spectral purity to the photon output, provides coherent photons, and offers high speed performance. *Thus the key difference between the LED and the laser diode arises from the difference between spontaneous and stimulated emission.*

Let us develop an understanding of this difference using Fig. 10.2. Consider an electron with wave vector k and a hole with a wave vector k in the conduction and valence bands, respectively, of a semiconductor. In the case shown in Fig. 10.2a, initially there are no photons in the semiconductor. The electron and hole recombine to emit a photon as shown, and this process is the spontaneous emission. The spontaneous emission rate was discussed in the context of the LED.

In the case of Fig. 10.2b, we show the electron-hole pair along with photons of energy $\hbar\omega$ equal to the electron-hole energy difference. In this case, in addition to the spontaneous emission rate, one has an additional emission rate called the stimulated emission process. The stimulated emission process is proportional to the photon density (of photons with the correct photon energy to cause the *e-h* transition). *The photons that are emitted are in phase (i.e., same energy and wave vector) with the incident photons.* As discussed in Chapter 4, the rate for stimulated emission is

$$\boxed{W_{em}^{st}(\hbar\omega) = W_{em}(\hbar\omega) \cdot n_{ph}(\hbar\omega)} \qquad (10.1)$$

where $n_{ph}(\hbar\omega)$ is the photon occupation number and W_{em} is the spontaneous emission rate discussed earlier. In the LED, when photons are emitted by spontaneous emission, they are lost either by reabsorption or simply leave the structure. Thus $n_{ph}(\hbar\omega)$ remains extremely small and stimulated emission cannot get started.

Consider now the possibility that when photons are emitted via spontaneous emission, an optical cavity is designed so that photons with a well defined energy are selectively confined in the semiconductor structure. If this is possible, two important effects occur: i) the photon emission for photons with the chosen energy becomes stronger due to stimulated emission; ii) the e-h recombination rate increases as can be seen by Eqn. 10.1. These two effects are highly desirable since they produce an optical spectrum with very narrow emission lines and the light output can be modulated at high speeds as outlined in Fig. 10.1.

The challenge for the design of the laser is, therefore, to incorporate an optical cavity that ensures that the photons that are emitted are allowed to build up in the semiconductor device so that stimulated emission can occur.

10.3   THE LASER STRUCTURE: THE OPTICAL CAVITY

$\longrightarrow$ While both the LED and the laser diode use a forward biased *p-n* junction to inject electrons and holes to generate light, the laser structure is designed to create an "optical cavity" which can "guide" the photons generated. The optical cavity is

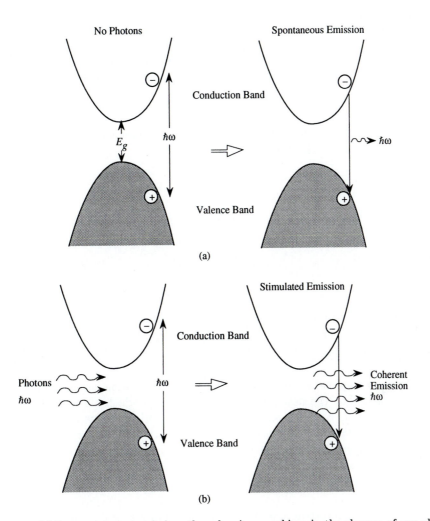

Figure 10.2: (a) In spontaneous emission, the *e-h* pair recombines in the absence of any photons present to emit a photon. (b) In simulated emission, an *e-h* pair recombines in the presence of photons of the correct energy $\hbar\omega$ to emit coherent photons. In coherent emission the phase of the photons emitted is the same as the phase of the photons causing the emission.

essentially a resonant cavity in which the photons have multiple reflections. Thus, when photons are emitted, only a small fraction is allowed to leave the cavity. As a result, the photon density starts to build up in the cavity. A number of important cavities are used for solid state lasers. These are the Fabry-Perot cavity, cavities for distributed feedback lasers containing periodic gratings, surface emitting laser cavities containing specially designed reflectors, etc.. For semiconductor lasers, the most widely used cavity is the Fabry-Perot cavity shown in Fig. 10.3a. The important ingredient of the cavity is a polished mirror surface which assures that resonant modes are produced in the cavity as shown in Fig. 10.3b. These resonant modes are those for which the wavelengths of the photon satisfy the relation

$$L = q\lambda/2 \tag{10.2}$$

where q is an integer, L is the cavity length, and λ is the light wavelength in the material and is related to the free space wavelength by

$$\lambda = \frac{\lambda_o}{n_r} \tag{10.3}$$

where n_r is the refractive index of the cavity. The spacing between the stationary modes is given by

$$\Delta k = \frac{\pi}{L} \tag{10.4}$$

As can be seen from Fig. 10.3a, the Fabry-Perot cavity has mirrored surfaces on two sides. The other sides are roughened so that photons emitted through these sides are not reflected back and are not allowed to build up. Thus only the resonant modes are allowed to build up and participate in the stimulated emission process.

While the optical cavity can confine the photons with certain characteristics, it must be noted that the active region of the laser in which electron-hole pairs are recombining may only occupy a small fraction of the optical cavity. *It is important that a large fraction of the optical waveform overlap with the active region since only this fraction will be responsible for stimulated emission. As a result, it is important to design the laser structure so that the optical wave has a high probability of being in the region where e-h pairs are recombining.*

If a planar heterostructure of the form shown in Fig. 10.3c is used to confine the optical wave in the z-direction, the optical equation has the form (this is the Helmholtz equation discussed in Appendix E)

$$\frac{d^2 F_k(z)}{d^2 z} + \left(\frac{\epsilon(z)\omega^2}{c^2} - k^2 \right) F_k(z) = 0 \tag{10.5}$$

where F is the electric field representing the optical wave. The dielectric constant $\epsilon(z)$ is chosen to have a z-direction variation so that the optical wave is confined in the z-direction as shown in Fig. 10.3c. This requires the cladding layers to be made from a large bandgap material. This leads to a structure similar to the one discussed for the heterostructure LED in Section 9.4.1.

Many advances in laser physics are being driven by superior optical cavities. In the above discussion the optical confinement is improved by the heterostructure cladding layers. This is straightforward to do in an epitaxial process. It is somewhat difficult to produce dielectric constant variation in the plane of the laser, i.e., in the y-direction. Thus, usually, the laser is fabricated as a stripe of width $\sim 10~\mu$m$-50~\mu$m. The strip is produced by etching. It is also possible to produce "buried" lasers where the y-direction optical confinement is produced by doping or defect introduction since these processes can also change the dielectric constant. In section 10.6 we will discuss the advanced cavity structure being used for lasers.

An important parameter of the laser cavity is the optical confinement factor Γ, which gives the fraction of the optical wave in the active region,

$$\Gamma = \frac{\int_{\text{active region}} |F(z)|^2 dz}{\int |F(z)|^2 dz} \qquad (10.6)$$

This confinement factor is almost unity for "bulk" double heterostructure lasers where the active region is $\gtrsim 1.0~\mu$m, while it is as small as 1% for advanced quantum well lasers. However, in spite of the small value of Γ, quantum well lasers have superior performance because of their superior electronic properties owing to their 2- dimensional density of states.

EXAMPLE 10.1. Consider a GaAs laser with a cavity length of 200 μm. Calculate the frequency separation of the resonant modes.

The frequency separation is given by

$$\Delta \nu = \frac{c}{2 n_r L} = \frac{3 \times 10^{10}}{2(3.66)(200 \times 10^{-4})} = 2 \times 10^{11}~Hz$$

The energy separation of the modes is

$$h \Delta \nu = \frac{(6.64 \times 10^{-34})(2 \times 10^{11})}{1.6 \times 10^{-19}} = 0.83~meV$$

The linewidth of each of these lines under lasing conditions is smaller than this separation, so that one can see single modes.

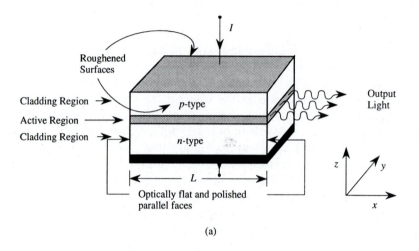

(a)

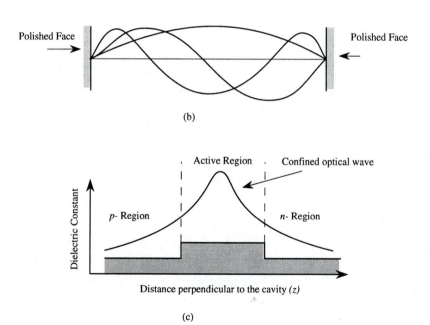

(b)

(c)

Figure 10.3: (a) A typical laser structure showing the cavity and the mirrors used to confine photons. The active region can be quite simple as in the case of double heterostructure lasers or quite complicated as in the case of quantum well lasers. (b) The stationary states of the cavity. The mirrors are responsible for these resonant states. (c) The variation in dielectric constant is responsible for the optical confinement. The structure for the optical cavity shown in this figure is called the Fabry-Perot cavity.

10.3.1   Optical Absorption, Loss and Gain

$\longrightarrow$ The photon current associated with an electromagnetic wave traveling through a semiconductor is described by

$$I_{ph} = I_{ph}^0 \, exp\,(-\alpha x) \tag{10.7}$$

where α (the absorption coefficient) is usually positive, and I_{ph}^0 is the incident photon current at $x = 0$. The optical intensity which is the photon current multiplied by the photon energy $\hbar\omega$ falls as the wave travels if α is positive. However, if electrons are pumped in the conduction band and holes in the valence band, the electron-hole recombination process (photon emission) can be stronger than the reverse process of electron-hole generation (photon absorption). In general, as discussed in Section 4.7, the gain coefficient is defined by gain = emission coefficient−absorption coefficient. If $f^e(E^e)$ and $f^h(E^h)$ denote the electron and hole occupation, the emission coefficient depends upon the product of $f^e(E^e)$ and $f^h(E^h)$ while the absorption coefficient depends upon the product of $(1-f^e(E^e))$ and $(1-f^h(E^h))$. Here the energies E^e and E^h are related to the photon energy by the condition of vertical k-transitions. In Section 9.3.1 we had found that (see Eqns. 9.9 and 9.10)

$$
\begin{aligned}
E^e &= E_c + \frac{m_r^*}{m_e^*}(\hbar\omega - E_g) \\
E^h &= E_v - \frac{m_r^*}{m_h^*}(\hbar\omega - E_g)
\end{aligned}
\tag{10.8}
$$

The occupation probabilities f^e and f^h are determined by the quasi-Fermi levels for electrons and holes as discussed in Section 4.7 (it would be very useful for the reader to review Section 4.7).

The gain which is the difference of the emission and absorption coefficient is now proportional to (see Eqn. 4.118)

$$g(\hbar\omega) \sim f^e(E^e) \cdot f^h(E^h) - \{1 - f^e(E^e)\}\{1 - f^h(E^h)\} = \{f^e(E^e) + f^h(E^h)\} - 1 \tag{10.9}$$

The optical wave has a general spatial intensity dependence

$$I_{ph} = I_{ph}^0 \, exp\,(g(\hbar\omega)x) \tag{10.10}$$

and *if g is positive, the intensity grows because additional photons are added by emission to the intensity.* The condition for positive gain requires "inversion" of the semiconductor system, i.e., from Eqn. 10.9,

$$\boxed{f^e(E^e) + f^h(E^h) > 1} \tag{10.11}$$

The quasi-Fermi levels must penetrate their respective bands for this condition to be satisfied. In Section 4.7, we had given the expression for the gain in a bulk semiconductor

$$g(\hbar\omega) = \frac{\pi e^2 \hbar}{m_0^2 c n_r \epsilon_0} \frac{1}{\hbar\omega} |a \cdot p_{cv}|^2 N_{cv}(\hbar\omega)[f^e(E^e) + f^h(E^h) - 1] \qquad (10.12)$$

Note that if $f^e = 0 = f^h$, the gain is just $-\alpha(\hbar\omega)$, the negative of the absorption coefficient.

If one uses the values of the density of states and a value of the momentum matrix element typical of direct gap semiconductors, one finds that the gain is given for GaAs for unpolarized light by ($\hbar\omega$, E_g are in eV)

$$\boxed{g(\hbar\omega) \cong 5.6 \times 10^4 \frac{(\hbar\omega - E_g)^{1/2}}{\hbar\omega} \left[f^e(E^e) + f^h(E^h) - 1\right]\ cm^{-1}} \qquad (10.13)$$

The prefactor for a different semiconductor A can be obtained by multiplying the ratio $(m_r^*(A)/m_r^*(GaAs))^{3/2}$. To evaluate the actual gain in a material as a function of carrier injection $n(=p)$, one has to find the electron and hole quasi-Fermi levels and the occupation probabilities $f^e(E^e)$ and $f^h(E^h)$, where E^e and E^h are related to $\hbar\omega$ by Eqn. 10.8. The procedure is described in more detail through Examples 10.3 and 10.4.

It must be noted that the laser operates under conditions where f^e and f^h are larger than 0.5. In this high injection limit, the occupation probabilities are not given accurately by the Boltzmann statistics. A useful approach is to use the Joyce-Dixon approximation for the position of the Fermi levels. For a given injection density $n(=p)$, the position of the quasi-Fermi levels is given by

$$E_{Fn} = E_c + k_B T \left[\ell n \frac{n}{N_c} + \frac{1}{\sqrt{8}} \frac{n}{N_c}\right] \qquad (10.14)$$

$$E_{Fp} = E_v - k_B T \left[\ell n \frac{p}{N_v} + \frac{1}{\sqrt{8}} \frac{p}{N_v}\right] \qquad (10.15)$$

where N_c and N_v are the effective density of states at the conduction and valence bands.

With these expressions the gain can be calculated as a function of photon energy for various levels of injection densities $n(=p)$. At low injections, f^e and f^h are quite small and the gain is negative. However, as injection is increased, for electrons and holes near the bandedges, f^e and f^h increase and gain can be positive. However, even at high injections, for $\hbar\omega \gg E_g$, the gain is negative. The general form of the gain-energy curves for different injection levels is shown in Fig. 10.4.

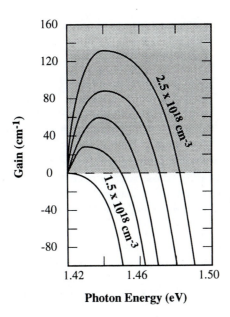

Figure 10.4: Gain vs. photon energy curves for a variety of carrier injections for GaAs at 300 K. The electron and hole injections are the same. The injected carrier densities are increased in steps of 0.25×10^{18} cm^{-3} from the lowest value shown.

The gain discussed above is called the material gain and comes only from the active region where the recombination is occurring. Often this active region is of very small dimensions. In this case, one needs to define the cavity gain which is given by

$$\boxed{\text{Cavity gain} = g(\hbar\omega)\Gamma} \qquad (10.16)$$

where Γ is the fraction of the optical intensity overlapping with the gain medium as discussed in Section 10.3. The value of Γ is almost unity for double heterostructure lasers and ~0.01 for quantum well lasers. In quantum well lasers, the overall cavity gain can still be very high since the gain in the quantum well is very large for a fixed injection density when compared to bulk semiconductors.

It is important to note that the δ-function in energy conservation of $\hbar\omega = E_e + E_h + E_g$ is broadened by a 1 to 2 meV Lorentian function to represent the lifetime of the carrier states. Thus the material gain is not "sharp" in 2D or 1D systems. The effect of the broadening is to "smoothen" out the gain curves. This is done by convoluting the gain given by Eqn. 10.13 with a broadening function.

In order for the laser oscillations to start, it is essential that when photons are emitted in the laser cavity, the gain associated with the cavity is able to surmount the loss suffered by the photons. The photon loss consists of two parts: i) loss because of absorption of the photons in the cladding regions and contacts of the laser; ii) loss due to the photons emerging from the cavity.

The cavity loss α_{loss} is primarily due to free carrier absorption of the light. As noted in Chapter 4, Section 4.6.1, this is a second order process and in high quality materials this loss can be as low as 10 cm^{-1}. It must be noted that the loss is dependent upon doping and defects in the material and, therefore, the material quality should be very good, especially in regions where the optical wave is confined.

To study the photon losses by reflection and transmission from the cavity, let us consider a Fabry-Perot cavity whose reflection coefficient and transmission coefficient is shown in Fig. 10.5a. Let us consider a wave with field F_0 incident on one edge of the cavity as shown in Fig. 10.5b and let us follow this wave as it moves through the cavity.

It is straightforward to show that the transmitted and the reflected fields are given by (refer to Fig. 10.5b)

$$F_{\text{trans}} \;=\; F_4 + F_{10} + \ldots = \frac{t_1 t_2 A}{1 - A^2 r_1^2} F_0$$

$$F_{\text{ref}} \;=\; F_2 + F_7 + \ldots = \left(r_2 + \frac{r_1 t_1 t_2 A^2}{1 - A^2 r_1^2} \right) F_0 \qquad (10.17)$$

The gain of the wave when it moves a distance L is given by

$$A = exp \left[\left(\frac{g_{tot}}{2} + ik \right) L \right] \qquad (10.18)$$

where g_{tot} consists of gain in the cavity and any loss term $\alpha_{\text{loss}}(g_{tot} = \Gamma g - \alpha_{\text{loss}})$.

Laser action occurs when non-zero F_{trans} and F_{ref} exist when F_0 is zero, i.e., photon generation in the cavity is sufficient to create photons outside the cavity. This requires a certain value of $g_{tot} = \Gamma g_{th}$. For lasings to start, we must have

$$A^2 r_1^2 = 1 \qquad (10.19)$$

The real part of this condition gives (using Eqn. 10.18)

$$g_{tot}(th) = \Gamma g_{th} - \alpha_{\text{loss}} = \frac{1}{L} \ell n r_1^{-2} \qquad (10.20)$$

or $(R = r_1^2)$

$$\Gamma g_{th} = \alpha_{\text{loss}} - \frac{1}{L} \ell n R \qquad (10.21)$$

The phase part of the lasing condition (Eqn. 10.19) requires that

$$k = \frac{m\pi}{L} \qquad (10.22)$$

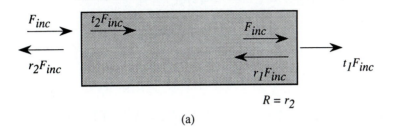

r_1 : Amplitude reflected at the semiconductor - air boundary

t_1 : Amplitude transmitted at the semiconductor + air boundary

r_2 : Amplitude reflected at the semiconductor - air boundary

t_2 : Amplitude transmitted at the semiconductor + air boundary

F_{inc} t_2F_{inc} F_{inc}

r_2F_{inc} r_1F_{inc} t_1F_{inc}

$R = r_2$

(a)

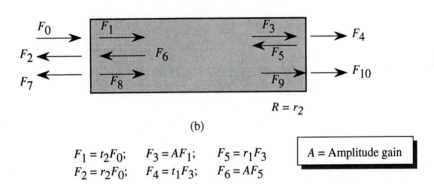

F_0 F_1 F_3 F_4

F_2 F_6 F_5 F_{10}

F_7 F_8 F_9

$R = r_2$

(b)

$F_1 = t_2F_0;$ $F_3 = AF_1;$ $F_5 = r_1F_3$ $A = $ Amplitude gain

$F_2 = r_2F_0;$ $F_4 = t_1F_3;$ $F_6 = AF_5$

Figure 10.5: (a) A schematic of the Fabry-Perot cavity showing the reflectance and transmittance of waves. (b) The path of a light wave as it moves through the cavity.

where m is an integer. This is the result we discussed from a less rigorous viewpoint in the previous section for the Fabry-Perot cavity. The Fabry-Perot cavity is not the only optical cavity used in lasers. In a later section we will discuss the distributed feedback cavity for which the lasing conditions are somewhat different. A comparison of various optical cavities will be done later in Section 10.6.

For the GaAs-air interface, the value of the reflection coefficient is

$$R = \frac{(n_r(GaAs) - 1)^2}{(n_r(GaAs) + 1)^2} \sim 0.33 \tag{10.23}$$

since the refractive index n_r of GaAs is 3.66.

EXAMPLE 10.2 Consider a GaAs Fabry-Perot laser cavity. The absorption loss in the cavity is given by an absorption coefficient of 20 cm^{-1}. Calculate the cavity length at which the absorption loss and the mirror loss become equal.

The length of the cavity is given by using the relation

$$\alpha_R = \alpha_{loss} \quad = \quad -\frac{1}{L} \, \ell n R$$

$$\text{or } L \quad = \quad \frac{-1}{20} \, \ell n(0.33) = 554 \ \mu m$$

10.4   THE LASER BELOW AND ABOVE THRESHOLD

$\longrightarrow$ In Fig. 10.6 we show the light output as a function of injected current density in a laser diode. If we compare this with the output from an LED shown in Fig. 9.12 we notice an important difference. The light output from a laser diode displays a rather abrupt change in behavior below the "threshold" condition and above this condition. The threshold condition is usually defined as the condition where the cavity gain overcomes the cavity loss for any photon energy, i.e., when

$$\boxed{\Gamma g(\hbar\omega) = \alpha_{\text{loss}} - \frac{\ell n R}{L}} \tag{10.24}$$

In high quality lasers $\alpha_{\text{loss}} \sim 10$ cm^{-1} and the reflection loss may contribute a similar amount. Another useful definition in the laser is the condition of transparency when the light suffers no absorption or gain, i.e.,

$$\boxed{\Gamma g(\hbar\omega) = 0} \tag{10.25}$$

When the p-n diode making up the semiconductor laser is forward biased, electrons and holes are injected into the active region of the laser. These electrons and holes recombine to emit photons. It is important to identify two distinct regions of operation of the laser. Referring to Fig. 10.7, when the forward bias current is small, the number of electrons and holes injected are small. As a result, the gain in the device is too small to overcome the cavity loss. The photons that are emitted are either absorbed in the cavity or lost to the outside. Thus, in this regime there is no buildup of photons in the cavity. However, as the forward bias increases, more carriers are injected into the device until eventually the threshold condition is satisfied for some photon energy. As a result, the photon number starts to build up in the cavity. As the device is further biased beyond threshold, stimulated emission starts to occur and dominates the spontaneous emission. The light output in the photon mode for which the threshold condition is satisfied becomes very strong.

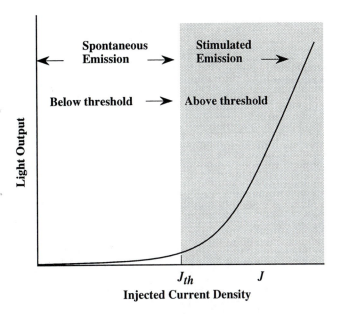

Figure 10.6: The light output as a function of current injection in a semiconductor laser. Above threshold, the presence of a high photon density causes stimulated emission to dominate.

Below the threshold the device essentially operates as an LED except that there is a higher cavity loss in the laser diode since photons cannot escape from the device due to the mirrors. Let β_{loss} be the fraction of photons that cannot escape from the device. The photon current output is given by

$$
\begin{aligned}
I_{ph} &= (1 - \text{loss})\,(\text{total } e{-}h \text{ recombination per second}) \\
&= (1 - \text{loss})\,(\text{electron particle current})
\end{aligned}
\tag{10.26}
$$

i.e.,

$$
I_{ph} = (1 - \beta_{\text{loss}})\,(R_{\text{spon}} A d_{las}) = (1 - \beta_{\text{loss}})\frac{I}{e}
\tag{10.27}
$$

where A is the laser cavity area, d_{las} is the thickness of the active layer where the recombination is occurring, and I is the injected current. The light output I_{ph} is lower than a corresponding output in an LED due to the high value of the photon loss term β_{loss}. This situation is shown schematically in Fig. 10.7a.

In order to understand the laser operation above threshold we will consider the interaction of the photons and the electrons via the rate equations. In this chapter, we will not discuss the dynamic response of the laser as this is examined in the next chapter. We will also postpone a detailed discussion of the spectral purity of the laser. Here we will focus on the light output as a function of injection for the various modes of the laser. We will write the equations relevant to two dimensional (areal) quantities. Thus, in the following S_m and N_{2D} are the photon population per unit area in the mode

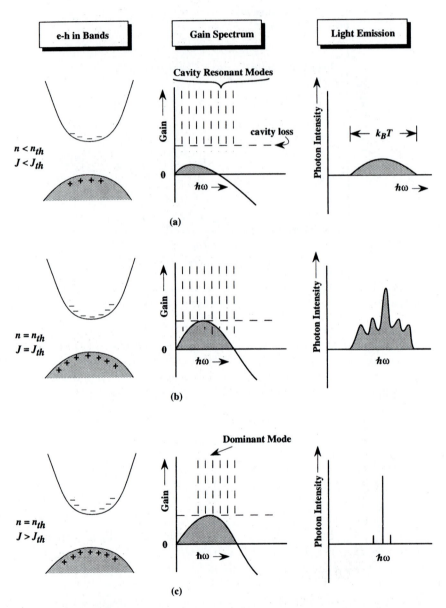

Figure 10.7: (a) The laser below threshold. The gain is less than the cavity loss and the light emission is broad as in an LED. (b) The laser at threshold. A few modes start to dominate the emission spectrum. (c) The laser above threshold. The gain spectrum does not change but, due to the stimulated emission, a dominant mode takes over the light emission.

m and the areal carrier density, respectively. The rate equation for the photons is (E_M is the energy of the mode m)

Rate of change of photon density $=$ Stimulated emission $-$ cavity loss rate

$+$ spontaneous emission

This gives

$$\frac{dS_m}{dt} = [\Gamma g(n_{2D}, E_M) - \alpha_c] \frac{c}{n_r} S_m + \beta R_{sp}(N_{2d}) \tag{10.28}$$

where the cavity loss α_c is

$$\alpha_c = \alpha_{loss} + \frac{1}{L} \ell n R \tag{10.29}$$

The parameter β is the spontaneous emission factor which represents the fraction of total spontaneous emission photons that are emitted into a particular mode. Since the total spontaneous emission is over an energy range of $\sim k_B T$ while the mode linewidth is only a few μeV, the factor β is typically 10^{-4} to 10^{-5} for Fabry-Perot cavities. The various factors in the rate equation come as follows:

- Stimulated emission:

$$R_{\text{stim}} = \text{cavity gain} \times \text{velocity of light} \times \text{photon density}$$
$$= \frac{\Gamma g \, c \, S_m}{n_r}$$

- Photon loss by cavity loss (absorption + photon loss by escape from the cavity):

$$\text{Loss rate} = \frac{\alpha_c c S_m}{n_r}$$

We note that the photon time in the cavity is given by

$$\frac{1}{\tau_{ph}} = \frac{\alpha_c c}{n_r}$$

- Spontaneous emission rate βR_{spon}

The rate equation for the carrier density is (we will first consider only the radiative part of the rate equations; the non-radiative part will be discussed in Section 10.4.1)

$$\frac{dn_{2D}}{dt} = \frac{J_{rad}}{e} - R_{sp}(n_{2D}) - \frac{c}{n_r} \sum_m \Gamma g(n_{2D}, E_m) S_m \tag{10.30}$$

Here J_{rad} the radiative part of the current density, i.e.,

$$J_{rad} = J - J_{nr} \tag{10.31}$$

The carrier loss rate is due to spontaneous recombination processes and stimulated recombination processes. These rate equations will be discussed in greater detail in the next chapter when we examine the dynamic response of the lasers. Here we will examine the steady state response where the time derivatives are zero. This gives us (from Eqns. 10.28 and 10.30)

$$S_m = \frac{\beta R_{sp}(n_{2D})}{\frac{c}{n_r}[\alpha_c - \Gamma g(n_{2D}, E_m)]} \tag{10.32}$$

$$\frac{J_{rad}}{e} = R_{sp}(n_{2D})\left\{1 + \sum_m \frac{\beta \Gamma g(n_{2D}, E_m)}{[\alpha_c - \Gamma g(n, E_m)]}\right\} \tag{10.33}$$

In general, one can solve these two coupled equations iteratively (self-consistently) to obtain the relation between J and the photon density in different modes. In performing these calculations, one usually makes the following assumptions: i) the photon and carrier densities are assumed to be uniform over the cavity length. In principle, there is spatial variation in the stationary modes of the longitudinal and transverse modes. However, the effect of these spatial variations is not so important for semiconductor lasers; ii) the refractive index is assumed to be uniform in space; iii) the confinement factor Γ and the spontaneous emission factor β is assumed to be independent of current injection; iv) gain compression effects are ignored. This effect will be discussed in the next chapter.

Noting that at threshold $g_{th} - \alpha_c = 0$, we have for the photons density in a mode m,

$$S_m = \frac{\beta R_{sp}(n_{2D})n_r}{c\Gamma}[g_{th} - g(n_{2D}, E_m)]^{-1} \tag{10.34}$$

As we know from the gain expression and Fig. 10.4, the gain versus energy curve has a peak at some energy. As the gain increases, the mode closest to this peak position will start to have a very high fraction of the photon density. Modes that are further away from the peak mode will have a lower photon density. If p is the mode corresponding to the peak in the gain curve, we may write ($dg/d\lambda = 0$ at the peak), for an energy E_s close to the maximum gain energy E_p

$$g(E_s) = g(E_p) - \frac{(\lambda_s - \lambda_p)^2}{2}\frac{d^2g}{d\lambda^2}$$

As the injection current is increased, the difference $g_{th} - g(E_p)$ decreases towards zero so that S_p increases rapidly. For other modes the difference $g_{th} - g(E_s)$ reaches the limit

$$g_{th} - g(E_s) \rightarrow \frac{(\lambda_s - \lambda_p)^2}{2}\frac{d^2g}{d\lambda^2} \tag{10.35}$$

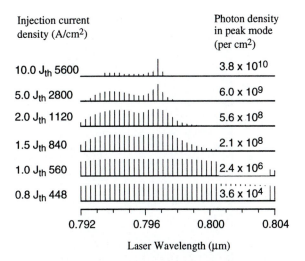

Figure 10.8: The spectral output of a quantum well laser as a function of injected current. Results are for a 50 Å GaAs/Al$_{0.3}$Ga$_{0.7}$As laser with threshold current density of 560 A/cm^2. (After Y. Lam, Loehr, and Singh, *IEEE J. Quant. Electron.*, **QE**-28, 1248 (1992).)

At or around threshold current, a large number of modes of the laser cavity participate in the photon emission. *However, as the laser is driven harder, the main peak starts to dominate and the other modes become relatively weak.* This is shown in Figs. 10.7 and 10.8.

Several important observations must be made regarding the solutions of the steady state rate equations:

i) The carrier density in the active region increases initially as the laser is pumped from the zero current, but as the carrier density is simply

$$n_{2D} = \frac{J_{rad}\tau_r}{e} \tag{10.36}$$

once stimulated emission starts and τ_r starts to rapidly decrease, the value of n_{2D} tends to a saturation value. A typical result is shown in Fig. 10.9. Note that the carrier concentration dependence on current is to some extent controlled by the factor β. A large value of β means that more e-h recombination is being coupled to the laser mode.

ii) The total photon output versus current has a kink as shown in Fig. 10.6 when the lasing starts to occur. For a laser cavity with high β, coherent emission starts to occur at very low pumping since there is a less fraction of wasted photons. It is possible to increase β by designing special optical cavities as discussed later in this chapter. However, the value of β is $\sim 10^{-4} - 10^{-5}$ for present state of the art lasers.

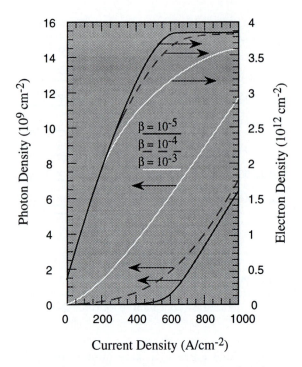

Figure 10.9: A typical dependence of photon density and electron (hole) density in a Fabry-Perot laser. The results are shown for an 80 Å GaAs/Al$_{0.3}$Ga$_{0.7}$As quantum well laser. (Courtesy of Y. Lam).

iii) The peak mode of emission shifts with current level since the gain curve's peak shifts slightly as the device is pumped. Additionally, the effective bandgap of the semiconductor changes slightly as e-h pairs are injected. In Fig. 10.10 we show a typical shift in the peak mode wavelength as the device is injected with current.

iv) The Fabry-Perot cavity provides no preference for one or another of the allowed longitudinal modes. All of the modes can, in principle, lase. The mode selectivity is provided by the gain spectrum, i.e., from the electronic properties of the active region. Thus, the envelope of the spectral output of the Fabry-Perot laser has several modes. The envelope width depends upon the current injection level as we have discussed already. We will discuss later how by a proper design of the optical cavity, a greater mode selectivity can be produced.

The performance of the semiconductor laser depends upon both the properties of the active region making up the laser and the optical cavity causing photon selection. In the next sections, we will discuss how these two issues can be optimized to produce state of the art lasers.

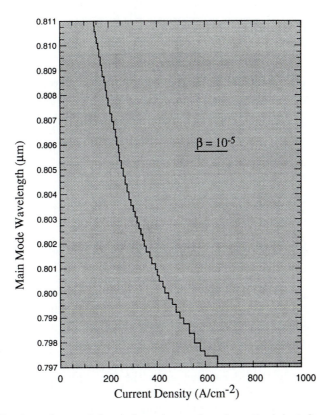

Figure 10.10: The dependence of the peak mode wavelength on the injected current density for a Fabry-Perot laser. (After Y. Lam, J. Loehr, and J. Singh, *IEEE J. Quant. Electron.*, **QE**-28, 1248 (1992).)

10.4.1 Non-Radiative Current

$\longrightarrow$ In the discussion above, the current that appears in the rate equations is the radiative current J_{rad}. The total current is given by

$$J = J_{rad} + J_{nr} \tag{10.37}$$

where the non-radiative current is due to *e-h* recombination where no photons are emitted. The non-radiative current can be attributed to two sources. The first source is due to defect (traps) related *e-h* recombination is discussed in Section 4.8, while the second source is due to Auger recombination as discussed in Section 4.9. The reader should review these sections to understand these two sources. The defect related recombination is proportional to the defect density and is usually quite low for lasers based on mature technologies (e.g., GaAs or InP based). However, the Auger recombination depends upon the intrinsic bandstructure of the active material and is controlled by the bandgap, carrier masses and operating temperature. The Auger rate can be written as

$$R_{\text{Auger}} = Fn^3 \qquad (10.38)$$

where F is the Auger coefficient and is given for several semiconductors in Appendix B. Note the very strong dependence of the Auger recombination on the carrier density. We have discussed above that the carrier density in the laser saturates at a value $n_{th}(= n_{2D}(th) \cdot d_{las})$. When the carrier density reaches n_{th}, the cavity gain is equal to the cavity loss, so that n_{th} *does not depend upon the non-radiative processes.* The threshold current can be written as

$$J_r(th) \quad = \quad \frac{en_{th}d_{las}}{\tau_r} \qquad (10.39)$$

$$J_{nr}(th) \quad = \quad eFn_{th}^3 d_{las} \qquad (10.40)$$

Here d_{las} is the width of the active region of the laser. The total threshold current is

$$J_{th} = J_r(th) + J_{nr}(th) \qquad (10.41)$$

The Auger part of the threshold current depends upon the bandgap of the material and the operating temperature. In GaAs based lasers, this part is quite negligible, but for 1.3 μm and 1.55 μm lasers, the Auger current can be significant.

We note finally that depending upon the laser design, a certain fraction of the injected electrons and holes will leak across the active region and recombine in the neutral p and n regions. This leakage current was discussed for the LEDs in the previous chapter. This current has to be added to the total current. The leakage current becomes quite large as temperature increases and is an important factor in the temperature dependence of the laser threshold current.

It may be noted from the discussion of this section that, for a low threshold current, it is useful to use a p-doped active region. This is due to the much heavier hole masses which cause the value of f^h to be much smaller than f^e for the same carrier injection. By p-doping, one can increase f^h and thus lower the injection density needed for threshold. In Example 10.7, we examine the difference that p-doping can make in this regard.

We also point out that in a semiconductor laser under lasing conditions the injected carrier density is $\sim 10^{18}$. At such a density there is a bandgap renormalization effect as discussed in chapter 4, Section 4.11. This effect causes the effective bandgap to be about 25 *meV* smaller than the low carrier density bandgap. Also at such densities there are essentially no excitonic effects.

EXAMPLE 10.3 According to the Joyce-Dixon approximation, the relation between the Fermi level and carrier concentration is given by Eqns. 10.14 and 10.15. Calculate the carrier density needed for the transparency condition in GaAs at 300 K and 77 K. The transparency condition is defined at the situation where the maximum gain is zero (i.e., the optical beam propagates without loss or gain). Calculate the transparency condition at $\hbar\omega = E_g$.

At room temperature the valence and conduction band effective density of states is

$$N_v = 7 \times 10^{18} \ cm^{-3}$$
$$N_c = 4.7 \times 10^{17} \ cm^{-3}$$

The values at 77 K are

$$N_v = 0.91 \times 10^{18} \ cm^{-3}$$
$$N_c = 0.61 \times 10^{17} \ cm^{-3}$$

In the semiconductor laser, an equal number of electrons and holes are injected into the active region. We will look for the transparency conditions for photons with energy equal to the bandgap. The approach is very simple: i) choose a value of n or p; ii) calculate E_{Fn} and E_{Fp} from the Joyce-Dixon approximation; iii) calculate $f^e + f^h - 1$ and check if it is positive at the bandedge. The same approach can be used to find the gain as a function of $\hbar\omega$.

For 300 K we find that the material is transparent when $n \sim 1.1 \times 10^{18}$ cm^{-3} at 300 K and $n \sim 2.5 \times 10^{17}$ cm^{-3} at 77 K. Thus a significant decrease in the injected charge occurs as temperature is decreased.

EXAMPLE 10.4 Consider a GaAs double heterostructure laser at 300 K. The optical confinement factor is unity. Calculate the threshold carrier density assuming that it is 20% larger than the density for transparency. If the active layer thickness is 2.0 μm, calculate the threshold current density.

From Example 10.3 we see that at transparency

$$n = 1.1 \times 10^{18} \ cm^{-3}$$

The threshold density is then

$$n_{th} = 1.32 \times 10^{18} \ cm^{-3}$$

The radiative recombination time is approximately four times τ_o, i.e., ~ 2.4 ns. The current density then becomes

$$J_{th} = \frac{e \cdot n_{th} \cdot d_{las}}{\tau_r} = \frac{(1.6 \times 10^{-19} \ C)(1.32 \times 10^{18} \ cm^{-3})(2 \times 10^{-4} \ cm)}{2.4 \times 10^{-9} \ s}$$
$$= 1.76 \times 10^4 \ A/cm^2$$

EXAMPLE 10.5 In Chapter 2 we had discussed the density of states for electrons. Calculate the density of states for photons using similar arguments.

Just like the electron states, the photon states also have a phase dependence $exp\,(ik\cdot r)$. However, the dependence of the energy of the k-vector is different. The photon "dispersion relation" is $(k = 2\pi/\lambda)$

$$\hbar\omega = \frac{hv}{\lambda} = \hbar v k$$

where v is the light velocity.

As in the case of the electrons, the k-space occupied by each photon state is $V/(2\pi)^3$, where V is the volume of the space in which photons are confined. The number of states having energy between $\hbar\omega$ and $\hbar\omega + d(\hbar\omega)$ is then ($\rho(\hbar\omega)$ is the density of photon states)

$$\begin{aligned}
\rho(\hbar\omega)d(\hbar\omega)V &= \frac{4\pi k^2\,dk}{(2\pi)^3}V \\
&= \frac{\omega^2\,d(\hbar\omega)V}{2\pi^2 v^3\hbar}
\end{aligned}$$

The density of states is then

$$\rho(\hbar\omega) = \frac{\omega^2}{2\pi^2 v^3\hbar}$$

In general, there are two different polarization modes for the light (both the modes are transverse, i.e., if the light is travelling in the z direction, the polarization can be in x or y direction). Thus the total photon density of states is twice the result obtained.

EXAMPLE 10.6 Consider a GaAs optical cavity which has a length of 200 μm and the reflectivity of the mirrors is 0.33. The absorption loss in the cavity is 10 cm^{-1}. Calculate the time spent by a photon in the cavity before it is absorbed or emitted. The time is called the photon lifetime τ_{ph}.

The loss coefficient for the photon is

$$\begin{aligned}
\alpha_{tot} = \alpha_{loss} + \alpha_R = \alpha_{loss} - \frac{1}{L}\ell n R &= 10 - \frac{\ell n(0.33)}{2\times 10^{-2}} \\
&= 65.43\ cm^{-1}
\end{aligned}$$

This represents the inverse distance travelled by the photon before it is either absorbed or emitted from the cavity. The lifetime is therefore (v = velocity of light)

$$\tau_{ph} = \frac{1}{v\alpha_{tot}} = \frac{3.6}{3\times 10^{10}\times 65.43} = 0.51\ ps$$

EXAMPLE 10.7 Consider two double heterostructure GaAs/AlGaAs lasers at 300 K. One laser has an undoped active region while the other one is doped p-type at 8×10^{17} cm^{-3}. Calculate the threshold current densities for the two lasers if the cavity loss is 50 cm^{-1} and the radiative lifetime at lasing is 2.4 ns for both lasers. The active region width is 0.1 μm.

This example is chosen to demonstrate the advantages of p-type doping in threshold current reduction. Since holes are already present in the active region, one does not have to inject as much charge to create gain. However, it must be noted that too much p-doping can cause an increase in cavity loss and even non-radiative traps (some dopants can be incorporated on unintended sites in the crystal).

To solve this problem, a computer program should be written. This program should calculate the quasi-Fermi levels for the electrons and holes and then evaluate the gain.

We have

$$
E_{Fn} = E_c + k_B T \left[\ell n \frac{n}{N_c} + \frac{1}{\sqrt{8}} \frac{n}{N_c} \right]
$$

$$
E_{Fp} = E_v - k_B T \left[\ell n \frac{p_{tot}}{N_v} + \frac{1}{\sqrt{8}} \frac{p_{tot}}{N_V} \right]
$$

where n is the electron (and hole) density injected and

$$
p_{tot} = p + p_A
$$

where p_A is the acceptor density. For the undoped laser, one finds that at approximately 1.1×10^{18} cm^{-3} the laser reaches the threshold condition. For the doped laser we get a value of $n = p = 8.5 \times 10^{17}$ cm^{-3}. The threshold current densities in the two cases are

$$
J(\text{undoped}) = \frac{(1.1 \times 10^{18} \ cm^{-3})(0.1 \times 10^{-4} \ cm)(1.6 \times 10^{-19} \ C)}{(2.4 \times 10^{-9} \ s)}
$$

$$
= 733 \ A/cm^2
$$

$$
J(\text{doped}) = \frac{(8.5 \times 10^{17} \ cm^{-3})(0.1 \times 10^{-4} \ cm)(1.6 \times 10^{-19} \ C)}{(2.4 \times 10^{-9} \ s)}
$$

$$
= 566 \ A/cm^2
$$

10.5 ADVANCED STRUCTURES: TAILORING ELECTRONIC STRUCTURE

One of the most important applications of semiconductor lasers is in the area of optical communications, where a key driving force for superior laser design is low threshold current and high modulation bandwidth. Other motivations include lasers with emission frequencies that are important for particular applications. These include long wavelength lasers for communication, short wavelength lasers for optical memory applications, etc. It is also very important to have lasers with very narrow emission linewidth if wavelength division multiplexing or coherent detection is to be feasible for optical communication.

In this chapter we will address the issues of low threshold and spectral purity in lasers. The issue of high bandwidth will be discussed in the next chapter. The laser performance is controlled by both electronic and optical properties. In Fig 10.11 we show the approaches being pursued to enhance laser performance. As can be seen, these approaches involve novel designs for both the electronic states and the photonic states. We will discuss the physics behind these approaches in this and the next section.

10.5.1 Double Heterostructure Lasers

$\longrightarrow$ The first successful lasers involved the use of active layers of thickness $d_{las} \gtrsim 1$ μm. The optical confinement factors in these layers were very high ($\Gamma \sim 1.0$). *It is important to note that the 3D density of electrons (holes) required to produce the transparency condition is a value n (transparency) which is independent of the active layer thickness if $\Gamma \sim 1.0$.* The value of n (transparency) is very close to n_{th} in high quality structures. *The current density needed at threshold is, however, related to the 2D carrier density (see Eqn. 10.36) as*

$$J_{th} = \frac{en_{th}d_{las}}{\tau_r(J_{th})} = \frac{en_{2D}(\text{threshold})}{\tau_r(J_{th})} \qquad (10.42)$$

Thus, in devices with thick active layers, the threshold current is proportional to the active layer thickness. This is indeed seen in actual devices (see Fig. 10.12).

Once d_{las} becomes much smaller than the emission wavelength, the value of the optical confinement starts to decrease. Also for very small values of d_{las} (~ 100 Å) quantum effects become important. The advantages of quantum well lasers are discussed next.

10.5.2 Quantum Well Lasers

$\longrightarrow$ In quantum well lasers, the active region where *e-h* recombination takes place is only about a hundred Angstroms. A typical quantum well laser structure is shown in Fig. 10.13. A narrow bandgap region is surrounded by a wider gap region to form the quantum well. Surrounding the quantum well is the wide gap bandgap cladding layer. Often the quantum well is surrounded by a region which "funnels" the electrons and holes into the well. Such structures are called graded index separate (GRIN) confinement structures.

Quantum well lasers have gained wide acceptance as high speed low threshold lasers. Their advantage in low threshold applications comes from the special density of

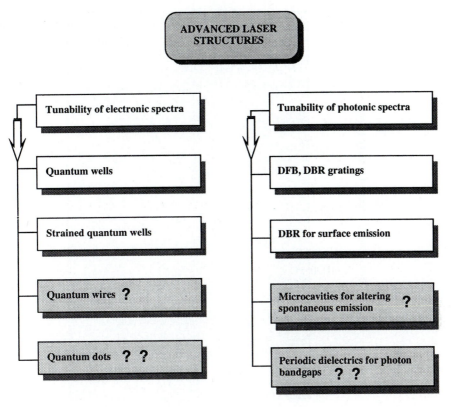

Figure 10.11: Approaches used to fabricate advanced semiconductor lasers. Question marks are placed after approaches where considerable technological challenges remain and whose merit is not yet established.

states that quasi-2D systems have. We have discussed the density of states of 3D, 2D and 1D systems in Chapter 2. In Fig. 10.14 we show the density of states for parabolic bands in a quantum well. The important point to note is that the density of states goes to zero at the bandedge in a 3D system, while for the 2D (and 1D system), the density of states is non-zero. As a result, when carriers are injected into a quantum well system, the product of occupation number and density of states at the bandgap increases much more rapidly than in 3D system. Even though the confinement factor Γ is small in a quantum well, the cavity gain given by the product Γg can be improved by going to a quantum well system.

Let us consider a narrow quantum well case where only the ground states of the conduction band and valence band subbands are occupied. Let E_1^e, E_1^{hh} and $E_1^{\ell h}$ represent the ground states of the electron, heavy hole and light hole as shown in Fig.

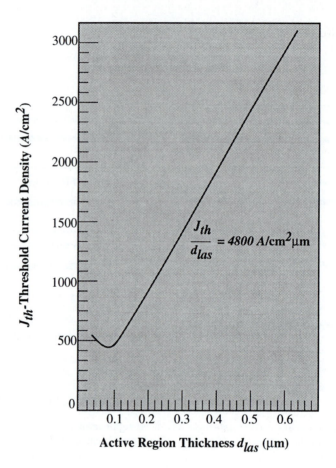

Figure 10.12: Dependence of threshold current in double heterostructure lasers on width of the active region. The threshold current density decreases with the active layer thickness because the 2-dimensional sheet charge density of the injected charge needed for the threshold condition decreases inversely with the active layer thickness. At very small active layer thicknesses $d \stackrel{<}{\sim}$ 50 Å, the threshold current density increases because the optical wave confinement factor goes towards zero so that the cavity gain is almost zero.

10.14. In the limit of single band occupation we have

$$
\begin{aligned}
n &= \frac{m_e^*}{\pi \hbar^2} \int_{E_1^e}^{\infty} \frac{dE}{exp\left(\frac{E - E_{Fn}}{k_B T}\right) + 1} \\
&= \frac{m_e^* k_B T}{\pi \hbar^2} \left[ln\left\{ 1 + exp\left(\frac{E_{Fn} - E_1^e}{k_B T}\right) \right\} \right]
\end{aligned} \tag{10.43}
$$

or

$$
E_{Fn} = E_1^e + k_B T \; ln \left[exp\left(\frac{n \pi \hbar^2}{m_e^* k_B T}\right) - 1 \right] \tag{10.44}
$$

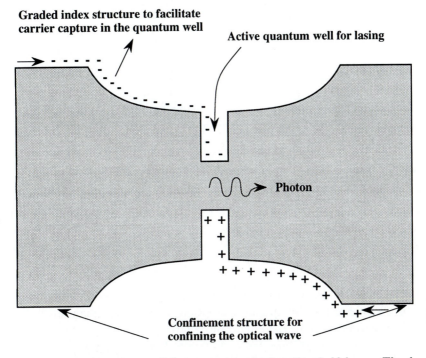

Graded index structure to facilitate carrier capture in the quantum well

Active quantum well for lasing

Photon

Confinement structure for confining the optical wave

Figure 10.13: A typical quantum well laser structure for low threshold lasers. The density of states in the 2D quantum well allows one to achieve the condition for inversion of bands at a lower injection density. This results in a lower threshold current. A cladding layer with a high bandgap surrounds the quantum well so that the optical wave is confined as much as possible near the quantum well to get a high confinement factor.

For the hole density we have (considering both the HH and LH ground state subbands)

$$p = \frac{m_{hh}^*}{\pi\hbar^2} \int_{E_1^{hh}}^{-\infty} \frac{dE}{exp\left(\frac{E_{Fp}-E}{k_BT}\right)+1} + \frac{m_{\ell h}^*}{\pi\hbar^2} \int_{E_1^{\ell h}}^{-\infty} \frac{dE}{exp\left(\frac{E_{Fp}-E}{k_BT}\right)+1} \qquad (10.45)$$

where m_{hh}^* and $m_{\ell h}^*$ are the in-plane density of states masses of the HH and LH subbands. Note that these masses depend upon the quantum well size and, in fact, the HH and LH bands are not parabolic for most quantum wells. However, to avoid a numerical calculation requiring a computer, one can use this simpler approximation. We have then

$$
\begin{aligned}
p &= \frac{m_{hh}^* k_B T}{\pi\hbar^2} \left[\ell n \left\{ 1 + exp \frac{\left(E_1^{hh} - E_{Fp}\right)}{k_B T} \right\} \right] \\
&+ \frac{m_{\ell h}^* k_B T}{\pi\hbar^2} \left[\ell n \left\{ 1 + exp \left(\frac{E_1^{\ell h} - E_{Fp}}{k_B T} \right) \right\} \right] \qquad (10.46)
\end{aligned}
$$

If

$$E_1^{hh} - E_1^{\ell h} > k_B T \qquad (10.47)$$

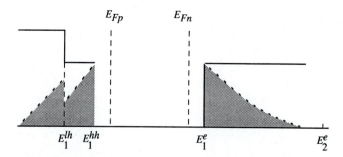

Figure 10.14: Fermi levels and carrier occupation in quantum well lasers.

we can make the approximation that the LH subband is not occupied and obtain an analytical expression for E_{Fp} in terms of p just as we have done for the electron subband. Otherwise, to obtain E_{Fp} from p we need to do an iterative calculation which is quite straightforward.

To find the material gain in the quantum well structure, we use the expressions derived in Chapter 4 for the optical absorption processes in quantum wells. At this point, it is important to note the important polarization selection rules that come into picture in quantum well systems. We remind ourselves that the heavy-hole to electron transition has a stronger coupling to the x-y polarized light (TE polarized) where z is the growth direction of the quantum well while the LH state only couples to the z-polarized light (TM polarized). We summarize the strength of the coupling to the different polarization by examining the momentum matrix elements (see Section 4.4).

TE polarized

$$HH \rightarrow C - band: \qquad |p_{if}|^2 = \frac{1}{2}| < p_x|p_x|s > |^2 = \frac{m_0 E_p}{4}$$

$$LH \rightarrow C - band: \qquad |p_{if}|^2 = \frac{1}{6}| < p_x|p_x|s > |^2 = \frac{m_0 E_p}{12}$$

TM polarized:

$$HH \rightarrow C - band: \qquad |p_{if}|^2 = 0$$

$$LH \rightarrow C - band: \qquad |p_{if}|^2 = \frac{2}{3}| < p_x|p_x|s > |^2 = \frac{m_0 E_p}{3} \qquad (10.48)$$

where E_p is an energy term with an approximate value (see Section 4.4)

$$E_p \cong 22eV$$

for most semiconductors.

The polarization detection rules have important effects on quantum well lasers where the HH and LH bands are not degenerate and, therefore, their occupations are different. *In general, since the HH band is closer to E_{Fp} than the LH band, there is a stronger inversion in the HH band. As a result, the light output is primarily TE polarized, i.e., the polarization is perpendicular to the growth direction taken to be z-axis.* In the presence of strain, this can be altered as will be discussed in the next subsection.

The TE and TM polarization gain in the quantum well can now be written for the mth valence band subband and n^{th} conduction band subband as

$$g_{nm}(\hbar\omega) = \frac{\pi e^2 \hbar}{n_r c \, m_0^2 W \epsilon_0 \hbar\omega} N_{nm}(\hbar\omega)|p_{nm}|^2 \left[f_e(E_n^e(k)) + f_h(E_m^h) - 1 \right] \qquad (10.49)$$

where $N_{nm}(\hbar\omega)$ is the joint density of the electron-hole bands. The momentum matrix element is

$$p_{nm} = \int g_c^{*n}(z) g_v^m(z) dz \sum_\nu <s|p_a|u_v^m> \qquad (10.50)$$

where $g_c^{*n}(z)$ and $g_v^m(z)$ are the electron and hole envelope functions and u_v^m is the central cell part of the hole state. The overlap between the envelope functions of the bands is essentially unity for $n = m$ (i.e., $CB1 \to HH1$ or $CB1 \to LH1$) and zero for $n \neq m$ (e.g., $CB1 \to HH2$ or $CB2 \to LH1$, etc.,) unless there is a strong electric field applied the quantum well. The momentum matrix elements of p_a have been discussed in detail in Chapter 4 and are given in Eqn. 10.48.

The important point to note is the following when considering the advantage of quantum well lasers over double heterostructure lasers. In double heterostructure lasers where the confinement factor $\Gamma \sim 1.0$, the cavity gain can easily approach $10^4 \ cm^{-1}$. However, in a high quality cavity, the losses are only $\sim 30-50 \ cm^{-1}$, so that one does not really need or use such a high potential gain. In a quantum well laser, the cavity gain Γg can only approach $100 \ cm^{-1}$ at the most. However, this is quite adequate. Thus one can reduce the threshold current greatly by using quantum well lasers.

It is important to note that the quantum well structure may not always be better than a wider double heterostructure laser. This is especially true in narrow gap materials where the Auger effects dominate. Due to the smaller optical confinement factor, the quantum well laser requires a higher material gain at threshold compared to a wide laser. As a result, the value of n_{th} is larger in a quantum well laser. In the presence of Auger processes, the threshold current is given by

$$J_{th} = \frac{e n_{th} d_{las}}{\tau_r} + e F n_{th}^3 d_{las} \qquad (10.51)$$

where the second term comes from the Auger processes. Due to the increase in n_{th} in quantum wells, it is not necessary that J_{th} will always decrease with reduced active layer thickness. Thus, in presence of strong Auger processes an optimum width exists for lasers which could be larger than the usual 50-100 Å for the case where Auger processes are unimportant.

10.5.3   Strained Quantum Well Lasers

↝ In Chapter 2 (section 2.10) we discussed the modification of bandstructure that is possible due to built-in strain. This strain can be produced by using an epitaxial layer with a lattice constant that is different from that of the substrate. The strain has a dramatic influence on the optical properties of the system. In Chapter 4 we have examined the effect of the built-in strain on the optical properties and the reader is advised to review that discussion. There are several motivations for using strain in quantum well lasers as outlined in Fig. 10.15. We will briefly discuss each of them.

Bandgap tunability
As noted in Chapter 2, the built-in strain (we will consider (001) growth) allows one to tailor the bandgap of the semiconductor. The strain causes a splitting of the HH and LH strate and the "bulk" bandgaps change as shown in Fig. 10.16a:

C-band → HH

$$E_g = E_{go} + 2a \left(\frac{C_{11} - C_{12}}{C_{11}} \right) \epsilon + b \left(\frac{C_{11} + 2C_{12}}{C_{11}} \right) \epsilon \tag{10.52}$$

C-band → LH

$$E_g = E_{go} + 2a \left(\frac{C_{11} - C_{12}}{C_{11}} \right) \epsilon - b \left(\frac{C_{11} + 2C_{12}}{C_{11}} \right) \epsilon \tag{10.53}$$

where ϵ is the strain (a_S, a_L are the substrate and overlayer lattice constants)

$$\epsilon = \frac{a_S}{a_L} - 1 \tag{10.54}$$

and C's are the force constants. The a and b are the deformation potentials. For compressive strain, ϵ is negative and the HH state is above the LH state. For tensile strain the reverse occurs. Note that effects of alloying and quantum confinement are over and above the values given above. For the InGaAs system grown on GaAs, the change in the bandgap is approximately 6ϵ eV.

Reduced Threshold Current
One of the most important motivations for using strain in quantum wells is the reduction in threshold current that is possible. An important consequence of strain is to lift the degeneracy of the HH and LH states which causes a reduction in the in-plane hole density of states mass. The HH, LH splitting can reach 100 meV for a strain of ∼ 2%, and the hole density of states mass can decrease by up to a factor of 3 as shown in Fig. 10.16b. The reduced hole mass allows the inversion condition to be reached at a lower injection density.

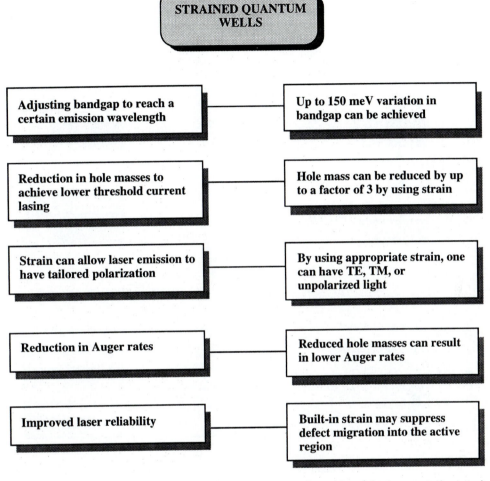

Figure 10.15: Some of the important advantages that can be achieved by incorporating strain in quantum well lasers. The issues of lower Auger recombination and laser reliability are still being researched.

Threshold currents are found to decrease for both compressive and tensile strained lasers since the bandedge mass decreases in both cases as discussed in Chapter 2. At a small tensile strain, the HH, LH degeneracy lifting due to quantization is restored so that the hole density of states becomes quite large. As a result, the threshold current increases slightly before decreasing in the presence of tensile strain.

Polarization Control

We had noted earlier that the HH state couples only to the TE mode. On the other hand, the LH state couples to the TM mode four times as strongly as it couples to the

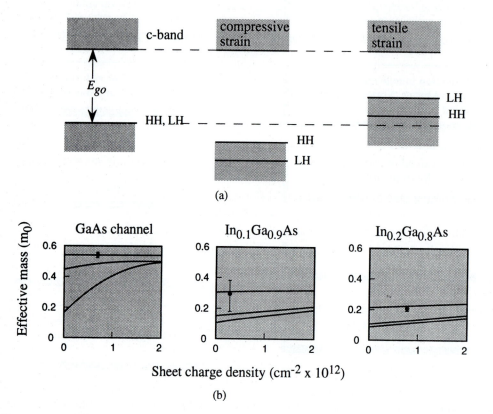

(a)

(b)

Figure 10.16: (a) The effect of strain on the bandedges (with reference to the conduction band). (b) The effect of strain on the near bandedge hole masses. Results are for layers grown on a GaAs substrate. The top curve is for 300 K, the middle curve for 77 K and the bottom curve for 4 K. (After M. Jaffe, J. E. Oh, J. Pamulapati, J. Singh and P. Bhattacharya, *Applied Physics Letters*, 54, 2345 (1989).)

TE mode. Since the strain can alter the HH, LH separation, and hence their occupation at threshold, it is possible to tailor the emission polarization of the laser light. For many applications, it is important to have unpolarized light. In an unstrained quantum well since the HH is above the LH, the light is slightly more TE polarized than TM polarized. By incorporating a small tensile strain unpolarized light can be obtained.

Reduction in Auger Rates

In Chapter 4, section 4.9, we had discussed the Auger rates in semiconductors. An important parameter in the Auger rates is the minimum threshold energy that carriers must have to instigate the Auger process. This minimum energy comes about due to momentum and energy conservation as discussed in Section 4.9. The strain in the system alters the hole masses and, as a result, the minimum energy for the Auger process also is changed. The threshold energy was calculated for the CHHS process and CCHC processes in Chapter 2, Section 2.9. There we found that the threshold energy depends upon the hole and electron masses as well as the difference between E_g and the split-off energy Δ.

If Δ is unaltered by strain, as the hole mass decreases E_{th} increases which means that the probability of electrons and holes at these energies will be low and the Auger coefficients will be low. However, if Δ approaches the bandgap, the Auger rates could increase. Of course, since the Auger rate is Fn_{th}^3, the decrease in n_{th} with strain certainly suppresses the Auger rates.

Laser Reliability

In the initial period of strained epitaxy, it was believed that the pseudomorphic devices would have a poor long term reliability due to the built-in strain energy. However, almost a decade of testing has shown that both electronic and optoelectronic devices exploiting strain seem to have better reliability than lattice matched devices!

It appears that defect propagation through a strained region is suppressed and defects trying to enter a strained region are repelled away. This unexpected benefit of strain is perhaps one of the most important benefits since device reliability is such an important issue in systems.

From the discussion of this subsection it is clear that strain provides a remarkable degree of tailorability in semiconductor lasers. It is, therefore, not surprising that strained quantum well lasers are being researched and developed by a very large number of optoelectronics groups around the world.

EXAMPLE 10.8 Consider two 80 Å quantum well lasers: an unstrained $GaAs/Al_{0.3}Ga_{0.7}As$; and another in which 20% excess In is added in the well region (i.e., with a $In_{0.2}Ga_{0.8}As$). The following parameters define the devices

GaAs/Al$_{0.3}$Ga$_{0.7}$As laser:

$$
\begin{aligned}
E_e^1 - E_c(GaAs) &= 40 \ meV \\
E_v(GaAs) - E_{HH}^1 &= 10 \ meV \\
E_v(GaAs) - E_{LH}^1 &= 30 \ meV \\
m_e^* &= 0.067 \ m_0 \\
m_{hh}^* &= 0.5 \ m_0 \\
m_{\ell h}^* &= 0.15 \ m_0
\end{aligned}
$$

In$_{0.2}$Ga$_{0.8}$As/Al$_{0.3}$Ga$_{0.7}$As laser:

$$
\begin{aligned}
E_e^1 - E_c(InGaAs) &= 40 \ meV \\
E_v(InGaAs) - E_{HH}^1 &= 10 \ meV \\
E_v(InGaAs) - E_{LH}^1 &= 0.12 \ meV \\
m_e^* &= 0.067 \ m_0 \\
m_{hh}^* &= 0.25 \ m_0
\end{aligned}
$$

Calculate the carrier densities needed for transparency for the two lasers at 300 K.

We note that the energy difference between the HH and LH states in the unstrained quantum well is 20 meV which is smaller than $k_B T$. Thus there will be a small contribution from the light hole occupation. For the strained laser, there will be no light hole occupation.

To find the transparency density we use Eqns. 10.44 and 10.47 to calculate E_{Fn} and E_{Fp}. Note that E_{Fp} has to be obtained iteratively if LH states are occupied. Once E_{Fn} and E_{Fp} are known as a function of carrier injection, it is straightforward to find the density at which $f^e + f^h = 1$ at the effective bandedge energy. The following values are found:

$$
\begin{aligned}
n_{2D}(GaAs \ well) &\sim 10^{12} \ cm^{-2} \\
n_{2D}(InGaAs \ well) &\sim 6.5 \times 10^{11} \ cm^{-2}
\end{aligned}
$$

We see from this example the importance of reducing hole masses to reduce the carrier injection needed for transparency (or the threshold condition). This translates into a reduced threshold current density.

10.5.4 Quantum Wire and Quantum Dot Lasers

⤳ From the discussions of the previous subsections, the reader may have started thinking about what an ideal bandstructure or density of states would be for low threshold lasers. For low threshold, the term $f_e + f_h - 1$ should be positive for the lowest possible injection. This places the following demands on the density of states: i) the density of states should be high at the lasing energy and low at other energies so that all the carriers injected into the diode can be used to improve f_e and f_h at the lasing energy; ii) the electron and hole density of states should be symmetric so that both f_e and f_h can be high at low injection. These considerations were the motivating forces for the quantum well and then the strained quantum well lasers. The next logical step is the quantum wire laser and even the quantum dot laser.

The quantum wire structure has a singular density of states if the energy momentum relation is parabolic. The density of states for each subband is (E_n is the subband energy)

$$N(E) = \sqrt{2}\frac{m^{*1/2}}{\pi\hbar}(E - E_n)^{-1/2} \qquad (10.55)$$

Expressing m^* in units of m_0 and E in eV, we have

$$N(E) = \left(\frac{m^*}{m_0}\right)^{1/2}(E - E_n)^{-1/2}(1.626 \times 10^7)eV^{-1}\ cm^{-1} \qquad (10.56)$$

The conduction bandedge can be described quite well by parabolic bands, but the valence band density of states is highly non-parabolic as shown in Fig. 3.20.

For a parabolic band the carrier concentration in a given subband is

$$n = \int_{E_n}^{\infty} N(E)f(E)dE \qquad (10.57)$$

If we assume that the occupation function is given by the Boltzmann approximation, we have

$$n = \sqrt{\frac{2}{\pi}}\frac{m^{*1/2}(k_BT)^{1/2}}{\hbar}exp\left(\frac{E_{Fn} - E_n}{k_BT}\right) \qquad (10.58)$$

Due to the bandedge singularity in the density of states, the gain near the bandgap energy is much greater than that in the quantum well case for a similar 3-dimensional injection density. As a result, it is possible to obtain a lower threshold current density.

In the quantum dot, the density of states are due to the discrete energy levels in the structure. As a result, one has δ-function density of states. It is, therefore, possible to get inversion with a very small charge injection. Of course, each dot may only have 2 to 3 electrons and holes at threshold!

From the discussion above, it would appear that quantum wire and quantum dot lasers would be fabricated and marketed by every optoelectronic house. However, the rosy picture painted above has some important hinderances. In Fig. 10.17 we show

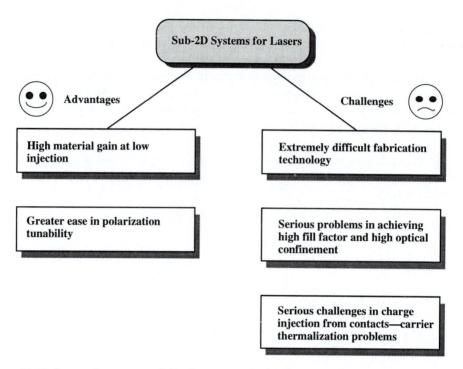

Figure 10.17: Some advantages and disadvantages of sub-2D systems (quantum wires and dots) for lasers.

some advantages and disadvantages of these systems. The most important problem arises from the fabrication process itself. The fabrication of the quantum well structure by MBE or MOCVD growth simply involves opening and closing of shutters or valves (see Chapter 1, Section 1.10). However, to create the lateral confinement of electrons and holes needed for quantum wires and quantum dots, one needs to use additional techniques which are very difficult to control. There are several approaches used to create lateral confinement: i) Growth on patterned substrates: In this approach one designs the substrate by either creating a V-groove or by cutting the substrate at an angle to create a staircase like arrangement of atoms. The lateral confinement is produced by depending upon growth kinetics of the different atoms during epitaxy. Thus, typically, atoms of the narrow bandgap materials (say Ga in a GaAs-AlAs structure) migrate more rapidly than the atoms of the larger gap material. As a result, in the growth on patterned substrates, lateral gradients in bandgap composition are created which can be exploited to make quantum wires. The lateral confinement is, however, rather weak and the technique is very fragile. ii) Etching and Regrowth: In this approach a quantum well is first grown and the sample is then etched and a high bandgap material is then regrown to create quantum confinement. This approach is also quite difficult since the etching and regrowth processes are rather poorly understood and controlled.

Quantum wire lasers have been fabricated by both approaches outlined above, but their performance has not been comparable to quantum well lasers so far.

An important concern in sub-2D systems has to do with carrier injection and thermalization. We have discussed this issue in Section 3.11 and it's consequences on laser dynamics will be discussed in the next chapter. However, the slow relaxation process creates a serious problem in the use of sub-2D lasers for high speed applications.

10.6 ADVANCED STRUCTURES: TAILORING THE CAVITY

10.6.1 Issues in a Fabry-Perot Cavity

A commonly used laser cavity structure is the Fabry-Perot cavity having a waveguiding structure of length L and width d_T as shown in Fig. 10.3. In this cavity, there are a number of optical modes which can form stationary states as discussed before. As a laser is pumped, the gain becomes positive and eventually the cavity loss is overcome. At this point, a number of the optical modes can start to have stimulated emission. As a cavity is pumped harder, a few modes, whose frequency lies close to the peak energy in the gain curve, start to get much stronger as discussed before and shown in Fig. 10.7. Nevertheless, several modes are usually emitted during the laser operation. Thus the modal purity of the Fabry-Perot laser is not very good. The spacing between these Fabry-Perot modes is given by (see Section 10.3)

$$\Delta k = \pi/L \qquad \Delta \omega = \frac{v}{4\pi L} \tag{10.59}$$

where v is the velocity of light. The photon modes that the *e-h* recombination can emit in a Fabry-Perot cavity are essentially the same as in a bulk semiconductor. This is because of the large size of the cavity compared to the wavelength of light.

The Fabry-Perot cavity is one of the simplest optical cavities used for semiconductor lasers. It is simply produced by cleaving a semiconductor wafer along the cleavage planes (110) or ($\bar{1}$10) for (001) direction grown wafer as shown in Fig. 10.18. The cleavage produces high quality mirrors with reflectivity of ~ 0.3 to 0.4 depending upon the semiconductor. Typical lengths of the cavity range from 150 μm to 1 mm depending upon the application of the laser diode.

The Fabry-Perot cavity also has a certain lateral dimension, d_T, which determines the transverse modes of the light that is emitted. As a result, the output of the Fabry-Perot mode not only has a number of longitudinal modes present, but in general may also have several transverse modes present as shown in Fig. 10.19.

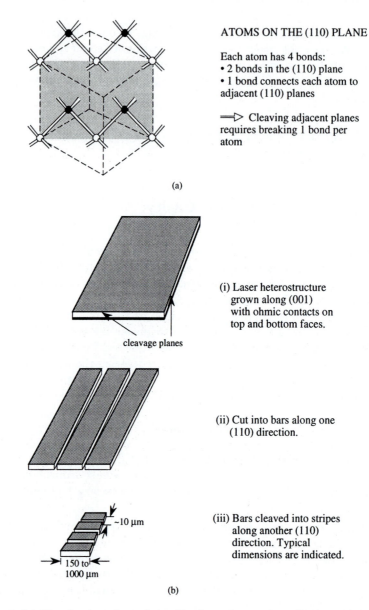

ATOMS ON THE (110) PLANE

Each atom has 4 bonds:
• 2 bonds in the (110) plane
• 1 bond connects each atom to adjacent (110) planes

⟹ Cleaving adjacent planes requires breaking 1 bond per atom

(a)

(i) Laser heterostructure grown along (001) with ohmic contacts on top and bottom faces.

cleavage planes

(ii) Cut into bars along one (110) direction.

~10 μm

150 to 1000 μm

(iii) Bars cleaved into stripes along another (110) direction. Typical dimensions are indicated.

(b)

Figure 10.18: (a) The cleaving plane of zinc-blende structures has adjacent planes connected by a single bond. (b) The approach used to produce a Fabry-Perot optical cavity involves cleaving a wafer containing the laser diode structure.

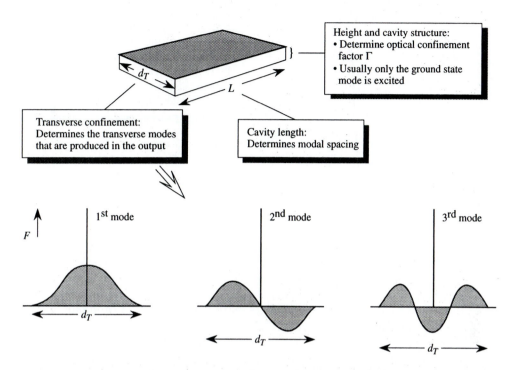

Figure 10.19: The various geometric parameters of a Fabry-Perot laser and their importance for the laser emission. A schematic of the various lateral modes are also shown.

As the laser injection current is increased, the relative strength of the various modes changes. The longitudinal mode which has the highest photon density is the one which has a photon wavelength closest to the peak in the gain spectra. Since the peak in the gain spectra shifts with carrier injection, the peak mode shifts. This is shown for a quantum well laser in a Fabry-Perot cavity in Fig. 10.10.

Additionally, if the transverse confinements of the optical wave is weak (i.e., d_T is large), there may be several transverse modes whose frequency is closely spaced. The laser output may involve different such modes as the injection current is increased. This results in a "kink" in the output power as shown in Fig. 10.20. To avoid the "kink" which produces noise in the optical transmitter, it is important to ensure a strong transverse confinement in the structure. This confinement can be achieved by two approaches: i) gain guided cavities, and ii) index guided cavities. We will now discuss these two approaches.

Gain Guided Cavities

The most common technique to confine the optical wave and to guide it along the cavity length is to fabricate a stripe geometry laser as shown in Fig. 10.21a. The semiconductor structure is covered with a thin layer of oxide (SiO_2) into which a thin strip of width

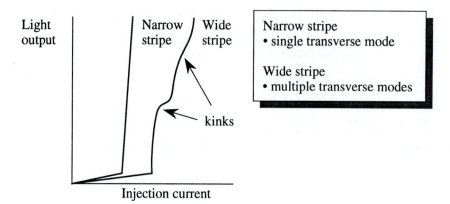

Figure 10.20: The shift in transverse modes participating in the optical output of a laser produce kinks in the light output-current curves.

$d_T = 2 - 10$ μm is etched. The contact is now made through this stripe as shown. This allows the injected current to be confined into a very narrow region. A similar effect is created in the ridge laser shown in Fig. 10.21b.

The current is injected through a narrow opening of width, d_T, as shown in Fig. 10.21c, but the current spreads out under the stripe as shown in Fig. 10.21d. This spreading is due to the diffusive nature of the current flow and is controllable by the device design. The carrier concentration under the stripe has a nonuniform behavior as shown in Fig. 10.21e. Since the refractive index of a material is dependent upon the carrier density, the index profile also becomes nonuniform as shown in Fig. 10.21f. Finally, the gain of the device has a nonuniform behavior shown schematically in Fig. 10.21g.

The important point to note from Fig. 10.21 is that not only is the real part of the refractive index nonuniform in the y-direction, the gain is highly nonuniform. In fact, the index step produced is very small and is not the main factor in determining the transverse mode behavior. The real part of the refractive index changes due to the carrier injection under the stripe and the change is such that an anti-guiding effect is produced. However, due to the strongly nonuniform gain profile, the gain produces a guiding effect.

It is difficult to produce single transverse mode operation in a gain guided laser unless one goes to a very narrow stripe (~ 2 μm). However, at such narrow stripes, the threshold current of the device increases significantly since a large fraction of the current spreads out and is wasted. For single transverse mode output index guided optical cavities are designed.

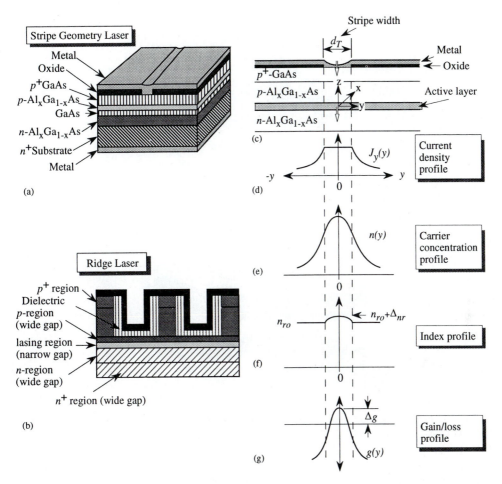

Figure 10.21: (a) The stripe geometry laser, (b) the ridge laser, (c) the current injection into the laser, (d) the current density profile, (e) the electron (hole) density profile in the active region, (f) The refractive index profile, and (g) the gain and loss profile.

Index Guided Cavities

Index guided cavities rely on a step in the index profile in the lateral direction. An example is shown in Fig. 10.22. Such a device is also called a buried heterostructure laser. The device fabrication is much more complex. To produce a lateral index step, one requires to first grow an epitaxial layer with a structure similar to the normal laser. The structure is then etched down leaving a few micron regions. Regrowth is then carried out to surround the active region by a large bandgap material.

The buried heterostructure laser, if fabricated correctly, does not suffer from kinks in the light-current curve. The output is single mode and the threshold current is very small. Of course, to take full advantage of the electronic properties, the active

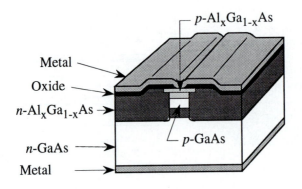

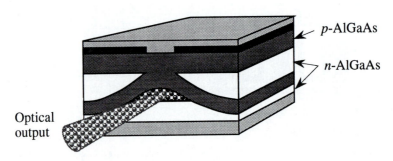

Figure 10.22: Index guided laser cavities. Etching and regrowth techniques are employed to produce buried active regions.

region must contain quantum well structures.

10.6.2 The Distributed Feedback Lasers

⤳ The Fabry-Perot cavity laser, although easy to fabricate, suffers from a number of drawbacks. Since a simple mirror is used to create the stationary states, there is no special preference given to a particular optical mode as far as the optical cavity is concerned. The determination as to which modes will dominate is left entirely up to the gain spectra determined by the electronic properties of the active region. Since the mode spacing is only 4-5 Å, and the gain spectra is rather flat on this scale, several modes end up lasing in the Fabry-Perot cavity. Of course, as discussed earlier, at high powers the side modes are suppressed relative to the peak mode. Nevertheless, at most operating conditions the laser linewidth is $\sim$ 20 Å for Fabry-Perot cavity, even though each mode is extremely narrow. The question naturally arises: Can the cavity itself

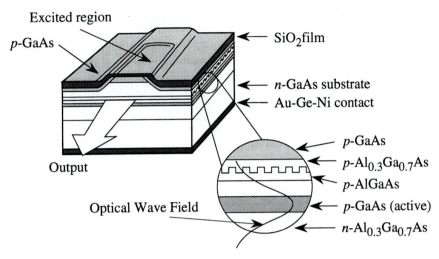

Figure 10.23: The distributed feedback structure incorporates a periodic grating in the laser structure. The confined optical wave senses the periodic grating as shown. (After K. Aiki, *IEEE Journal of Quantum Electronics*, **QE**-12, 601 (1976).)

provide mode selection? After all, in electronic circuits, resonant cavities can be designed with tremendous mode selectivity. Indeed, where would the microwave field be without resonant cavities? While, in semiconductor optoelectronics, one cannot design mode selective optical cavities with as much ease as in electronics, there are some solutions.

An important approach towards design of a mode selective optical cavity is the distributed feedback (DFB) structure. The DFB approach to create strong mode selectivity is based upon the propagation of waves in periodic structures. We know that in periodic structures, special effects occur when the wavelength of the wave approaches the wavelength of the periodic structure. In semiconductor crystals, this leads to bandgaps and Bragg reflections. Similar effects occur for optical waves, as we shall discuss below.

In the DFB structure, a periodic grating is incorporated into the laser structure as shown in Fig. 10.23. The fabrication process is by no means trivial and involves growing the basic laser structure, etching a periodic structure and regrowing the top layer. The grating should be as close as possible to the active region so that the optical wave interacts strongly with the grating. However, since the placement of the grating creates defects, the grating cannot be too close to the active region. The difficulty in the fabrication of the DFB lasers is manifested in a cost that is a thousand times greater than that of a Fabry-Perot laser.

The wave equation for the optical wave field is given in general by (we will focus on the z-direction propagation in the analysis given below)

$$\frac{\partial^2 F}{\partial z^2} + k^2 F = 0 \tag{10.60}$$

where k is determined by the complex refractive index N and the free space wavevector k_o as

$$k^2 = N^2 k_o^2 \tag{10.61}$$

The real part of the refractive index has a uniform part and a periodic part

$$n_{rr}(z) = n_{ro} + \sum_G n_G e^{iGz} \tag{10.62}$$

where

$$G = \frac{2\pi n}{a} \tag{10.63}$$

and a is the grating periodicity. Out of all the Fourier coefficients of the expansion, the dominant one will be the 1st order harmonic coefficient n_{G_1} corresponding to $G_1 = 2\pi/a$.

In addition to the real part of the refractive index, due to the gain in the structure, there is an imaginary part given by

$$n_{ri} = \frac{-g}{2k_o} \tag{10.64}$$

Substituting the real and imaginary parts of the refractive index for N^2, and making the approximations

$$\left(\frac{n_G}{n_{ro}}\right)^2 \sim 0$$

$$\left(\frac{g}{2k_o}\right)^2 \sim 0 \tag{10.65}$$

we get for the wave equation

$$\frac{\partial^2 F}{\partial z^2} + (\beta^2 + i\beta g)F = -k_o\beta \sum_G n_G e^{iGz} F \tag{10.66}$$

where

$$\beta = k_o n_{ro} \tag{10.67}$$

If the right hand side were zero, i.e., if there is no periodic variation of the refractive index, the solution of the wave equation is a wave with a wave vector β whose amplitude grows in space due to the gain term. However, due to the periodic variation, the general solution to the equation has a form

$$F(z) = \sum_{\alpha} F_{\alpha o} exp(i\alpha z) \tag{10.68}$$

where $F_{\alpha o}(z)$ is an (slowly varying) amplitude describing the propagating wave.

We are interested in the case where the wavevector is close to the half Bragg wavelength (this the 1st order Bragg reflection point):

$$\beta_B = \frac{2\pi n_{ro}}{\lambda_o} \tag{10.69}$$

The field at the Bragg zone edge consists of two counter propagating waves given by ($\alpha = \beta_B$ and $\alpha = -\beta_B$ in Eqn. 10.68)

$$F(z) = F_+(z)exp(i\beta_B z) + F_-(z)exp(-i\beta_B z) \tag{10.70}$$

with essentially little mixture from other terms of Eqn. 10.68. Substituting this solution in the wave equation (Eqn. 10.66) and comparing terms of equal phase, we get two coupled equations (we ignore second derivative terms of $F_+(z)$ and $F_-(z)$)

$$\frac{\partial F_+}{\partial z} + (\frac{g}{2} - i\delta)F_+ = iK_C F_-$$

$$\frac{-\partial F_-}{\partial z} + (\frac{g}{2} - i\delta)F_- = iK_C F_+ \tag{10.71}$$

where δ defines the shift in frequency from the Bragg reflection frequency

$$\delta = \frac{\beta^2 - \beta_B^2}{2\beta_B} \cong \beta - \beta_B \equiv \frac{n_{ro}}{c}(\omega - \omega_B) \tag{10.72}$$

and K_C defines the coupling factor for the 1st order refractive index of the periodic grating $n_{G_1} = \Delta n_{r_1}$

$$K_C = \frac{\pi \Delta n_{r_1}}{2\lambda o} \tag{10.73}$$

The system of the two coupled equations now describes the propagation of waves in the DFB structure. As shown in Fig. 10.24a, the wave F_+ and F_- build up from each edge of the structure. The boundary conditions for the waves are

$$F_-\left(\frac{L}{2}\right) = F_+\left(-\frac{L}{2}\right) = 0 \tag{10.74}$$

To solve the coupled mode equations we assume a general form for the wave amplitudes

$$F_- = C_1 e^{\gamma z} + C_2 e^{-\gamma z}$$

$$F_+ = D_1 e^{\gamma z} + D_2 e^{-\gamma z} \tag{10.75}$$

From the symmetry of the problem we can note that either

$$F_+(z) = -F_-(-z)$$

or

$$F_+(z) = F_-(-z) \tag{10.76}$$

Thus we have the impositions

$$C_1 = \pm D_2, C_2 = \pm D_1 \tag{10.77}$$

Also from the boundary conditions,

$$\frac{C_1}{C_2} = \frac{D_2}{D_1} = -e^{\gamma L} \tag{10.78}$$

This yields the following form for F_- and F_+

$$
\begin{aligned}
F_-(z) &= sinh\gamma\left(z + \frac{L}{2}\right) \\
F_+(z) &= \pm sinh\gamma\left(z - \frac{L}{2}\right)
\end{aligned}
\tag{10.79}
$$

Substituting the spatial dependence in the coupled equations we get from Eqn. 10.71

$$
\begin{aligned}
-\gamma sinh\frac{\gamma L}{2} + \left(\frac{g}{2} - i\delta\right)cosh\frac{\gamma L}{2} &= \pm\, i\, K_C\, cosh\, \frac{\gamma L}{2} \\
-\gamma cosh\frac{\gamma L}{2} + \left(\frac{g}{2} - i\delta\right)sinh\frac{\gamma L}{2} &= \mp\, i\, K_C\, sinh\, \frac{\gamma L}{2}
\end{aligned}
\tag{10.80}
$$

Equivalently, taking the difference and sum of the two equations, we get

$$
\begin{aligned}
\gamma + \left(\frac{g}{2} - i\delta\right) &= \pm\, i\, K_C\, e^{\gamma L} \\
\gamma - \left(\frac{g}{2} - i\delta\right) &= \mp\, i\, K_C\, e^{-\gamma L}
\end{aligned}
\tag{10.81}
$$

Multiplying these equations we get

$$\gamma^2 = K_C^2 + \left(\frac{g}{2} - i\delta\right)^2 \tag{10.82}$$

Remember that in this solution γ represents the rate at which the optical waves build up as they propagate, δ is the normalized separation of the optical frequency from

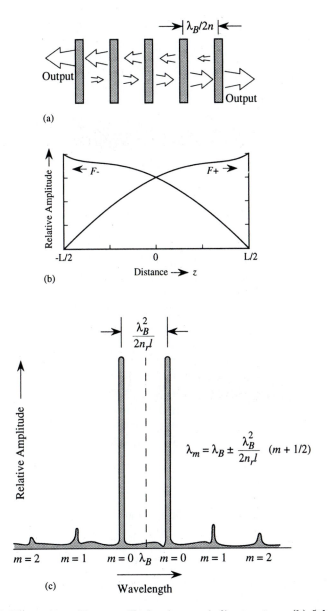

Figure 10.24: (a) Illustration of laser oscillation in a periodic structure; (b) field amplitudes of a left-traveling wave F_- and a right-traveling wave F_+ versus distance; and (c) optical output vesus wavelength for a DFB laser.

the Bragg frequency, and g is the gain in the cavity. We are interested in developing an understanding of the threshold conditions for various lasing modes. In general, we need a computer to solve for this information, but if we assumed that the gain is large, i.e., $g >> K_C$, we can write from Eqn. 10.82

$$\gamma \simeq \frac{g}{2} - i\delta \tag{10.83}$$

and from Eqn. 10.81, we get

$$2\left(\frac{g}{2} - i\delta\right) \simeq \pm i \ K_C \ exp\left(\frac{g}{2} - i\delta\right) L \tag{10.84}$$

Multiplying this by its complex conjugate, we get the condition

$$\frac{K_C^2}{4\left(\frac{g^2}{4} + \delta^2\right)} = exp(-gL) \tag{10.85}$$

This equation represents the condition for laser oscillation in the DFB structure. Comparing this equation with Eqn. 10.21 for the Fabry Perot structure, we note that the term $(\Gamma_g - a_{loss})$ of the Fabry-Perot laser corresponds to g in this equation and the mirror reflectivity corresponds to an effective reflectivity

$$\frac{K_C^2}{4\left(\frac{g^2}{4} + \delta^2\right)} \tag{10.86}$$

The important point to note is that the effective reflectivity decreases sharply as δ^2 increases, i.e., as one moves away from the Bragg condition. The allowed modes of the system are obtained by equating the phase of the terms in Eqn. 10.84. This gives

$$\left(m + \frac{1}{2}\right)\pi + tan^{-1}\frac{2\delta}{g} = \delta L \tag{10.87}$$

where m = 0, $\pm$ 1 .... For small values of δ, we get (near the Bragg condition)

$$\delta = \frac{1}{L}\left(m + \frac{1}{2}\right)\pi \tag{10.88}$$

and the frequency of the mode is

$$\nu = \nu_B \pm \frac{c}{2n_{ro}L}\left(m + \frac{1}{2}\right) \tag{10.89}$$

There are no allowed waves at the Bragg frequency ν_B. Around the Bragg condition, the modes are arranged symmetrically (in our simple formalism). *The mode spacing is the same as in a Fabry-Perot laser, but the gain needed for laser oscillation is lowest for the m=0 modes, and increases rapidly for the other modes as shown in Fig. 10.24c. As a result, the side modes are severely suppressed in the DFB laser.*

For an optimum design, the gain peak of the active region of the laser must coincide in energy (or frequency) with the main mode of the DFB structure. This would produce a single mode output. It may be noted that higher order harmonics of the grating would similarly produce a strong mode selective laser. For short wavelength lases, the 2nd or 3rd order reflection is used since it is difficult to fabricate very fine gratings. For longer wavelength lasers (1.55 μm) the 1st order effect is usually exploited.

The DFB laser is the device of choice for high performance communication systems. As discussed in the next chapter, the single mode output of this laser allows one to reach linewidths of a few MHz without the need of external feedback. In principle, such lasers can be used for coherent detection applications.

EXAMPLE 10.9 Consider a 1000 μm DFB laser tuned to a 1.55 μm light emission. The effective (i.e., weighted with the optical intensity) first order periodic variation of the refractive index is $\Delta n_{r1} = 0.001$. Calculate the effective reflectivity for the main mode and the first side mode at lasing. Also calculate the gain needed for these two modes to lase.

For this device we have

$$K_c = \frac{\pi \Delta n_{r1}}{2\lambda_o} = 10.13 \ cm^{-1}$$

$$\delta(m = 0) = \frac{\pi}{(1000 \times 10^{-4} \ cm) \times 2} = 15.7 \ cm^{-1}$$

$$\delta(m = 1) = 3 \times \delta(m = 0) = 47.1 \ cm^{-1}$$

The effective reflectivity depends upon the gain in the device as can be seen from Eqn. 10.86. For the $m = 0$ mode to lase, we find that the cavity gain is ~ 28 cm^{-1}. The effective reflectivity is then

$$R(m = 0) = 5.8\%$$
$$R(m = 1) = 1.06\%$$

Thus a strong difference exists for the two adjacent modes. This relative difference is created by the small periodic perturbations in the structure and not due to any external mirrors. The gain needed to lase in the $m = 0$ and $m = 1$ states can be obtained from the Eqns. 10.85 and 10.87

$$g(m = 0) \cong 28 \ cm^{-1}$$
$$g(m = 1) \cong 47 \ cm^{-1}$$

Thus, the DFB structure produces a strong enhancement in the gain needed if the side mode at $n = 1$ is to lase in the structure. The Fabry Perot cavity does not produce any such selection.

10.6.3   The Surface Emitting Laser

⤳ In the discussion of semiconductor lasers, so far, we have discussed "edge emitting" lasers where the laser output comes out from the edge and is parallel to the substrate on which the device is grown. As noted earlier the lasing condition has the form

$$(\Gamma g_H - \alpha_c) = -\frac{1}{L}\ell_n R \tag{10.90}$$

where L is the cavity length. Usually R has a value of ~ 0.3, so that to ensure a reasonably low threshold, the laser length L has to be $\gtrsim 100$ μm. Thus the edge emitting laser is a fairly large device on the scale of other microelectronic devices such as transistors. Additionally, it is difficult to produce a high density array of edge emitting lasers on a single chip. While this may not be a serious problem for the usual optical communication applications, it places a severe limitation on the use of lasers for high density optical interconnects for computer chips.

The surface emitting laser (SEL) or the vertical cavity surface emitting laser (VCSEL) overcomes the problems outlined above. These lasers can be produced in very high density on a single wafer and the light output emerges perpendicular to the wafer. The devices can also be made very small so that threshold currents can be in the range of sub-milliamperes.

In the SEL, the mirrors are placed not at the edges of the laser, but at the top and the bottom of the device. However, since it is difficult to epitaxially grow very thick layers, the thickness of the cavity is limited to ~ 10 μm. If the reflectivity is maintained at around 0.3, the reflection losses will approach 10^3 cm^{-1} or higher. This would cause the threshold current to be extremely high. This problem is overcome by developing mirrors with very high reflection coefficient. For example, if the reflectivity of the mirrors can be made to be 0.99, the mirror loss becomes only 10 cm^{-1} which is quite acceptable for low threshold current density lasers. Thus the success of the SEL depends upon incorporating very high reflectivity mirrors in the structure.

In the previous section we have discussed how a periodic structure can be used to create an effective reflectivity for light with a wavelength that matches the spatial periodicity. This concept of distributed Bragg reflectors (DBRs) is used in the SEL to produce high quality mirrors. The DBR provides the very high reflectivity needed to compensate for the short cavity length. A typical DBR structure is shown in Fig. 10.25a. The active region of the SEL is essentially similar to that of a Fabry Perot cavity. However, the cladding layers are chosen as DBRs which consist of a periodic arrangement of layers with thicknesses d_1 and d_2 and refractive indices n_{r1} and n_{r2} as shown in Fig. 10.25a. The periodicity is chosen so that

$$n_{r1}d_1 + n_{r2}d_2 = \lambda/2 \tag{10.91}$$

which corresponds to the Bragg reflection condition for an optical beam with wavelength

$\lambda = \frac{\lambda_o}{\bar{n}_r}$ where $\bar{n}_r$ is the effective refractive index. The wavelength λ is chosen to coincide with the peak in the gain curve.

The DBR structure can be formed from crystalline materials like GaAs/AlAs or from amorphous materials such as Si/SiO_2 as shown in Figs. 10.25b and 10.25c. As seen in our discussion of the DFB structure, the reflectivity of a DBR structure is highly dispersive. The reflectivity is very high at the Bragg condition and drops off strongly when the wavelength is different from the Bragg condition wavelength. A schematic of the reflectivity is shown in Fig. 10.25a.

To carry out an accurate analysis of the SEL, the reflectivity of the DBRs has to be carefully calculated. Numerous methods have been developed and the interested reader is referred to the references at the end of this chapter. For wide SEL with device lateral diameters greater than $\sim 10~\lambda$, the reflectivity is given by

$$R = \left(\frac{1 - \left(\frac{n_{r1}}{n_{r2}} \right)^{2N}}{1 + \left(\frac{n_{r1}}{n_{r2}} \right)^{2N}} \right)^{2} \tag{10.92}$$

where n_{r1} is smaller than n_{r2} and we have chosen $d_1 = d_2$. Here N is the number of periods used in the DBR. The reflectivity can easily reach 99% if a proper combination of N and n_{r1}/n_{r2} is used. The reflectivity is somewhat lower if the diameter of the cavity starts to approach a few times λ.

Laser arrays can be fabricated using the SEL concept and such arrays have been shown to have output optical power approaching 1 watt. However, the SEL structure does face some difficulties, especially when compared to the edge emitting device. These difficulties arise from the following considerations:

i) Charge injection: The presence of the DBR mirrors causes considerable difficulty in charge injection into the SEL. Great care has to be taken in developing a low resistance path to the active region. If the charge is to be injected through the DBR, this structure is to be doped. The thick DBR offers significant resistance to the current flow. If ring contacts are to be used to inject charge directly into the active region, by passing the DBR, current crowding effects become significant. In this case the edges of the device carry the current while the center has very low current flow. Improvements in SEL technology are likely to come from innovative techniques for DBR fabrication and charge injection.

ii) Device heating: The high resistance in the SEL current path causes considerable device heating. As a result, it becomes difficult to operate the device at high drive current. At high drive current, device heating causes loss in the device efficiency due to current leakage and Auger recombination (for low bandgap materials) and the optical output tends to saturate.

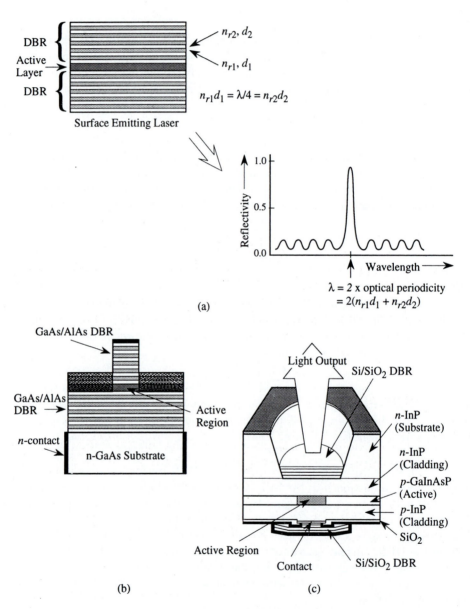

Figure 10.25: a) A schematic of the placement of DBR reflectors used in a surface emitting laser. A schematic of the reflectivity of a DBR stack is shown. (b) A typical structure using crystalline (GaAs/AlAs) DBR. (c) A structure using amorphous DBR stacks. (Figures such as ones shown can be seen in the special issue of *IEEE J. Quantum Electronics*, **QE**-29 (1993).)

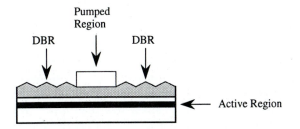

Distributed Bragg Reflector Structure

Figure 10.26: The DBR laser, where a Bragg reflector is placed outside the active region. The optical periodicity of the DBR ($n_{r1}d$) can be altered electronically.

We have mentioned above the use of SELs in laser arrays. The arrays are important not only for inter-chip interconnects, but for high power optical devices. It is difficult to make a single laser structure with large enough area to give, say, 1 watt of optical power. Structural non-uniformities and technology limitations make it difficult for a single laser to emit more than ~ 10 mW of power. However, with laser arrays the power level can approach watts. While the output power of laser arrays can be high, it is difficult to control the relative phase of the different lasers on an array. Considerable research is still ongoing to develop techniques to control the phase of the individual optical emitters.

10.6.4 The DBR Laser

$\mathcal{R}$ In the periodic structure based cavities discussed so far, the periodicity of the structure is fixed. As a result, the wavelength of emission that is given selective preference is also fixed. *In the DBR laser, the periodic reflecting stack is placed outside the active lasing region. Also, the refractive index of the stack is alterable by, say, current injection.* In Fig. 10.26 we show a typical DBR laser. The DBR is responsible for providing a reflection to the light emitted from the active region. The wavelengths that get the highest feedback must satisfy

$$\lambda_B = 2qa \qquad (10.93)$$

where q is a positive integer and a is the optical periodicity of the structure

$$a = n_{r1}d_1 + n_{r2}d_2 \qquad (10.94)$$

The values of n_{r1} and n_{r2} can be altered electronically and, therefore, the laser can have a certain degree of wavelength tunability (say ~ 30 Å).

10.6.5 Cleaved Coupled Cavity Laser

$\mathcal{R}$ Another important structure satisfying the need for spectrally pure and tunable laser output is the cleaved coupled cavity (C^3) laser. A schematic of the C^3 laser is shown in Fig. 10.27a and consists of two independent gain or index guided laser diodes optically coupled with each other by a space less than 5 μm. Each laser diode has a Fabry-Perot cavity with slightly different length so that the longitudinal modes of each laser are not exactly overlapping as shown in Fig. 10.27b.

To fabricate the C^3 laser, a regular Fabry-Perot laser is formed through cleaving. This laser is covered with a thin Au layer and recleaved near the middle. The Au film keeps the two parts connected and the two diodes can now be mounted on a Cu heat sink with a small separation between them.

The C^3 laser can be operated in a tuning or nontuning mode. One of the lasers is operated above threshold while the second one is kept below threshold for creating a tunable light source. The longitudinal modes of the second laser are controlled by carrier injection which alters the refractive index of the laser and thus the wavelength (free space) of the photons that get feedback. If n_{r1} and n_{r2} are the refractive indices of the two lasers and L_1 and L_2 their lengths, we have for the spacing of the modes

$$
\begin{aligned}
\Delta\lambda_1 &= \frac{\lambda_o^2}{2n_{r1}L_1} \\
\Delta\lambda_2 &= \frac{\lambda_o^2}{2n_{r2}L_2}
\end{aligned}
\tag{10.95}
$$

As shown in Fig. 10.27b, the modes that coincide will get a constructive feedback while the others will be suppressed. The spectral spacing of the C^3 laser is then

$$
\Delta\lambda_C = \frac{\Delta\lambda_1\Delta\lambda_2}{|\Delta\lambda_1 - \Delta\lambda_2|} = \frac{\lambda_o^2}{2\,|n_{r1}L_1 - n_{r2}L_2|}
\tag{10.96}
$$

Thus the mode spacing is much larger than that of a single laser and only one allowed mode falls in the energy range where the gain is high. As a result, a single mode emission is observed. By changing the value of n_{r2} by current injection, the peak mode can shift by one Fabry-Perot mode of the lasing diode. A tunability of $\sim$ 150 Åcan be achieved by this approach.

10.6.6 The Microcavity Laser—Thresholdless Laser?

$\mathcal{R}$ We have seen in the previous section, how by using quantum wells, quantum wires and quantum dots, the electronic density of states can be altered. The obvious question

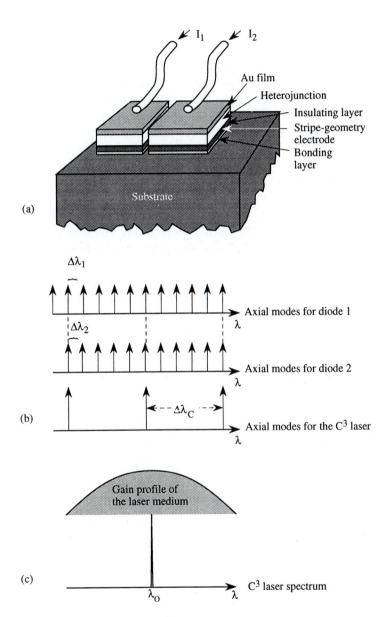

Figure 10.27: (a) A schematic of the C^3 laser. (b) The longitudinal modes of the two diodes and the output spectrum of the laser.

is: Can the same thing be done to photons? Of course, one has to know if there are any advantages to tailoring the photon density of states. As we will see from the following discussion, there are, indeed, many advantages that can be reaped if the photon density can be tuned.

In Fig. 10.28a we show the gain profile of a laser along with the density of photon states in a usual laser cavity. Note that the density of photons function includes not only the lasing modes but all other modes into which a photon can be emitted. Most of the photons emitted by *e-h* recombination do not go into the lasing mode but are wasted. As a result, the laser does not start to emit coherent photons (i.e., lase) until a fairly large current is injected. In the ideal case, there should be only one possible photon mode so that the moment emission occurs, the photons build up in the cavity and the device lases. This would create the "thresholdless" laser.

A number of schemes have been suggested to produce the single photon mode cavity. A straightforward possibility is to simply shrink the lateral dimensions of a surface emitting laser structure. In Fig. 10.28b we show the photon density of states for a 0.5 μm×0.5 μm SEL with DBRs with 20 periods each. As can be seen, the density of states does not quite go to the ideal case since there are a lot of "leaky modes" in the system. However, there is a significant improvement over the bulk laser (dimensions > 4.0 μm). A strong reduction in threshold current density is expected for this structure.

10.7   TEMPERATURE DEPENDENCE OF LASER OUTPUT

$\longrightarrow$ As in the case of the LED, the temperature dependence of the laser diode output is extremely important for most applications. We will see in the next chapter that for high speed applications the laser has to be driven at a very high input power. As a result, the device gets heated even with good heat sinking. Issues of importance in the study of the temperature dependence are: i) Effect of temperature on the threshold current and the optical intensity; ii) Effect of temperature on the frequency of emission. We will briefly discuss these issues.

10.7.1   Temperature Dependence of the Threshold Current

$\longrightarrow$ As the temperature of the laser diode increases, the threshold current increases and at a given injection, the photon output decreases. One can ascribe three reasons for this to occur:

i) The increase in temperature causes the quasi-Fermi functions f_e and f_h to

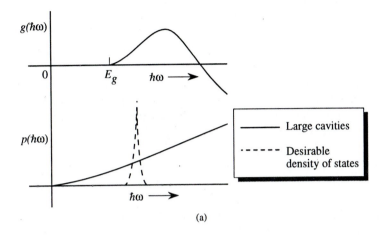

(a)

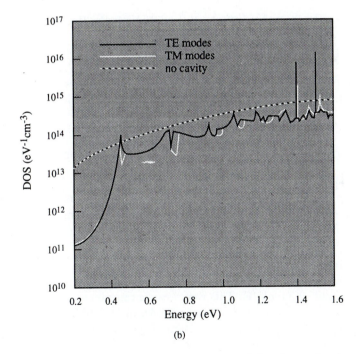

(b)

Figure 10.28: (a) A schematic of the gain profile in a laser and the photon density of states. The ideal desirable density of states of a cavity. (b) Calculated density of states in a 0.5 μm×0.5 μm GaAs/AlGaAs microcavity.(After I. Vurgaftman and J. Singh, *Applied Physics Letters*, 64, 1472 (1994).)

"smear" out more. As a result, the condition for inversion $f_e + f_h > 1$ requires a higher injection carrier density to be fulfilled. This increases the threshold current. This effect arises in all semiconductor lasers.

ii) The increase in the temperature causes the electrons and holes distribution to be spread out into higher energies. As a result, a greater fraction of the injected charge can cross over the active region and end up in the cladding or contact region. This leakage current wad described for the LED in the previous chapter. The leakage current depends upon the design of the laser and can be suppressed by using a wide active region or a GRIN structure for quantum well lasers.

iii) Increase in temperature causes more electrons and holes to possess energies greater than the threshold energy needed for Auger recombination. This, coupled with the increase in the threshold carrier density, causes the Auger recombination to increase exponentially with temperature. The Auger processes are particularly important for narrow gap materials.

As a result of the three effects discussed above the threshold current density of a laser can, in general, be described by the form

$$J_{th}(T) = J_{th}^o exp\,(T/T_o) \tag{10.97}$$

A large value of T_o is desirable. Typical values in GaAs lasers are ~ 120 K. For long wavelength lasers ($\lambda = 1.55$ μm), T_o values are smaller (~ 50 K).

10.7.2   Temperature Dependence of the Emission Frequency

$\longrightarrow$ For many applications it is important that the laser frequency remain stable over the operating conditions. If the temperature changes, the emission frequency shifts. There are two effects that control the shift of the laser frequency:

i) Change in the semiconductor bandgap causes the entire gain spectrum to shift to lower energies as the temperature increases. The bandgap change in most semiconductors is about -0.5 meV/K. This would cause the gain spectra to change by ~ 3 to 4Å/K if there were no additional factors as shown in Fig. 10.29a. However, in the laser the emission depends not just on the position of the gain peak, but on the Fabry-Perot mode that is closest to the gain peak. This brings us to the second effect.

ii) As temperature changes, the thermal expansion of the laser cavity and the change in the refractive index alters the position of the resonant modes. The resonant modes are given by (q is an integer)

$$q\lambda_q = 2L; \lambda_q = \frac{\lambda_{qo}}{n_r} \tag{10.98}$$

If the effective cavity length increases due to the temperature, the positions of the modes will shift relative to the gain spectra which is itself shifting with temperature. This is shown schematically in Fig. 10.29b. For most semiconductors, the overall effect is to produce a shift in the resonant wavelengths of about 1 Å K^{-1}.

As a result of the two effects, the emission wavelength of a Fabry-Perot laser has a behavior schematically shown in Fig. 10.29b. The emission wavelength shifts by $\sim$ 1 Å K^{-1} until an adjacent mode becomes closer to the gain peak. Thus mode hopping occurs. However, in the case of a DFB laser where the lasing mode is not selected by the gain peak but by the spacing of the grating, the mode hopping does not occur and the lasing wavelength shifts by only about 1 Å K^{-1} over a wide temperature range as shown in Fig. 10.29b.

10.8 CHAPTER SUMMARY

$\mathcal{R}$ In this chapter we have discussed light emitting devices using semiconductors. The summary tables (Tables 10.1-10.3) highlight the important areas discussed in this chapter.

10.9 PROBLEMS

Section 10.3

10.1 Consider a GaAs laser with a cavity length of 75 μm. Calculate the number of allowed longitudinal modes in an energy width $E_g \pm \frac{1}{2}k_B T$ where $E_g = 1.43$ eV and $T = 300$ K.

10.2 Consider two optical cavities, one a Fabry-Perot cavity with a length of 100 μm and a mirror reflectivity of 0.33 and another with a length 10 μm. What should be the mirror reflectivity of the second cavity to ensure that a photon emitted inside spends the same time in the two cavities before escaping? Cavity loss due to absorption is 10 cm^{-1}.

10.3 Consider a Fabry-Perot cavity of length 100 μm. The mirror reflectivity is 0.33 and the absorption loss in the cavity is 20 cm^{-1}. Calculate the photon lifetime τ_{ph}. The refractive index in the cavity is 3.6.

Section 10.4

10.4 Consider the semiconductor alloy InGaAsP with a bandgap of 0.8 eV. The electron and hole masses are 0.04 m_0 and 0.35 m_0, respectively. Calculate the injected electron and hole densities needed at 300 K to cause inversion for the electrons and holes at the bandedge energies. How does the injected density change if the temperature is 77 K?

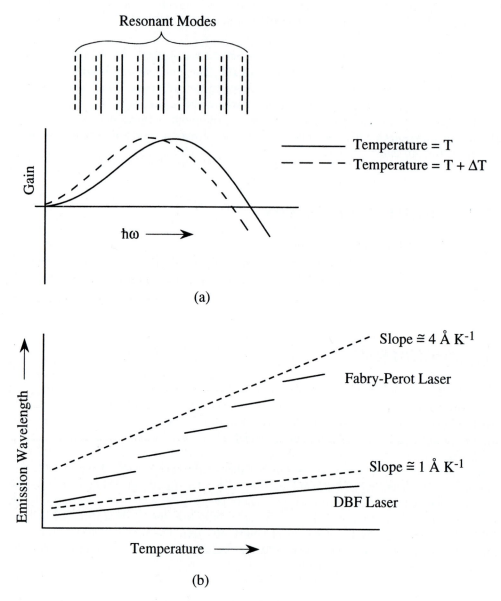

Figure 10.29: (a) The shift of the gain spectra and the resonant modes of a cavity with temperature. (b) Shift in the emission wavelength of a laser with temperature.

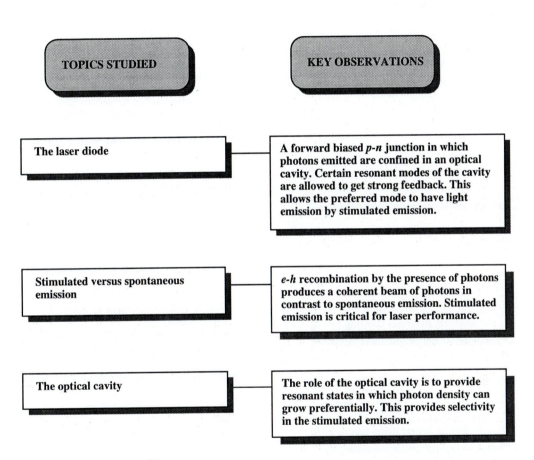

Table 10.1: Summary table.

TOPICS STUDIED	KEY OBSERVATIONS
Gain and threshold in a laser	In the presence of carrier injection, some energies photon emission coefficients minus the absorption coefficient (= gain) can be positive. If the gain times optical confinement in the active region exceeds the cavity loss, stimulated emission or lasing starts.
The laser below and above threshold	Below threshold, the emission of photons is by spontaneous emission—incoherent photons are emitted over a broad spectral range. Above threshold, some modes start to dominate the photon output. The spectra becomes sharp and coherent.
Laser structures for superior performance: electronic tailoring	Laser threshold current is reduced by using narrow active regions, quantum wells, strained quantum wells, and quantum wires. Essentially, the improvements results from improvements in inversion due to the density of states.
Laser structures for superior performance: optical cavity tailoring	Laser performance is improved by going from gain guided cavities to index guided cavities. The use of DFB lasers allows suppression of side modes. Surface emitting lasers are useful for vertical emission. DBR and C^3 cavities are excellent tunable output. Microcavity lasers and other "photonic bandgap" lasers promise threshold-less lasing.

Table 10.2: Summary table.

Use the Joyce-Dixon approximation.

10.5 Consider a GaAs based laser at 300 K. Calculate the injection density required at which the inversion condition is satisfied at i) the bandedges; ii) at an energy of $\hbar\omega = E_g + k_B T$. Use the Joyce-Dixon approximation.

10.6 Consider a GaAs based laser at 300 K. A gain of 30 cm^{-1} is needed to overcome cavity losses at an energy of $\hbar\omega = E_g + 0.026$ eV. Calculate the injection density required. Also, calculate the injection density if the laser is to operate at 400 K.

10.7 Consider the laser of the previous problem. If the time for e-h recombination is 2.0 ns at threshold, calculate the threshold current density at 300 K and 400 K. The active layer thickness is 2.0 μm and the optical confinement is unity.

10.8 Two GaAs/AlGaAs double heterostructure lasers are fabricated with active region thicknesses of 2.0 μm and 0.5 μm. The optical confinement factors are 1.0 and 0.8, respectively. The carrier injection density needed to cause lasing is 1.0×10^{18} cm^{-3} in the first laser and 1.1×10^{18} cm^{-3} in the second one. The radiative recombination times are 1.5 ns. Calculate the threshold current densities for the two lasers.

10.9 Consider a laser in which the carriers masses could be tuned. Assume that the hole density of states mass is 0.5 m_0 while the electron density of states mass changes from 0.02 m_0 to 0.2 m_0. Calculate and plot the transparency density needed at an energy of $E_g + k_B T$ where E_g is 1.4 eV. The temperature is 300 K.

10.10 In the previous problem, the electron mass was tuned. Consider now the case where the electron mass is fixed at 0.067 m_0 and the hole mass varies between 0.1 m_0 and 0.5 m_0. Calculate the corresponding transparency density.

10.11 Calculate the temperature dependence of the transparency condition injection density from 77 K to 400 K. Use the energy $\hbar\omega = E_g + k_B T$ to define the point where $f^e + f^h = 1$. The electron mass is 0.067 m_0 and the hole mass is 0.45 m_0.

10.12 Consider a 1.55 μm laser operating at 77 K and 300 K. A cavity loss of 40 cm^{-1} exists in this laser. The other laser parameters are:

$$\Gamma = 0.1$$
$$m_e^* = 0.04 \, m_0$$
$$m_h^* = 0.4 \, m_0.$$

Auger coefficient

$$F = 10^{-28} \, cm^6 \, s^{-1} \, at \, 300 \, K$$
$$F = 10^{-30} \, cm^6 \, s^{-1} \, at \, 77 \, K$$
$$d_{las} = 0.1 \, \mu m$$

Calculate the threshold carrier density needed at 77 K and 300 K. If the radiative lifetime at threshold is 2ns calculate the radiative and non-radiative current densities at threshold.

10.13 Consider the previous problem. Repeat the problem if the active region is doped p-type at 8×10^{17} cm^{-3}. Assume that the Auger process involves CHHS transitions.

10.14 Consider a GaAs/Al$_{0.3}$Ga$_{0.7}$As laser with an active region thickness of 0.1 μm at 300 K. The active region is doped p-type at 1.0×10^{18} cm^{-3}. The cavity loss is 50 cm^{-1} and $\Gamma = 0.12$. Calculate the transparency density for the bandgap energy and the threshold density for this device.

10.15 Consider the device of the previous problem. Calculate the threshold current density if the radiative lifetime at threshold is 2.4 ns. What would the threshold current density be if the device active region was undoped and the lifetime was unchanged? Consider only the radiative current.

10.16 Consider a GaAs/Al$_{0.3}$Ga$_{0.7}$As laser with an active region thickness of 0.1 μm at 300 K. Calculate the gain curves versus energy for injection densities of 0; 8.0×10^{17} cm^{-3}; 1.0×10^{18} cm^{-3} and 1.30×10^{18} cm^{-3}.

10.17 Consider a laser with an active region bandgap of 0.8 eV and a thickness of 0.1 μm with $\Gamma = 0.1$. Calculate the threshold current for the device at 300 K. The device is described by the following parameters:

Spontaneous emission lifetime at threshold		$= 3.0\ ns$
Laser length		$= 300\ \mu m$
Laser width		$= 10\ \mu m$
Auger coefficient	F	$= 10^{-28}\ cm^6\ s^{-1}$
electron mass	m_e^*	$= 0.04\ m_0$
hole mass	m_h^*	$= 0.4\ m_0$
cavity loss	α_c	$= 40\ cm^{-1}$

10.18 Consider a laser emitting at 1.3 μm with a threshold current of 1 mA. The laser is biased at 10 mA so that you can assume that all the photons are being emitted in a single mode. Calculate the photon density in the peak mode in the cavity which has a dimension 5 μm$\times$150 μm and a loss $\alpha_c = 40$ cm^{-1}. What is the output power of the device at this bias? The cavity refractive index is 3.3.

10.19 Consider a GaAs/Al$_{0.2}$Ga$_{0.4}$As laser of active layer thickness of 0.2 μm and $\Gamma = 0.2$. Calculate the ratio between the photon densities in the peak mode and the next closest mode when the device is biased at 1.5 J_{th} and 5.0 J_{th}. The device has the following parameters:

$$m_e^* = 0.067\ m_0$$
$$m_h^* = 0.5\ m_0$$
$$\beta = 10^{-5}$$
$$\tau_r = 2.4\ ns \text{ at threshold density}$$
$$\alpha_c = 50\ cm^{-1}$$
$$\text{cavity length } L = 200\ \mu m$$

Section 10.5

10.20 Show that for the allowed quantum well transitions, the gain can be written as follows for $E_p = 22\ eV, n_r = 3.6$, (W is in Å; $\hbar\omega$ in eV):

TE polarized: HH $\rightarrow$ C-band

$$g(\hbar\omega) = 1.44 \times 10^5 \left(\frac{m_r^*}{m_0}\right) \left(\frac{1}{\hbar\omega}\right) \left(\frac{100}{W}\right) \left[f_e(E^e(k)) + f_h(E^h) - 1\right] \ cm^{-1}$$

The values for the gain of the other transitions can be obtained from Eqn. 10.48.

10.21 Calculate the carrier density needed to reach a peak TE gain of 1000 cm^{-1} in a 100 Å GaAs/AlGaAs quantum well at 300 K. Assume the following parameters: ($m_e^* = 0.067 \ m_0; m_{HH}^* = 0.45 \ m_0; m_{LH}^* = 0.45 \ m_0$):

$$
\begin{aligned}
E_{e1} - E_c(GaAs) &= 33 \ meV \\
E_{e2} - E_c(GaAs) &= 120 \ meV \\
E_v(GaAs) - E_{HH1} &= 8 \ meV \\
E_v(GaAs) - E_{LHI} &= 20 \ meV
\end{aligned}
$$

Use a Gaussian broadening function of halfwidth 1.3 meV. What is the peak TM gain at the same injection?

10.22 Consider a 80 Å GaAs/Al$_{0.3}$Ga$_{0.7}$As quantum well laser with $\Gamma = 0.02$. Calculate the threshold carrier density and threshold current for the following device parameters.

$$
\begin{aligned}
E_{e1} - E_c(GaAs) &= 40 \ meV \\
E_v(GaAs) - E_{HH1} &= 12 \ meV \\
E_v(GaAs) - E_{LH} &= 20 \ meV \\
m_e^* &= 0.067 \ m_0 \\
m_{HH}^* &= 0.45 \ m_0 \\
m_{LH}^* &= 0.4 \ m_0 \\
\alpha_c &= 50 \ cm^{-1}
\end{aligned}
$$

The radiative lifetime is 2.5 ns. Assume a halfwidth of 1.0 meV for a Gaussian broadening for the gain.

10.23 Calculate the gain and dg/dn in a 100 Å GaAs/Al$_{0.3}$Ga$_{0.7}$As quantum well and compare it to the case of a "bulk" GaAs/AlGaAs quantum well. Plot the results versus $\hbar\omega$ for injection densities of 8×10^{17} cm$^{-3}; 10^{18}$ cm$^{-3}; 2 \times 10^{18}$ cm$^{-3}; 3 \times 10^{18}$ cm^{-3} and 4×10^{18} cm^{-3}. (You can convert the 3D density to a 2D density by multiplying by the well width).

10.24 Design an InGaAs 80 Å quantum well laser which would emit at 1.55 μm. The laser has to be grown on an InP substrate. If the system has any strain include the strain effects using the deformation potentials from the Appendix B.

10.25 Design a 100 Å quantum well semiconductor laser grown on an InP substrate which has an almost equal mixture of the TE and TM polarized light. Calculate the effective bandgap of the structure. Assume that $m_e^* = 0.067 \ m_0; m_{hh}^* = 0.045 \ m_0$; and $m_{\ell h}^* = 0.01 \ m_0$.

10.26 Design the composition of an 80 Å GaAsP quantum well laser which has a pure TM polarized light emission. The substrate to be used is GaAs. The barrier region is $Al_{0.3}Ga_{0.7}As$. Calculate the effective bandgap of the structure. Assume that $m_e^* = 0.067\ m_0; m_{hh}^* = 0.45\ m_0$; and $m_{\ell h}^* = 0.1\ m_0$.

10.27 Consider a 100 Å $GaAs/Al_{0.3}Ga_{0.7}As$ quantum well laser and a 100 Å $In_{0.3}Ga_{0.7}As/Al_{0.3}Ga_{0.7}As$ laser. Compare the gain and differential gain curves for the two lasers near threshold. What can you say about threshold current and side mode suppression of the two lasers? The devices are defined by the following parameters. GaAs well (cavity loss for both lasers is 40 cm^{-1} and confinement factor is 0.002):

$$
\begin{aligned}
m_e^* &= 0.067\ m_0 \\
m_{HH}^* &= 0.5\ m_0 \\
m_{LH}^* &= 0.5\ m_0 \\
E_e^1 - E_c(GaAs) &= 33\ meV \\
E_v(GaAs) - E_{HH}^1 &= 8\ meV \\
E_v(GaAs) - E_{LH}^1 &= 20\ meV
\end{aligned}
$$

InGaAs well:

$$
\begin{aligned}
m_e^* &= 0.067\ m_0 \\
m_{HH}^* &= 0.2\ m_0 \\
E_e^1 - E_c(InGaAs) &= 33\ meV \\
E_v(InGaAs) - E_{HH}^1 &= 8\ meV \\
E_v(InGaAs) - E_{LH}^1 &= 120\ meV
\end{aligned}
$$

Assume that the gain is broadened by a Gaussian function with halfwidth of 1.5 meV. Temperature is 300 K.

Note that for the compressive strain

$$
E_v(InGaAs) = E_v(HH)
$$

and

$$
\begin{aligned}
E_{gap}(GaAs) &= 1.43\ eV \\
E_{gap}(In_{0.3}Ga_{0.7}As) &= 1.25\ eV
\end{aligned}
$$

Section 10.6

10.28 In the discussion on the DFB laser we have seen that two symmetric modes around the Bragg peak have equal reflectivities. In a real DFB laser, the output is single mode. Discuss the reasons for this.

10.10 REFERENCES

- **General**

  - H. Kressel and J. K. Butler, *Semiconductor Lasers and Heterojunction LEDs*, Academic Press, New York (1977).
  - R. Baets, "Heterostructures in III-V Optoelectronic Devices," *Solid State Electronics*, 30, 1175 (1987).
  - G. P. Agrawal and N. K. Datta, *Long-Wavelength Semiconductor Lasers*, Van Nostrand Reinhold, New York (1986).
  - A. E. Siegman, *Lasers*, University Science books, Mill Valley, CA (1986).
  - Articles in *Semiconductors and Semimetals*, ed. W. T. Tsang, Volume 22, Parts B and C, Academic Press, New York (1985).
  - Special issues of *IEEE Journal of Quantum Electronics* are regularly produced to address issues in semiconductor lasers. For example, see the issues **QE**-29, June (1993) and **QE**-30, February (1994).

- **Advanced Laser Structures: Quantum Well and Strained Quantum Well**

  - W. T. Tsang, *Applied Physics Letters*, 39, 786 (1981).
  - N. Holonyak, Jr., R. M. Kolbas, W. D. Laidig, B. A. Vojak, K. Hess, R. D. Dupuis, and P. D. Dapkus, *J. Applied Physics*, 51, 1328 (1980).
  - Y. Arakawa and A. Yariv, *IEEE J. Quantum Electronics*, **QE**-22, 1887 (1986).
  - W. D. Laidig, P. J. Caldwell, Y. F. Lin, and C. K. Peng, *Applied Physics Letters*, 44, 653 (1984).
  - N. Chand, E. E. Becker, J. P. Vander Ziel, S. N. G. Chu, and N. K. Dalta, *Applied Physics Letters*, 58, 1704 (1991).
  - E. Yablanovitch and E. O. Kane, *J. Lightwave Technology*, 4, 504 (1986).
  - U. Koren, M. Oron, M. G. Young, B. I. Miller, J. L. D. Miguel, G. Raybon, and M. Chien, *Electron. Lett.*, 26 465 (1990).
  - J. P. Loehr and J. Singh, *IEEE J. Quantum. Electron.*, 27 708 (1991).
  - I. Suemune, L. A. Coldren, M. Yamanishi, and Y. Kan, *Appl. Phys. Lett.*, 53 1378 (1988).
  - S. D. Offsey, W. J. Schaff, P. J. Tasker, and L. F. Eastman, *IEEE Photon. Tech. Lett.*, 2, 9 (1990).

- **Advanced Laser Structure: Optical Cavities**

  - Articles in special issue of *IEEE J. Quantum Electronics*, **QE**-29 (June 1993). Excellent discussion on surface emitting lasers appear in this issue.

– H. Kogelnik and C. V. Shank, *J. Applied Physics*, 43, 2327 (1972). This is a classic paper on DFB structures.

– Articles in *Semiconductors and Semimetals*, Volume 22B, Academic Press, New York (1985).

CHAPTER
11

SEMICONDUCTOR LASERS: DYNAMIC PROPERTIES

11.1 INTRODUCTION

In the previous chapter we have discussed important steady state properties of the semi-conductor laser diode. In this chapter the dynamics of the laser diode will be discussed. The semiconductor laser is an important device in the information processing age with the special feature that it can code and send information in the form of spectrally pure optical waves. However, no matter how pleasing the words "information processing by photons" sound, the laser must compete with alternate electronic devices. An important consideration in this competition is the temporal response of the device. In sheer speed, the laser cannot compete with a state of the art compound semiconductor FET or an HBT which can operate at hundreds of gigahertz (a laser is hard pressed at 50 GHz). However, the parallelism offered by optics allows a potential bandwidth which is in terrahertz. In this chapter we will address the following issues:

(i) What limits the laser time response for large and small signal modulation?

(ii) What controls the linewidth of laser emission under modulation? The line-width issue is closely tied to the laser dynamics.

11.2   IMPORTANT ISSUES IN LASER DYNAMICS

$\mathcal{R}$ The laser diode is used to produce an optical output which must be modulated to be useful for information transmission. The most important and simplest approach to modulation is direct modulation in which the current through the laser is modulated. Depending upon the application, one can divide the modulation techniques into the three categories outlined below and shown in Fig. 11.1.

Large Signal Modulation
In this approach the laser is turned ON and OFF, i.e., the current goes from well above the threshold value to a value below threshold. This kind of modulation may be used for "optical interconnects," or for certain logic applications. The response of the laser to such a modulation is extremely slow ($\sim$ 10 ns) as we shall see later. Large signal modulation is not used for optical communication due to the slow response and due to the spectral width of the output. In fact, the large signal response of a laser is not much better than that of a LED.

Small Signal Modulation
In the small signal modulation approach the laser is biased at a point well above threshold and a small ac signal is applied to it. This approach produces the highest frequency response for the laser. It also provides insight into the fundamental limitations of the laser. Modulation bandwidth approaching 50 GHz can be achieved in a small signal modulation approach.

Pulse Code Modulation
Pulse code modulation is the most widely used scheme for modern optical communication. The approach falls between the large signal and small signal modulation approaches. The laser is biased well above threshold and a current (voltage) pulse is applied so that the current goes to a high value and a low value as shown in Fig. 11.1. The important point to note is that even in the low state, the diode current remains above the threshold current. In the pulse code modulation scheme the laser can operate at up to a few tens of GHz.

The overall modulation response of semiconductor lasers is controlled by both extrinsic and intrinsic factors as shown schematically in Fig. 11.2. The extrinsic limits arise from a variety of sources as shown in Fig. 11.2. An important restriction is due to the laser heating produced when the laser is biased at a high current. The high current biasing is necessary to operate the device at high speeds as will be shown later. The heat produced causes a deterioration in the device parameters such as gain spectra, threshold current, etc..

Another important issue in lasers biased at high power is the catastrophic degeneration produced by facet damage. This produces a catastrophic failure of the laser

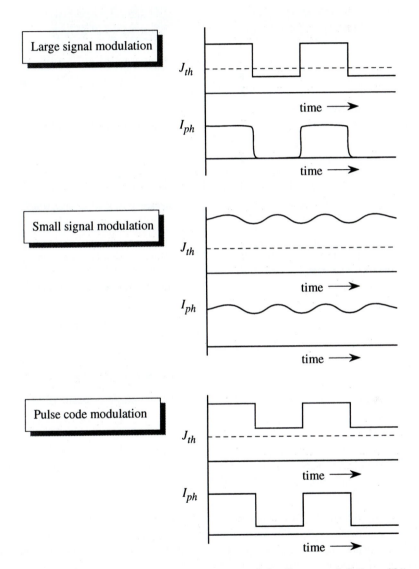

Figure 11.1: Three different modulation approaches used for direct modulation of lasers.

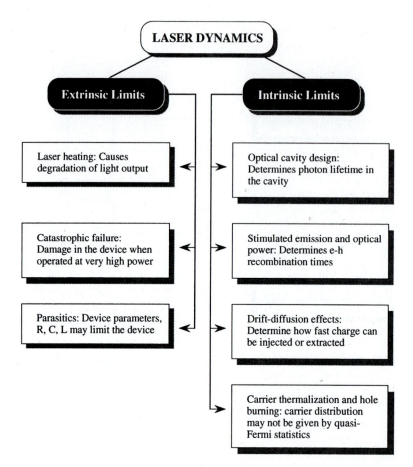

Figure 11.2: Extrinsic and intrinsic factors limiting the temporal response of semiconductor lasers.

due to the damage produced in the mirror of the cavity. Thus the laser has an upper limit on injection, at which it can be operated safely.

A final and important extrinsic limit to the laser speed is due to the extrinsic parasitics of the laser diode. The laser must be designed so that the resistance, capacitance and the inductance of the device do not limit the device response. We will address these extrinsic issues in a later section.

The intrinsic modulation limits of the laser come from several different sources outlined in Fig. 11.2. These include limits arising from the cavity design, carrier drift and diffusion as well as "gain compression" effects which limit the gain and speed of small signal modulation. We will discuss these important limits in the following sections. We start with the large signal response of the laser.

11.3 LARGE SIGNAL SWITCHING OF A LASER

$\longrightarrow$ As discussed in the previous section, the large signal switching of a laser involves changing the diode current from a value below threshold to a value above threshold. The initial value of the current may be zero. Before carrying out a mathematical analysis of the problem, let us see what happens physically when a laser diode is subjected to a step pulse of current.

Before the current step, the carrier density in the active region of the laser is essentially zero. As the current pulse is turned on, the carrier density increases. As a result, the gain in the device also starts to increase. However, until the gain reaches the cavity loss value, there are very few photons emitted out of the cavity. Thus, for a time t_d, no photons emerge from the device. Once the carrier concentration reaches n_{th}, stimulated emission starts. However, the carrier concentration overshoots the value n_{th} thus increasing the photon output beyond the steady state value. The high photon output, in turn, reduces the carrier density through higher *e-h* recombination. Thus oscillations are produced in the carrier density and the photon output. The entire process is shown schematically in Fig. 11.3.

Let us calculate the delay time t_d by examining the simple rate equation for the carrier density. This equation has been discussed in Chapter 10, and can be written as (See Eqn. 10.30)

$$\frac{dn_{2D}}{dt} = \frac{J}{e} - \frac{n_{2D}}{\tau} - R_{stim} \tag{11.1}$$

where τ is the total *e-h* recombination time. If the current density is changed from 0 to a value J, during the time $n_{2D} < n_{2D}(th)$, no photons are present in the cavity and $R_{stim} \sim 0$. Integrating this equation from $t = 0$ to $t = t_f$, and $n_{2D} = n_{2D}(i)$ to $n_{2D}(f)$, we get

$$t_f = \tau \, ln \left(\frac{J - en_{2D}(i)/\tau}{J - en_{2D}(f)/\tau} \right) \tag{11.2}$$

The photon density starts to change when $n_{2D}(f) = n_{2D}(th)$. Thus, the delay time is (for $n_{2D}(i) = 0$)

$$t_d = \tau \, ln \left(\frac{J}{J - en_{2D}(th)/\tau} \right)$$
$$= \tau \, ln \left(\frac{J}{J - J_{th}} \right) \tag{11.3}$$

The time τ is due to radiative and non-radiative processes below threshold. If the non-radiative processes are negligible, $\tau = \tau_r$, and thus there is a delay of several nanoseconds between the time the current is switched and the photons emerge from the cavity. This and the relaxation oscillations produced after photons start to emerge is

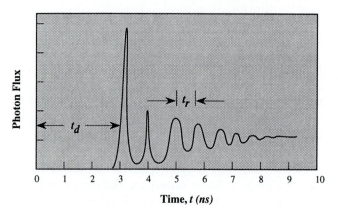

Figure 11.3: A schematic of the photon output when a large signal pulse is applied to a semiconductor laser at time zero.

a serious handicap of lasers for many applications. One must compare the total time of ~ 10 ns it takes for the laser output to reach steady state with switching of Si n-MOSFETs which is now ~0.1 ns and GaAs/AlGaAs MODFETs or HBTs which can switch in ~10 ps.

   To understand further the details of the large signal switching, let us return to the rate equations which were presented in Chapter 10. The coupled multimode rate equations are (see Eqns. 10.28 and 10.30)

$$\frac{dS_m}{dt} = [\Gamma g(n_{2D}, E_m) - \alpha_c]\frac{c}{n_r}S_m + \beta R_{sp}(n_{2D}) \qquad (11.4)$$

$$\frac{dn_{2D}}{dt} = \frac{J}{e} - R_{sp}(n_{2D}) - \frac{c}{n_r}\sum_m \Gamma g(n_{2D}, E_m)S_m \qquad (11.5)$$

Here we have written the rate equations for areal quantities. The origin of the various terms is discussed in Chapter 10, Section 10.4. The quantities Γ, $g(n_{2D}, E_m)$, $\alpha_c, n_r, \beta, R_{sp}(n_{2D})$ and J are all known, as discussed in Chapter 10. The equations can be solved in the time domain by the Runge-Kutta algorithm.

   We will show the results of a calculation done for an $In_{0.2}Ga_{0.8}As/Al_{0.3}Ga_{0.7}As$, 50 Å single quantum well laser with a 300 μm cavity. The cavity loss is taken to be 48 cm^{-1}. The 300 K threshold current density was calculated to be 180 A/cm^2. In Fig. 11.4, results are shown for a large signal step in which the current density goes from 0 to 360 A/cm^2 (= 2 J_{th}) at time zero. The electron concentration and the photon concentration are shown. The quantum well is under biaxial compressive strain so that the light hole state is pushed deeper into the valence band. Since the heavy hole state has no interaction with TM polarized light, only TE polarized light is observed. Note that both the electron density and the photon density take about 5 ns to stabilize. It is also important to note that while the photon output has very large swings during the

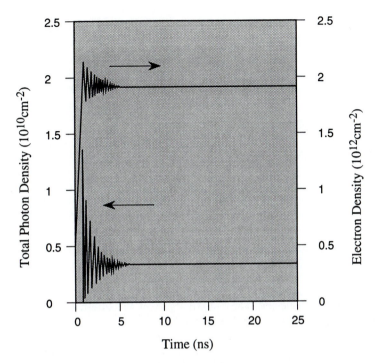

Figure 11.4: Calculated transient response of the electron and photon densities for a strained quantum well laser subjected to step function modulation from zero to twice threshold current density (360 A/cm^2). (After Y. Lam, J. Loehr and J. Singh, *IEEE J. Quantum Electronics*, **QE**-28, 1248 (1991).)

relaxation oscillations (up to 3 times the final steady state value), the electron density changes by only $\sim 10\%$ during the oscillations. This is because of the extremely strong dependence of the photon output on $g - g_{th}$.

In Fig. 11.4, we see how the total photon output behaves. It is useful to examine how the photons in individual longitudinal modes behave during the large signal switching. This effect is shown for the same laser considered above during switching (in Fig. 11.5). It is important to note that it takes a much longer time for the individual modes to reach steady state. For example, even though the total photon density is seen to stabilize in 5 ns for this laser, it takes ~ 25 ns for the individual modes to stabilize. This, of course, seriously limits the switching performance for applications where the spectral output is of importance.

The large signal switching is somewhat improved if there are strong Auger effects in the laser. The Auger rates tend to suppress the relaxation oscillations and can also reduce the delay time, t_d. Of course, this is not a very desirable way to improve the laser performance.

It is important to note that large signal switching can be exploited to produce narrow optical pulses. This approach is called gain switching and involves injecting a current pulse into the laser. The current goes from zero to a large value above threshold and back to zero. As a result only the first peak in the laser relaxation oscillations is generated. In semiconductor lasers this peak is about 20 to 30 picoseconds. A disadvantage of the pulse produced by gain switching is that the pulse has a strong chirp, a feature that will be discussed later in this chapter.

EXAMPLE 11.1 A 1.55 μm laser has a threshold current density of 400 A/cm^2 at 300 K. From a theoretical analysis of the cavity loss and gain curves it is found that $n_{th} = 2.5 \times 10^{18}$ cm^{-3} and $\tau_r = 2.5$ ns. Estimate the Auger coefficient for the laser. Also calculate the time delay produced in the photon output when the laser is switched from zero current to a current density equal to twice the threshold value. The width of the active region is 100 Å.

The threshold current density is given by

$$J_{th} = \frac{e n_{th} \cdot d_{las}}{\tau_r} + e F n_{th}^3 \cdot d_{las}$$

where F is the Auger coefficient. The radiative part of the threshold current density is 160 A/cm^2. This gives for the Auger coefficient

$$
\begin{aligned}
F &= \frac{(240 \ A/cm^2)}{(2.5 \times 10^{18} \ cm^{-3})^3 \cdot (100 \times 10^{-18} \ cm)(1.6 \times 10^{-19} \ C)} \\
&= 9.6 \times 10^{-29} \ cm^6 \ s^{-1}
\end{aligned}
$$

The non-radiative lifetime can be defined as

$$
\begin{aligned}
\tau_{nr}^{-1} &= \frac{F n_{th}^3}{n_{th}} \\
&= (9.6 \times 10^{-29} \ cm^{-6} \ s^{-1})(2.5 \times 10^{18} \ cm^{-3})^2
\end{aligned}
$$

or

$$\tau_{nr} = 1.67 \ ns$$

The total carrier lifetime is now given by

$$\frac{1}{\tau} = \frac{1}{\tau_r} + \frac{1}{\tau_{nr}}$$

or

$$\tau = 1.0 \ ns$$

The time delay is now

$$
\begin{aligned}
\tau_d &= \tau \ell n 2 \\
&= 0.7 \ ns
\end{aligned}
$$

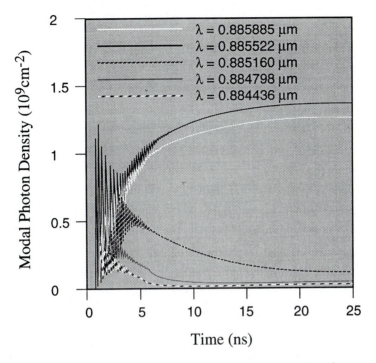

Figure 11.5: Calculated transient response of the modal photon densities for a strained quantum well laser subjected to step function modulation from zero to twice threshold current density (360 A/cm^2). Only 5 modes from the full spectrum are plotted. (After Y. Lam, J. Loehr and J. Singh, *IEEE J. Quantum Electronics*, **QE**-28, 1248 (1992).)

11.4 SMALL SIGNAL MODULATION OF A LASER

$\longrightarrow$ The small signal modulation of a laser has become a key benchmark for laser performance. Though, at present, small signal modulation is rarely used in real life systems, it provides important insight into the physics of the laser and how the laser design can be improved.

In the small signal modulation, the laser is biased at a high injection current and a small time dependent current signal is applied to it. The issue of importance here is the transfer function between the current and the photon output. Before we begin the analysis of the small signal response, we will briefly discuss some important regimes of laser operation. When the laser is biased at a high injection level, the photon output is very high and *the e-h recombination time at the energy corresponding to the peak mode energy is very short.* This time can approach a few picoseconds at very high biasing. In the laser, because of the need of the large bandgap cladding regions for optical confinement, electrons and holes are injected at high energies into the active

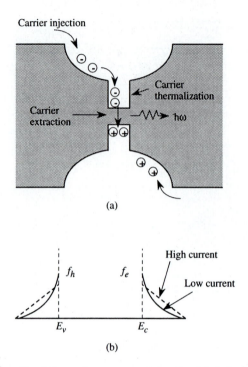

Figure 11.6: (a) A schematic of the various processes involved in injecting charge into the active region of a laser. (b) The electron and hole distribution functions at low and high drive current.

region. This is schematically shown in Fig. 11.6 for a quantum well laser.

Before electrons and holes can recombine, they have to drift and diffuse across the cladding regions and thermalize into the active region. The thermalization process is an important one since the electron and holes are injected quite "hot" into the active region. In all our discussions on the material gain, we have assumed that the carriers are described by quasi-Fermi distribution functions. *This in turn is based on the assumption that e-h recombination rates are slower than the rates of optical phonon emission which is responsible for carrier thermalization.* The reader should re-examine Sections 3.11 and 3.12 for a discussion of these issues.

At low injection current levels, the *e-h* recombination rate is slow ($\tau_{stim} \gtrsim 50$ ps) and the quasi-Fermi approximation is valid. However, at high injection, the carriers are unable to relax down and thermalize fast enough and the distribution function develops a "hole" as shown in Fig. 11.6b. Since electron thermalization times are slower than those of holes due to the small electrons density of states, the "hole" is expected to be produced in the electron distribution. The result of the spectral hole in the carrier distribution is to cause a reduction of the gain at a given carrier concentration. This phenomenon is called gain compression and results in rate equations that are somewhat modified from the ones discussed so far.

We will first discuss the laser small signal response under conditions of linear response, i.e., gain is independent of optical power and only depends upon the carrier concentration. Then we will discuss the effect and importance of gain compression.

11.4.1   Small Signal Response: Linear Theory

$\longrightarrow$ To understand the response of the laser to a small signal with frequency ω, let us return to the photon and carrier density rate equations for the mode m

$$\frac{dS_m}{dt} = [\Gamma g(n_{2D}, E_m) - \alpha_c]\frac{c}{n_r}S_m + \beta R_{sp}(n_{2D}) \tag{11.6}$$

$$\frac{dn_{2D}}{dt} = \frac{J}{e} - \frac{J_a}{e} - R_{sp}(n_{2D}) - \frac{c}{n_r}\sum_m \Gamma g(n_{2D}, E_m)S_m \tag{11.7}$$

Here we have explicitly included the Auger current J_a. We will initially assume that J_a is zero and later comment on the influence of the Auger processes. In the small signal theory we apply a current signal

$$J = \bar{J} + \tilde{J}exp(i\omega t) = \bar{J} + \Delta J \tag{11.8}$$

which causes a variation in the carrier density and the photon density

$$n_{2D} = \bar{n}_{2D} + \tilde{n}_{2D}exp(i\omega t) = \bar{n}_{2D} + \Delta n_{2D} \tag{11.9}$$

$$S_m = \bar{S}_m + \tilde{S}_m exp(i\omega t) = \bar{S}_m + \Delta\tilde{S}_m \tag{11.10}$$

We also assume that the gain and spontaneous rate can also be linearized:

$$g(\bar{n}_{2D} + \Delta n_{2D}, E_m) \simeq g(\bar{n}_{2D}, E_m) + \frac{\partial g(\bar{n}_{2D}, E_m)}{\partial n_{2D}}\Delta n_{2D} \tag{11.11}$$

$$R_{sp}(\bar{n}_{2D} + \Delta n_{2D}) \simeq R_{sp}(\bar{n}_{2D}) + \frac{\partial R_{sp}(\bar{n}_{2D})}{\partial n_{2D}}\partial n_{2D} \tag{11.12}$$

In the small signal modulation, the laser is usually biased well above threshold for maximum available bandwidth (we will see this below), and at such high biasing, the main mode of the laser dominates. Thus it is sufficient to carry out a single mode study for the small signal response. Substituting the small signal variations into the rate equations and retaining only the first order terms in the small signals, we get the equations

$$\tilde{S}\left[i\omega - \Gamma g\frac{c}{n_r} + \alpha_c\frac{c}{n_r}\right] = \tilde{n}_{2D}\left[\Gamma\frac{\partial g}{\partial n_{2D}}\frac{c}{n_r}\bar{S} + \beta\frac{\partial R_{sp}(n_{2D})}{\partial n_{2D}}\right] \tag{11.13}$$

$$\tilde{n}_{2D}\left[i\omega + \frac{\partial R_{sp}(n_{2D})}{\partial n_{2D}} + \frac{c}{n_r}\Gamma\frac{\partial g}{\partial n_{2D}}\bar{S}\right] = \frac{\tilde{J}}{e} - \frac{c}{n_r}\Gamma\bar{g}\tilde{S} \tag{11.14}$$

Eliminating $\tilde{n}_{2D}$ we get a relation between the photon signal $\tilde{S}$ and the current signal $\tilde{J}$

$$\tilde{S}\left[\Gamma\frac{cg(\bar{n}_{2D})}{n_r} + \frac{i\omega + \gamma}{\zeta}\{i\omega - \Gamma g(\bar{n}_{2D}, E_p) - \alpha_c\}\frac{c}{n_r}\right] = \frac{\tilde{J}}{e} \qquad (11.15)$$

where

$$\gamma = \frac{\partial R_{sp}(\bar{n}_{2D})}{\partial n_{2D}} + \frac{\Gamma c}{n_r}\frac{\partial g(\bar{n}_{2D}, E_p)}{\partial n_{2D}}\bar{S} \qquad (11.16)$$

$$\zeta = \beta\frac{\partial R_{sp}(\bar{n}_{2D})}{\partial n_{2D}} + \frac{\Gamma c}{n_r}\frac{\partial g(\bar{n}, E_p)}{\partial n_{2D}}\bar{S} \qquad (11.17)$$

we may also denote the photon lifetime τ_{ph} by (see section 10.4)

$$\tau_{ph} = \frac{1}{\Gamma\frac{c}{n_r}g(\bar{n}_{2D})} = \frac{n_r}{\alpha_c c} \qquad (11.18)$$

The laser response is given by the transfer function relating $\tilde{S}$ to $\tilde{J}$. We get

$$\left|\frac{\tilde{S}}{\tilde{J}}\right| = R(\omega) = \frac{\omega_r^2}{(\omega_r^2 - \omega^2) + i\omega\gamma} \qquad (11.19)$$

where

$$\omega_r^2 = \frac{\beta}{\tau_{ph}}\frac{\partial R_{sp}(\bar{n}_{2D})}{\partial n_{2D}} + \frac{c\Gamma\bar{S}}{n_r\tau_{ph}}\frac{\partial g(\bar{n}_{2D})}{\partial n_{2D}} \qquad (11.20)$$

At high biasing where stimulated emission dominates, the resonance frequency is

$$\boxed{f_r = \frac{\omega_r}{2\pi} = \frac{1}{2\pi}\sqrt{\frac{c\Gamma\bar{S}}{n_r\tau_{ph}}\frac{\partial g(n_{2D})}{\partial n_{2D}}}} \qquad (11.21)$$

The term γ defined above is the damping rate and has the value in terms of the resonance frequency and photon lifetime

$$\gamma = \frac{\partial R_{sp}(\bar{n}_{2D})}{\partial n_{2D}} + \omega_r^2\tau_{ph} \qquad (11.22)$$

The general response of the laser is shown in Fig. 11.7. The response function has a peak at ω_r (or f_r) as shown and the magnitude of the resonance peak is

$$\frac{R(\omega_r)}{R(0)} = \frac{\omega_r}{\gamma} \qquad (11.23)$$

Having derived the response function, let us briefly examine the factors that affect the frequency response of lasers.

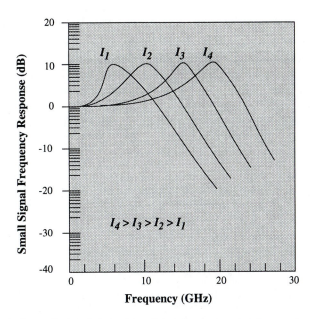

Figure 11.7: A typical small signal frequency dependent response for semiconductor lasers.

- **The Injection Current and Threshold Current**

It is clear from the transfer function calculation that a laser's high frequency response will improve as the photon density $\bar{S}$ becomes higher. Equivalently, if the laser is biased at high injection current value, the device will operate better. *A low threshold current laser is important since the photon density will be higher at the same injection current in a lower threshold laser as discussed in the previous chapter.*

It must be appreciated that one cannot simply drive the laser at a very high current value due to extrinsic effects briefly mentioned in Section 11.1. At high injection heating and high photon density induced effects will degrade the laser performance. These effects will be discussed in Section 11.6.

- **Auger Effects**

The Auger effects have two important effects on the laser performance. Since a fraction of the current is not available for creating photons, one has to drive the laser at a higher current to reach a certain value of $\bar{S}$ (compared to the case of $J_a = 0$). In addition, the damping factor becomes

$$\gamma(Auger) = \gamma(J_a = 0) + \frac{1}{e}\frac{\partial J(\bar{n}_{2D})}{\partial n_{2D}} \tag{11.24}$$

As a result, the damping factor increases and the device response suffers.

● **Photon Lifetime**

The expression for the frequency response suggests that the photon lifetime should be made as small as possible. However, one has to be very careful since the photon lifetime is strongly coupled to the gain and the threshold current. The photon lifetime can be decreased by using a shorter length cavity, but this increases the cavity loss and, as a result, a higher $n_{2D}(th)$ is needed to reach the same gain. Thus, there is an optimum cavity length for a given laser. Typical optimum lengths for quantum well lasers are around 100 μm.

● **Differential Gain**

The appearance of the differential gain $\frac{\partial g(\bar{n}_{2D})}{\partial n_{2D}}$ shows that the device response can be improved by choosing the active region of the laser to have a high differential gain. In Chapter 10 we had discussed how this can be done by using quantum well structures and strained quantum well structures. In Fig. 11.8, we show calculated high frequency response of unstrained and strained quantum well lasers. The effect of cavity length is also shown.

From the discussion of this section, it may appear that small signal frequency response (say, the 3dB cutoff frequency) can be pushed to a very high value. The obvious question is: how high is the cutoff frequency in real laser diodes? Unfortunately, the cutoff frequency has not reached very high values, in spite of a tremendous amount of effort on pushing this limit. Lasers have been demonstrated to perform up to ~ 40 GHz. State of the art quantum well lasers have been shown to operate up to ~ 35 GHz. While these numbers are impressive, they are rather embarrassingly low when compared to state of the art microwave devices. Cutoff frequencies of up to 300 GHz have been demonstrated in modulation doped FETs based on InGaAs technology.

What is causing the laser diodes to come up against this high frequency performance wall, while electronic devices continue to march forward? The answer to this question lies in a variety of intrinsic and extrinsic factors. We will now examine some of the intrinsic factors.

11.4.2 Gain Compression Effects

$\rightsquigarrow$ At the end of the previous subsection, we made an important observation regarding the cutoff frequencies of semiconductor lasers and FETs. To understand one important reason for the difference in the performance of the two devices, let us examine the importance of carrier distribution function in the device response. In Chapter 3 we discussed the transport properties of charged carriers in semiconductors. We find that *the carrier drift velocity at high fields is essentially independent of the distribution function.* For example, for GaAs the distribution function changes enormously between, say, a field of 10 KV/cm and 100 kV/cm, but the drift velocity is $\sim 10^7$ cm/s for these fields. The

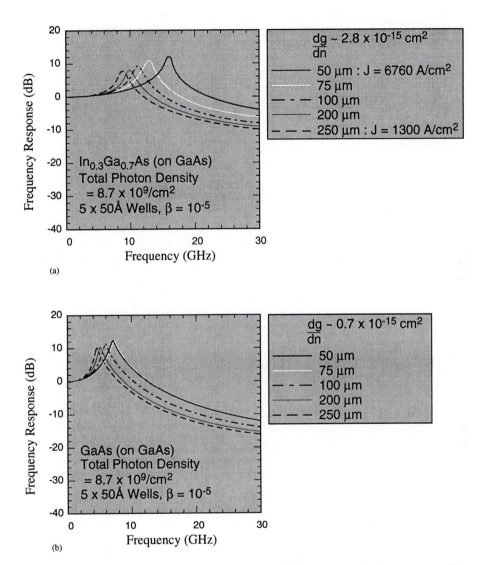

Figure 11.8: Small signal response of (a) a strained and (b) an unstrained quantum well laser. The results are given for the same output photon density. The improved performance in the strained laser is due to an improvement in $\frac{dg}{dn}$. (After Y. Lam, J. Loehr, and J. Singh, *IEEE J. Quantum Electronics*, **QE**-28, 1248 (1992).)

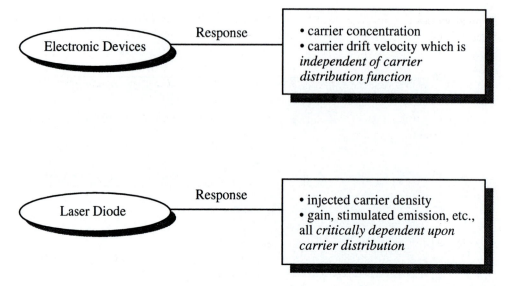

Figure 11.9: The response of electronic devices is essentially independent of how electrons (holes) are distributed in energy space. However, the carrier distribution is central to the laser response.

device temporal response then depends upon the carrier transit time which is simply controlled by the device dimensions. This issue is illustrated in Fig. 11.9.

The semiconductor laser performance does not directly depend upon the total injected charge, but depends more on how these carriers are distributed in energy. This is reflected in the occupation values f_e and f_h which are assumed to be given by the quasi- Fermi functions. We have discussed the limitations under which the distribution function is given by the quasi-Fermi function (see Sections 3.10 and 3.11). At high drive current when the *e-h* recombination times at the lasing energy E_p approach a few picoseconds, carriers are unable to relax down from the higher energies to "fill in the hole" created and the values of $f_e(E_p)$ start to decrease below that given by the quasi-equilibrium. This causes a reduction in the gain, a phenomenon referred to as gain compression.

An approach used to include the effect of gain compression is to introduce a gain compression coefficient ϵ so that

$$g = g_o(1 - \epsilon S) \tag{11.25}$$

where now g_o is the gain in the absence of any stimulated emission (i.e., for quasi- Fermi statistics). Under a small drive current when S is small, $g \simeq g_o$. The gain compression is defined through the partial derivative

$$\epsilon = -\frac{\Delta g}{\Delta S}\frac{1}{g_o} \tag{11.26}$$

Typical values of ϵ are found to be $1 - 5 \times 10^{-17}$ cm^3 from experiments. Monte Carlo calculations have been carried out to obtain ϵ from the carrier thermalization times on quantum wells and are found to be $\sim 1.1 \times 10^{-17}$ cm^3. We can see that once the photon density in the laser cavity approaches 10^{15} cm^{-3}, the gain of the laser starts to decrease and it becomes increasingly harder to generate a higher power from the device.

If the gain compression effects are included, it can be shown that the maximum -3dB bandwidth is given by

$$f_{-3dBmax} = \frac{2^{3/2}\pi}{K} \tag{11.27}$$

where the K factor is given by

$$K = 4\pi^2 \left(\tau_{ph} + \frac{\epsilon}{v_g dg/dn} \right) \tag{11.28}$$

The differential gain dg/dn is in the range of 7×10^{-16} cm^2 under high biasing conditions. Using a value of $\tau_{ph} = 1.0$ ps and $\epsilon = 1.0 \times 10^{-17}$ cm^3, we find

$$f_{-3dBmax} \simeq 90 \; GHz \tag{11.29}$$

Such a value has not yet been observed in lasers, perhaps because of limitations due to drift-diffusion effects and due to extrinsic considerations to be discussed in Section 11.6.

It may be noted that quantum wire lasers where carrier thermalization is quite a bit slower than for quantum wells, the gain compression coefficient is expected to be quite large so that $f_{-3dBmax}$ maybe no more than 20 GHz.

11.4.3 Drift-Diffusion Considerations in Laser Modulation

$\longrightarrow$ A final topic of importance in the small signal modulation of lasers is the time required for electrons and holes to diffuse or drift from the contacts into the active region of the device. This does not involve the parasitics of the device which will be discussed in Section 11.6. The Fabry-Perot laser has a total thickness of ~ 1 to 2 μm over which carriers have to be transported.

The laser is a forward-biased p-n junction so that the electric fields over the structure are rather small. The carrier transport is thus mainly governed by low field transport properties. The time taken for carriers to diffuse across ~ 0.5 μm is around 20 ps which is a significant time. High speed lasers have to be designed to minimize the distance over which carriers are to diffuse.

EXAMPLE 11.2 In this example we will explore some of the important time scales involved in the laser operation.

A semiconductor laser has a threshold current density of 200 A/cm^2 at 300 K. The spontaneous emission time is 2.0 ns. The laser is driven at an output power of 2.0 milliwatt (in a single mode operation). Calculate the average stimulated emission time for the e-h system. Also calculate the stimulated emission time for those electrons and holes whose energies correspond to the laser emission energy of 1.41 eV.

$$
\begin{aligned}
\text{Device area, A} &= 2.0 \times 200 \ \mu m^2 \\
\text{Laser linewidth,} \ \Delta(\hbar\omega) &= 10^{-4} \ eV \\
\text{Width of optical mode} &= 1.0 \ \mu m \\
\text{Photon lifetime,} \ \tau_{ph} &= 1.5 \ ps \\
\text{Radiative efficiency} &= 100\%
\end{aligned}
$$

We first calculate the threshold density $n_{th}(2D)$,

$$
\begin{aligned}
n_{th}(2\text{D}) &= \frac{(200A/cm^2)(2.0 \times 10^{-9} \ s)}{(1.6 \times 10^{-19} \ C)} \\
&= 2.5 \times 10^{12} \ cm^{-2}
\end{aligned}
$$

Assuming that stimulated emission starts after threshold,

$$
\text{Power} = \frac{(J - J_{th}) \cdot \hbar\omega \cdot A}{e}
$$

or

$$
\begin{aligned}
(J - J_{th}) &= \frac{(1.6 \times 10^{-19} \ C)(2 \times 10^{-3} \ W)}{(1.41 \times 1.6 \times 10^{-19} \ J)(2 \times 200 \times 10^{-8} \ cm^2)} \\
&= 709.2 \ A/cm^2
\end{aligned}
$$

The average stimulated emission time for the e-h ensemble is (of course, only a very small fraction of this ensemble is actually participating in the stimulated emission process)

$$
\begin{aligned}
\frac{1}{\tau(\text{average})} &= \frac{(J - J_{th})}{en_{th}(2\text{D})} \\
&= \frac{(709.2 \ A/cm^2)}{(1.6 \times 10^{-19} \ C)(2.5 \times 10^{12} \ cm^{-2})} \\
\tau(\text{average}) &= 0.564 \ ns
\end{aligned}
$$

One simple way to calculate the stimulated emission time for e-h that have the correct energy corresponding to 1.41 eV is to say that the density of such electrons is approximately

$$
\begin{aligned}
n(\text{stimulated}) &\cong n_{th}\frac{\Delta(\hbar\omega)}{k_B T} \\
&= (2.5 \times 10^{12} \ cm^{-2})\frac{(10^{-4} \ eV)}{(26 \times 10^{-3} \ eV)} \\
&= 9.6 \times 10^9 \ cm^{-2}
\end{aligned}
$$

The stimulated emission time is then

$$\tau(\text{stimulated}) = \frac{(1.6 \times 10^{-19} \ C)(9.6 \times 10^9 \ cm^{-2})}{(709.2 \ A/cm^2)} = 2.17 \ ps$$

Another way to estimate this time is to calculate the photon occupation number for the lasing mode. The photon density in the cavity is

$$
\begin{aligned}
S(2D) &= \frac{\text{Power} \cdot \tau_{ph}}{A \cdot \hbar\omega} \\
&= \frac{(2 \times 10^{-3} \ W)(1.5 \times 10^{-12} \ S)}{(2 \times 200 \times 10^{-8} \ cm^2)(1.41 \times 1.6 \times 10^{-19} \ J)} \\
&= 3.32 \times 10^9 \ cm^{-2}
\end{aligned}
$$

The photon 3D density is (the optical mode is confined to 1.0 μm)

$$S(3D) = \frac{3.32 \times 10^9 \ cm^{-2}}{(1.0 \times 10^{-4} \ cm)} = 3.32 \times 10^{13} \ cm^{-3}$$

To calculate the photon occupation n_{ph} we note that

$$n_{ph} = \frac{S(3D)}{\rho(\hbar\omega) \cdot \Delta(\hbar\omega)}$$

where $\rho(\hbar\omega)$ is the photon density of states. At 1.41 eV, we have

$$\rho(\hbar\omega) \cong 4.1 \times 10^{14} \ cm^{-3} \ eV^{-1}$$

This gives

$$
\begin{aligned}
n_{ph} &= \frac{3.32 \times 10^{13} \ cm^{-3}}{(4.1 \times 10^{14} \ cm^{-3} \ eV^{-1})(10^{-4} \ eV)} \\
&= 810
\end{aligned}
$$

From the discussions of Chapter 4 ($W_{st} = W_{spon} \ n_{ph}$)

$$
\begin{aligned}
\tau_{st} &= \frac{\tau_{spon}}{n_{ph}} \\
&= \frac{2.0 \times 10^{-9} \ s}{810} \\
&= 2.96 \ ps
\end{aligned}
$$

This value is quite close to the value calculated earlier. We see from this example how the build-up photons in a laser affects the *e-h* recombination.

EXAMPLE 11.3 A semiconductor laser is driven at high injection so that it operates in single mode and emits a power of 5 milliwatts. Calculate the resonant frequency for the dynamical response of the laser, using the following parameters:

$$
\begin{aligned}
\text{photon energy}, \hbar\omega &= 1.41 \ eV \\
\text{photon lifetime}, \tau_{ph} &= 1.5 \ ps \\
\text{confinement factor}, \Gamma &= 10^{-2} \\
\text{refractive index}, n_r &= 3.6 \\
\text{differential gain}, \frac{dg}{dn} &= 2 \times 10^{-15} \ cm^2 \\
\text{active region thickness}, d_{las} &= 100 \ \text{Å} \\
\text{laser area}, A &= 2 \ \mu m \times 200 \ \mu m
\end{aligned}
$$

The value of $dg/dn_{(2D)}$ is

$$
\begin{aligned}
\frac{dg}{dn_{2D}} &= \frac{1}{d_{las}} \frac{dg}{dn} \\
&= \frac{2.0 \times 10^{-15} \ cm^2}{(10^{-6} \ cm)} = 2 \times 10^{-9} \ cm
\end{aligned}
$$

The photon density is related to optical power by the relation

$$
\begin{aligned}
S &= \frac{\text{Power} \times \tau_{ph}}{A\hbar\omega} \\
&= \frac{(5 \times 10^{-5} \ W)(1.5 \times 10^{-12} \ s)}{(2 \times 200 \times 10^{-8} \ cm^2)(1.41 \times 1.6 \times 10^{-19} \ J)} \\
&= 2.58 \times 10^9 \ cm^{-2}
\end{aligned}
$$

The resonant frequency is now

$$
\begin{aligned}
f_r &= \frac{1}{2\pi} \left[\frac{(3 \times 10^{10} \ cm/s)(0.01)(2.58 \times 10^9 \ cm^{-2})(2 \times 10^{-9} \ cm)}{3.6(1.5 \times 10^{-12} \ s)} \right]^{1/2} \\
&= 2.69 \ GHz
\end{aligned}
$$

11.5   PULSE CODE MODULATION

$\longrightarrow$ As we have noted several times, modern day optical communication systems depend upon intensity modulation with direct detection. We had shown in Chapter 8 that the bit error rate of such transmission-detection systems depends upon the ratio of the optical power in the "zero" and "one" states. This ratio should be as close to zero as possible. In small signal modulation, this ratio is almost unity so that a very large error rate would be produced if direct detection was carried out with a small signal modulated laser.

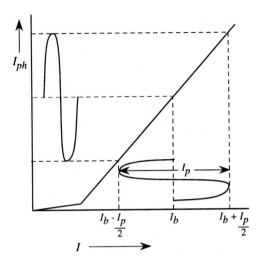

Figure 11.10: A schematic of a laser biased at a dc value I_b on which a large signal is imposed.

Ideally, one would like to use the large signal modulation approach discussed in Section 11.3, where the photon output is zero in the "off" state. However, as discussed in that section, the laser response is extremely slow (~ 10 ns) so that the laser cannot be used for any high speed transmission.

In the pulse code modulation approach, the laser is biased at a current I_b well above threshold and pulse signals of amplitude $I_p/2$ are applied to the laser. The low current reached in the modulation is $I_b - I_p/2$ as shown in Fig. 11.10 and at this point the photon output is essentially zero in steady state. However, the lowest current reached is still above the threshold current. By keeping the bias points above threshold at all times, the time delay produced in large signal switching from below threshold is avoided.

To fully understand the device response in pulse modulation, one has to solve the rate equations numerically. Here we will summarize the key findings of experimental and theoretical studies.

Maximum Frequency
The maximum resonance frequency suffers in a pulse modulation scheme, when compared to a small signal modulation around the bias point. An example of theoretical studies on this issue is shown in Fig. 11.11. Here the maximum frequency is shown to degrade as the modulation depth increases.

Distortion of the Output
In the small signal modulation, a sine wave input leads to a sinusoidal optical output. In pulse modulation, the output signal is significantly distorted. The output becomes more pulselike and depends quite significantly on the structure of the input signal. An

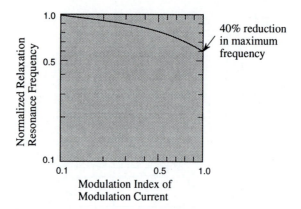

Figure 11.11: Normalized relaxation resonance frequency versus optical modulation depth ($I_b = 2J_{th}$). The reduction factor is defined as the ratio of the relaxation oscillation frequency obtained under large-signal conditions to that under small-signal conditions. (After T. Ikegami and Y. Suematsu, *Electronics Communications*, B51, 51, Japan (1970).)

example of an optical output of a 2.3 Gbps data stream is shown in Fig. 11.12. As can be seen, a serious distortion of the output is produced.

Spectral Output

Finally the spectral output of a pulse modulation is quite broad in Fabry Perot lasers since a number of modes participate in the response. Thus, while in a small signal modulation the linewidth may approach ~ 100 MHz, in a large signal pulse modulation of the overall output could be as high as 100 GHz.

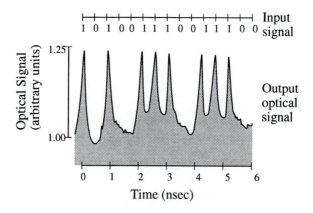

Figure 11.12: Experimental results of a 2.3 Gbps modulation of a GaAs based laser (After P. Russer and S. Schultz, *AEU*, 27, 193 (1973).)

11.6 EXTRINSIC EFFECTS ON LASER MODULATION

$\longrightarrow$ So far in this chapter we have discussed the intrinsic limits to laser modulation. We have noticed in Fig. 11.2 that there are important extrinsic limits to laser modulation that can seriously handicap the laser performance. The dominant extrinsic limitations on laser performance are due to: i) laser heating at high biasing; ii) damage of facets at high photon output; iii) parasitic effects due to the capacitance and resistance of the device. We will briefly discuss these issues.

Laser Heating
We have seen that in order to be able to modulate the laser at high speeds, the laser must be biased at as high a photon output as possible. This requires a very large current in the forward biased p-n diode. As a result, a large amount of heat is produced in the device and it is essential that the heat be extracted. If the laser starts to heat, essentially every performance parameter starts to degrade. Thus heat sinking of the laser is a critical technology component of every high speed transmitter. However, in spite of heat sinking, there is a limit to how much heat can be extracted from a laser. This limits the current at which the laser can be biased which, in turn, may limit the high speed response of the laser.

Catastrophic Failure
An important limitation of a laser biased at a very high drive current is due to the catastrophic failure of the laser structure. This is an important issue for the lifetime and reliability of the laser. At a high drive current, damage can occur in the laser facets which can reduce the reflectivity R so that the cavity loss increases. This changes the threshold current and lowers the photon output. Eventually, failure can occur. Additional failure mechanisms may involve diffusion of non-radiative centers into the active region which may further reduce the laser lifetime. Thus, while high speed performance demands that the laser be biased at a very high photon output, the laser lifetime will be significantly reduced.

Parasitic Effects
The semiconductor laser is an optoelectronic device whose design has to be carefully optimized if it is to perform at high speeds. One has to consider both the intrinsic circuit parameters of the device and the extrinsic parameters arising from wire bonding, contact resistances and device capacitance. The laser diode is a forward biased p-n diode and normally a forward biased p-n diode has a very slow response because of the rather slow e-h recombination time. However, as we have seen, this recombination time is dramatically reduced because of the stimulated emissions that occurs in the laser cavity. The intrinsic laser is described by the rate equations which describe the relationship between current and photon output. An equivalent circuit designed to describe these equations has been given by Morishita et. al., (1979) and is shown in Fig. 11.13. The laser diode can be modelled by a parallel RLC circuit as shown. The various elements

of the circuit are (for a bias current of I)

$$R_i = R_d \left[\frac{I_{th}}{I} - 1 \right] \tag{11.30}$$

$$L_i = \frac{R_d \tau_{ph}}{\left(\frac{I}{I_{th}} - 1 \right)} \tag{11.31}$$

$$C_i = \frac{\tau}{R_d} \tag{11.32}$$

where

$$R_d = \frac{2 k_B T}{e I_d} \tag{11.33}$$

and

$$I_d = \frac{n}{e \tau A d_{las}} \tag{11.34}$$

This equivalent circuit has a response that is described by an impedance that peaks at the resonance frequency and is essentially zero elsewhere.

To the intrinsic equivalent circuit one usually adds another resistance R_{sp} due to the spontaneous emission process as shown in Fig. 11.13a. This extra resistance is extremely small in comparison to other effects and can be ignored especially at high biasing.

To consider the effect of extrinsic parasitics one constructs an equivalent circuit of the laser as shown in Fig. 11.13b. Important additional parasitics are generated due to the bond wire inductance to contact the laser contact and the capacitance between the top and the back plane. A good bond wire inductance should not be more than 1 nH. The real parasitic limit arises from the capacitance. This capacitance depends upon whether the laser is fabricated on a conducting substrate or a semi-insulating (SI) substrate.

The fabrication of lasers on conducting substrates is technologically simpler than on a SI substrate. This is due to the ease of contact formation on the backside. However, one has to pay the prices due to the additional capacitance associated with the substrate. In simple stripe geometry lasers, the parasitic capacitance can be ~ 100 pF which is too large for high speed applications. In buried channel lasers, the capacitance is reduced by including a p-type blocking layer that reduces the capacitance between the top and bottom contacts to ~ 10 pF.

For high frequency response the lasers have to be fabricated on SI substrates. A typical buried channel device is shown in Fig. 11.14. With such lasers it is possible to reach modulation frequencies approaching 50 GHz.

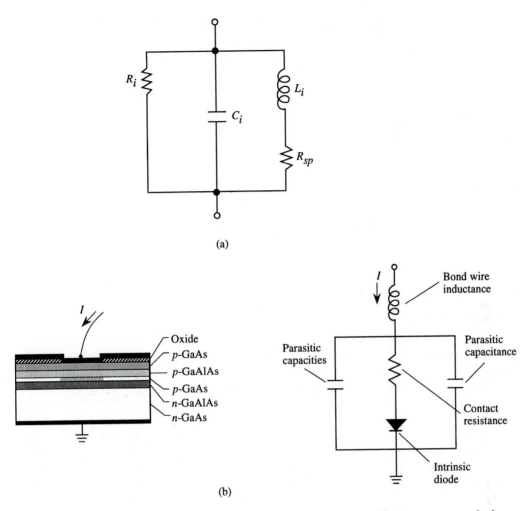

(a)

(b)

Figure 11.13: (a) The equivalent circuit of an intrinsic laser diode. (b) The geometry of a laser diode and a simple equivalent circuit.

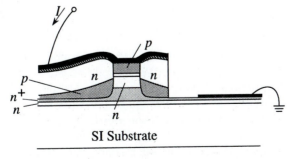

SI Substrate

Figure 11.14: A typical buried channel laser fabricated on a SI substrate to reduce parasitic effects.

11.7 SPECTRAL OUTPUT AND LASER LINEWIDTH

⤳ Since one of the most important applications of semiconductor lasers is in the area of communication, the linewidth of the optical output (along with stability of the frequency) is, perhaps, the most important feature. This is of special importance for applications involving coherent detection using AM or FM schemes. While at present most system applications involve direct detection in which the intensity modulation is used to code information, advances in the linewidth control are making it feasible for coherent detection to be possible.

The linewidth of the LED discussed in Chapter 9 was found to be $\sim k_B T$ which, at 300 K, is ~ 26 meV or 39 THz! Semiconductor lasers can do much better as we shall discuss in this section. While in this section we will focus on the linewidth of the laser, it must be kept in mind that it is important that the laser frequency remain stable as well. One of the important challenges in this regard is temperature control. Semiconductor lasers depend upon band to band transitions and 1 K change of temperature effects most semiconductors bandgaps by ~ 0.5 meV (~ 0.75 THz). Thus temperature stability is essential.

As we discuss the various factors responsible for laser linewidth, it is useful to examine the demands placed on lasers for various optical communication systems. These demands and their underlying reasons will be discussed in Chapter 13. Fig. 11.15 gives an overview of the demands placed on laser linewidths. As can be seen, the demands become more stringent as one tries to exploit the full potential of optical communication.

The various factors controlling the laser linewidth are outlined in Fig. 11.16. The factors range from the cavity quality to modulation schemes used. We will now discuss these controlling factors.

11.7.1 The Optical Cavity

The optical cavity is a critical component of the semiconductor laser and without the cavity, the laser will perform like a LED. We have discussed the various optical cavities used for lasers in the previous chapter (Section 10.6). The optical cavity plays a role in two aspects of the laser output. The first one involves the number of laser modes that the output has. The second one concerns the width of each mode. We have discussed in Chapter 10 how the Fabry-Perot cavity allows a number of longitudinal modes to lase while the DFB laser usually transmits a single longitudinal mode. In the Fabry-Perot laser, the mode separation is typically 4-5 Å and if, say, 4 modes participate, the output could have a width of ~15 Å. Additionally, when the laser is modulated, the

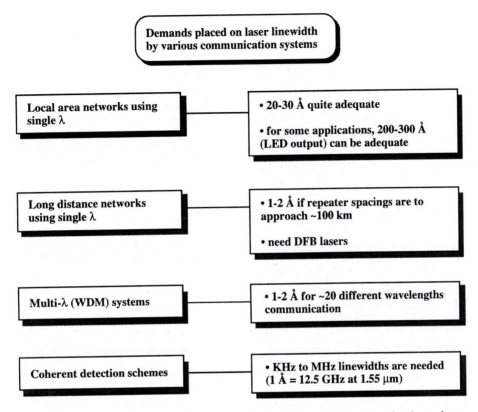

Figure 11.15: Important linewidth requirements for various optical communication schemes.

envelope width could be even higher. Clearly, the Fabry-Perot laser cannot be used for wavelength division multiplexing or coherent detection. One has to use a DFB laser for these applications.

In this chapter we will be concerned with the linewidth of a single mode of a laser output. This corresponds to either a DFB laser or a Fabry-Perot laser driven at very high injection current. The width of the transmission spectra of an optical cavity is given by ($\Delta\omega\Delta t \sim 1$)

$$\Delta\omega \cong \frac{1}{\tau_{ph}} \sim \frac{c\,|\ell n R|}{n_r\ L} \tag{11.35}$$

where τ_{ph} is the photon time in the cavity. In gas lasers where L, the cavity length, can be as large as several meters, the transmission resonance width is extremely narrow and can dominate to control and determine the laser linewidth. In semiconductor lasers, the optical cavity is quite short and the transmission resonance width is in the range of THz and, therefore, provides no selectivity to produce a narrow linewidth output. The intrinsic quality of a resonant cavity is often described by the quality factor, Q, defined as $\omega/\Delta\omega$. For the semiconductor lasers, the Q-factor is not very high so that the laser

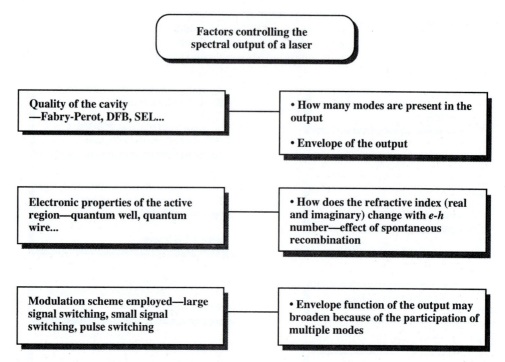

Figure 11.16: A list of important factors controling the linewidth of semiconductor laser output.

linewidth is dominated by the optical processes occurring in the semiconductor.

11.7.2   Spontaneous Emission and the Linewidth Enhancement Factor

In our discussion of spontaneous and stimulated emission, we have noted that the photons produced by stimulated emission are coherent (same frequency and phase) with the photons responsible for the emission. The photons emitted by spontaneous emission, on the other hand, have no coherence with the other photons in the cavity. It is, thus, intuitively apparent that the photons emitted through spontaneous emission will be an important source of incoherence or line broadening.

In order to understand the intrinsic limits on the laser linewidth due to the optoelectronic properties of the active region of the laser, let us examine the effect of carrier injection into the region. The carrier injection causes a change in both the real and imaginary part of the refractive index. The imaginary part of the refractive index is related to the absorption or gain in the material while the real part determines the phase velocity of the light. As discussed in Chapter 4, Section 4.2, the two are related by the Kramers-Kronig relation.

If $\Delta n_r'$ and $\Delta n_r''$ are the changes in the real and imaginary parts of the refractive index, we have

$$\Delta g = \frac{-2\omega \Delta n_r''}{c} \tag{11.36}$$

where Δg is the change in the gain due to the carrier injection. For small changes in the refractive index, the Kramer's-Kronig relations give

$$\Delta n_r'(E) = \frac{2}{\pi} P \int_o^\infty \frac{E' \Delta n_r''(E') dE'}{E'^2 - E^2} \tag{11.37}$$

where P is the principal part of the integral. Since the change in the gain in the cavity is known as a function of carrier injection, it is straightforward to calculate $\Delta n_r'(E)$ also. A typical value of $\Delta n_r'$ and $\Delta n_r''$ is shown in Fig. 11.17 when a laser is taken from zero current to the threshold value.

The change that occurs in the real part of the refractive index alters the wavelength of the lasing mode and thus causes a broadening of the emission line. An important parameter is the rate of the real and imaginary part of the change in the refractive index. This ratio is called the linewidth enhancement factor α_{enh} evaluated at the lasing energy $\hbar\omega_{las}$

$$\alpha_{enh} = \frac{\Delta n_r'(\hbar\omega_{las})}{\Delta n_r''(\hbar\omega_{las})} \tag{11.38}$$

This factor plays an important role in the spectral width of a single lasing mode. It is important to note that, in general, the peak in the gain curves and the peak in the dg/dn curve will not coincide. The value of $\Delta n_r'$ goes to zero at the peak position of the dg/dn curve ($\Delta n_r'' \propto \Delta g$). Thus α_{enh} will be low at the point were the dg/dn peak occurs. *A reduction in $\alpha_{enh}(\hbar\omega_{las})$ thus involves choosing a structure for which the peaks in g and dg/dn coincide.*

The spontaneous emission that occurs in the laser (even in the high injection lasing condition) causes fluctuations of the phase and amplitude of the optical field. These fluctuations have an important effect on the coherence or linewidth of the laser output. As the ratio of the spontaneous emission to total emission decreases, the effect of these fluctuations is reduced, so that the laser linewidth (of a single mode) decreases with output power.

We will present a simple treatment for the single mode laser linewidth for a semiconductor laser. So far, in our treatment of the rate equations of the photon density and the carrier density, we have considered only the average values of these quantities. For the laser linewidth or coherence we need to examine the fluctuations in the laser photon field. We write the photon field as (ω_o is the frequency of the laser mode)

$$F = (P_{op})^{1/2} exp(-i\phi) exp(i\omega_o t) \tag{11.39}$$

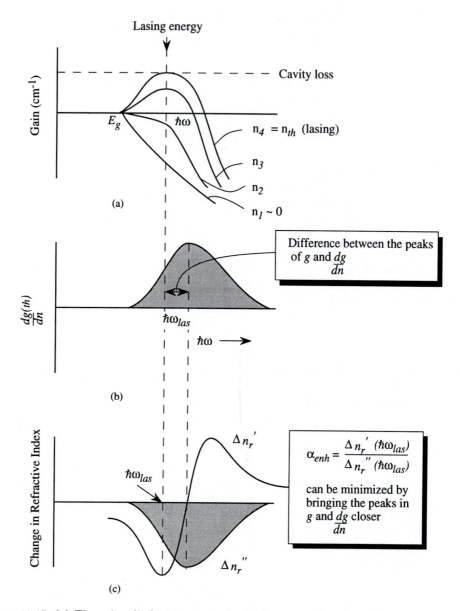

Figure 11.17: (a) The gain $g(\hbar\omega)$ of a semiconductor laser as a function of injection density. (b) The value of the change in the gain as the carrier density is changed. (c) The changes in the real and imaginary part of the refractive index.

where I is the photon intensity, and ϕ is the phase of the field. We normalize the intensity so that the average value of P_{op} gives S the photon density in the laser mode. We have already discussed in Chapter 10 the rate equation for the photon density. In the case of a transient phenomenon, the carrier concentration and, hence, the gain of the laser will change. As discussed earlier, this will produce a change in the refractive index which, in turn, will alter the phase of the electric field associated with the optical field. The rate of change of the phase is given by

$$\frac{\delta\phi}{\delta t} = k\Delta v \tag{11.40}$$

where Δv is the change in the velocity of the light due to the refractive index variation away from the steady state value. This gives us the expression

$$\begin{aligned}
\frac{d\phi}{dt} &= \frac{\omega}{n_r}\delta n'_r \\
&= \frac{\omega}{n_r}(\Gamma g - \alpha_c)\frac{\Delta n'_r}{\Delta g} \\
&= \frac{v\alpha_{enh}(\Gamma g - \alpha_c)}{2}
\end{aligned} \tag{11.41}$$

$$\frac{dS}{dt} = v(\Gamma g - \alpha_c)S + \beta R_{spon} \tag{11.42}$$

The second equation is just the rate equation we have used to describe the photon density earlier. To complete the equations we must add the noise sources, due to fluctuations in the rate equations. We refer to Fig. 11.18 where we show what happens when the photon number (or intensity) changes by unity due to spontaneous emission in the lasing mode. The fluctuation in the photon field intensity (normalized to photon density) and the photon field phase by the event i is

$$\Delta S_i = 1 + 2S^{1/2}\cos\theta_i \tag{11.43}$$

$$\Delta\phi_i = \frac{\sin\theta_i}{S^{1/2}} \tag{11.44}$$

Note that on average the fluctuations add up to zero. However, they introduce important phase incoherences that lead to the laser linewidth.

We note that the phase is affected by intensity fluctuations in addition to the fluctuation given by Eqn. 11.44. This fluctuation called delayed phase fluctuation is given from Eqns. 11.41, 11.42 and 11.43 as

$$\Delta\phi(delay) = \frac{\alpha_{enh}}{S^{1/2}}\cos\theta_i \tag{11.45}$$

This additional phase fluctuation is a characteristic of the semiconductor laser due to the presence of α_{enh} and is not present in gas lasers.

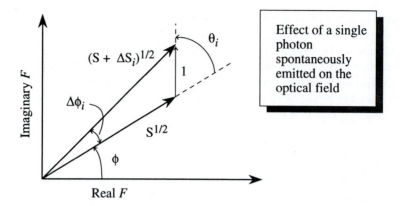

Figure 11.18: The effect of a spontaneous event on the optical intensity (normalized to equal photon density) and the field phase of the laser mode. (After C. H. Henry, *Semiconductors and Semimetals*, ed. W. T. Tsang, volume 22B, Academic Press, New York (1985).)

We are now in a position to calculate the laser linewidth. We are interested in the laser power spectrum $G(\omega)$ which is given by

$$G(\omega) = \int dt < F(t)F(0) > e^{i\omega t} \qquad (11.46)$$

Thus $G(\omega)$ is the Fourier transform of the correlation of the laser field at different times. If the phase coherence is maintained over all times the linewidth is simply zero. The photon field is given by Eqn. 11.39. We will calculate the laser linewidth assuming that the linewidth is dominated by phase fluctuations with little effect from direct intensity fluctuations. This is a reasonable assumption except that certain details in the laser lineshape are not accurate. We have from Eqns. 11.39 and 11.46

$$G(\omega) = \int dt \; exp\left[i(\omega - \omega_o)t\right] < exp(i\Delta\phi(t)) > \qquad (11.47)$$

where

$$\Delta\phi(t) = \phi(t) - \phi(0) \qquad (11.48)$$

The phase fluctuation $\Delta\phi$ is a Gaussian variable because of the random nature of the spontaneous emission. Thus it is reasonable to assume that the probability function $\rho(\Delta\phi)$ is a Gaussian, an assumption that is verified experimentally. For a Gaussian function we have

$$< exp(-i\Delta\phi) >= exp\left(\frac{\Delta\phi^2}{2}\right) \qquad (11.49)$$

To evaluate $\Delta\phi^2(t)$ we calculate $\Delta\phi(t)$ from Eqns. 11.44 and 11.45 after summing over n spontaneous emissions where $n = \beta R_{spont}t$ is the number of random events occurring in time t

$$\Delta\phi(t) = \frac{1}{S^{1/2}} \sum_{i=1}^{n} [sin\theta_i + \alpha_{enh}cos\theta_i] \qquad (11.50)$$

Taking a square of this and averaging, we get

$$<\Delta\phi^2(t)> = \frac{\beta R_{spon}}{2S}(1 + \alpha_{enh}^2)t \qquad (11.51)$$

This gives, for the power spectrum (from Eqns. 11.48 and 11.51)

$$G(\omega) = \int dt exp \left[\left\{ i(\omega - \omega_o) - \frac{\beta R_{spon}(1 + \alpha_{enh}^2)}{4S} \right\} t \right] \qquad (11.52)$$

This results in a Lorentzian power spectrum with a full width at half maximum (FWHM) of

$$\boxed{\Delta f = \frac{\beta R_{spon}}{4\pi S}(1 + \alpha_{enh}^2)} \qquad (11.53)$$

The importance of the linewidth enhancement factor α_{enh} is quite apparent. The linewidth enhancement factor can be tailored to some extent by tailoring the density of states of the active region. For double heterostructure lasers the value of α_{enh} is found to be ~ 7. However, lower values have been suggested for quantum well and quantum wire lasers.

The inverse power dependence of laser linewidths has been seen in a number of lasers. Laser linewidths as low as kilohertz have been observed in high power DFB lasers where the combination of high power and superior optical cavity have been exploited. Such lasers can be used for coherent detection schemes, although the stability of the laser line is still a challenge. It is observed that the intercept of the laser linewidth at infinite power is not zero, but a finite value. This additional effect is traced to a 1/f noise related to defects and traps in the laser structure.

11.7.3   Spectra Under Modulation

$\mathcal{R}$ When the laser is under modulation, the spectral output of the laser tends to broaden. The level of broadening depends upon the modulation scheme that is being used. In large signal modulation where the laser is taken from below threshold to above threshold, a large number of modes participate in the photon output and the linewidth is dominated by the envelope function of the various modes. The linewidth is in the range of hundreds of terrahertz (several ten Angstroms).

In pulse code modulation, the laser linewidth can be maintained at a relatively low value. However, due to the nonlinear response of the laser, sidebands may appear in the laser output (see Chapter 8, Section 8.2). The output broadens and the emission frequency shifts. In addition, during the switching process, the phase of the optical field shifts, as can be seen from Eqn. 11.41. An associated change in frequency is then produced

$$\Delta\nu(t) = \frac{1}{2\pi}\frac{d\phi}{dt} = \frac{v\alpha_{enh}}{4\pi}(\Gamma g - \alpha_c) \tag{11.54}$$

This phenomenon is called "chirp" and produces a spread of $\sim$ 10-20 GHz during pulse code modulation. In the next chapter we will discuss the effect of chirp on the broadening or compression of optical pulses.

It must be noted that in this section, we have considered the properties of the laser as it is modulated by direct current injection. If the modulation is external, as will be discussed in the next chapter, the issues are quite different.

EXAMPLE 11.4 Consider a Fabry Perot cavity of length 400 μm and mirror reflectivity of 0.33. Calculate the bandwidth of the resonant states of the cavity. The refractive index of the cavity is 3.6. The linewidth of a typical laser mode is 100 MHz for the laser. What does the cavity length have to be if the optical cavity is to set the limit on the laser linewidth?

The bandwidth of the resonant state is

$$\Delta\omega = \frac{c|\ln R|}{n_r L}$$
$$= \frac{(3\times 10^{10}\ cm/s)(1.109)}{(3.66)(400\times 10^{-4}\ cm)}$$
$$= 2.27\times 10^{11}\ Hz$$

This bandwidth is much larger than the laser linewidth. If the resonant mode bandwidth is to be smaller than the laser linewidth, the cavity length has to be 91 cm. Such a length cannot be achieved in semiconductor technology unless external feedback approaches are used.

EXAMPLE 11.5 The linewidth enhancement factor in a laser is found to be 7. Calculate the laser linewidth if the laser is emitting a power of 10 milliwatts at a photon energy of 1.41 eV. The laser parameters are given below:

Stripe width	5 μm
Laser length	200 μm
Threshold current density	200 A/cm^2
Photon lifetime	1 ps
Spontaneous emission factorβ	10^{-5}

The spontaneous emission rate (areal) is

$$R_{spon} \cong \frac{J_{th}}{e} = \frac{(200\ A/cm^2)}{(1.6\times 10^{-19}\ C)}$$
$$= 1.25\times 10^{20}\ cm^{-2}\ s^{-1}$$

To calculate the photon density (also areal) we have

$$\text{Power} \ = \ A\frac{S\hbar\omega}{\tau_{ph}}$$

or

$$S \ = \ \frac{(10^{-2}W)(10^{-12}\ s)}{(1.41 \times 1.6 \times 10^{-19}\ J)(5 \times 200 \times 10^{-8}\ cm)}$$

$$= \ 4.43 \times 10^{9}\ cm^{-2}$$

The laser linewidth is now

$$\Delta f \ = \ \frac{\beta R_{spon}(1 + \alpha_{enh}^{2})}{4\pi S}$$

$$= \ \frac{10^{-5}(1.25 \times 10^{20}\ cm^{-2}\ s^{-1})(1 + 49)}{4\pi(4.43 \times 10^{9}\ cm^{-2})}$$

$$= \ 1.12\ MHz$$

11.8 ADVANCED STRUCTURES

$\mathcal{R}$ As we have seen in this and the previous chapter, the semiconductor laser is a complex device which depends upon both the electronic and optical properties of the semiconductor structure. As shown in Fig. 11.19, advances in semiconductors are occurring from the use of structures and materials that can optimize both the optical and electronic properties. Advances are also occurring due to the desire to use new materials for either short wavelength or long wavelength lasers. Finally, the desire to have OEICs is leading to research and development in the new lasers.

11.8.1 The Quest for the Dream Laser

$\mathcal{R}$ The ideal laser would have zero threshold current and a very high bandwidth. Conventional lasers do not satisfy these requirements for several reasons, all related to the electronic and photonic properties of the semiconductor laser structure. *The electronic states of the semiconductor form a continuum so that when carriers are injected they occupy energies not only corresponding to the lasing mode but also to the non-lasing modes.* Emission in the non-lasing modes is a drain on the laser current. To avoid this drain, one has to either design a semiconductor structure with discrete energy states (quantum dot) or use a cavity where non-lasing photon states are forbidden. As shown in Fig. 11.19, this has led to research into sub-3D semiconductor structures for electronic bandstructure. Also research involving microcavity lasers is motivated by the desire to alter the photon bandstructure.

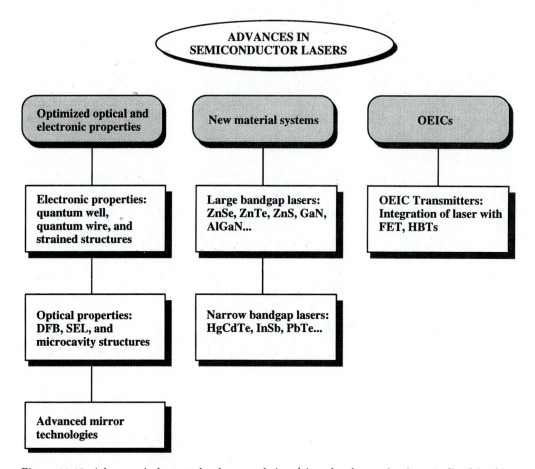

Figure 11.19: Advances in laser technology are being driven by the motivations outlined in this figure.

The single photon state, single *e-h* pair state laser is far from being realized at present. Challenges that have to be overcome are not only technological (growth and fabrication) but also fundamental. For example, how is charge injected rapidly into a discrete level? Conventional lasers depend upon phonon emission to get charge into the lasing levels. This process becomes painfully slow for quantum dots.

11.8.2   Tailoring the Emission Energy

$\mathcal{R}$ Semiconductor lasers emitting at $\sim 0.8~\mu$m, 1.55 μm and 1.3 μm have received the greatest attention over the last several decades due to their importance in communication systems. However, as the application of lasers broaden, other wavelength become important as well. For example, the desire for high density optical memories, which can be read by lasers, is fueling research into short wavelength lasers. Similarly, the desire to send free space optical signals through fog, clouds, etc., is encouraging work in the area of long wavelength lasers.

The challenges in short wavelength ($\lambda \stackrel{<}{\sim} 0.5~\mu$m) or longwavelength ($\lambda \stackrel{>}{\sim} 2.5~\mu$m) lasers are primarily technological. The short wavelength materials are ZnS, ZnSe, ZnTe, GaN, AlGaN, etc., where bandgaps are above 2.0 eV. These materials are quite difficult to grow because of their very high melting points. More importantly, it is extremely difficult to dope the materials reliably. Since ohmic contacts depend upon very high doping, the ohmic contact technology in these materials is quite undeveloped. However, great strides have been made recently in the ZnSe, ZnTe system for applications as green or blue lasers.

The technology of the narrow bandgap materials is also very challenging. These materials are inherently "soft" and fragile and it is very easy to create defects and damage. Another important challenge for these materials is the very high Auger recombination in the lasers fabricated from them. As a result, these lasers can only operate at low temperatures (77 K) and in pulsed mode. It is possible that quantum well structures and especially strained quantum well structures will benefit this area. Also, the potential of using intersubband transitions discussed in the context of longwavelength detectors (Section 7.10) for light emission is quite strong.

11.8.3 OEIC Laser Transmitters

$\mathcal{R}$ The present day optical transmitters involve hybrid technology where a high speed transistor (a bipolar device or a FET) drives a laser diode. There are many advantages in going towards an integrated technology where the laser and the driver are fabricated on the same chip. The advantages include better manufacturing yield, better potential performance due to the reduction in parasitics, and perhaps a transition to an OEIC where thousands (or even millions) of lasers and transistors are placed on the same chip. Such chips could find use in imaging technologies as well as in future "optical computers."

Of course, along with the advantages of integrated technologies, there are some serious challenges. Consider the structure of a typical edge emitting laser. The laser is typically 3-4 μm "high" and has a material structure which is quite incompatible with a transistor structure. If the laser is a vertical cavity laser, the structure is even higher ($\sim 10~\mu$m) with Bragg reflectors. Also, for the edge emitting laser, high quality mirrors have to be defined at the edges. This may require cleaving of the structure. Also, the lateral structure of the laser requires one to define a narrow stripe or buried region so that a single transverse mode is excited. Finally, the laser performance is extremely sensitive to the temperature of the device so that a very efficient heat sinking mechanism has to be used.

From the discussion above, it is clear that all the difficulties of etching and regrowth have to be surmounted for OEIC transmitters. The presence of defects and dislocations has a much more disastrous effect on lasers than on photoreceivers so that the demands are more severe on OEIC transmitters than on OEIC receivers.

A great deal of work on integrated transmitters has focussed on the fabrication of mirrors. Microcleaving techniques are being developed along with the use of dry etching such as ion beam milling to produce the mirrors. The etched mirrors do not have the performance that can be reached in cleaved mirrors due to the lack of atomic abruptness at the edge. However, rapid progress is being made in this area.

As the reader can surmise, the laser fabrication process is not as simple as the fabrication of transistors or photodiodes. The need for high quality mirrors makes the process cumbersome. The problem becomes even more serious if the laser is the DFB laser. The problems of including gratings into a laser are quite serious and the manufacturing yield of DBF lasers is very poor. As a result, a DFB laser can cost several thousand dollars (1993), while a Fabry-Perot laser costs a few dollars.

11.9 SUMMARY

In this chapter we have examined the dynamic response of semiconductor lasers. The dynamic response is critical for applications of lasers in optoelectronic information processing systems. We also examined the issue of laser linewidth. In Tables 11.1-11.2 we summarize the findings of this chapter.

11.10   PROBLEMS

Unless otherwise stated, assume that the temperature is 300 K and radiative efficiency is unity.

Section 11.3

11.1 A GaAs/AlGaAs laser has a threshold current density of 200 A/cm^{-2}. In a large signal switching, the laser is switched from zero current to 4 times the threshold current. Calculate the time delay before photon emission if the carrier lifetime is 2 ns.

11.2 A 1.55 μm quantum well laser has a threshold carrier density of 2×10^{12} cm^{-2}. The radiative lifetime is 2.5 ns and the Auger coefficient is 10^{-28} cm^6 s^{-1}. Calculate the

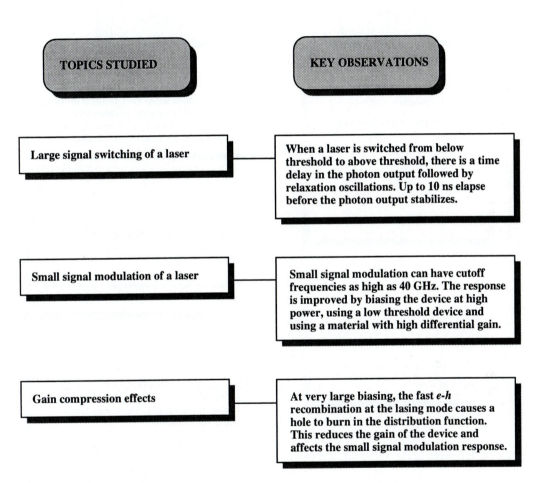

TOPICS STUDIED

KEY OBSERVATIONS

Large signal switching of a laser

When a laser is switched from below threshold to above threshold, there is a time delay in the photon output followed by relaxation oscillations. Up to 10 ns elapse before the photon output stabilizes.

Small signal modulation of a laser

Small signal modulation can have cutoff frequencies as high as 40 GHz. The response is improved by biasing the device at high power, using a low threshold device and using a material with high differential gain.

Gain compression effects

At very large biasing, the fast *e-h* recombination at the lasing mode causes a hole to burn in the distribution function. This reduces the gain of the device and affects the small signal modulation response.

Table 11.1: Summary table.

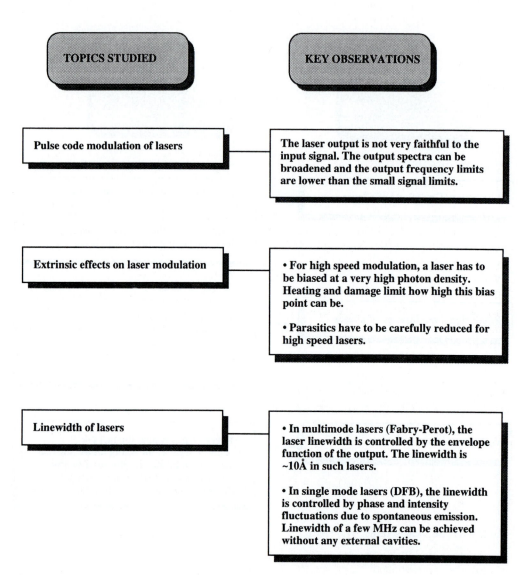

Table 11.2: Summary table.

time delay for a large signal switching in which the current density is switched from 0.5 J_{th} to 3.0 J_{th}. Assume that the Auger rate can be evaluated using the carrier density n_{th}. The active layer thickness is 100 Å.

11.3 Describe in mathematical detail how the Auger coefficient can be evaluated by measuring the time delay in a large signal switching in a laser. Assume that τ_r is known as a function of carrier density.

11.4 Write a computer program to study the large signal switching of a laser. Write the program so it can calculate the photon density evolution in 5 modes of a Fabry-Perot laser (as shown in Fig. 11.5).

11.5 The switching time of a 2.5 μm laser is found to be much faster than that of a 0.8 μm laser. Discuss the possible reasons for this.

Section 11.4

11.6 Consider a GaAs/AlGaAs laser at 77 K and 300 K with the optical confinement factor $\Gamma = 1.0$. Calculate dg/dn for the laser near threshold if the cavity loss is 100 cm^{-1}. What can you say about the effect of temperature on the small signal modulation?

11.7 Calculate the differential gain near threshold of an 80 Å GaAs/Al$_{0.3}$Ga$_{0.7}$As quantum well laser. The cavity loss is 30 cm^{-1}. Temperature is 300 K and the optical confinement factor is 0.01.

11.8 A GaAs/AlGaAs quantum well laser has a threshold density of $n_{2D}(th) = 2 \times 10^{12}$ cm^{-2}. The laser is driven well above threshold so that the photon density in the peak mode is 4×10^9 cm^{-2}. The factor β for the laser is 10^{-5}. Estimate the stimulated emission lifetime for the electron-hole pairs that have energies appropriate for the lasing mode if the spontaneous emission lifetime is 2.4 ns.

11.9 Discuss the effect of decreasing the cavity length of a Fabry-Perot laser on the modulation speed of the laser.

11.10 Consider a GaAs/AlGaAs quantum well laser with a cavity loss of 40 cm^{-1}. The device is defined by the following parameters at 300 K:

$$
\begin{aligned}
m_e^* &= 0.067 \, m_0 \\
m_{hh}^* &= 0.45 \, m_0 \\
E_e^1 - E_c(GaAs) &= 33 \, meV \\
E_v(GaAs) - E_{HH}^1 &= 8 \, meV \\
\tau_r &= 2.4 \, ns \\
d_{las} &= 2 \times 100 \, \text{Å}
\end{aligned}
$$

You may ignore the LH band contribution. Calculate the resonant frequency f_r of the laser at an injection current density of $2J_{th}$ and $5J_{th}$. You may assume a single mode operation and an optical confinement of 0.025.

11.11 Consider the previous problem. What does the injection current density have to be to push the resonance frequency to reach 25 GHz? Assume no gain compression effects. Estimate the stimulated e-h lifetime at the lasing energy. If the carrier thermalization time is 10 ps, is the assumption of no gain compression valid? What is the source of the poor high speed performance of this laser?

11.12 Compare the differential gain at threshold for a lattice matched and strained quantum well laser with the following parameters at 300 K (well size in each case is 100 Å).

Lattice matched laser:

$$
\begin{aligned}
m_e^* &= 0.04\ m_0 \\
m_{HH}^* &= 0.45\ m_0 \\
E_e^1 - E_{HH}^1 &= 0.8\ eV \\
\tau_r &= 2.0\ ns \\
\alpha_c &= 50\ cm^{-1} \\
\Gamma &= 0.015
\end{aligned}
$$

Strained laser:

$$
\begin{aligned}
m_e^* &= 0.04\ m_0 \\
m_{HH}^* &= 0.2\ m_0 \\
E_e^1 - E_{HH}^1 &= 0.8\ eV \\
\tau_r &= 2.0\ ns \\
\alpha_c &= 50\ cm^{-1} \\
\Gamma &= 0.015
\end{aligned}
$$

You only need to consider the HH states for the valence band. What does this problem tell you about the high speed performance of lattice matched versus strained lasers?

11.13 Discuss the qualitative reasons for gain compression and why it suppresses the high speed performance of lasers.

11.14 The carrier thermalization time in a quantum well laser is found to be 5.0 ps. At what photon density do you expect gain compression effects to be important? Assume that the laser mode width is 10^{-4} eV.

11.15 Calculate the time taken for electrons to diffuse across a 1.0 μm region where the mobility is 8000 cm^2/V-s.

11.16 Discuss the reasons why as one goes from 3D to 2D to 1D to 0D systems, the carrier thermalization times increase.

Section 11.5

11.17 Write a computer program to solve the rate equations of a laser for a pulse code modulation scheme.

Section 11.8

11.18 Consider a GaAs/AlGaAs double heterostructure laser at 300 K. If $\alpha_c = 50$ cm^{-1} and $\Gamma = 1.0$, calculate the peak energy positions of the gain and differential gain at threshold. Also calculate the linewidth enhancement factor for the laser.

11.19 Calculate the separation in the positions of the peak of the gain and differential gain in a bulk GaAs and an 80 Å quantum well GaAs/AlGaAs laser at 300 K. For the quantum well, assume that only the ground states of the electron and the heavy hole are to be included. The material gain needed for lasing is 50 cm^{-1} for the bulk laser, and 3000 cm^{-1} for the quantum well laser.

11.11 REFERENCES

- **Large Signal Response**

  - P. J. Herre and V. Barabas, *IEEE Journal of Quantum Electronics*, 25, 1794 (1989).
  - H. Kressel and J. K. Butler, *Semiconductor Lasers and Heterojunction LEDs*, Academic, Orlando (1977).
  - Y. Lam, J. P. Loehr, and J. Singh, *IEEE Journal of Quantum Electronics*, 28, 1248 (1992).
  - D. Marcuse and T. Lee, *IEEE Journal of Quantum Electronics*, 19, 1397 (1983).

- **Small Signal Response**

  - G. P. Agrawal and N. K. Datta, *Long-Wavelength Semiconductor Lasers*, Van Nostrand Reinhold, New York (1986).
  - Y. Lam, J. P. Loehr, and J. Singh, *IEEE Journal of Quantum Electronics*, 29, 42, (1993).
  - K. Y. Lau and A. Yariv, *IEEE Journal of Quantum Electronics*, 21, 121 (1985).
  - Lau, K. Y. and A. Yariv, *Semiconductors and Semimetals*, Volume 22, part B, ed. W. Tsang, Academic Press, New York (1985).
  - P. A. Morton, R. A. Logan, T. Tanbun-Ek, R. F. Sciortino, Jr., A. M. Sergent, R. K. Montgomery, and B. T. Lee, *Electronic Letters*, 28, 1256 (1992).
  - K. Petermann, *Laser Diode Modulation and Noise*, Kluwer Academic, Tokyo (1988).
  - I. Suemune, L. A. Coldren, M. Yamanishi, and Y. Kan, *Applied Physics Letters*, 5a3, 1378 (1988).
  - S. Weisser, J. D. Ralston, E. C. Larkins, I. Esquivias, P. A. Tasker, J. Fleissner, and J. Rosenweig, *Electronic Letters*, 28, 2141 (1992).

- **Gain Compression Effects**

  - G. P. Agrawal and N. K. Datta, *Long Wavelength Semiconductor Lasers*, Von Norstrand Reinhold, New York (1986).

- J. E.Bowers, *Solid State Electronics*, 30, 1 (1987).

- Y. Lam, *Static and Dynamic Properties of Pseudomorphic Quantum Well Lasers*, Ph.D. Thesis, University of Michigan, Ann Arbor, MI (1993).

- A. Tomita, *IEEE Photonics Technology Letters*, 4, 342 (1992).

- I. Vurgaftman and J. Singh, *Journal of Applied Physics*, 74, 6451 (1993).

- **Drift Diffusion Effects in Lasers**

  - S. Kan C., D. Vassilovski, T. C. Wu, and K. Y. Lau, *Applied Physics Letters*, 61, 751 (1992).

  - R. Nagarajan, M. Ishikawa, T. Fukushina, R. S. Geels, and J. E. Bowers, *IEEE Journal of Quantum Electronics*, 28, 1990 (1992).

- **Pulse Modulation of Lasers**

  - W. Harth and D. Siemsen, *AEU*, 28, 391 (1974).

  - K. Y. Lau and A. Yariv, *Semiconductors and Semimetals*, Volume 22B, ed. W. Tsang, Academic Press, New York (1985).

  - P. Russer and S. Schultz, *AEU*, 27, 193 (1973).

  - R. Tell and S. T. Eng, *Electronics Letters*, 16, 497 (1980).

- **Spectral Width of Lasers**

  - G. P. Agrawal, W. B. Joyce, R. W. Dixon, and M. Lax, *Applied Physics Letters*, 43, 11 (1983).

  - B. Diano, P. Spano, M. Tambarini, and S. Piazolla, *IEEE Journal of Quantum Electronics*, **QE**-19, 226 (1983).

  - C. H. Henry, *Semiconductors and Semimetals*, Volume 22B, ed. W. Tsang, Academic Press, New York (1985).

  - C. H. Henry, *IEEE Journal of Quantum Electronics*, **QE**-18, 259 (1982).

  - C. H. Henry, R. A. Logan, and K. A. Bertness, *Journal of Applied Physics*, 52, 4453 (1981).

  - D. Welford and A. Mooradian, *Applied Physics Letters*, 40, 865 (1982).

- **Extrinsic Limits to Laser Performance**

  - J. Katz, S. Margalit, C. Harder, D. P. Wilt, and A. Yariv, *IEEE Journal of Quantum Electronics*, **QE**-17, 4 (1981).

  - M. Morishita, T. Ohmi, and J. Nishizawa, *Solid State Electronics*, 22, 951 (1979).

  - K. L. Yu, U. Koren, T. R. Chen, P. C. Chen, and A. Yariv, *Electronic Letters*, 17, 790 (1981).

- **Advanced Structures: Wide Gap Lasers**

  - M. A. Haase, J. Oui, J. M. DePuydt, and H. Cheng, *Applied Physics Letters*, 59, 1272 (1991).
  - H. Hagerolt, H. Jeon, A. V. Nurmikko, W. Xie, D. E. Grillo, M. Kobayashi, and R. L. Gunshor, *Applied Physics Letters*, 60, 2825 (1992).
  - I. Suemune, H. Masato, N. Kakanishi, K. Yamada, Y. Kan, and M. Yamanishi, *Journal of Crystal Growth*, 101, 754 (1990).

- **Advanced Structures: Narrow Gap Lasers**

  - S. J. Eglesh and H. K. Choi, *Applied Physics Letters*,
  - N. Koguchi, T. Kiyowasa, and S. Takahashi, *Journal of Crystal Growth*, 81, 400 (1987).
  - J. Singh and R. Zucca, *Journal of Applied Physics*, 72, 2043 (1992).
  - M. Tacke, B. Spanger, A. Lambrecht, P. R. Norton, and H. Bottner, *Applied Physics Letters*, 53, 2260 (1988).
  - M. Zandian, J. M. Arias, R. Zucca, R. V. Gil, and S. H. Shin, *Applied Physics Letters*, 59, 1022 (1991).

- **Advanced Structures: OEIC Transmitters**

  - M. Kuno, T. Sanada, H. Nobuhara, M. Makiuchi, T. Fugii, O. Wada, and T. Sakurai, *Applied Physics Letters*, 49, 1575 (1986).
  - O. Wada, "Recent Progress in OEICs," *SPIE Proceeding*, No. 797, 224 (1987).

CHAPTER
12

MODULATION AND
AMPLIFICATION
DEVICES

12.1   INTRODUCTION

In the last several chapters we have discussed semiconductor devices that can detect and generate light. While these devices are of obvious importance in the information processing age we live in, they are not "intelligent" devices. Intelligence comes from manipulating information—switching, modulating and carrying out arithmetic and logic operations. In the electronic domain, the flip-flops, registers, and logic gates, all based on transistor technology provide the intelligent devices that run today's computers, microprocessors and even optoelectronic systems.

Consider an optical signal as it carriers a data bit from New York and heads to Tokyo. There is obviously no single one to one optical fiber connection between the two end points. The data bit has to go through several repeaters and exchanges as it goes along. The repeaters are there to compensate for the optical power loss suffered in transmission. At present the repeaters convert the optical pulse to an electronic pulse, amplify it and reconvert it to a boosted optical pulse. The same is done in the switching exchanges. Thus all the data manipulation is done by the same electronic devices that run our personal computers (PC's). All this seems rather embarrassing to the proponents of optics and "optical computing," but at present there are few optoelectronic devices that can compete with the transistor in data manipulation. However, intelligent optoelectronic devices are improving rapidly and are beginning to make a serious impact on technology. It may still be a long time before these devices find their way into our

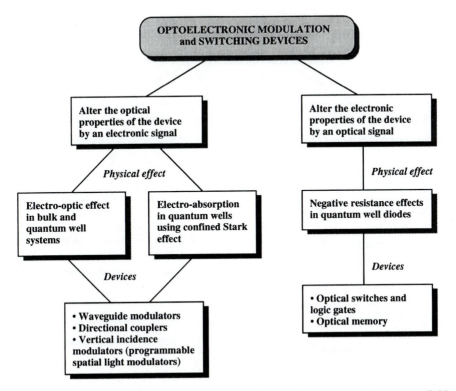

Figure 12.1: An overview of the physical effects and the resulting devices that are available for modulation and switching.

PC's, but they will impact high speed communication systems, in the near future.

In order to develop devices that can modulate or switch light, one needs to exploit some physical effect that alters the optical and electronic property of the semiconductor. In Fig. 12.1, we show how one can alter the optical properties of a semiconductor by exploiting the electro-optic effect or the quantum confined Stark effect (QCSE) discussed in Chapter 5. It is also possible to alter the electronic properties of a material by an optical signal. This can be used to create a variety of memory and switching devices. We will examine these devices in this chapter.

It is also possible to use "all optical" effects in semiconductors where one light beam can be used to switch another beam. Some of the effects used for this approach are shown in Fig. 12.2. For example, exciton quenching in quantum well structures discussed in Section 5.9 can be used to create all optical logic gates. Another useful effect is the photorefractive effect in which the intensity modulation produced by two interfering beams can produce a refractive index pattern. This pattern can persist even after the write beams are turned off, and can be read out by another beam.

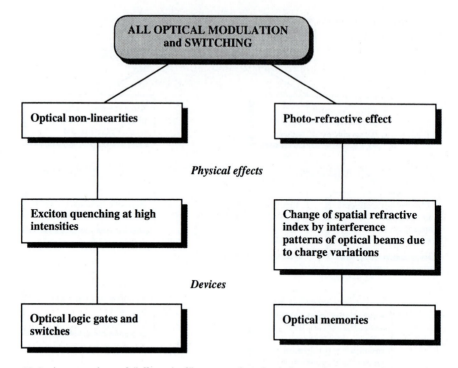

Figure 12.2: An overview of "all optical" approaches for information processing devices.

In this chapter we will focus on the optoelectronic devices that use both light and electronic signals to produce switching, modulation and amplification.

12.2   OPTICAL MODULATORS

→   As noted in Chapter 8, information content is coded in optoelectronic devices as a variation of some aspect of the optical output from a LED or a laser diode (LD). The property of light that is modulated depends upon the particular application and technology available and could involve the amplitude of the optical signal, the phase of the signal, widths of pulses being sent, etc. Regardless of the coding scheme used, it is clear that modulation of light is a critical ingredient of optoelectronic technology. In this regard the optoelectronic devices are somewhat lacking when compared to electronic devices.

In the electronic devices (MOSFET, MESFET, MODFET, BJT, HBT, etc.), the electrical signal is readily modulated by a gate signal or a base signal. The gate or the base is an integral part of the device. The device dimensions are quite small and allow modulation frequencies of up to 20 GHz in advanced silicon technology and up to several hundred gigahertz in advanced heterostructure technologies. When compared with electronic devices, the size of optoelectronic sources is rather huge. We have already discussed the limitations for LEDs in high speed applications. The laser diode can, in principle, overcome some of the problems but, as discussed in Chapter 11, is still limited to 20-30 GHz. Additionally, the laser diode does not have a simple "gate" from which one can control the light output.

There are two schemes used to modulate the optical signals in LEDs or LDs. The first one is direct modulation in which an electronic circuit is designed to simply modulate the current injected into the device. Since the light output is controlled by the injected current, one has the desired modulation. The "driver" for this direct modulation may be an FET or an HBT. Because of the different structural nature of the LD and the electronic device, it is not a trivial matter to fabricate these devices on the same chip. Thus such circuits are usually based on hybrid technology. The goal to build the driver and the source on the same substrate and develop an OEIC technology for the transmitter is being pursued actively in a number of labs.

Direct modulation of the laser is simple, which is the biggest attraction of this scheme as shown in Fig. 12.3. Note that, compared to electronic devices, the circuit is by no means simple or easily "manufacturable." It is simple when compared to other modulation approaches we will discuss later. However, direct modulation has several problems including the upper modulation frequencies (which are ~ 40 GHz) and shift in emission frequency.

External modulation of lasers offers the advantages and disadvantages listed in Fig. 12.3. The key disadvantage is that the modulator is usually large on the scale of microelectronic devices and is usually not a part of a simple integrated circuit. Not only that, it is usually fabricated from materials that are not compatible with semiconductor technology. Recently, however, this is changing with the use of quantum well systems based upon the same semiconductors that are used in laser diodes.

In the external modulation scheme, the light output passes through a material whose optical properties can be modified by an external means. Depending upon the means used, one can have electro-optic, acousto-optic, or magneto-optic modulators. The electro-optic effect is most widely used for high speed applications and is most compatible with modern electronics.

In this section we will discuss briefly the electro-optic modulators. We will also discuss the use of quantum wells in enhancing the electro-optic effect as well as in producing electro- absorption modulators. Finally, we will discuss the liquid crystal display (LCD) devices. The LCDs are not used for high speed modulation, but are an integral part of the modern display technology. Also, they may have been married so well with semiconductor technology that it is important to examine the underlying physics.

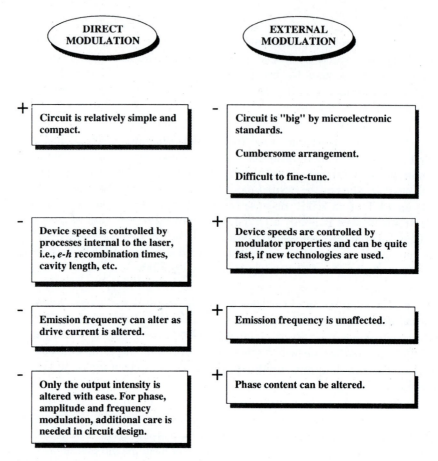

Figure 12.3: A comparison of the advantages (+) and disadvantages (−) of internal (or direct) and external modulation of laser diodes.

The reader should review Section 5.7 and 5.8 where the basic effects which are exploited for electro-optic modulation are discussed. Here we will simply discuss how the effects are exploited in devices.

12.2.1   Electro-Optic Modulators

Electro-optic modulators can produce amplitude, frequency or phase modulation in an optical signal by exploiting the electro-optic effect in which the optical properties of a crystal can be altered by an electric field. A number of crystals exist which have desirable response behavior. These include potassium dihydrogen phosphate (KDP), ferroelectric peroskites such as $LiNbO_3$ and $LiTaO_3$ as well as semiconductors such as GaAs and

CdTe.

In general, the refractive index of a crystal is not isotropic and is described in terms of an index ellipsoid (or indicatrix). The equation for the indicatrix, along the principle axis x_i, is (see Section 5.7 and Appendix E.2)

$$\sum_{i=1}^{3} \frac{x_i^2}{n_{ri}^2} = 1 \qquad (12.1)$$

where n_{ri} are the principle refractive indices. When an electric field is applied along a particular direction, the refractive indices are affected. However, because of anisotropy, different indices are affected differently. In uniaxial crystals, there are two axis, say x_1 and x_2, along which the refractive index is the same, say n_{ro}, and one along which the index is n_{re}. These are called the ordinary and extraordinary indices, respectively.

When a linearly polarized light wave enters the modulator as shown in Fig. 12.4, it resolves into two components. In general, as discussed in Appendix E.2, the two directions have a different refractive index and a phase ϕ develops between them after they propagate a distance L. Consider an input signal that is linearly polarized and given by

$$F_x = \frac{F_o}{\sqrt{2}} exp(i\omega t)$$

$$F_y = \frac{F_o}{\sqrt{2}} exp(i\omega t)$$

After transmission through the modulator, the wave emerges with a general polarization given by

$$F_x = \frac{F_o}{\sqrt{2}} exp(i\omega t + i\theta_1) \qquad (12.2)$$

$$F_y = \frac{F_o}{\sqrt{2}} exp(i\omega t + i\theta_2) \qquad (12.3)$$

with the phase difference given by $\phi = \theta_2 - \theta_1$. If ϕ is $\pi/2$, the output beam is circularly polarized and if it is π, it is linearly polarized with polarization 90° with respect to the input beam. If the output beam passes through a polarizer at 90° with respect to the input beam polarizer as shown in Fig. 12.4, the modulation ratio is given by

$$\frac{I_{out}}{I_{in}} = sin^2 \frac{\phi}{2} \qquad (12.4)$$

Thus if ϕ can be controlled by an electric field, the intensity can be modulated. For GaAs, the electric field dependent phase is given by (see Section 5.7)

$$\phi = \frac{2\pi}{\lambda} L n_r^3 r_{41} \frac{V}{d} \qquad (12.5)$$

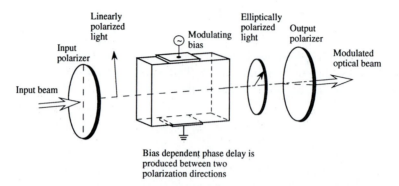

Figure 12.4: The use of an electro-optic device to modulate an optical signal. The applied bias introduces a phase change between light travelling in two polarization directions and the output light is modulated.

where λ is the wavelength of light, L the device length, n_r the GaAs refractive index, r_{41} the electro-optic coefficient for GaAs, V the transverse applied bias, and d the thickness of the modulator. Similar results are obtained for other electro-optic materials.

It may be noted that the electric field can be applied in a transverse or longitudinal way to the modulator. When a transverse field is applied, the effect is called Kerr effect. On the other hand, when the field is longitudinal the effect is called Pockel effect.

It is clear from Eqn. 12.5 that a high electro-optic coefficient can allow one to achieve a modulation using a smaller interaction length for the same applied field. However, the electro-optic coefficients of most materials are rather small ($\sim 10^{-12}$ m/V), so that for realistic bias values, the length required is quite long (millimeters or more).

As noted above, the small values of the electro-optic coefficients force one to make external modulators from long devices that are incompatible with modern microelectronics. Thus there is an intense search for materials with improved physical properties. An important advance in this direction is due to quantum well structures. In Chapter 5 we have discussed the effect of the e-h Coulombic interaction on the optical absorption spectra. The e-h pair is the exciton and it results in sharp peaks in the absorption (or transmission) spectra as discussed in Chapter 5. If an electric field is applied perpendicular to the quantum wells, the exciton peak is shifted. This phenomenon, shown in Fig. 12.5 and discussed in detail in Section 5.8, is called the Quantum Confined Stark Effect (QCSE) and can be used for optical modulation. The modulation scheme is shown in Fig. 12.5, where a MQW structure is placed in a p-i(MQW)-n structure. The p-i-n structure is reverse biased and an optical signal can be modulated by applying a modulated electrical signal to the structure. If an optimum structure is designed in which the carriers can be swept out rapidly, the modulation speeds could reach ~ 100 GHz.

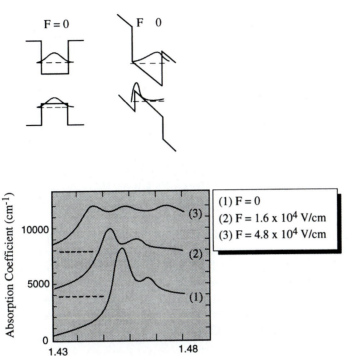

Figure 12.5: The quantum confined Stark effect in which a transverse electric field alter the optical absorption spectra in a quantum well.(After D.S. Chemla, *Nonlinear Optics: Materials and Devices*, Springer-Verlag, New York (1986).)

Modulators based on QCSE can be used in two modes, as discussed in Chapter 5, Section 5.8. In the electro-optic mode the optical frequency is below the exciton energy so that no absorption occurs. In this case, the electric field shifts the exciton position which alters the refractive index. The benefit of using quantum wells is that the electro-optic coefficient can increase by up to two orders of magnitude over the bulk value. Thus devices can be of the order of 100 μm, instead of the order of a centimeter. The other mode of operation is based on actual light absorption. Such modulators are called electro absorption modulators.

The excitonic absorption for the heavy hole exciton is given approximately by (see Section 5.8)

$$\alpha_{ex}(\hbar\omega) \cong \frac{2.9 \times 10^3}{W\sigma} exp\left\{-\frac{(E^{ex} - \hbar\omega)^2}{\sqrt{2}\sigma^2}\right\} \ cm^{-1} \qquad (12.6)$$

where W is the well in size Å, σ is the exciton linewidth in eV, and E^{ex} is the position of the exciton peak energy. For example, the peak absorption is $\sim 1.45 \times 10^4$ cm^{-1} if $W = 100$ Å and $\sigma \sim 2$ meV.

The shift in the exciton energy with applied field is approximately

$$\Delta E^{ex} \cong -3 \times 10^{-20} \left(m_e^* + m_{hh}^* \right) F^2 W^4 (eV) \tag{12.7}$$

where F is the applied field in V/cm and m_e^*, m_{hh}^* are the electron and heavy- hole masses in units of free electron mass.

The shift in the excitonic absorption resonance results in a change in the real part of the dielectric constant of the device. This effect can be calculated by using the Kramers- Kronig relationship and was discussed in Section 5.8. As can be seen from the quadratic nature of the shift in the exciton resonance, the change in the refractive index is also quadratic.

In order for the electro-optic effect to be as strong as possible, it is necessary for the optical beam to be as close to the exciton resonance as possible. However, it cannot be too close, otherwise the light will be absorbed which is not desirable. Thus, an optimization has to be done between the closeness of the photon energy to the exciton resonance, the voltage applied for switching and the interaction length of the device. Due to the strong quadratic electro-optic effect, it is possible to obtain a good modulation ratio in devices as small as 100 μm.

In Section 5.9 we have discussed how the exciton resonance can be quenched due to screening of the Coulombic interaction when charge is injected into the device. This approach can be exploited to produce an effective optical modulator. A device using this effect is called the barrier reservoir and quantum well electron transfer (BRAQWET) modulator. By introducing charge in the channel well, the optical properties are modified.

12.2.2 Electro-Absorption Modulators

In the electro-absorption mode changes in the imaginary part of the refractive index are exploited. The energy of the incoming photon is close to the exciton resonance and by applying an electric field, the transmittance of the device is changed.

In Fig. 12.6 we show a typical arrangement for using QCSE in optical modulation. The light signal could impinge vertically or in a waveguide mode. The device itself is a p- i(MQW)-n reverse biased diode. Depending upon the energy of the photons, the device could be normally ON (i.e., high transmittance at no bias) or normally OFF as shown.

The modulation depth that can be reached is quite large (see Example 12.2) even when the active region is only a few microns. The modulation depth may be defined

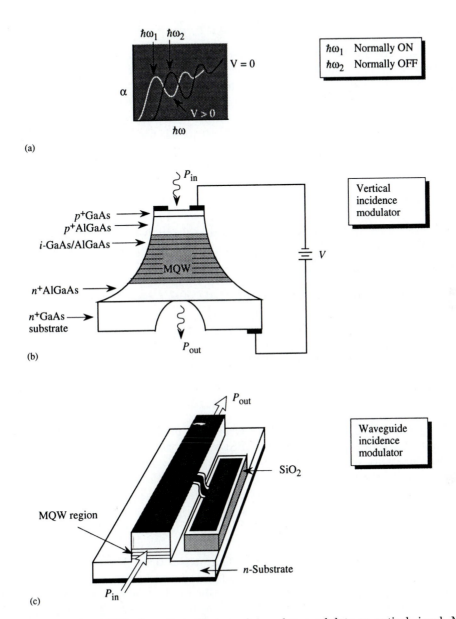

Figure 12.6: (a) The QCSE absorption effect can be used to modulate an optical signal. Normally ON or normally OFF devices can be produced. The QCSE can be used in quantum well based modulators either for (b) vertical incidence light or for (c) light incidence in a waveguide geometry. Both GaAs based and InP based devices have been demonstrated.

as

$$m = \frac{I_{trans}(V_{ON})}{I_{\text{trans}}(V_{OFF})} \qquad (12.8)$$

where $I_{trans}(V_{ON})$ is the transmitted light when the absorption is the least (at voltage V_{ON}) and $I_{\text{trans}}(V_{OFF})$ is the transmitted light when the absorption is high (at voltage V_{OFF}). The modulation depends upon the position of the photon energy relative to the exciton resonance, the interaction length of the device with the optical beam, the bias swing applied and the exciton linewidth. For vertical incident light, the interaction length is $\sim 2~\mu$m and a voltage signal of $\sim$ 5- 10 volts is needed to produce an ON/OFF ratio of $\sim$ 10:1. For the waveguide geometry, the interaction length can be quite large ($\sim 100~\mu$m) and much smaller voltage levels are needed to get a high ON/OFF ratio.

The intrinsic time response of the QCSE modulator is determined by the mechanism of carrier extraction from the quantum well. Electron hole pairs are generated during absorption and these need to be extracted, otherwise the charge build up will render the device inoperable. When the applied electric field is small the electrons and holes generated during absorption will be unable to escape from the quantum wells and will have to recombine as shown in Fig. 12.7a. Since the *e-h* recombination time is $\sim$ 1-2 ns, the device cannot operate beyond a GHz in this case.

If the applied bias is large, as shown in Fig. 12.7b, the carriers (which are in quasi-equilibrium in the quantum well) can escape by either above the barrier thermionic emission, or by tunneling. At high fields, the device can respond at times approaching a few picoseconds. Thus, if the device is designed with small parasitics, it can operate up to 100 GHz.

The *p-i-*(MQW)-*n* modulator can be used in the vertical incidence mode as a programmable spatial light modulator (PSLM), i.e., as a programmable transparency. The transmittance of each pixel can be controlled by an electrical signal. In such a case, if an optical beam passes through the PSLM as shown in Fig. 12.8a, it emerges after being multiplied by the transmittance of the pixels. This allows a highly parallel approach to information processing. As shown in Fig. 12.8b, this approach can be used for vector- matrix multiplication.

In the vector-matrix architecture, light passes through a vector with transmittance T_i for the pixel i. This light is spread out via a lens to impinge upon an array representing a matrix with elements T_{ij}. Finally, the light is collected by another lens onto a vector photodetector array. If the incident light is I_o on each pixel, the light level on the detector pixel j is

$$I_j = I_o \sum T_i T_{ij} \qquad (12.9)$$

which is the vector-matrix product. Other architectures have also been proposed to carry out a variety of important algorithms. While many of these architectures have been demonstrated in labs, the transition to usable technology has been slow. For this

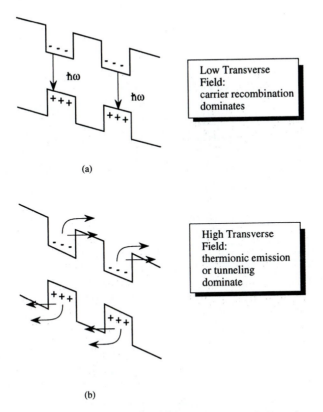

Figure 12.7: A schematic description of the extraction of *e-h* pairs during absorption in quantum wells. (a) At low fields, the carriers must recombine to be extracted. (b) At high fields, the carriers can tunnel out or be thermionically emitted.

include reliability and non-uniformity problems in quantum well based structures, high cost of devices compared to Si devices, difficulties in using multi-chip modules where alignment of light beams can be a serious problem. It is also important to point out that in the GaAs/AlGaAs technology, the substrate has to be etched away if the modulator is to be used in the transmission mode. This poses an additional complication for the technology.

12.2.3 Liquid Crystal Devices

$\mathcal{R}$ Liquid crystal devices are not semiconductor devices, but they have been so well integrated with semiconductor technology that it is important to understand their operating principles. These devices are used as display devices (liquid crystal displays or LCDs), programmable spatial modulators (programmable transparencies), display

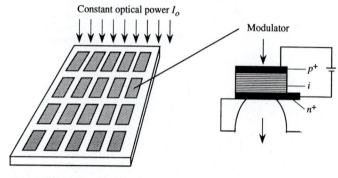

(a)

Output from pixel (i,j): $O_{ij} = I_o T_{ij}$

T_{ij} = transmittance of pixel (i,j)

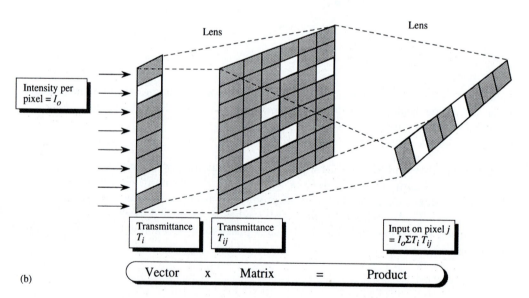

(b)

Figure 12.8: (a) A programmable spatial light modulator (PSLM) in which the transmittance of a pixel can be tuned electronically. (b) The use of PSLMs to calculate a vector-matrix product in parallel by using appropriate lenses.

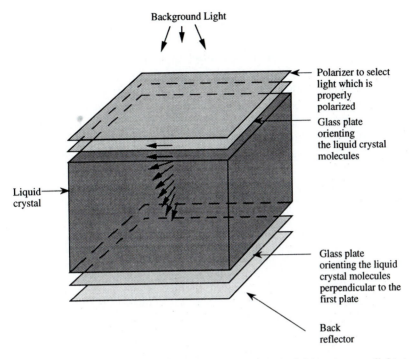

Background Light

Polarizer to select
light which is
properly
polarized

Glass plate
orienting
the liquid crystal
molecules

Liquid
crystal

Glass plate
orienting the liquid
crystal molecules
perpendicular to the
first plate

Back
reflector

Figure 12.9: A schematic of an LCD showing how an electric field can cause light to reflect back or scatter. The arrows show the orientation of the molecules of the liquid crystal. In the absence of a field, the incoming light is reflected back. An ac electric field disturbs the orientation sequence, causing no reflection and the device looks dark.

screens for laptop computers, and a host of other applications.

Liquid crystals are materials which have rod-like molecules that are free to twist in the crystal around certain axis. In the temperature range of ~ -10 to 50°C the molecules form a nice ordered sequence. Since the molecules have strong dielectric anisotropics, the ordered sequence can be disturbed by a strong enough electric field. The molecules also have optical anisotropics so that when light passes through them, the molecules can cause a change in the polarization of the light.

LCDs are fabricated by placing a crystal in between two glass plates, as shown in Fig. 12.9 The glass plates have a preferential direction (produced by rubbing or some other lithographic technique) so that these directions orient the liquid crystal molecules. The plates are oriented so that the preferential directions are perpendicular to each other causing a 90° (or even a 270°) twist in the liquid crystal molecules.

If light polarized along the molecules on the top impinges on the LCD, it changes its polarization by 90° as it goes to the back plate which is reflecting. As a result the light is reflected back.

If an ac electric field (ac field is used to avoid electromigration) is applied to the LCD, the orientation of the molecules is disturbed so that the light is scattered instead of being reflected back. Thus the device can appear bright or dark and can be used for a variety of applications. Since in this text we have focused on semiconductor optoelectronic devices, we will not discuss liquid crystals in any detail. It is important, however, to realize that these materials are the basis for many new technologies and, therefore, hold the key to explosive growth of new information processing systems.

EXAMPLE 12.1 Calculate the length required for a GaAs electro-optic modulator for maximum modulation depth. The parameters for the device are the following:

Refractive index,	n	=	3.6
Electro-optic coefficient,	r_{41}	=	1.2×10^{-12} m/V
Voltage swing,	V	=	10 V
Device thickness,	d	=	1.0 μm
Optical wavelength,	λ	=	1.0 μm

The length is given by

$$L = \frac{\lambda d}{2n^3 r_{41} V} = \frac{(10^{-4})(10^{-3})}{2(3.6)^3(1.2 \times 10^{-10})(10)}$$
$$= 0.893 \ cm$$

This is a long device on the scale of microelectronics where one could have thousands of devices on the same size GaAs chip.

EXAMPLE 12.2 Consider an electro-absorption modulator based on 100 Å GaAs/ $Al_{0.3}Ga_{0.7}As$ quantum well structures. The active layer thickness consists of 1 μm of GaAs wells. In absence of the electric field, the optical signal energy coincides with the exciton energy. Calculate the ON/OFF ratio if a field of 70 kV/cm is applied. The exciton linewidth σ is 3 meV.

In the absence of the field, the absorption coefficient is ($\hbar\omega = E^{ex}$)

$$\alpha(0) = \frac{2.9 \times 10^3}{(100)(3 \times 10^{-3})} = 0.97 \times 10^4 \ cm^{-1}$$

When a field is applied, the shift in the exciton energy is

$$\Delta E^{ex} = 3 \times 10^{-20} (m_e^* + m_{hh}^*) F^2 W^4$$
$$= -(3 \times 10^{-20})(0.51)(7 \times 10^4)^2(100)^4 = 7.5 \ meV$$

The new absorption coefficient is

$$\alpha(F) = \frac{2.9 \times 10^3}{(100)(3 \times 10^{-3})^2} \ exp\left[-\frac{(\Delta E^{ex})}{\sqrt{2\sigma^2}}\right]$$

$$= 1.16 \times 10^2 \; cm^{-1}$$

The ON/OFF ratio in the transmitted light is

$$\frac{I(F = 70 \; kV/cm)}{I(F = 0)} = \frac{exp(-\alpha(F)L)}{exp(-\alpha(0)L)}$$

$$= \frac{exp\left\{-(1.16 \times 10^2)(10^{-4})\right\}}{exp\left\{-(0.97 \times 10^4)(10^{-4})\right\}} = \frac{0.988}{0.38} = 2.6$$

Thus a fairly large modulation can be obtained with just a micron of material, using reasonable applied biases.

12.3 SWITCHING DEVICES: DIRECTIONAL COUPLERS

⤳ In the previous section we have seen how the electro-optic effect can be used to modulate an optical signal. By changing the refractive index of a material, one can introduce a phase modulation, polarization modulation or, through interference of two beams, an intensity modulation. It is straightforward to use the same concept for switching applications. An important application in switching is satisfied by the directional coupler.

The directional coupler in its basic form couples two optical inputs I_1 and I_2 to two outputs O_1 and O_2 as shown in Fig. 12.10. The device should be able to connect an entering optical signal to either O_1 or O_2. Such a device is extremely important in communication applications, where it can be used to route signals.

To understand the basic operation of the directional coupler, consider two single mode waveguides on the same substrate as shown in Fig. 12.10. The two guides are parallel to each other and are separated by a gap g over a length L. Outside the interaction length L, the guides separate out as shown. The reason the two waveguides are brought close to each other is to provide a small coupling between the optical modes of the two guides. As the waves progress through the structure, the optical energy transfers back and forth between the two guides much like the problem of two coupled pendulums. This feature together with the electro-optic effect can be exploited to make switching devices.

In the discussion of the DFB laser, we had considered the equations of motion for two coupled optical modes. We will use the same equations to understand the operation of the directional coupler. Let A_1 and A_2 be the amplitudes of the optical modes in the guide 1 and 2. We assume that the propagation constants in the two guides are the same. Also let K represent the coupling coefficient for the two guides. The coupled mode equations are

$$\frac{dA_1}{dz} = -i\beta A_1(z) - iKA_2(z)$$
$$\frac{dA_2}{dz} = -i\beta A_2(z) - iKA_1(z) \tag{12.10}$$

The propagation constant is

$$\beta = \beta_r - \frac{i\alpha_{\text{loss}}}{2} \tag{12.11}$$

where α_{loss} is the loss coefficient in the guide and β_r is the real part of the propagation constant.

Let us assume that initially, light is coupled only into the guide 1 at $z = 0$, so that

$$A_1(0) = 1$$
$$A_2(0) = 0 \tag{12.12}$$

It can be seen that the solutions to the coupled mode equations are

$$A_1(z) = cos\ (Kz)e^{i\beta z}$$
$$A_2(z) = -isin\ (Kz)e^{i\beta z} \tag{12.13}$$

The optical power in the guides at a point z is

$$P_1(z) = A_1 A_1^* = cos^2(Kz)e^{-\alpha_{\text{loss}} z}$$
$$P_2(z) = A_2 A_2^* = sin^2(Kz)e^{-\alpha_{\text{loss}} z} \tag{12.14}$$

As can be seen from these equations and as shown in Fig. 12.10b, the power sloshes back and forth between the two guides as the signal propagates. A complete transfer of power occurs when the signal propagates a distance

$$L = \frac{\pi}{2K} \tag{12.15}$$

If the guides are not phase matched ($\beta_1 \neq \beta_2$), the entire light is not coupled from one guide to another. If we write

$$\Delta = |\beta_1 - \beta_2| \neq 0 \tag{12.16}$$

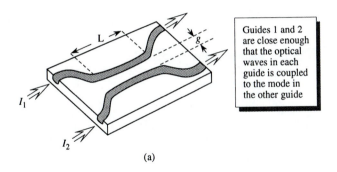

(a)

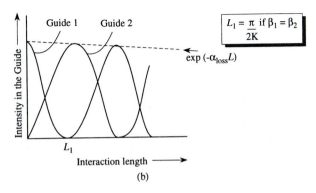

Interaction length ⟶

(b)

Figure 12.10: (a) A schematic of the directional coupler where two waveguides are separated by a gap g over a length L. (b) The transfer of optical energy from one guide to another when initially the light signal enters Guide 1.

It can be shown that if $A_1(0) = 1; A_2(0) = 0$, one gets

$$\frac{P_2(z)}{P_1(0)} = \frac{K^2}{K^2 + \Delta^2} \sin^2\left[(K^2 + \Delta^2)^{1/2} z\right] \qquad (12.17)$$

Clearly, if $\Delta^2 >> K^2$, there is no coupling of light.

The working of the directional coupler depends upon altering the value of Δ by an electric field. Note that if $\Delta = 0$, the length over which the first transfer of energy occurs is

$$\boxed{L_1 = \frac{\pi}{2K}} \qquad (12.18)$$

Now if one applied a voltage so that a change Δ_{sw} occurs so that

$$\boxed{(K^2 + \Delta_{sw}^2)^{1/2} L_1 = \pi} \qquad (12.19)$$

then there will be no light emerging from guide 2 and the light will re-emerge from the guide 1 as shown in Fig. 12.11a.

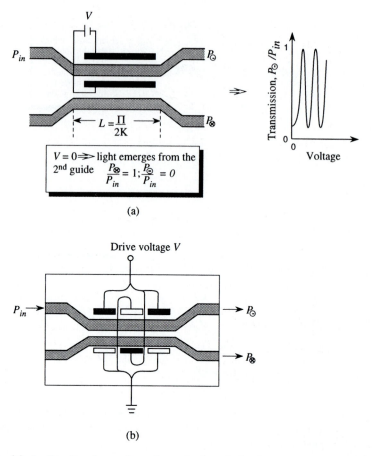

Figure 12.11: (a) A directional coupler where the length is chosen so that complete transfer occurs at zero applied bias. When the bias is changed, the light coupling can be changed. (b) A "push-pull" electrode arrangement for a directional coupler to double the phase change between the guides.

The switching is controlled by the coupling coefficient K and the strength of the electro- optic effect and the bias applied (through Δ). In order that the voltage level needed is small, it is important that the material have a large electro-optic effect. The coupling coefficient depends exponentially on the separation g, since the coupling is due to the evanescent waves in each guide.

In order to require less total applied potential, it is common to use a "push-pull" electrode arrangement as shown in Fig. 12.11b. The total shift Δ is then doubled. The shift Δ has been discussed in Section 5.9 and is given by the expression

$$\Delta = \frac{\pi n_{ro}^3}{2\lambda} \left[r_{eff} E_y + (S_{12} - S_{11}) E_y^2 \right] \tag{12.20}$$

where r_{eff} is the effective electro-optic coefficient. For GaAs the value to be used is of r_{41}. The quadratic terms S_{12} and S_{11} are essentially zero for bulk devices but can be large for quantum well devices.

For directional coupler devices based on bulk materials (L_iNbO_3, GaAs, L_iTaO_3, etc.,) the size of the device is fairly large, since the electro-optic coefficient is not so large. Typically the devices require an interaction length of $\sim$ 1 cm and a bias voltage of $\sim$ 5-10 V. The use of quantum well structures are allowing smaller device structures due to the improved electro-optic effect related to the excitonic features.

12.4 QCSE BASED SWITCHING DEVICES

$\rightsquigarrow$ We have discussed in the previous section how QCSE can be exploited to modulate an optical signal. Can the effect be exploited to make switches that can be controlled electronically or optically? The answer is yes, and over the last decade a very large variety of arithmetic and logic devices have been demonstrated using this effect. To understand the use of QCSE in switching devices, it is important to examine the photocurrent produced when light shines on the p-i(MQW)-n diode. The device photocurrent is the same as a p-i-n detector discussed in Chapter 7, Section 7.6 *except that the absorption coefficient depends upon the applied field.* As a result, the photocurrent has a strong dependence upon the electric field or the voltage across the device.

In order for the photocurrent to flow in the p-i-n device with multiquantum wells in the intrinsic region, the e-h pairs that are produced must be extracted. As discussed in the previous section, and as shown in Fig. 12.12, there are two regions of operation: i) at low applied fields, the carriers are unable to escape from the wells and recombine with each other. The collection efficiency, and hence the photocurrent, is extremely small in this case; ii) At high fields ($F > 20$ kV/cm) the carriers are able to escape by thermionic emission and tunneling and the photocurrent is simply proportional to the optical flux and the absorption coefficient. At high biasing, the photocurrent is simply given by

$$I(ph) = Ae \int_o^W GL \cdot dx \qquad (12.21)$$

where

$$G_L(x) = \frac{\alpha P_{op}(x)}{\hbar \omega} \qquad (12.22)$$

Assuming a uniform carrier generation rate G_L, we have

$$I(ph) = \frac{Ae\alpha(V)P_{op}W}{\hbar \omega} \qquad (12.23)$$

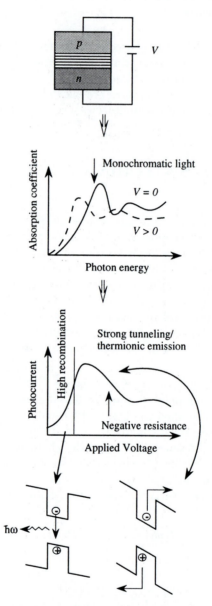

Figure 12.12: The excitonic spectra of a reverse biased p-i(MQW)-n modulator. The photocurrent produced essentially follows the absorption-voltage relation once the applied field is strong enough for e-h collection by tunneling a thermionic emission.

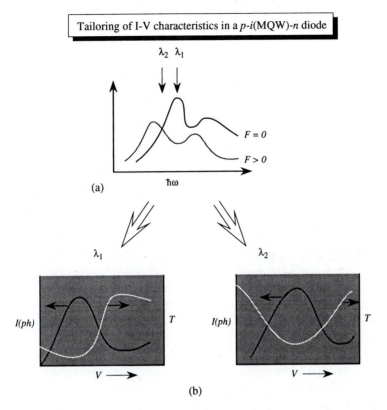

Figure 12.13: (a) Excitonic absorption spectra in the absence and presence of a field. (b) Current-voltage and transmittance-voltage curves at different wavelengths.

where W is the width of the active region (i.e., the well parts of the region). Since α is field dependent, the photocurrent is also field dependent.

In Fig. 12.12 we show how the device can have a photocurrent that has a negative differential resistance in it. In fact, by a suitable choice of position of the incidents photon energy, one can have an $I(ph) - V$ relationship that is tailorable as shown in Fig. 12.13. Such a tailorability is not possible in the conventional electronic devices and this can be exploited for a variety of devices.

The Self Electro-Optic Effect Device (SEED)
The first use of QCSE device as an optical switch was the SEED where the strong negative resistance in the p-i-(MQW)-n device is used for low power switching. In Fig. 12.14a we show a schematic of the device in which two parallel p-i(MQW)-n diodes are connected in series to a load resistance R. One of the diodes is a modulator through which the light can transmit while the other one can be called a controller. The equivalent circuit is shown in Fig. 12.14b.

In Figure 12.14c we show the operation of the device. The photocurrent-voltage curves are shown for two different inputs of optical power $P_{op}(\ell)$ and $P_{op}(h)$ into the controller. Also shown is the load line. When the optical power is low $P_{op}(\ell)$, the voltage across the diode is given by the point B where the transmittance is high and the diode voltage is high. When the power goes to a high value $P_{op}(h)$ the stable point switches over to A where the transmittance is low and the diode voltage is low. Thus a control beam can cause the switching of the modulator beam.

A very large number of device operations have been demonstrated using SEED and similar concepts. These include essentially all of the important logic operations. In its simplest form, the SEED has no gain and it is, therefore, not possible for one device to control multiple devices. Gain can be introduced by incorporating QCSE in a transistor instead of a diode.

The Multiquantum Well HBT

In order to make the SEED more compatible with OEIC technology, it is important to have built-in electronic gain in the device. Gain is also essential for larger tolerance in the devices as well as large fan-out and cascadability. Such gain can be realized by using an HBT with a MQW region in the base collector region. This device provides the following advantages:

i) Since the HBT is a vertical device, a large uniform potential can be developed across the base collector region to cause QCSE.

ii) The third terminal, i.e., the base, allows one to efficiently coupled optical and electronic features.

iii) The entire structure of the n^{+}-p-i(MQW)-n HBT and the p-i(MQW)-n modulator can be grown epitaxially in one step.

A schematic of the integrated MQW-HBT is shown in Fig. 12.15a. The MQW-HBT is shown in a circuit where it is controlling a modulator. The important point to realize is the presence of amplification of the photocurrent which allows low optical power switching. In a usual SEED the load resistances required are $\sim$ 100 KΩ with an applied bias of 10-15 volts. The load resistance is $\leq$ 5 KΩ for the MQW-HBT's. An important asset of the MQW-HBT is that the I-V curves can be shifted either by optical power or through base current and as long as some optical power is present, the negative resistive region is maintained. In Fig. 12.15 we have shown schematically how the switching operation can be carried out using as input either the base current or the optical intensity.

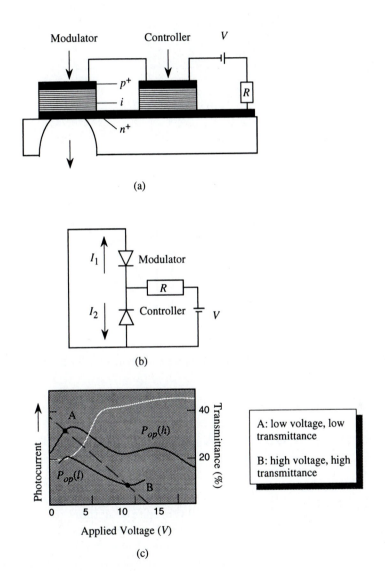

Figure 12.14: A schematic for the operation of the SEED. The I-V curve and the load line illustrate the switching concept.

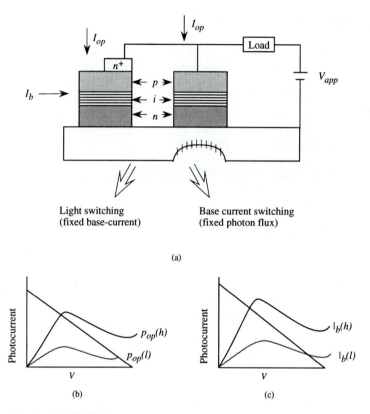

Figure 12.15: (a) An MQW-HBT structure along with the schematic of the switching operation by (b) optical power and (c) base current.

12.5   OPTICAL AMPLIFICATION: THE LASER AMPLIFIER

$\longrightarrow$ In the previous two sections, we have discussed a number of optoelectronic switching devices. The reader may be wondering why, in spite of all these great devices, general purpose computer designers are not rushing to use these devices. There are many reasons why a new device technology takes time to reach the market. Some of the reasons have to do with the enormous inertia present in the existing way of doing things. Keep in mind that silicon (and to lesser extent, GaAs) technology has invested billions of dollars into chip fabrication plants and process control software. It is highly unlikely that all this knowledge and capital investment will be cast aside simply because QCSE was found to exist. A second class of reasons for a device technology to take time to reach the consumer is that there may be some hidden pitfalls in the technology. One of the important drawbacks in the devices discussed above is related to signal amplification or gain.

Apart from special purpose applications, most switching applications require a signal to cause the switching of not one but many devices. This property, called the fan-out, is especially important for general purpose switching devices. In electronic devices the ability to provide gain or amplification ensures that fan-out is quite high. If optical signals are to be a part of a switching circuit, it is necessary that the optical signal not suffer attenuation as it passes through a device. In Fig. 12.16a we show that in an electronic transistor, the input signal is significantly amplified (either in voltage or current level). On the other hand, in the optoelectronic devices considered so far, the output signal is attenuated.

How can one amplify an optical signal so that the signal does not attenuate as it passes through a device? Conceptually, the solution is quite simple. In fact, in our discussion of semiconductor lasers, we have discussed the concept of gain. If the gain of a medium is positive an optical wave grows as

$$I_{ph}(z) = I_{ph}(z = 0)exp(gz) \tag{12.24}$$

This concept can be carried further in a laser cavity to produce the laser amplifier. The laser amplifier is essentially a laser without feedback. The structure is biased with positive gain, as shown in Fig. 12.17a. In this state there are essentially no photons emanating from the cavity, i.e., $S \sim 0$. When an optical signal enters the cavity as shown in Fig. 12.17b, the presence of the photons causes an onset of stimulated emission, so that a *coherent and amplified signal emerges from the cavity.*

The laser amplifier can operate as either a travelling wave amplifier (TWA) or a Fabry-Perot amplifier. In the TWA mode, the optical signal simply makes a single pass through the cavity and the signal grows during this pass. In the Fabry-Perot cavity, the signal goes through multiple passes since it is reflected at the mirrors. The TWA mode is specially attractive for high speed applications where pulses with subpicosecond width can be amplified.

In Fig. 12.18 we show the important performance parameters of an amplifier and how they impact performance. Important issues that need to be addressed are the width of the gain spectra which determines how suitable the amplifier is for WDM applications and for ultrashort pulse amplification. The bandwidth of several terrahertz of semiconductor amplifiers is adequate for many applications, but needs to be enhanced for some optical communication applications. The issues of gain saturation, noise and chirp of the amplifier will be discussed below.

We divide the amplifier study into two areas: i) amplification of CW signals or pulses of width greater than a nanosecond; ii) amplification of short pulses (1ns to subpicosecond). For CW applications we can assume that the gain and the injection carrier density in the amplifier are time independent, while for short pulses, the time dependence is significant and plays an important role.

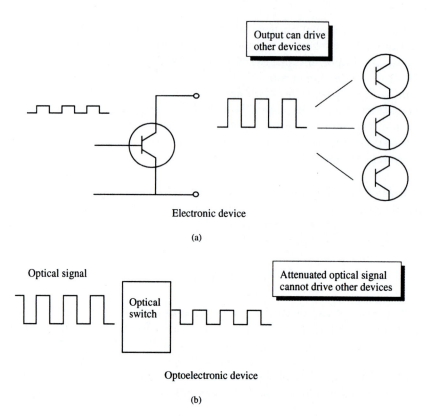

Figure 12.16: (a) In a conventional electronic device (a transistor), a small input electrical signal produces a large output signal. The amplification occurs by using the electrical energy from an attached battery. (b) In optoelectronic devices, the signal decays as it emerges. Amplifiers are needed to boost the signal.

12.5.1 Amplifications of CW Signals

$\longrightarrow$ To describe the performance of the semiconductor laser amplifiers, we make use of the rate equations for the electron and photon system. We write the amplifier gain in terms of the carrier concentration as

$$g = \frac{dg}{dn}(n - n_o) \qquad (12.25)$$

where Γ, the optical confinement factor, is included so that g is the cavity gain and n_o is the carrier concentration at transparency, i.e., at zero gain. The rate equation for the injected carrier concentration is (writing in terms of optical power P instead of photon density)

$$\frac{dn}{dt} = \frac{J}{ed_{las}} - \frac{n}{\tau} - \left(\frac{dg}{dn}\right)\frac{(n - n_o)P}{A_m \hbar \omega} \qquad (12.26)$$

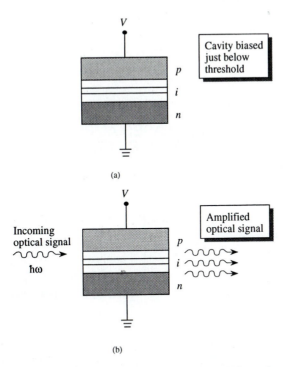

(a)

(b)

Figure 12.17: (a) The laser amplifier biased just under threshold so that there is no photon output. (b) An input signal causes stimulated emission so that the output signal is amplified.

where A_m is the area of the optical mode. For amplification of CW optical signals or pulses with widths longer than the carrier recombination time τ, we can assume steady state conditions. Using the value of n at $dn/dt = 0$, we get for the amplifier gain

$$g = \frac{g_o}{1 + P/P_s} \tag{12.27}$$

where g_o is the low power gain and P_s is called saturation power. These quantities are given by

$$g_o = \frac{dg}{dn}\left(\frac{J\tau}{ed_{las}} - n_o\right) \tag{12.28}$$

$$P_s = \hbar\omega \ A_m/\left(\tau\frac{dg}{dn}\right) \tag{12.29}$$

The above equations show that when the optical signal impinges on the amplifier, the gain decreases for a fixed injection current. This is quite understandable and arises from the increased net e-h recombination rate due to the stimulated emission. As a result, the injected carrier density for a fixed current decreases causing the gain saturation effect. This effect plays an important role in the amplification and alteration of pulse shapes of optical pulses as will be discussed later.

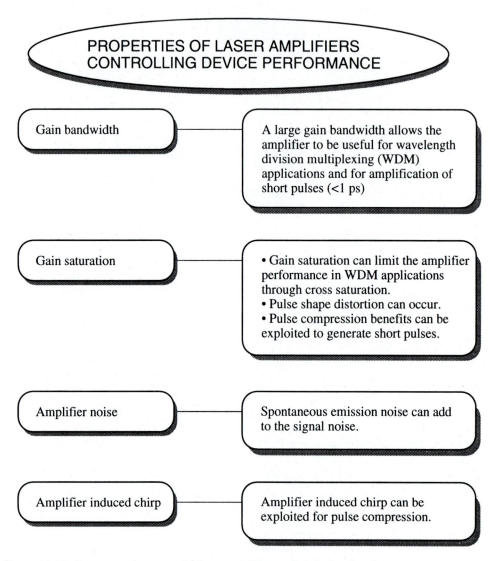

Figure 12.18: Important features of a laser amplifier and their implications.

It may be noted that if the amplifier is being used for amplifying a number of pulses at different frequencies (say, in a wavelength division multiplexing system), the gain saturation effect for a particular wavelength pulse is determined by the power in all the channels. This causes cross-saturation, an effect that is quite undesirable in lightwave systems.

To find the gain g, a pulse propagating through an amplifier, we note that the power variation in space is given by

$$\frac{dP}{dz} = gP$$

$$= \frac{g_o P}{1 + P/P_s} \tag{12.30}$$

Integration of this equation over the length L of the amplifier gives an implicit equation for the amplifier factor G

$$G = \frac{P_{out}}{P_{in}} = G_o \; exp \left(-\frac{G-1}{G} \frac{P_{out}}{P_s} \right) \tag{12.31}$$

where

$$G_o = exp \left(g_o L \right) \tag{12.32}$$

We see that the amplification factor G decreases as P_{out} approaches P_s. Depending upon G_o, the amplification factor falls by a factor of ~ 2.5 when P_{out} is equal to P_s. The decrease of the amplification factor with power output is shown in Fig. 12.19. Note that for semiconductor laser amplifiers P_s is about 5-10 mW.

12.5.2 Pulse Amplification and Pulse Shaping

$\rightsquigarrow$ For most practical applications, the laser amplifier is used to amplify optical pulses. The large bandwidth of the semiconductor amplifiers (several terrahertz) allow them to amplify subpicosecond optical pulses. However, to study this process we need to consider the time dependence of the amplifier gain and injection density. From Eqns. 12.25 and 12.26 we can write

$$\frac{\partial g}{\partial t} = \frac{g_o - g}{\tau} - \frac{gP}{E_s} \tag{12.33}$$

where we have defined

$$E_s = \hbar\omega \; A_m/(\partial g/\partial n) \tag{12.34}$$

If we write the amplitude of the electric field of the pulse as

$$F = P^{1/2} \; exp \left(i\phi \right) \tag{12.35}$$

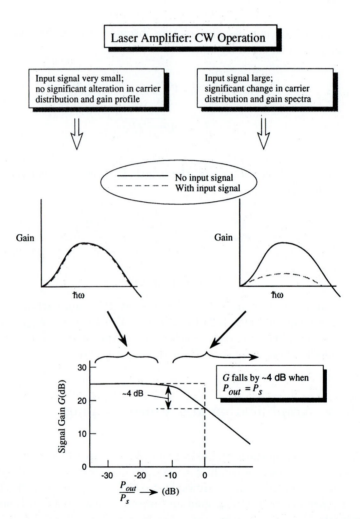

Figure 12.19: The operation of a laser amplifier under CW conditions. The amplification factor G falls by about 4dB when $P_{out} = P_s$. The exact decrease depends upon G_o.

we get as discussed in Chapter 11, Section 11.7.2, for the optical power and the phase

$$\frac{\partial \phi}{\partial t} = \frac{v \alpha_{enh} g}{2} \tag{12.36}$$

$$\frac{\partial P}{\partial t} = v g P \tag{12.37}$$

Transforming to the reduced time given by

$$\tau = t - \frac{z}{v} \tag{12.38}$$

and ignoring the first term in Eqn. 12.33 we get (we can transform from $\partial/\partial t$ to $\partial/\partial z$)

$$\frac{\partial P}{\partial z} = g(z, \tau) P(z, \tau) \tag{12.39}$$

$$\frac{\partial \phi}{\partial z} = -\frac{1}{2} \alpha_{enh} \, g(z, \tau) \tag{12.40}$$

$$\frac{\partial g}{\partial \tau} = -\frac{g(z, \tau) P(z, \tau)}{E_s} \tag{12.41}$$

Note that the gain has a position dependence and a τ dependence. Integrating Eqn. 12.39 over space we get

$$P_{out}(\tau) = P_{in}(\tau) exp\left[h(\tau)\right] \tag{12.42}$$

where

$$h(\tau) = \int_0^L g(z, \tau) dz \tag{12.43}$$

To obtain $h(\tau)$ we integrate Eqn. 12.41 over the amplifier length after replacing gP by $\partial P/\partial z$. This gives

$$\frac{dh}{d\tau} = -\frac{1}{E_s} \left[P_{out}(\tau) - P_{in}(\tau)\right]$$

$$= -\frac{P_{in}(\tau)}{E_s} (e^h - 1) \tag{12.44}$$

Integration of this equation gives $h(\tau)$ and one gets for the gain of the structure, $G(\tau) = exp\ h$

$$G(\tau) = \frac{G_o}{G_o - (G_o - 1) exp\left[-E_o(\tau)/E_s\right]} \tag{12.45}$$

where

$$E_o(\tau) = \int_{-\infty}^{\tau} P_{in}(\tau') d\tau' \tag{12.46}$$

$$G_o = exp\ (g_o L) \tag{12.47}$$

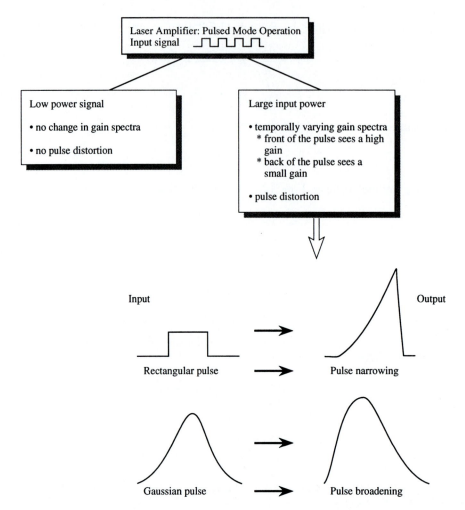

Figure 12.20: The operation of a laser amplifier under pulsed mode. At high rep rate the average power of the optical signal determines the power at which gain saturation effects become important.

Note that $E_o(\tau)$ is the partial energy of the pulse passing up to a certain point in time ($E_o(0) = 0; E_o(\infty) = E_{in}$).

An important outcome of the result for the amplifier gain is that the gain seen by different parts of the pulse are different. For example, the leading edge of the pulse sees the unsaturated gain G_o, while the trailing end sees a lower gain. This process, arising due to the gain saturation, causes pulse distortion as shown in Fig. 12.20. Depending upon the initial pulse shape, the pulse could be narrowed or broadened. This could have serious effects on the performance of a lightwave system.

In addition to pulse distortion, gain saturation can have an effect on the chirp of the pulse. As discussed in Chapter 11, chirp arises from the frequency shifting of an optical signal during a time dependent process. The time dependence results in a change in the gain which alters the refractive index through the linewidth enhancement factor. This in turn causes a phase modulation which produces chirp. The optical pulse modulates its own phase as it moves through the amplifier due to the gain saturation effect. This phenomenon is called saturation-induced self-phase modulation. The frequency chirp is given by

$$\Delta \nu_c = -\frac{1}{2\pi} \frac{d\phi}{d\tau} \tag{12.48}$$

From Eqns. 12.40 and 12.44 we get

$$\Delta \nu_c = \frac{\alpha_{enh}}{4\pi} \frac{dh}{d\tau}$$

$$= -\frac{\alpha_{enh} P_{in}(\tau)}{4\pi E_s} (G(\tau) - 1) \tag{12.49}$$

The chirp imposed by the amplifier can be exploited for pulse compression in a properly designed lightwave system. To see how this process works we will summarize some key features of wave propagation. We will consider a Gaussian pulse which has a certain spread in time and frequency. We consider the initial pulse amplitude to be of the form

$$F(0,t) - F_o\, exp \left[-\frac{1+iC}{2} \left(\frac{t}{T_o}\right)^2\right] = F_o\, exp \left(-\frac{t^2}{2T_o^2}\right) exp\, (-i\phi) \tag{12.50}$$

Here F_o is the field amplitude, T_o describes the pulse halfwidth and C is a parameter called chirp parameter. A value of $C \neq 0$ causes the frequency of the pulse to change with time. The frequency change is given by

$$\delta\omega = -\frac{\partial \phi}{\partial t} = \frac{Ct}{T_o^2} \tag{12.51}$$

The Fourier transform of the field is given by

$$\tilde{F}(0,\omega) = F_o \left(\frac{2\pi T_o^2}{1+iC}\right)^{1/2} exp \left[-\frac{\omega^2 T_o^2}{2(1+iC)}\right] \tag{12.52}$$

and the spectral width (at 1/e intensity point) is given by

$$\Delta\omega_o = \frac{(1+C^2)^{1/2}}{T_o} \tag{12.53}$$

In the absence of any chirp, the spectral width satisfies the relation

$$\Delta\omega_o T_o = 1 \tag{12.54}$$

Such a pulse is called transform limited and has the narrowest spectrum associated with a pulse of temporal width given by T_o. The chirp enhances the spectral width by a factor of $(1 + C^2)^{1/2}$. The propagation of the pulse in a medium can, in general, be described by

$$\tilde{F}(z, \omega) = \tilde{F}(0, \omega)\, exp\,(i\beta z) \qquad (12.55)$$

and, in general, we can write the propagation vector as an expansion around the peak frequency ω_o

$$\beta(\omega) = \frac{\bar{n}_r \omega}{c} = \beta_o + \beta_1 \Delta\omega + \frac{1}{2}\,\beta_2\,\Delta\omega^2 + \frac{1}{6}\,\beta_3 \Delta\omega^3 \dots \qquad (12.56)$$

where

$$\beta_m = \frac{d^m \beta}{d\omega^m}\,|_{\omega=\omega_o} \qquad (12.57)$$

The parameter β_1 is the inverse of the group velocity, β_2 is related to the material dispersion, etc.

The propagation of the pulse under consideration is given by

$$
\begin{aligned}
F(z, t) &= \frac{1}{2\pi} \int_{-\infty}^{\infty} \tilde{F}(z, \omega) e^{-i\omega t}\, d\omega \\
&= \frac{1}{2\pi} \int \tilde{F}(0, \omega)\, exp\left[\frac{i\beta_2 \omega^2 z}{2} + \frac{i\beta_3 \omega^3 z}{6} - i\omega t\right] d\omega \qquad (12.58)
\end{aligned}
$$

Ignoring the β_3 term which is acceptable as long as one is not near the zero dispersion point where $\beta_2 = 0$, we get

$$F(z, t) = \frac{F_o T_o}{[T_o^2 - i\beta_2 z(1 + iC)]^{1/2}}\, exp\left(-\frac{(1 + iC)t^2}{2\,[T_o^2 - i\beta_2 z(1 + iC)]}\right) \qquad (12.59)$$

From the above equation we see that a Gaussian pulse maintains its Gaussian shape, although the width of the pulse changes as

$$\frac{T_1(z)}{T_o} = \left[\left(1 + \frac{C\beta_2 z}{T_o^2}\right)^2 + \left(\frac{\beta_2 z}{T_o^2}\right)^2\right]^{1/2} \qquad (12.60)$$

where $T_1(z)$ is the half width after a propagation over a distance z.

One defines a distance called dispersion length L_d given by

$$L_d = \frac{T_o^2}{|\,\beta_2\,|} \qquad (12.61)$$

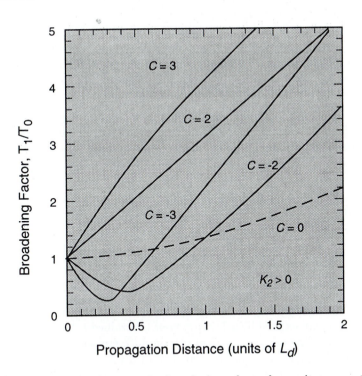

Figure 12.21: The effect of chirp on the broadening of a pulse as it propagates through a medium with positive β_2.

As can be seen from Eqn. 12.60 an unchirped pulse broadens by a factor of $\sqrt{2}$ at $z = L_d$. The chirped pulse may broaden or compress depending upon the relative signs of β_2 and C. If β_2 and C have the same sign, the chirped Gaussian pulse broadens monotonically with propagation at a rate faster than the unchirped pulse as shown in Fig. 12.21. However, if β_2 and C have opposite sign (i.e., $\beta_2 C < 0$), the pulse initially shrinks and then broadens. The pulse width is minimum at

$$z_{min} = \frac{|C| L_d}{1 + C^2} \qquad (12.62)$$

The minimum width is given by

$$T_1^{min} = \frac{T_o}{(1 + C^2)^{1/2}} \qquad (12.63)$$

and we get a transform limited pulse, since

$$\Delta\omega_o T_1^{min} = 1 \qquad (12.64)$$

This feature of pulse compression can be exploited in lightwave systems. In Chapter 13 we discuss the properties of the optical fibers used for optical communication.

These fibers are characterized by the parameters $\beta_1, \beta_2, \beta_3$, etc., discussed above. For long distance communication at 1.55 μm, the value of β_2 is about -18 ps/km. The combination of chirp produced in a laser amplifier and the dispersion produced in the fiber has been exploited to produce optical pulses as short as 2 ps.

It may be noted that gain saturation is not the only means to produce pulse compression. Due to the finite bandwidth of the gain spectra, the group velocity has a frequency dependence which can be exploited for pulse compression. This approach is useful only for ultrashort pulses of width $\sim$ 1 ps.

In this section we have discussed some of the important issues of semiconductor amplifiers. In this context it is important to point out an extremely important development in fiber amplifiers—the erbium doped fiber amplifier. The physics of the fiber amplifiers is qualitatively similar to that of the semiconductor laser amplifier. In the erbium doped (and other rare earth ion doped) fiber amplifiers (EDFAs), erbium ions are placed inside the fiber core during the fabrication process. The doped ions have associated with them energy bands at different wavelengths. In Fig. 12.22 we show the energy spectra of the erbium ions in a fiber. The levels that are exploited are the $4\ell_{15/2}$ and $4\ell_{13/2}$. Carriers are excited by a 1.48 μm laser (a strained InGaAs laser can be used) by optical pumping. The relaxing carriers can produce emission and gain over a broad spectrum ranging from 1.5 μm to 1.56 μm. Use of other rare earth ions can produce amplifiers at other spectral regions.

The development of EDFAs has revolutionized the optical communication systems. The loss of the optical signal as it travels can be compensated by these fiber amplifiers thus eliminating the need to convert to electrical signals and reconvert to optical signals.

EXAMPLE 12.3 Calculate the saturation power P_s for a GaAs/AlGaAs amplifier. The optical mode area is 1.0 μm $\times 1.0 \mu$m, the e-h recombination time is 2.0 ns and the differential gain is 10^{-15} cm^2.

The saturation power is ($\hbar\omega = 1.43$ eV)

$$
\begin{aligned}
P_s &= \frac{(1.43 \times 1.6 \times 10^{-19} \ J)(10^{-8} \ cm^2)}{(2.0 \times 10^{-9} \ s)(10^{-15} \ cm^2)} \\
&= 1.14 \ mW
\end{aligned}
$$

EXAMPLE 12.4 Consider a GaAs/AlGaAs amplifier biased at a current density of 10^4 A/cm^2. Calculate the amplification factor if two long optical pulses with $\hbar\omega = 1.5$ eV and initial power 0.1 mW and 1.0 mW pass through the traveling wave amplifier. The device is described by the following parameters:

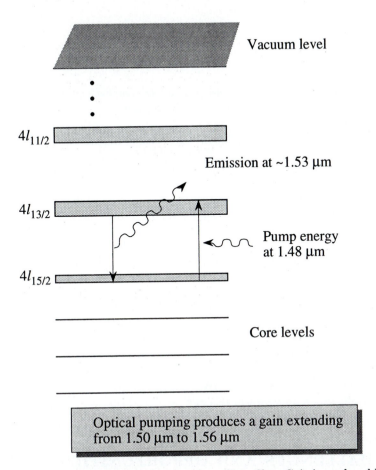

Figure 12.22: The energy diagram of erbium ions in a silicon fiber. Gain is produced by optically pumping electrons from the $4\ell_{15/2}$ level to the broader $4\ell_{13/2}$ band.

$$
\begin{aligned}
\text{Device length, } L &= 100 \ \mu m \\
\text{Pulse mode area, } A_m &= 5 \times 10^{-8} \ cm^2 \\
\text{Differential gain, } \frac{dg}{dn} &= 10^{-15} \ cm^2 \\
\text{Optical confinement factor, } \Gamma &= 1.0 \\
\text{e-h recombination time, } \tau &= 2.0 \ ns \\
\text{Active region width, } d_{las} &= 0.8 \ \mu m
\end{aligned}
$$

The injection density (at no input optical power) is given by

$$
n = \frac{J\tau}{ed_{las}}
$$

$$= \frac{(10^4 \ A/cm^2)(2 \times 10^{-9} \ s)}{(1.6 \times 10^{-19} \ C)(0.8 \times 10^{-4} \ cm)}$$
$$= 1.56 \times 10^{18} \ cm^{-3}$$

The gain can be obtained directly by using a simple computer program. It is found to be 610 cm^{-1} at $\hbar\omega = 1.5$ eV. The saturation power is given by

$$P_s = \frac{1.43 \times 1.6 \times 10^{-19} \ J)(5 \times 10^{-8} \ cm^2)}{(2 \times 10^{-9} \ s)(10^{-15} \ cm^2)}$$
$$= 5.7 \ mW$$

The small signal amplification factor is

$$G_o = exp \ (gL)$$
$$= exp \ \left[(610 \ cm^{-1})(100 \times 10^{-4} \ cm)\right]$$
$$= 403.4$$

Using the equation

$$G = G_o \ exp \ \left(-\frac{G-1}{G}\frac{P_{out}}{P_s}\right)$$

and solving it iteratively ($P_{out} - GP_{in}$), we find that for

i) $P_{in} = 0.1$ mW; $P_{out} = 86 \times 0.1 = 8.6$ mW
ii) $P_{in} = 1.0$ mW; $P_{out} = 18.5 \times 1.0 = 18.5$ mW.

12.6   CHALLENGES AND OPPORTUNITIES IN OPTICAL SWITCHING

$\mathcal{R}$ In this chapter we have examined some of the optoelectronic modulation and switching elements that are presently being actively studied. Will these elements find their way into tomorrow's information processing systems? There is no doubt that in certain specialized applications within optical communication, these elements are going to play an important role. In the area of general purpose computing, the outlook is not as optimistic at present, since important challenges remain. While many of these devices seem to be based on elegant effects and are, thus, aesthetically pleasing, they must compete with the good old FET or BJT.

In the area of communication, optical modulation, switching and amplification is of great importance since there is a great premium on maintaining the optical character of the signal as it travels hundreds or even thousands of kilometers. This is most essential if coherent communication systems are to be realized. The key challenges here are in the area of device integration and manufacturability.

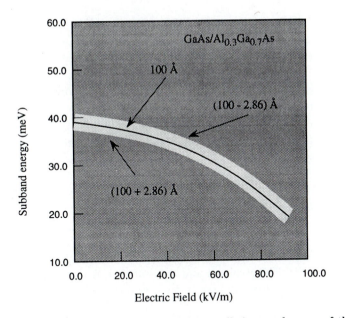

Figure 12.23: The effect of one monolayer variation in well size on the sum of the conduction and HH subbands as a function of the electric field. The solid line is for a 100 Å GaAs/ $Al_{0.3}Ga_{0.7}As$ well. The upper and lower limits are for the cases where the well size changes by a monolayer distance ($= 2.83$ Å).

Let us consider the case of a quantum well modulator or switch which exploits excitonic resonances and their dependence on applied field. We point out three key critical device parameters that determine the performance of the device: i) exciton linewidth; ii) fluctuations in quantum well size; iii) temperature fluctuations. The following observations are relevant to the excitonic properties :

• Peak absorption coefficient is inversely proportional to the exciton linewidth.

• Peak position (approximately) is inversely proportional to the square of the well size.

• Peak position shifts by ~ 0.5 meV for each degree change of the temperature around 300 K.

In Fig. 12.23 we show the variation of the quantum well subband energy (exciton energy = E_g + subband energy - exciton binding energy) as a function of field for a one monolayer variation in well size for a GaAs/$Al_{0.3}Ga_{0.7}As$ well. Remembering that the laser line that we are trying to modulate or switch is less than 1 microelectron volt, it is easy to see the need of extreme precision in the device fabrication.

The exciton linewidth is another serious concern. As discussed in Chapter 5 (section 5.6), the exciton linewidth is controlled by homogeneous effects (temperature dependent) and inhomogeneous effects (usually dominated by interface roughness and alloy disorder). Even small changes in the structural quality produce a significant effect on the device characteristics. Changes in the exciton linewidth will affect the I-V characteristics of devices such as SEED and also affect performance characteristics of devices such as directional couplers based on quantum wells.

The issues mentioned above point out that extreme precision is needed in device structure if commercial scale use of quantum well switching devices is to be viable. In such applications, system designers do not have the luxury enjoyed by researchers in the laboratories. In the laboratory, if the exciton linewidth is not quite where it should be, the laser is slightly tweaked to compensate.

Another challenge for the technology is that of device integration. The challenges for OEIC's from etching and regrowth have been discussed in Section 1.15 and also in the context of OEIC receivers and transmitters. A tremendous degree of control is needed in the area of etching and regrowth as well as integrated waveguide fabrication.

Coming to the question of optical switching for general purpose computing, it is important to recognize some important challenges to the devices discussed. If one were to simply develop a "digital optical computer" (DCO) along the lines of existing electronic computers, the DCO will not be able to compete. In electronic computers, simple operations like addition, multiplication, subtraction, etc., require a number of steps in which bits of electrical signals are fetched, stored, shifted in registers, gated and re-stored. While all this can be, in principle, done with optoelectronic switches and photons replacing electrical signals, at present several factors weight against this. These are: i) most optical switches do not have gain. While a laser amplifier can boost a signal, the size of the amplifier is $\sim 10~\mu m \times 300 - 500~\mu m$ which is huge, compared to the size of FETs ($\sim 5~\mu m \times 10~\mu m$). ii) Electrical signals can be easily sent down hardwired paths to different transistors and a power supply can easily provide the requisite power. There is no equivalent "photon source." Photons will have to be generated by lasers which requires a large chip area and high power.

In fact, the use of optoelectronic devices in DCO to replace electronic devices while fashionable a decade or so ago, is not considered seriously by optoelectronic scientists. Why should "expensive" photons be made to do things that "cheap" electrons can do? It seems that optics and optoelectronic devices should pursue paths that are extremely difficult for electronics to tread on. The success of one such path— communication, shows that in certain areas optics has an edge over electronics. In Fig. 12.24 we show some of the other areas where a similar edge is present. There are areas where the optical or optoelectronic systems have advanced to a stage where they are commercially viable. The area of optical communication (including LANs) will be examined in Chapter 13.

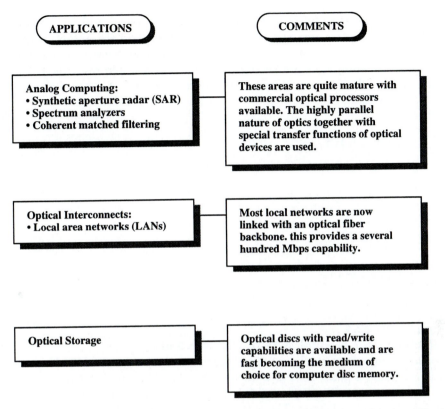

Figure 12.24: Areas of information processing where optoelectronics is competing effectively with electronics.

An area where optical processing has been viable for several decades is analog processing of a signal using certain natural transfer functions of optical elements. The most widely used one is the capability of a lens to do a Fourier transform. Optical storage has also developed greatly over the last several years so that optical read/write discs are now available at competitive prices.

There are other application areas where optoelectronics is likely to make significant impact in near future. These include the area of massive interconnection of devices and chips using lasers, holograms and detectors. Also, optical switching devices are expected to allow optical "on the fly" recognition of headers (addresses) on a data stream so that a decision can be made at a local node if the data is for that node. We have briefly discussed in this chapter the use of PSLMs for vector-matrix multiplication. Similar architectures have been designed for matrix-matrix multiplication and matrix inversion which are key computations for several classes of important problems.

Another important problem is that of associative memory where a partial input is supposed to generate a full stored data entry. For example, in a library, one would get a full description of this text even if the user entered under the author: SNGH. Once again, the parallelism of optics is expected to be of great benefit in this application where electronic computers are severely limited. Considerable research in optics and optoelectronics is focusing on bringing these ideas to reality.

12.7   SUMMARY

In this chapter we have examined optoelectronic modulation and switching devices which are necessary if optoelectronics is to develop into a truly versatile technology. The summary tables provide an overview of this chapter.

12.8   PROBLEMS

Section 12.2 - 12.4
See Problems 5.7 through 5.14 in Chapter 5 also

12.1 An electro-optic material is needed which has the property that a device of length 150 μm can cause a phase shift of π when a bias of 10 V is applied. Calculate the value of the electro-optic coefficient r_{41} needed. The refractive index of the material is 3.5 and the device thickness is 1.0 μm. The optical wavelength is 1.55 μm. Compare the desired value of r_{41} to the value for bulk GaAs. Such high values of r_{41} are possible in quantum well systems.

12.2 An excitonic transition in a quantum well is defined by a peak height of 2×10^4 cm^{-1} and a linewidth $\sigma = 1.5$ meV. The peak position is at 7300 Å at zero bias. A transverse field of 100 kV/cm is applied to shift the exciton peak by 25 meV. Assume that the linewidth is unchanged. Calculate the effect of the exciton shift on the refractive index change for a light beam with a wavelength of i) 8000 Å; ii) 7500 Å. Assume that the exciton resonance has a Lorentzian shape.

12.3 A bulk GaAs device is used to produce an electro-optic modulator with an ON/OFF ratio of 2:1 when an electric field of 10^5 V/cm is applied. Calculate the minimum length of the device needed assuming that the entire optical field sees the electro-optic effect. The wavelength of light is 1.0 μm.

12.4 Consider a vertical incidence 100 Å GaAs/AlGaAs MQW stack that is used to produce an electro-absorption modulator with an ON/OFF ratio of 2:1 when a field of 6×10^4 V/cm is applied. The incident wavelength is 10 meV below the zero field exciton peak and is 8000 Å and the excitonic halfwidth is $\sigma = 2.0$ meV. Calculate the minimum total width of the wells needed for the modulator. The exciton peak shifts by 10 meV at 6×10^4 V/cm and $\sigma = 3.0$ meV.

12.5 A vertical incidence 100 Å GaAs/AlGaAs multiquantum well modulator is to be designed with an ON/OFF ratio of 2:1. The electric field is to switch between 0 and

TOPICS STUDIED

KEY OBSERVATIONS

Electro-optic modulators

These devices depend upon the change of the real part of the refractive index due to an electrical signal. Bulk and quantum well devices can be used, although in quantum wells, the effect is enhanced and smaller size devices can be used. the device is based on a phase shift created by the electro-optic effect.

Electro-absorption modulators

The devices depend upon the change of the imaginary part of the refractive index (or absorption coefficient) due to an electrical signal. The most widely used devices are based on the excitonic transition in quantum wells. The excitonic resonance is altered by either an applied field or by charge injection.

Liquid crystal displays

These devices create light modulation by altering the polarization of a light pulse by an applied bias.

Directional couplers

The devices consists of two waveguides in which there is a small overlap of the optical field. The optical energy couples from one guide to another over a certain distance. The fraction of light coupled depends upon the length of the guide and the propagation constant differences between the guides. The propagation constant can be altered by an applied bias, thus the optical output can be switched.

Table 12.1: Summary table.

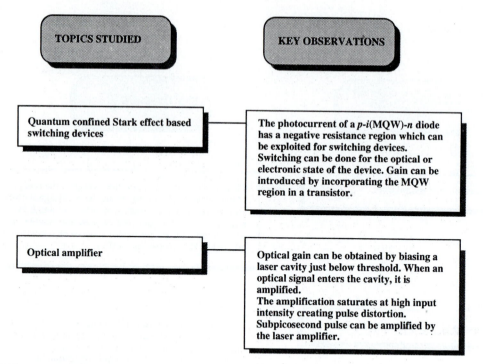

Table 12.2: Summary table.

70 kV/cm. The exciton linewidth is 5.0 meV. Calculate the thickness of the active region needed for this device after optimizing the structure for the choice of input light energy.

Section 12.5

12.6 An InGaAsP/InP amplifier is designed to amplify 1.55 μm optical signals. Calculate the saturation power for the following device parameters:

$$
\begin{aligned}
\text{InGaAsP bandgap, } E_g &= 0.76 \ eV \\
\text{Optical mode area, } A_m &= 2 \times 10^{-8} \ cm^2 \\
\text{e-h recombination time, } \tau &= 2 \ ns \\
\text{Differential gain for the amplifier, } &\quad 5 \times 10^{-15} \ cm^2
\end{aligned}
$$

12.7 Consider the amplifier discussed in the previous problem. The device is biased to give a cavity gain of 400 cm^{-1} and the device length is 200 μm. A CW optical signal of 0.01 mW power enters the amplifier. Calculate the power of the output signal.

12.8 Discuss the differences between gain saturation of an amplifier discussed in this chapter and gain compression discussed in the previous chapter on high speed laser response.

12.9 An amplifier is designed to amplify a 1 μW signal to 1 mW level. Calculate the output signal power when a 2 mW signal is incident on this amplifier. The value of P_s

for the amplifier is 10 mW.

12.10 A laser amplifier is used in a three channel amplification system. The channels are separated by 1 GHz. The amplifier is able to amplify each signal by 30 dB in isolation when $P_{in}/P_s = 10^{-3}$. Calculate the amplification when all channels are carrying optical signals. The *e-h* recombination time is 1.5 ns.

12.9 REFERENCES

- **Electro-optic Effect and the Directional Coupler**

  - Buckman, A. B., *Guided Wave Photonics*, Saunders College Publishing, Harcourt Brace Javanovich College Publisher, Orlando, FL (1992).

  - Hammer, J. M., "Modulation and Switching in Dielectric Waveguides," *Integrated Optics*, ed. T. Tamir, Springer-Verlag, New York (1982). Also see other articles in this book.

  - Wegener, M., T. Y. Chang, I. Bar-Joseph, J. M. Kuo, and D. S. Chemla, *Applied Physics Letters*, 55, 583 (1989).

  - Zucker, J. E., I. Bar-Joseph, G. Sucha, U. Koren, B. I. Miller, and D. S. Chemla, *Electronic Letters*, 24, 458 (1988).

- **QCSE Based Switching Devices**

  - Hinton, H. S. and J. E. Midwinter, eds., "Photonic Switching; Progress in Lasers and Electro-Optics," Series ed. P. W. E. Smith, *IEEE Press*, New York (1990).

  - Li, W-Q, S. Goswami, P. K. Bhattacharya, and J. Singh, "A Programmable Memory Cell Using Quantum Confined Stark Effect in a Multi-Quantum Well Heterojunction Bipolar Transistor," *Electronics Letters*, 27, pp. 31-32 (1991).

  - Miller, D. A. B., D. S. Chemla, T. C. Damen, T. H. Wood, C. A. Burris, A. C. Gossard, and W. Wigemann, "The Quantum WEll Self-Electro-optic Effect Device: Optoelectronic Bistability and Oscillation and Self-Linearized Modulation," *IEEE Journal of Quantum Electronics*, **QE**-21, 1462 (1985).

  - Miller, A. B., D. S. Chemla, T. C. Damen, T. H. Wood, C. A. Burrus, A. C. Gossard, and W. Wigemann, "Novel Optical Level Shifter and Self-Linearized Optical Modulator Using a Wuantum Well Self Electroptic Effect Device," *Optics Letters*, 9, 567 (1984).

  - Wheatley, P., P. J. Bradley, M. Whitehead, G. Parry, J. E. Midwinter, P. Mistry, M. A. Pate, and J. S. Roberts, "Novel Nonresonant Optical Logic Device," *Electronics Letters*, 92, (1985).

- **Laser Amplifiers**

    - Agrawal, G. P., *Fiber-Optic Communication Systems*, John Wiley and Sons, New York (1992).

    - Lowery, A. J., "Pulse Compression Mechanisms in Semiconductor Laser Amplifiers," *IEEE Proceedings*, Part J., vol. 136, p. 141 (1989).

    - Saileh, T. and T. Mules, "Gain Saturation Characteristics of Travelling Wave Semiconductor Laser Amplifiers in Short Optical Pulse Amplification," *IEEE Journal of Quantum Electronics*, **QE**-26, 2086 (1990).

    - Wiesenfeld, J. M., G. Eisenstein, R. S. Tucker, G. Rayban, and P. B. Hansen, "Distortionless Picosecond Pulse Amplification and Gain Compression in a Travelling Wave InGaAsP Optical Amplifier," *Applied Physics Letters*, 53, 1239 (1988).

    - Yamamoto, Y. and T. Kimura, "Coherent Optical Fiber Transmission Systems," *IEEE Journal of Quantum Electronics*, **QE**-17, 919, (1981).

- **Optical Information Processing**

    - Abu, Y. S. and D. Psaltis, "Optical Neural Computers," *Scientific American*, 256, 66 (1987).

    - Athale, R. A. and W. C. Collins, "Optical Matrix-Matrix Multiplier Based on Outer Product Decomposition," *Applied Optics*, 21, 2089 (1982).

    - Casasent, D., "Coherent Optical Pattern Recognition: Review," *Optical Engineering*, 24, 26 (1985).

    - Feitelson, D. G., *Optical Computing: A Survey for Computer Scientists*, MIT Press, Cambridge (1988).

    - Miller, D. A. B., "Quantum Wells for Optical Information Processing," *Optical Engineering*, 26, 368 (1987).

    - J. Singh, S. Hong, P. Bhattacharya, R. Sahai, C. Lastufka, and H. Sobel, "Systems Requirements and Feasibility Studies for Optical Modulators Based on GaAs/AlGaAs Multiquantum Well Structures for Optical Processing," *Journal of Lightwave Technology*, 6, p. 818 (1988).

CHAPTER 13

OPTICAL COMMUNICATION SYSTEMS: DEVICE NEEDS

13.1 INTRODUCTION

The fall of Troy was communicated to Queen Clytemnestra in 1184 B.C. by a series of line of sight beacons of fires covering 900 km! Today light has again become a medium of choice for communication of information. Optical communication systems provide a shining example of an area where optoelectronic devices are providing superior overall performance than electronic devices. Optical communication networks have given a major boost to the compound semiconductor technology and have already established themselves as the high performance networks of choice.

In the modern information age, the era of optical communication was launched with the invention of the LED and the laser and the availability of low loss optical fibers. Without the optical fiber, there would be no optical communication. Because of the central role of optical fibers in the communication system, it is essential that the optoelectronic devices be compatible with the demands placed by the fiber on light sources and detectors. It is fair to say that a large fraction of research in semiconductor optoelectronics is motivated by the desire to optimize and exploit the potential of the optical fiber. While optoelectronic devices are used in a variety of applications, the demands of a high performance communication system are extremely stringent.

In the previous chapters we discussed some important performance parameters of detectors and light sources. An important question in the reader's mind must be: What is an adequate performance for an optoelectronic device? Is an LED emitting a 150 Å spectral width optical beam at 8800 Å with a 100 μW power adequate for a 1 Gbps transmission system? Why does one need a laser source with a 0.001 Å spectral width for a coherent system? What should be the detectivity of a detector to qualify for an optical communication system? These questions cannot be answered without understanding the fundamentals of a communication system in general and the optical fiber in particular.

While optical communication systems have been used in local area network (LAN) applications (within a building) for several decades, the most important system that established the superiority of optical communication was the TAT-8, the eighth transatlantic link . This fiber optics based cable connected Tuckerton, New Jersey to Penmarch in France. The single mode fiber system provided a 280 Mbps capacity system using InGaAsP LDs ($\lambda = 1.3$ μm) and InGaAs p-i-n photodetectors. TAT-7, the predecessor of TAT-8, based on metal cables carried 5000 voice circuits while TAT-8 carried 37,800. The launch of TAT-8 was not without opposition. The chief opposition came from satellite companies which anticipated a big loss in business (rightly so) if an optical fiber link was available between North America and Europe. However, the satellite lobby was not able to stop the progress of optical fiber communication in spite of their shortlived legal battles against the optical communication companies. Optical communication has been growing by leaps and bounds and promises to turn individual homes into major information exchange centers.

13.2 A CONCEPTUAL PICTURE OF THE OPTICAL COMMUNICATION SYSTEM

$\mathcal{R}$ Optical communication is one of the fastest growing segments of optoelectronics. This is in spite of the fact that the ideal optoelectronic devices that can exploit the full potential of optical communication still do not exist. For example the potential of coherent communication is not yet exploited because optoelectronic devices are not yet as developed as electronic devices. Consequently, a very small capacity of the optical system is used at present. This, of course, suggests that the future of optical communication will be even brighter.

The chief advantage of the optical communication system comes from the properties of the optical fiber which is the medium used to convey information. In Fig. 13.1 we show a typical optoelectronic communication system layout. The information to be transmitted (data, voice, etc.), is first coded onto an optical signal. This requires a proper driver or modulator and an optical source. The signal is next coupled to an optical fiber, which is a key component which has made optical communication possible.

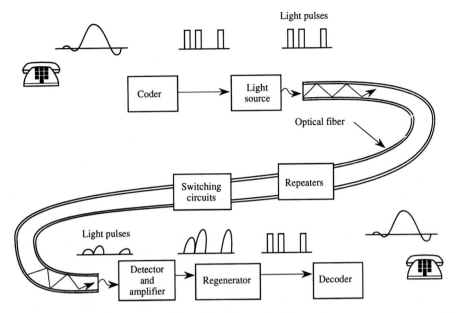

Figure 13.1: Components of an optoelectronic communication system.

As the signal passes along the fiber, at some point it may need to be switched to other channels. This requires appropriate switching elements. Once the data have reached the desired point, it is detected by an optical detector. The generated signal is amplified and received.

In real systems, the optical signal attenuates as it propagates. This requires the placement of "repeaters" to regenerate the optical signal or amplifiers which amplify the signal. In addition, the optical data may need to be switched from one channel to another. The switching circuitry can be quite complicated requiring lasers and detectors as well as silicon based decision making circuitry, etc. The basic devices involved in the communication system have already been discussed in the previous four Chapters. Here we are more interested in identifying the requirements for these devices.

The optical communication system is quite similar in concept to the electronic (microwave based) communication system. The key differences are in the wavelength of the radiation used and the use of an optical fiber instead of a metallic cable. The use of fibers provides four key areas of advantage for the optical system over the microwave based system: i) far greater repeater spacings are possible since the attenuation of an optical signal in a fiber is much smaller than that of a microwave signal in a cable; ii) a very high information capacity is available due to the use of the optical frequencies; iii) the system has low cost and low weight; and iv) the effects of electromagnetic interference are minimal. However, at present due to device capability limitations, the enormous bandwidth of the optical fiber is not fully utilized. In Fig. 13.2 we show a

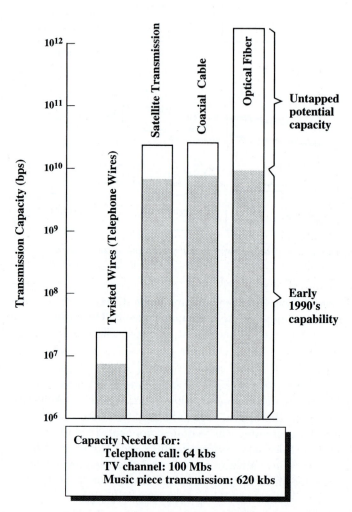

Figure 13.2: A comparison of the transmission capacity of some important transmission media. The current and potential capacities are shown. Uncompressed transmission rates needed for some day to day applications are also shown. Data compression can reduce the required rates by more than an order of magnitude. (The capacities shown are an updated version of a figure in *Computer Engineering*, (March 1987).)

comparison of the currently available and potentially available transmission capacity of several systems. Clearly the optical fiber has an enormous potential. However, to reach this potential it is essential that coherent detection systems be available. These issues will be discussed later in this chapter.

13.3 INFORMATION CONTENT AND CHANNEL CAPACITY

$\longrightarrow$ In modern optical communication systems, the information to be sent is coded onto the optical signal in a digital manner. Thus the optical pulse goes between a high and a low value. Once the transmission is complete, it should be possible to convert the optical signal to recover the original signal. An important question in this regard is the following: If an analog signal has a maximum frequency content f_m, at what rate must the signal be sampled for transmission so that it can be reproduced faithfully at the receiving end? *According to the sampling theorem the signal must be sampled at a rate that is at least twice the maximum frequency in order to recover the original signal.* Thus the transmission rate needed depends upon the information that has to be transmitted. A typical telephone call needs a transmission rate of 64 kbps, while the transmission of a high quality compact disc musical piece requires a rate of 620 kbps (see Example 13.1). The transmission rate required for television is $\sim$ 75 Mbps.

The capacity of usual optical fiber systems is, in principle, well over terrabit/s ($\sim$ 10% of the optical frequency), so that to maximize the system performance the information from different sources is multiplexed as shown schematically in Fig. 13.3. At present there are no practical optical systems which can use frequency division multiplexing (FDM) in which a number of frequency channels are used as carrier frequencies and the information is coded as to these frequencies. This would be the optimum system but at present it is difficult to develop transmission and detection systems that can send out and carry out "coherent" detection to selectively decode the information on the carrier frequencies. Research in the area of narrow linewidth stable laser sources is to a large part driven by the desire for coherent optical systems.

Another related multiplexing scheme is wavelength division multiplexing (WDM) in which sources emit the information coded in different "color" or "wavelength" light. These colors can travel down an optical fiber without interference. However, this scheme is also at a research level at the moment because there are no reliable tunable optical sources that can emit at different closely spaced wavelengths. Also, the standard optical detectors have broadband responsivity and cannot select a particular wavelength. With WDM it should be possible to reach upto a 100 Gbps transmission capacity.

Due to the lack of optoelectronic devices that can serve FDM or WDM schemes,

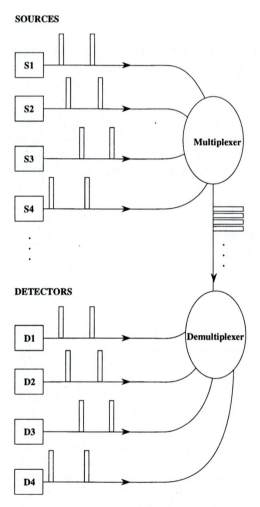

Figure 13.3: A schematic of a multiplexing (demultiplexing) scheme used to exploit the large capacity of a communication channel. At present the optical communication cannot exploit multiplexing of channels with different frequencies or wavelengths. As a result, the multiplexing is done "serially."

the massively parallel transmission capabilities of optical fibers are lost. In fact the status of the optical communication is similar to the status of microwave transmission in the 1930's. Thus at present, the advantages are mainly due to cost, reliability, and size reductions. Ofcourse, as advanced devices become available the full potential will be reached.

A commonly used multiplexing scheme is time division multiplexing (TDM) in which information is sent in a serial manner with information being sent in different time slots. Such a scheme does not exploit the full capabilities of optical communication but is compatible with existing devices.

Important requirements for optoelectronic devices are based upon the requirements placed by optical fibers. In the next section we will examine the properties of optical fibers.

EXAMPLE 13.1 In speech transmission the audio frequency range is 300 to 3400 Hz. Calculate the sampling rate needed for sending an audio signal in a telephone line.

According to the sampling theorem, the rate has to be 6800 s^{-1}. However, the international standard is to use a sampling rate of 8000 s^{-1} to improve the margin of error. Also one uses 8 bits per sample to transmit the speech in a clear manner. This requires a transmission rate of 64 kbps.

EXAMPLE 13.2 It is required to transmit an encyclopedia stored in a computer having 10,000 pages of written text. Estimate the information content of the encyclopedia. A communication channel is available with a transmission capacity of 500 Mbps. Calculate the time taken for transmission.

Let us assume that on an average a page has 45 lines, each with 12 words and each word has 6 letters. In the ASCII code, each letter is represented by 7 bits. The total information content of the encyclopedia is then

$$(10^4)(45)(12)(6)(7) = 2.268 \times 10^8 \text{ bits}$$

Using the transmission capability of the channel, the time of transmission is

$$t = \frac{2.268 \times 10^8 \text{ bits}}{500 \times 10^6 \text{ bps}} = 0.4536 \ s$$

13.4 SOME PROPERTIES OF OPTICAL FIBERS

$\longrightarrow$ The superior properties of the fiber when compared to the metallic cables have been the most important driving forces for the use of optoelectronics in telecommunications, computer links, industrial automation, medical technology, and military applications. The output of the optical sources (LDs and LEDs) must be coupled to the optical fiber for most applications. It is thus essential to understand some of the basic principles governing the optical fiber to appreciate the demands placed on optical sources and detectors.

13.4.1   Materials and Structure

Optical fibers come in three main categories depending upon the ingredients used—silica, glass, and plastic. The silica fibers are made from SiO_2 with appropriate addition of metal oxides to fine-tune the refractive index. Important dopants that are used to adjust the refractive index are TiO_2, Al_2O_3, GeO_2 and P_2O_5. The glass fibers are made from a variety of glasses which have very high chemical stability. Finally, plastics are used for many applications. At present the plastics have a higher attenuation than the silica or glass fibers, but continuous improvements in the field may change this.

The optical fiber is a cylindrical waveguide system through which the optical waves can propagate. The basic structure consists of a core at the center and a cladding layer on the outside as shown in Fig. 13.4. The core is a cylinder of transparent dielectric wire with a refractive index n_{r1}, and the cladding is a dielectric sheath with a lower index n_{r2}. The fibers have several classifications based on the index profile and the core size as shown in Fig. 13.4. The core size determines how many modes of the optical wave can propagate in the fiber. An optical cable consists of hundreds of such optical fibers.

13.4.2   Fiber Losses

In an ideal fiber, the optical waves should propagate in the core region due to the confinement from the cladding region without any loss. However, in real fibers, there are losses which arise from various sources. Some of the important losses are due to the following sources:

i) Absorption Loss: The light can excite several transitions in the material making up the fibers. The absorption loss is simple $\exp(-\alpha L)$ where α is the absorption coefficients and L is the optical path. The values of α are typically ~ 1 km^{-1}. The absorption loss is very important since it determines how much distance the optical signal can propagate before it needs to be regenerated by repeaters. In Fig. 13.5 we show a typical dependence of the loss on photon wavelength for silica fibers.

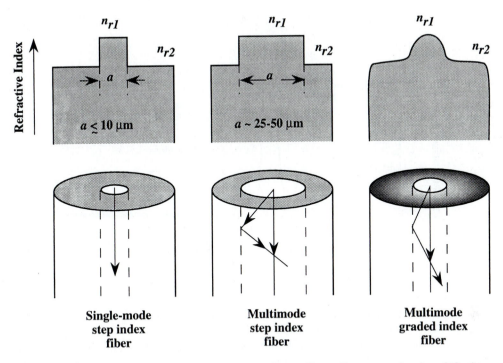

Figure 13.4: Structures of various kinds of optical fibers. Depending upon the size of the inner core, one or more modes of optical waves can propagate in the fiber. The single mode fibers are used for long haul high performance networks while the multimode fibers (which are cheaper) are used for LANs.

ii) Scattering Loss: In the fabrication process used for optical fibers, some irregularities are produced which can cause scattering losses. Improved techniques are reducing these losses.

iii) Bending Losses: If the optical fiber is bent too tightly (say, in circles of radius of a few millimeters), the light cannot bend along with the fiber and some of the lights can ooze out and be lost.

13.4.3 Acceptance Angle and Numerical Aperture

The optical fiber is a carrier of the optical information, and one of the important parameters of an optical fiber based system is the efficiency with which light can be coupled into a fiber from an optical source. This requires that the incident light arrive with such an angle that it suffers a total internal reflection at the core-cladding layer interface as shown in Fig. 13.6.

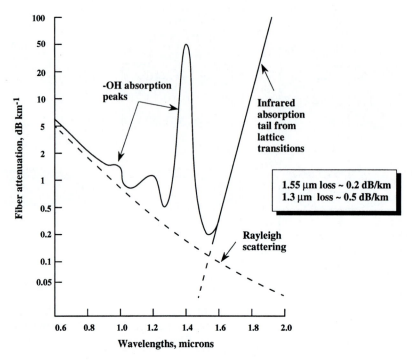

Figure 13.5: Attenuation in optical fibers as a function of wavelength. Low transmission losses occur at 1.3 μm and 1.55 μm. The loss figures are decreasing further as the fiber technology improves.

The critical angle for total internal reflection can be obtained by applying Snell's law to an optical beam propagating from a medium with index n_{r1} to a medium with index n_{r2}. The angles of incidence and refraction are given by

$$\frac{sin\phi_2}{sin\phi_1} = \frac{n_{r1}}{n_{r2}} \tag{13.1}$$

For total reflection $\phi_2 = 90°$ so that the critical angle $\phi_1 = \phi_{1c}$ is

$$\phi_{1c} = sin^{-1}\frac{n_{r2}}{n_{r1}} \tag{13.2}$$

Now let us consider the case of an optical fiber in which light is incident from a medium with index n_{ro} into the fiber as shown in Fig. 13.6. We wish to find the maximum angle of acceptance for the fiber. According to Snell's law,

$$n_{ro}sin\theta = n_{r1}sin\theta_1 = n_{r1}cos\phi_1 \tag{13.3}$$

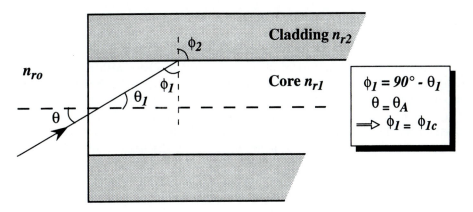

Figure 13.6: The geometry used to calculate the acceptance angle for an optical fiber. When the incident angle exceeds a value θ_A, the ray is not able to propagate within the optical fiber.

The total internal reflection between the core and the cladding layer interface occurs when

$$sin\phi_1 \geq \frac{n_{r2}}{n_{r1}} \qquad (13.4)$$

or

$$cos\phi_1 \leq \left(1 - \frac{n_{r2}^2}{n_{r1}^2}\right)^{1/2} \qquad (13.5)$$

Using Eqns. 13.3 and 13.5 we get for the limit on $sin\theta$

$$n_{ro}sin\theta < \left(n_{r1}^2 - n_{r2}^2\right)^{1/2} \qquad (13.6)$$

or

$$sin\theta \leq \frac{1}{n_{ro}}\left(n_{r1}^2 - n_{r2}^2\right)^{1/2} \qquad (13.7)$$

The maximum angle of acceptance is now

$$\boxed{\theta_A = sin^{-1}\left[\frac{1}{n_{ro}}\left(n_{r1}^2 - n_{r2}^2\right)^{1/2}\right]} \qquad (13.8)$$

The quantity $\left(n_{r1}^2 - n_{r2}^2\right)$ is called the numerical aperture (NA) of the fiber.

As can be seen from this analysis, the light gathering properties of the optical fiber depend critically on n_{ro}, n_{r1}, and n_{r2}. Improving the light gathering power and reducing losses are the key driving forces for research in fiber optics. As discussed in Chapter 9, in an LED, the light output is not very well collimated. As a result, a smaller fraction of light emitted is coupled into a fiber as compared to the light coupled from a laser diode.

13.4.4 Multipath Dispersion

A very important issue in optical fibers, especially in fibers with a wide core, is the problem of multipath dispersion. This problem arises from the fact that there are a number of paths that an optical beam can take to traverse a certain length of a fiber. Let us assume that light signals come at angles between $\theta = 0$ (axial ray) to $\theta = \theta_A$ (extreme ray). The two extreme paths are defined by the axial ray and the most oblique ray which enters at an angle corresponding to ϕ_{1c}, as discussed above. The rays travel with a velocity v given by

$$v = c/n_{r1} \tag{13.9}$$

and take a time

$$t_a = \frac{\ell}{v} = \frac{n_{r1}\ell}{c} \tag{13.10}$$

for the axial ray to travel across a fiber of length ℓ. The extreme ray effectively has to travel a distance $\ell/sin\phi_{1c}$. The time taken is (using the value of $sin\phi_{1c}$ from Eqn. 13.2)

$$t_e = \frac{n_{r1}\ell}{csin\phi_{1c}} = \frac{n_{r1}^2\ell}{n_{r2}c} \tag{13.11}$$

Thus, if a source sends in the two rays, the time difference between them when they emerge from the fiber is ($\Delta n_r = n_{r2} - n_{r1}$)

$$\tau_{mp} = t_a - t_e = \frac{n_{r1}\ell}{c}\left[1 - \frac{n_{r1}}{n_{r2}}\right] = \frac{n_{r1}\ell\Delta n_r}{n_{r2}c} \tag{13.12}$$

The multipath or model dispersion is given by the time spread per unit length,

$$\boxed{\frac{\tau_{mp}}{\ell} = \frac{n_{r1}\Delta n_r}{n_{r2}c}} \tag{13.13}$$

As shown in Fig. 13.7, the input pulse spreads in time as it passes through the fiber and emerges with a width τ_{mp}. The multipath dispersion is closely related to the bandwidth the fiber can support over a given distance. To a good approximation the bandwidth Δf limited by the multipath dispersion is simply

$$2\Delta f \sim \frac{1}{\tau_{mp}} \tag{13.14}$$

The bandwidth distance product is, therefore, (from Eqns. 13.13 and 13.14)

$$\boxed{\Delta f\ell = \frac{n_{r2}c}{2n_{r1}\Delta n_r}} \tag{13.15}$$

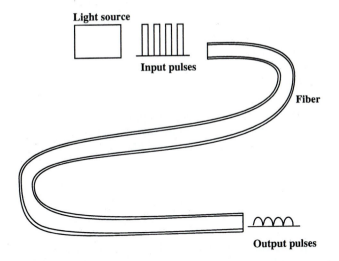

Figure 13.7: Due to multipath delays (and material dispersion delays) an optical pulse spreads as it passes through an optical fiber. The pulse spreading determines the maximum bandwidth the fiber can carry (in serial transmission) over a certain distance.

For a high bandwidth distance product, the value of Δn_r should be small. Since the numerical aperture is proportional to Δn_r, this means that for high frequency transmission, the numerical aperture should be small. However, it must be kept in mind that reducing τ_{mp} by reducing the numerical aperture can have a negative effect on the coupling of light into the fiber, especially if the light source is not highly collimated as is the case for an LED. Thus, to collect sufficient light, one needs a fiber with high numerical aperture and this in turn can reduce the bandwidth distance product. In semiconductor lasers, on the other hand, it is possible to get highly collimated beams and thus use fibers with high bandwidth-distance products.

To improve the bandwidth-distance product for a fiber, special approaches are taken. An important approach is the graded index fiber in which the index of the core is varied as shown in Fig. 13.4. The grading allows one to have a much higher bandwidth-distance product compared to the step index fiber.

A fiber that is increasingly dominating advanced optical systems is the single mode fiber. As the diameter of the core of the fiber starts to become comparable to the wavelength of light, the simple ray picture developed above breaks down. One has to consider the proper waveguide theory where the Maxwell equations are solved in the fiber. We will only give the relevant results here. *As the fiber dimensions shrink, one eventually reaches a stage where only one mode of radiation can propagate in the optical fiber.* Fibers having such parameters are called single mode fibers. The condition for a step index fiber of core diameter a as shown in Fig. 13.4 is

$$\boxed{\frac{2\pi a \left(n_{r1}^2 - n_{r2}^2\right)^{1/2}}{\lambda} < 2.405} \qquad (13.16)$$

By choosing a small core diameter, a single mode fiber can be fabricated. The cladding diameter is kept at 60-100 μm to maintain the strength of the fiber. *Since one has a single mode propagating in a single mode fiber, there is no multipath dispersion.*

13.4.5   Material Dispersion

In addition to the multipath delay problem discussed above, another limitation on the capacity of the optical fiber is placed by the material dispersion of the fiber. The material dispersion arises from the variation of the refractive index of the optical fiber with wavelength. Because of the variation, radiation with different wavelength travels a different optical path and thus a time delay is introduced. Of course, if an optical signal has a single wavelength, one would not be concerned with material dispersion. However, even the purest optical sources have some spread in the wavelength which makes material dispersion a serious problem.

In a material in which the dielectric constant is dependent upon the wavelength of light, one can define a phase velocity v_p which defines the phase change of a signal and the group velocity v_g which describes the rate of energy transfer by the signal. We have

$$v_p = \frac{\omega}{k} \qquad (13.17)$$

and

$$v_g = \frac{d\omega}{dk} \qquad (13.18)$$

If the material has no dispersion the group velocity and the phase velocity are the same. We can also define the ordinary refractive index n_r for the phase and the group refractive index N_r. These are related to the respective velocities by

$$n_r = \frac{c}{v_p} \qquad (13.19)$$

$$N_r = \frac{c}{v_g} \qquad (13.20)$$

In general n_r and N_r are a function of the wavelength λ. We can write (using Eqns. 13.17 through 13.20)

$$N_r = \frac{c}{v_g} = c\frac{dk}{d\omega} = c\frac{d}{d\omega}\left(\frac{\omega n_r}{c}\right) = n_r + \omega\frac{dn_r}{d\omega} \qquad (13.21)$$

We also have

$$\frac{dn_r}{d\omega} = \frac{dn_r}{d\lambda} \cdot \frac{d\lambda}{d\omega} \tag{13.22}$$

Since

$$\omega = \frac{2\pi c}{\lambda} \tag{13.23}$$

$$\frac{d\omega}{d\lambda} = -\frac{2\pi c}{\lambda^2} \tag{13.24}$$

and we get, from Eqns. 13.21, 13.22, 13.23, and 13.24,

$$N_r = n_r - \frac{2\pi c}{\lambda}\frac{dn_r}{d\lambda}\left(\frac{\lambda^2}{2\pi c}\right) = n_r - \lambda\frac{dn_r}{d\lambda} \tag{13.25}$$

This gives for the group velocity,

$$v_g = \frac{c}{N_r} = \frac{c}{n_r - \lambda dn_r/d\lambda} \tag{13.26}$$

An impulse of light at a fixed λ travels a distance ℓ of the fiber in a time

$$t = \frac{\ell}{v_g} = \left[n_r - \lambda\frac{dn_r}{d\lambda}\right]\frac{\ell}{c} \tag{13.27}$$

If the optical signal is made up of a spread of wavelengths there will be a spread in the time intervals taken by the different wavelengths to traverse the fiber. If $\Delta\lambda$ is the wavelength spread, the time spread is given by differentiating Eqn. 13.27 to get

$$\frac{\Delta t}{\Delta\lambda} = -\frac{\ell}{c}\lambda\frac{d^2n_r}{d\lambda^2} \tag{13.28}$$

An important property of an optical signal is, therefore, the spread in wavelengths $\Delta\lambda$ over which the optical has, say, 50% or higher spectral power. The relative spectral width of the source is defined by

$$\gamma = \left|\frac{\Delta\lambda}{\lambda}\right| = \left|\frac{\Delta\omega}{\omega}\right| \tag{13.29}$$

An impulse, after travelling a distance ℓ through the fiber, will have a half power spread in time τ_{disp} given by ($\tau_{disp} \equiv \Delta t$)

$$\tau_{disp} = -\frac{\ell}{c}\lambda\Delta\lambda\frac{d^2n_r}{d\lambda^2} = -\frac{\ell}{c}\gamma\,\lambda^2\left(\frac{d^2n_r}{d\lambda^2}\right) \tag{13.30}$$

An important parameter is the spread in time per unit length,

$$\frac{\tau_{disp}}{\ell} = \frac{\gamma}{c} |Y_m|$$ (13.31)

where the quantity Y_m represents the material dispersion

$$Y_m = \lambda^2 \frac{d^2 n_r}{d\lambda^2}$$ (13.32)

We can approximately define by $1/4\tau_{disp}$ the bandwidth of the signal the fiber system can support for a length ℓ. The exact relation depends somewhat on the spectral shape of the light output. We then have

$$\boxed{(\Delta f)\ell = \frac{\ell}{4\tau_{disp}} = \frac{c}{4\gamma |Y_m|}}$$ (13.33)

This relation emphasizes the importance of the optical source quality γ and the fiber dispersion quality in determining the channel capacity and the distance up to which the signal can be carried. A small spread in the optical source emission spectrum is critical and is a key reason for the use of semiconductor laser diodes instead of light emitting diodes in high performance optical communication systems. Typical dispersion values for silicon fibers are shown in Fig. 13.8. (Note that the values shown in Fig. 13.8 are not for Y_m, which can be obtained by multiplying the value given by λc).

It is interesting to compare the capacity of the fiber-source system at the wavelength of no dispersion versus the wavelength of minimum attenuation (e.g., 1.3 μm vs. 1.55 μm transmission). The signal can be transmitted farther at 1.55 μm, but the transmission bandwidth is not as high as at 1.3 μm. The optimum system has to consider the properties of the detector and the modal dispersion of the fiber as well. In Example 13.4 we will carry out a detailed comparison of various devices and emission wavelengths.

In this and the previous subsection we have examined two important sources that limit the transmission capacity of the fiber. The sources are modal dispersion and material dispersion. In general, both these limitations will be present in a fiber. The question then arises: What is the broadening factor τ_{tot} that describes the combined dispersion effects? In general, to answer this question one needs to know details of the optical pulse shape. However, a reasonable approximation that can be easily applied is the following:

$$\tau_{tot} = \left(\tau_{in}^2 + \tau_{mp}^2 + \tau_{disp}^2\right)^{1/2}$$ (13.34)

where τ_{tot} is the total pulse width, τ_{in} is the initial pulse width before entering the fiber, τ_{mp} is the width due to multipath or modal dispersion, and τ_{disp} is the width due to

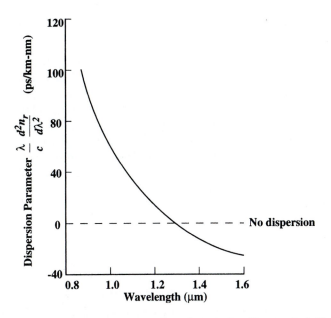

Figure 13.8: A typical dispersion curve versus wavelength for silica optical fibers. Notice that the dispersion goes to zero at $\sim 1.3\ \mu$m. Note also from Fig. 13.5 that the attenuation is lowest in the silica fiber at 1.55 μm, although the attenuation is not too high at 1.3 μm.

the material dispersion. As noted earlier, for a single mode fiber we can assume that $\tau_{mp} = 0$.

Soliton Communication Systems

$\mathcal{R}$ In this chapter we have seen that when optical pulses propagate in a fiber, they broaden and suffer distortion due to the fiber dispersion. The dispersion essentially causes different optical frequencies to propagate with different group velocities. In Chapter 11 we discussed the concept of self phase modulation in laser amplifiers. This effect occurs in fibers also where the refractive index of the fiber is dependent upon the optical intensity due to non-linear effects. The refractive index of silica fibers is given by

$$n'_r = n_r + \frac{n_{2r} P}{A_{eff}}$$

where A_{eff} is the effective cross-section of the optical beam and $n_{2r} \sim 3.2 \times 10^{-16}$ cm^2/W. It is possible to exploit the group velocity dispersion and self phase modulation to propagate optical pulses without distortion. The pulses, however, have to be generated in a special way in order to exploit this possibility. Optical pulses that do travel undistorted are called solitons. The optical soliton that is launched into a fiber must have the amplitude of the form

$$U(0, \tau) = sech(\tau)$$

to propagate undistorted. The pulse like soliton propagation occurs when the dispersion parameter β_2 (see Chapter 11) is negative, i.e., the material has anomalous dispersion.

While solitons have been known to exist for decades and their presence in optical fibers was shown in 1973, only recently have they received attention as something of practical use. Remarkable progress has been made in soliton studies and it has been shown that they can propagate over thousands of kilometers. This of course, makes them very attractive for optical communication. However, challenges remain before they can be used in practical systems. These challenges include generation of solitons from semiconductor lasers in a practical manner. Optical pulses produced by semiconductor lasers do not normally have the hyperbolic secant form required by solitons. Pulses have to be reshaped by passing them though fiber or amplifiers, an approach that requires significant improvements. Nevertheless, considerable improvements in transmission rate are expected when practical soliton networks become available.

EXAMPLE 13.3 An optical fiber has a minimum attenuation point at 1.55 μm. Calculate the maximum bandwidth-distance product for the fiber for an LED and a laser diode emitting at this wavelength. The spectral widths for the two sources are 300 Å and 30 Å, respectively, and the fiber dispersion is given by

$$Y_m = \lambda^2 \left. \frac{d^2 n_r}{d\lambda^2} \right|_{\lambda = 1.55 \mu m} = -0.01$$

The bandwidth-distance product is given by

$$\Delta f \cdot \ell = \frac{c}{4Y_m} \left(\frac{\lambda}{\Delta\lambda} \right)$$

This gives for the LED

$$\Delta f \cdot \ell = \frac{(3 \times 10^{10} \, cm/s)(1.55 \times 10^{-4} cm)}{4 \times (0.01) \times (300 \times 10^{-8} \, cm)} = 3.875 \times 10^{13} \, Hz \cdot cm$$
$$= 0.3875 \, GHz \cdot km$$

For the laser diode the value becomes ten times greater. Based upon this example, if this LED is to be used to send a signal over a 10 km fiber, the maximum bandwidth is 38.75 MHz.

13.4.6   Signal Attenuation and Detector Demands

An extremely important consideration in fiber optics is the attenuation suffered by light as it travels through the fiber. The success of the optical fiber communication system owes a great deal to the low attenuation suffered by the electromagnetic waves as they travel in the fiber. The original proposal to use fiber as a communication medium was made by Kao and Hockham, who estimated that an attenuation figure of 20 dB/km was needed if fibers were to be competitive. The attenuation in dB/km is given by

$$\text{Attenuation} = \frac{10 \log_{10}(P_i/P_f)}{L} \ dB/km \qquad (13.35)$$

where P_i is the initial optical power and P_f is the power left after travelling a distance L (in km). Note also that the absorption coefficient α also defines the ratio of P_i and P_f via the equation

$$\frac{P_f}{P_i} = exp\,(-\alpha L) \qquad (13.36)$$

Thus, for example, an attenuation of 20 dB/km corresponds to an absorption coefficient of 4.6 km^{-1}. Modern optical fibers are capable of attenuation as low as 0.2 dB/km at certain wavelengths. The attenuation curve for the most widely used silica fiber is shown in Fig. 13.5. It may be noted that fibers with high attenuation may be selected for certain special applications because they may have to be used in adverse environment, like under high radiation dosage. Under high radiation certain fibers do not degrade as much as other fibers and thus provide a better system.

The attenuation in the fiber is due to various scattering processes that occur causing light to be scattered or absorbed as it travels in the fiber. The improvements in the fiber attenuation have resulted from a detailed understanding of the absorption process. All fibers have minimas in the attenuation curve at some special wavelengths. The choice of the optical source and the detection system is greatly influenced by these special wavelengths. For example, as shown in Fig. 13.5, the most widely used silica fibers have an attenuation minimum at 1.55 μm. There is also a local minimum at 1.3 μm. The selection of an operating wavelength is affected by considerations such as fiber dispersion, attenuation, source spectral purity, as well as demands imposed upon the system. For example, if the communication system is for a local area network (LAN), a GaAs based source ($\lambda \sim 0.88$ μm) may be chosen due to the advanced level of GaAs based optoelectronics and electronics. However, for long distance communication, the concerns of dispersion and attenuation are quite important. Example 13.4 brings out some very important features that influence choice of devices for communication systems.

EXAMPLE 13.4 Consider the specifications of the components of an optoelectronic communication system shown in Table 12.1. The detector sesitivity is such that the minimum power needed to detect a signal is given by 0.1 nW/Mbps.

Calculate the repeater spacing needed if the system is to be used for transmission of data at i) 100 Mbps; ii) 1.0 Gbps. Assume that the cable coupling losses and splicing losses are included in the attenuation specifications provided.

Source	Emission Wavelength (μm)	Spectral Width (Å)	Power Coupled to the Fiber (μW)
LED	0.88	300	50
	1.3	300	50
	1.55	300	50
Laser diode (LD-1)	0.88	30	1000
	1.3	30	1000
	1.55	30	1000
Laser diode (LD-2)	0.88	3.0	1000
	1.3	3.0	1000
	1.55	3.0	1000

Fiber Type	Wavelength (μm)	Modal Dispersion (ns/km)	Material Dispersion $\left(\frac{\lambda}{c}\frac{d^2 n_r}{d\lambda^2}\right)$ (ps/km-nm)	Attenuation (dB/km)
Graded index	$\lambda = 0.88$	0.5	70	1.5
	$\lambda = 1.3$	0.5	2	0.6
	$\lambda = 1.55$	0.5	20	0.2
Single mode	$\lambda = 0.88$	0.0	70	1.5
	$\lambda = 1.3$	0.0	2	0.6
	$\lambda = 1.55$	0.0	20	0.2

Table 13.1: Specifications for the optical fibers and optical sources.

Given the system component specifications, the repeater spacing is determined by pulse spreading (dispersion) considerations or by attenuation considerations. The dispersion limit is determined by the source spectral purity and the dispersion in the fiber, while the attenuation limit is determined by the minimum power needed for detection. Let us consider the case of the LED first.

$$\lambda = 0.88 \ \mu m; 0.1 \ Gbps \text{ transmission}; P_i = 5 \times 10^{-5} \ W$$

$$P_f = \text{Detector minimum power required} = (10^{-10} \ W/Mbps)(100 \ Mbps) = 10^{-8} \ W$$

$$\frac{P_i}{P_f} = \frac{50 \times 10^{-6} \ W}{10^{-8} \ W} = 5 \times 10^3$$

From Eqn. 13.35 we get (attenuation is 1.5 dB/km)

$$L \equiv l_{att} = \frac{10 log(5 \times 10^3)}{1.5} = 24.7 \ km$$

A material dispersion of 70 ps/km-nm corresponds to a value of the dispersion parameter $Y_m = 0.018$. Taking the bit rate B as $2\Delta f$ we get for the single mode (SM) fiber (see Eqn. 13.33)

$$L \equiv l_{disp}(SM) = \frac{c\lambda}{2B\Delta\lambda Y_m} = 2.4 \ km$$

Source	Wavelength (μm)	Repeater Spacing for SM fiber (km)		Repeater Spacing for GI Fiber (km)	
		0.1 Gbps	1.0 Gbps	0.1 Gbps	1 Gbps
LED	0.88	2.4(D)	0.24 (D)	2.35 (D)	0.24 (D)
	1.33	61 (A)	45 (A)	10.0 (D)	1.0 (D)
	1.55	78 (D)	7.8 (D)	6.2 (D)	0.6 (D)
LD-1	0.88	24 (D)	2.4 (D)	10.0 (D)	1.0 (D)
	1.33	61 (A)	45 (A)	10.0 (D)	1.0 (D)
	1.55	78 (D)	7.8 (D)	10.0 (D)	1.0 (D)
LD-2	0.88	33 (A)	24 (D)	10.0 (D)	1.0 (D)
	1.33	61 (A)	45 (A)	10.0 (D)	1.0 (D)
	1.55	250 (A)	200 (A)	10.0 (D)	1.0 (D)

Table 13.2: Calculated results for repeater spacings for the design problem of Example 13.4. The symbol D is for dispersion limited; A is for attenuation limited.

which is smaller than the length obtained from attenuation considerations.

Thus for the single mode fiber, the dispersion limit is the one which sets the repeater spacing of 2.4 km for the LED at 0.88 μm. In the case of the graded index fiber, we must include the effect of the modal dispersion. The total broadening is

$$\frac{\tau_{tot}}{\ell} = \frac{(\tau_{mp}^2 + \tau_{disp}^2)^{1/2}}{\ell}$$

using $\tau_{mp}/\ell = 0.5$ ns/km and $\tau_{disp}/\ell = 2.07$ ns/km for the modal and material dispersion broadening, we get

$$\frac{\tau_{tot}}{\ell} = 2.13 \ ns/km$$

This gives for the repeater distance for the graded index (GI) fiber for bandwidth $B = 10^8$ bps

$$L(GI) = \frac{10^9}{2 \times 10^8 \times 2.13} = 2.35 \ km$$

Thus the repeater spacing is a little smaller in the graded index fiber.

Similar calculations for the other device combinations give the repeater spacings listed in Table 13.2.

This example shows the complicated interplay between the various component specifications in deciding the performance of the total system. The reader is urged to carefully examine the repeater distances given in Table 13.2. The reader can look for the following comparisons: *i) the LED has a poor performance for 0.88 μm transmission, but can perform reasonably well if it is used at 1.33 μm or 1.55 μm; ii) the 1.33 μm source does not perform as well as the 1.55 μm source in a single mode fiber at low transmission rates, but performs better at a high transmission rate.*

13.5   COHERENT COMMUNICATION SYSTEMS: DEVICE REQUIREMENTS

$\mathcal{R}$ As we have discussed the various optoelectronic devices in this text, we have repeatedly mentioned the limitations of these devices. These limitations arise from the difficulties in developing stable narrow bandwidth devices. Coherent communications is an area where these limitations are felt the most. The present status of optical communications is similar to the status of microwave communication before 1930. The transmitter simply produces a noisy output that is detected by the detector. In modern electronic communication, even the simplest radio receiver uses a local oscillator to detect the AM or FM signals sent over the airwaves. While optical communication is far from reaching this state, progress is being made in coherent optical communication. Coherent optical communication has been demonstrated in a number of labs and in some field trials. Coherent communication can improve the overall system performance by improving the receiver sensitivity and by eventually allowing FDM transmission.

The coherent system places extremely stringent limitations on the device performance of the lasers and detectors. For example the laser linewidth has to be very small so that the coded information can be detected without loss due to lack of coherence. A conservative rule of thumb says that the intermediate frequency Δf_{IF} should be given by

$$\Delta f_{IF} = 2 \times B + 12 \times \Delta f_{sl} \qquad (13.37)$$

where B is the bit rate and Δf_{sl} is the sum of the linewidths of the transmitter and local oscillator lasers. Thus if the different assigned channels are to be placed at separation of say 5 GHz each the laser linewidths have to be in the range of 20 MHz for a few gigabit per second transmission. The linewidth of a Fabry Perot laser is in the range of a few THz so that these lasers cannot be used at all. The DFB lasers are certainly capable of providing the required performance. However, the dificulty of fabricating DFB lasers causes their price to be almost a factor of thousand greater than the price of a Fabry Perot laser (a few dollars versus a few thousand dollars). Moreover, not all commercial DFB lasers have the high performance needed for a coherent system.

In addition to the narrow linewidths needed, automated circuitry is needed to stabilize the laser output and in the case of the local oscillator, to search and lock at the incoming optical signal. As a result, the system is extremely expensive and will remain so for some time to come. The various components needed are described below:

Low Linewidth Lasers

As we have noted already, linewidths of a few tens of MHz are needed in a coherent system. As discussed in Chapter 11, the DFB lasers operated at very high injection can produce such linewidths. It is now possible to buy commercial DFB lasers with the desired performance. In fact if external cavities are used in a feedback loop, linewidths of a few kilohertz can be achieved.

Polarization Maintaining Optical Fibers

It is important that the polarization state of the signal be maintained during transmission through the optical fiber. Ordinary optical fibers permit the transmission of two polarization modes of light. As a result, geometric variations and polarization dispersion cause fluctuations in the signal polarization. Special fibers are now available the allow propagation of a single polarization and maintain the polarization of the signal.

Polarization Independent Devices

It is important to develop optical transmitters and receivers which are polarization insensitive. Thus one would not have to worry about the polarization state of the signal. We have seen in Chapter 10 that quantum well lasers emit TE polarized light preferentially. However, the use of a small tensile strain can merge the HH and LH states and allow polarization insensitive response.

Improved Modulation Schemes

In Chapter 11 we discussed the intensity modulation of a laser by direct modulation of the current flow into the device. For coherent communication systems, the frequency, phase or amplitude of the light has to be modulated. In digital systems a shift in these quanities has to be produced. The frequency shift is produced by a weak current bias modulation. At high biasing the frequency of the laser shifts due to the fluctuation in the carrier density. The frequency shift in AlGaAs lasers is found to be ~ 500 MHz/mA at a modulation rate of 1 GHz. This effect can be used in FSK modulation.

Phase locking techniques are needed for phase and amplitude modulation schemes. This requires master-slave lasers which are coupled by a feedback loop. A photodetector serves as a phase detector. A frequency modulation of the slave laser is converted to phase modulation. Amplitude modulation is also possible in a phase-locked mode with no FM or PM provided the loop gain is high. Needless to say, compared to the IM-DD approach, the coherent modulation is quite complex.

Tunable Optical Filters

Coherent optical receivers offer two advantages over the incoherent systems. As discussed before these are: i) improved receiver sensitivity; and ii) super wideband FDM. In the FDM approach hundreds of channels, each ~ 10 GHz wide could be used for optical communication. At present even if single channel coherent systems become available, FDM is not possible because of non-availability of optical filters capable of selecting frequencies with a few GHz bandwidth. In absence of such filters, a large power loss is suffered if the FDM signal is shared equally by a number of detectors. There is considerable ongoing research in tunable optical filters but the performance is still inadequate.

To summarize this section coherent communication has a great promise and rapid advances are occuring in the field. However, the present cost is too high to compete with incoherent communication systems.

13.6   CHAPTER SUMMARY

$\mathcal{R}$ In this chapter we have discussed some of the driving forces behind modern opto-electronic information processing systems. The optical communication system is one of the most important systems which is exploiting the best of optics and electronics. We discussed the device demands placed by the communication systems on light sources, detectors, switches, etc. We have noted that many of the device needs remain unfulfilled so that the full potential of the optical communication system has not yet been achieved. Many of the device needs for the communication systems are quite generic and other optoelectronic systems also make similar demands.

13.7   PROBLEMS

Section 13.3

13.1 A CD is to be broadcast digitally over a communication channel. The disc has 60 minutes of music covering a bandwidth of 20 kHz. The recording has a dynamic range of 80 dB. Calculate the transmission rate and the total information content of the transmission. Dynamic range is given by

$$20 log \frac{A_{\text{signal}}}{A_{\text{noise}}}$$

For a digital transmission, a 80 dB dynamic range means that $A_{\text{signal}}/A_{\text{noise}} = 10^4$. This requires one to have 14 digital bits assigned to each sampling $(2^{14} = 1.6 \times 10^4)$.

13.2 A television channel is to be broadcast over a transmission system. The signal occupies a total of 5.5 MHz bandwidth and requires a dynamic range of 50 dB. What is the capacity needed if the signal is to be transmitted digitally? Recently, some telephone companies have suggested that by using data compression/decompression technologies they can send a TV signal over telephone wires. If the capacity of copper wires is ~ 10 Mbps, what level of compression is needed to send a TV signal?

13.3 Using the information given in Fig. 13.2 on the present (early 1990's) capacity of various transmission systems, estimate the number of TV channels that can be sent simultaneously over the four transmission media mentioned in Fig. 13.2. Assume that each TV channel needs ~ 100 Mbps capacity.

Section 13.4

13.4 Calculate the distance an optical signal will travel before it falls to a hundredth of its initial value in optical fibers with attenuations of i) 20 dB/km; ii) 2 dB/km; and iii) 0.2 dB/km.

13.5 Calculate the maximum acceptance angles θ_A for the following step index fibers:

i) $\qquad n_{r1} \; = \; 1.470; n_{r2} = 1.45; n_{ro} = 1.0$

ii) $\qquad n_{r1} \; = \; 1.46; \; n_{r2} = 1.4; \; n_{ro} = 1.0$

13.6 Calculate the multipath dispersion parameter τ_{mp}/ℓ for the two fibers of Problem 13.5. Also calculate the bandwidth-distance product.

13.7 In order to exploit the low loss, low dispersion window of silica fibers, an optical source emitting at 1.3 μm is selected. Why is it not possible to use silicon detectors for this technology?

13.8 An LED is used as an optical source for an LAN. The LED emits at 0.8 μm, and 15.0 μW of optical power is coupled into a fiber with attenuation of 2 dB/km. An InGaAs *p-i-n* detector with detection limits defined by a minimum power of 1.0 nW/Mbps receives the signal. If the source to detector spacing is 1 km and the system performance is governed by signal attenuation, calculate the maximum transmission rate possible.

13.9 Discuss the reasons why the simple detectors discussed in Chapter 10 cannot exploit the capacity of an optical fiber.

13.10 A GaAs/AlGaAs laser emitting at 0.88 μm is to be used in an LAN. Assume that the material dispersion of the fiber sets the limit of the system performance. The parameter Y_m describing the material dispersion has a value of 0.025. Calculate the maximum spectral width $\Delta\lambda$ of the laser if a 1 Gbps transmission is to occur with a repeater or detector spacing of 2 km.

13.8   REFERENCES

- General

  - G.P. Agrawal, *Fiber Optic Communication Systems*, John Wiley and Sons, New York (1992).

  - T. Edwards, *Fiber Optic Systems: Network Applications*, John Wiley and Sons, New York (1989).

  - J. Gowar, *Optical Communication System*, Prentice-Hall, Englewood Cliffs, NJ (1984).

APPENDIX

A

LIST OF SYMBOLS

a	lattice constant (edge of the cube for the semiconductor fcc lattice)
AM	amplitude modulation
ASK	amplitude shift keying
B	base transport factor in a bipolar transistor
B	bandwidth
BER	bit error rate in a transmission stream
c	velocity of light
C	chirp parameter
C_j, C_d	junction, diffusion capacitance in a p-n diode
C_{ij}	elastic force constants
d_{las}	active region thickness of a laser
D_n	electron diffusion coefficient
D_p	hole diffusion coefficient
D_b	diffusion coefficient in the base of a bipolar transistor
D_c	diffusion coefficient in the collector of a bipolar transistor
D_e	diffusion coefficient in the emitter of a bipolar transistor
D, D^*	detectivity, specific detectivity of a detector
DBR	distributed Bragg reflector
DFB	distributed feedback (laser)

e	magnitude of the electron charge

E	energy of a particle
E_F	Fermi level
E_{Fi}	intrinsic Fermi level
E_{Fn}	electron quasi-Fermi level
E_{Fp}	hole quasi-Fermi level
$E^e(E^h)$	energy of an electron (hole) in an optical absorption or emission measured from the bandedges
$E_c(E_v)$	conduction (valence) bandedge
E_{ex}	exciton binding energy

$f(E)$	occupation probability of an electron state with energy E at equilibrium. This is the Fermi-Dirac function
$f^e(E)$	occupation function for an electron in non-equilibrium state. This is the quasi-Fermi function
$f^h(E)$	occupation function for a hole $= 1 - f^e(E)$
f_T	cutoff frequency for unit current gain
f_{max}	available power gain is unity at this frequency

F	electric field
F	Auger coefficient
F_{ext}	external force such as an electric or magnetic force
F_f	fill factor of a solar cell
FDM	frequency division multiplexing
FM	frequency modulation
FSK	frequency shift keying

g, g_{th}	gain, threshold gain
g_m	transconductance of a transistor
g_D	output conductance of a transistor

G_L	electron-hole generation rate due to a light beam
G_{ph}	gain of a detector

$\hbar$	Planck's constant divided by 2π

H	magnetic field

$IMDD$	intensity modulation direct detection
I_{ph}	photon particle current

I_E, I_B, I_C	emitter, base, and collector current in a BJT
I_{En}, I_{Ep}	electron, hole part of the emitter current in an npn BJT
I_L	photocurrent due to an optical signal
I_D	drain current in an FET
I_o	reverse bias saturation current in a p-n diode
I_s	reverse bias saturation current in a Schottky diode
I_{GR}	generation recombination current in a diode
I_{GR}^o	prefactor for the generation recombination current
I_{sc}	short circuit current of a solar cell

J	current density
J_L	photocurrent density
J_{ph}	photon particle current density

k_B	Boltzmann constant

ℓ	mean free path between successive collisions

L_n	diffusion length for electron
L_p	diffusion length for holes

m_0	free electron mass
m_e^*	electron mass
m_h^*	hole mass
m_{dos}^*	density of states mass
m_σ^*	conductivity mass
m_{hh}^*	mass of the heavy hole
$m_{\ell h}^*$	mass of the light hole
m_r^*	reduced mass of the electron-hole system

M, M_e, M_h	multiplication factor, multiplication factor for electrons, mulitplication factor for holes

n	electron concentration in the conduction band
n_i	intrinsic electron concentration in the conduction band
n_d	electrons bound to the donors
n_r	refractive index
$n_{ph}(\hbar\omega)$	photon density of photons with energy $\hbar\omega$
$n_p(n_p)$	equilibrium electron density in the p-side (n-side) of a p-n junction
n_{th}	threshold carrier density for a laser
NA	numerical aperture

N_{cv}	joint density of states for electrons and holes
$N_e(E)$	density of states of electrons in the conduction band
$N_h(E)$	density of states of holes in the valence band
$N_c(E)$	effective density of states in the conduction band
$N_v(E)$	effective density of states in the valence band
N_d	donor density
N_a	acceptor density
$N_{2D}(E)$	2-dimensional density of states
N_t	density of impurity states (trap states)
N_{ab}	acceptor concentration in the base of an *npn* BJT
N_{de}	donor concentration in the emitter of an *npn* BJT
N_{dc}	donor concentration in the collector of *npn* BJT
NEP	noise equivalent power of a detector

p	momentum of a particle
p	hole concentration in the valence band
p_i	intrinsic hole concentration in the valence band
p_a	holes bound to acceptors
p_{cv}	momentum matrix element for an optical transition between the valence and conduction band
$p_n(p_p)$	equilibrium hole density in the *n*-side (*p*-side) of a *p-n* junction

PM	phase modulation
PSK	phase shift keying
P_{op}	optical power density (energy flow/sec/area)

$QCSE$	quantum confined Stark effect

R_{ph}	responsivity of a detector material
R_{spon}	total rate at which an electron-hole system recombines to emit photons by spontaneous recombination
R_{st}	total rate at which *e-h* recombine by stimulated emission of photons
R_s, R_G, R_D	parasitic resistances associated with the source, gate and drain of a transistor respectively
R_L	load resistance
R^*	Richardson constant in a Schottky barrier

$SEED$	self electro-optic effect device
SNR	signal to noise ratio

t_d	delay time in large signal laser response
t_{tr}	transit time of a carrier through a channel
T	tunneling probability
$U(r)$	position dependent potential energy
v	velocity of the electron
v_s	saturation velocity of the carrier (electron, hole)
V_{oc}	open circuit voltage of a solar cell
V_{bi}	built-in voltage
$V_r(V_f)$	reverse (forward) bias voltage in a diode
V_{BE}, V_{BC}	base to emitter, base to collector bias in a bipolar transistor
V_{pt}	punchthrough voltage
W_{em}	spontaneous emission rate for an electron-hole radiative recombination
W_{em}^{st}	stimulated emission rate for an electron-hole radiative recombination
$W_n(W_p)$	depletion region edge on the n-side (p-side) of a p-n junction
W	depletion region width
WDM	wavelength division multiplexing
W_b, W_{bn}	base width, neutral base width of a bipolar transistor
α	optical absorption coefficient
α	current transfer ratio in a bipolar transistor
α_{enh}	linewidth enhancement factor
α_R	reflection loss coefficient in an optical cavity
α_{imp}	impact ionization coefficient for electrons
β	base to collector current amplification factor in a BJT
β_2	dispersion parameter
β_{imp}	impact ionization coefficient for holes
γ_e	emitter efficienty of a bipolar transistor
γ_{inj}	injection efficiency of a p-n diode for electron (hole) current
Γ	Optical confinement factor in a semiconductor laser
ΔE_g	bandgap difference between two materials
$\Delta E_c, \Delta E_v$	band discontinuity in the conduction, valence band in a heterostructure
ϵ_o	free space permittivity
ϵ	product of the relative dielectric constant and ϵ_o
ψ	electron wavefunction
$\sigma_e(\sigma_n)$	electron (hole) captive cross section for an impurity

σ	conductivity of a material
μ	mobility of a material
$\mu_n(\mu_p)$	electron (hole) mobility
τ_{sc}	scattering time between successive collisions. Also called relaxation time
ω	frequency
τ_o	rate at which an electron recombines radiatively with a hole at the same momentum value
τ_r	radiative recombination time for *e-h* pair
τ_{nr}	non-radiative recombination time for a *e-h* pair
τ_n	lifetime of an electron to recombine with a hole
τ_p	lifetime of a hole to recombine with an electron
τ_{sd}	storage delay time in a diode
τ_{ph}	photon lifetime in a laser cavity
δn	excess electron density in a region. This is the density above the equilibrium density
δp	excess hole density in a region
ϕ_m	metal work function
χ_s	electron affinity of a semiconductor
ϕ_s	work function of a semiconductor
ϕ_{ms}	difference between a metal and semiconductor work function
ϕ_b	barrier height seen by electrons coming from a metal towards a semiconductor
λ_c	cutoff wavelength of a detector
η_Q	quantum efficiency of a detector
η_{conv}	power conversion efficiency of a solar cell
η_{det}	efficiency of a detector to convert an optical signal current to an electrical current
η_{Qr}	radiative quantum efficiency of electron-hole recombination
η_{fiber}	coupling efficiency for a optical fiber
θ_A	acceptance angle for an optical fiber

APPENDIX
B

IMPORTANT PROPERTIES OF SEMICONDUCTORS

The data and plots shown in this Appendix are extracted from a number of sources. A list of useful sources is given below. Note that impact ionization coefficient and Auger coefficients of many materials are not known exactly.

- S. Adachi, *J. Appl. Phys.*, 58, R1 (1985).

- H.C. Casey, Jr. and M.B. Panish, *Heterostructure Lasers*, Part A, "Fundamental Principles;" Part B, "Materials and Operating Characteristics," Academic Press, N.Y. (1978).

- Landolt-Bornstein, *Numerical Data and Functional Relationship in Science and Technology*, Vol. 22, Eds. O. Madelung, M. Schulz, and H. Weiss, Springer-Verlag, N.Y. (1987).

- S.M. Sze, *Physics of Semiconductor Devices*, Wiley, N.Y. (1981). This is an excellent source of a variety of useful information on semiconductors.

Semi-conductor	Type of Energy Gap	Experimental Energy Gap E_g (eV)		Temperature Dependence of Energy Gap (eV)
		0 K	300 K	
AlAs	Indirect	2.239	2.163	$2.239 - 6.0 \times 10^{-4} T^2/(T + 408)$
GaP	Indirect	2.338	2.261	$2.338 - 5.771 \times 10^{-4} T^2/(T + 372)$
GaAs	Direct	1.519	1.424	$1.519 - 5.405 \times 10^{-4} T^2/(T + 204)$
GaSb	Direct	0.810	0.726	$0.810 - 3.78 \times 10^{-4} T^2/(T + 94)$
InP	Direct	1.421	1.351	$1.421 - 3.63 \times 10^{-4} T^2/(T + 162)$
InAs	Direct	0.420	0.360	$0.420 - 2.50 \times 10^{-4} T^2/(T + 75)$
InSb	Direct	0.236	0.172	$0.236 - 2.99 \times 10^{-4} T^2/(T + 140)$
Si	Indirect	1.17	1.11	$1.17 - 4.37 \times 10^{-4} T^2/(T + 636)$
Ge	Indirect	0.66	0.74	$0.74 - 4.77 \times 10^{-4} T^2/(T + 235)$

Table B.1: Energy gaps of some semiconductors along with their temperature dependence.

Material	Electron Mass (m_0)	Hole Mass (m_0)
AlAs	0.1	
AlSb	0.12	$m_{dos} = 0.98$
GaN	0.19	$m_{dos} = 0.60$
GaP	0.82	$m_{dos} = 0.60$
GaAs	0.067	$m_{lh} = 0.082$ $m_{hh} = 0.45$
GaSb	0.042	$m_{dos} = 0.40$
Ge	$m_l = 1.64$ $m_t = 0.082$	$m_{lh} = 0.044$ $m_{hh} = 0.28$
InP	0.073	$m_{dos} = 0.64$
InAs	0.027	$m_{dos} = 0.4$
InSb	0.13	$m_{dos} = 0.4$
Si	$m_l = 0.98$ $m_t = 0.19$	$m_{lh} = 0.16$ $m_{hh} = 0.49$

Table B.2: Electron and hole masses for several semiconductors. Some uncertainty remains in the value of hole masses for many semiconductors.

Compound	Direct Energy Gap E_g (eV)
$Al_x In_{1-x} P$	$1.351 + 2.23x$
$Al_x Ga_{1-x} As$	$1.424 + 1.247x$
$Al_x In_{1-x} As$	$0.360 + 2.012x + 0.698x^2$
$Al_x Ga_{1-x} Sb$	$0.726 + 1.129x + 0.368x^2$
$Al_x In_{1-x} Sb$	$0.172 + 1.621x + 0.43x^2$
$Ga_x In_{1-x} P$	$1.351 + 0.643x + 0.786x^2$
$Ga_x In_{1-x} As$	$0.36 + 1.064x$
$Ga_x In_{1-x} Sb$	$0.172 + 0.139x + 0.415x^2$
$GaP_x As_{1-x}$	$1.424 + 1.150x + 0.176x^2$
$GaAs_x Sb_{1-x}$	$0.726 + 0.502x + 1.2x^2$
$InP_x As_{1-x}$	$0.360 + 0.891x + 0.101x^2$
$InAs_x Sb_{1-x}$	$0.18 + 0.41x + 0.58x^2$

Table B.3: Compositional dependence of the energy gaps of the binary III-V ternary alloys at 300 K. (After Casey and Panish (1978).)

Semiconductor	Bandgap (eV) 300K	Mobility at 300 K (cm^2/V-s) Electrons	Holes
C	5.47	800	1200
Ge	0.66	3900	1900
Si	1.12	1500	450
α-SiC	2.996	400	50
GaSb	0.72	5000	850
GaAs	1.42	8500	400
GaP	2.26	110	75
InSb	0.17	8000	1250
InAs	0.36	33000	460
InP	1.35	4600	150
CdTe	1.56	1050	100
PbTe	0.31	6000	4000

Table B.4: Bandgaps, electron and hole mobilities of some semiconductors.

Breakdown Electric Fields in Semiconductors

Material	Bandgap (eV)	Breakdown Electric Field(V/cm)
GaAs	1.43	4×10^5
Ge	0.664	10^5
InP	1.34	
Si	1.1	3×10^5
$In_{0.53}Ga_{0.47}As$	0.8	2×10^5
C	5.5	10^7
SiC	2.9	$2\text{-}3 \times 10^6$
SiO_2	9	10^7
Si_3N_4	5	10^7

Table B.5: Breakdown electric fields in some semiconductors.

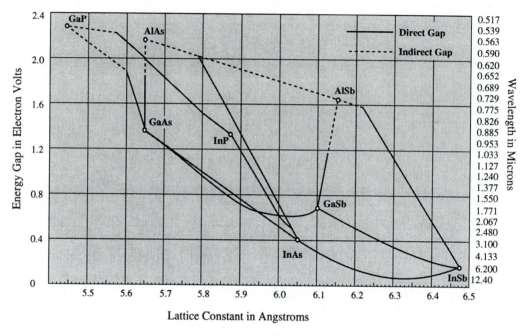

Figure B.1: Lattice constants and bandgaps of semiconductors at room temperature.

Material	b (eV)	d (eV)	dE$_g$/dP (10^{-12} eV cm^2/dyne)	Ξ_u (eV)
Si	-1.5 300 K	-3.4 300 K	-1.41 300 K	9.2 295 K
Ge	-2.2 300 K	-4.4 300 K	5. 300 K	15.9 297 K
AlSb	-1.35 77 K	-4.3 77 K	-3.50 77 K	6.2 300 K
GaP	-1.3 80 K	-4.0 80 K	-1.11 300 K	6.2 80 K
GaAs	-2.0 300 K	-6.0 300 K	11.7 300 K	
GaSb	-3.3 77 K	-8.35 77 K	14. 300 K	
InP	-1.55 77 K	-4.4 77 K	4.7 300 K	
InAs			10. 300 K	
InSb	-2.05 80 K	-5. 80 K	16.0 300 K	

Table B.6: Strain parameters of some semiconductors. The temperature is specified.

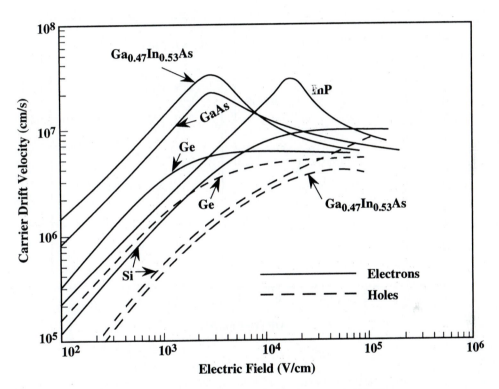

Figure B.2: Velocity-Field relations for several semiconductors at 300 K.

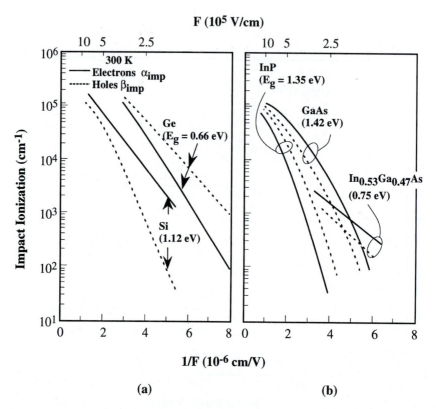

Figure B.3: Ionization rates for electrons and holes at 300 K versus reciprocal electric field for Ge, Si, GaAs, $In_{0.53}Ga_{0.47}$ and InP. (Si, Ge results are after S.M. Sze, *Physics of Semiconductor Devices*, John Wiley and Sons (1981); InP, GaAs, InGaAs results are after G. Stillman, *Properties of Lattice Matched and Strained Indium Gallium Arsenide*, ed. P. Bhattacharya, INSPEC, London (1993).

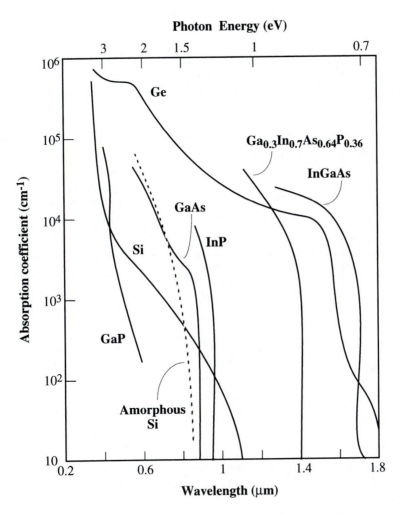

Figure B.4: Absorption coefficient as a function of wavelength for several semiconductors.

Material	Auger Coefficient (cm^6s^{-1})	Comments
$In_{0.53}Ga_{0.47}As$	~ 10^{-28} at 300 K	For a compilation of results in literature, see J Shah in "Indium Gallium Arsenide," ed. P. Bhattacharya, *INSPEC*, London (1992).
GaInAsP	~ 6×10^{-28} at $E_g = 0.8$ eV at 300 K ~ 1.2×10^{-27} at $E_g = 0.7$ eV at 300 K	Based on work of A. Sugimura, *Quantum Electronics*, QE-18, 352 (1982).
GaInAsSb	~ 6×10^{-27} at $E_g = 0.4$ eV at 300 K	

Table B.7: Auger coefficients of some semiconductors. Considerable uncertainty still exists in the Auger coefficients. The values given are only rough estimates.

APPENDIX

C

IMPORTANT QUANTUM MECHANICS CONCEPTS

C.1 THE HARMONIC OSCILLATOR PROBLEM: SPONTANEOUS AND STIMULATED EMISSION

The solution to the harmonic oscillator problem in quantum mechanics provides one of the most widely used results in semiconductor physics. The problems of scattering of electrons by phonons as well as absorption and emission of photons make direct use of our understanding of the harmonic oscillator problem. The Hamiltonian for the problem is

$$H = \frac{p^2}{2m} + \frac{1}{2}Kx^2 \tag{C.1}$$

In classical physics, the energy of the harmonic oscillator can start at zero and increase continuously. The different energy values correspond to different amplitudes. The frequency of the classical oscillator is

$$\omega_c = \sqrt{\frac{K}{m}} \tag{C.2}$$

In quantum mechanics, however, the energy of the oscillator does not start from zero and does not change continuously. The energy values allowed are given by

$$E_n = \left(n + \frac{1}{2}\right)\hbar\omega_c \text{ for } n = 0, 1, 2, \ldots \tag{C.3}$$

with the general state being denoted by $|n\rangle$.

One defines a raising operator $a^\dagger$ which operates on an energy state defined by the quantum number n and produces a state with a quantum number $n+1$. The lowering operator a operates on a state with quantum number n and produces a state with quantum number $n-1$. When the lowering operator operates on the vacuum state (quantum number 0) the outcome is zero.

We may write

$$H = \left(a^\dagger a + \frac{1}{2}\right)\hbar\omega_c \tag{C.4}$$

The operator $a^\dagger a$ has eigenvalues which are zero and positive integers which represent the number of "quanta" of energy present in a given state.

The mathematics of the harmonic oscillator problem is extremely useful in connection with quantization of the wavefields associated with particles (i.e., treating the wavefields or probability amplitudes as operators—just like we treat x and p as operators). In such a treatment, also, the operators $a^\dagger$ and a appear as combinations of wavefields and have the effect of raising or lowering the number of quanta or particles in a given state. In this context, these operators are called the creation and destruction operators. This concept is extremely useful in semiconductor physics when we discuss the scattering of electrons from phonons or photons. The photon field is represented by a particular state where the occupation (number) of photons of various modes is well-defined. During scattering, the electron absorbs or emits photons, thus changing the occupation of the final state.

It is useful to find the matrix representation of the operators $a^\dagger$, a. We note that the only nonvanishing elements are $\langle n-1|a|n\rangle$ for the operator a. Similarly, the only nonvanishing element for $a^\dagger$ is $\langle n|a^\dagger|n-1\rangle$ which is just the complex conjugate of $\langle n-1|a|n\rangle$. The diagonal element of $a^\dagger a$ is $\langle n|a^\dagger a|n\rangle = \langle n|a^\dagger|n'\rangle\langle n'|a|n\rangle = |\lambda_n|^2 = n$. Thus, apart from a phase factor, which we can choose to be unity

$$\begin{aligned} \langle n|a^\dagger|n-1\rangle &= \sqrt{n} \\ &= \langle n-1|a|n\rangle \end{aligned} \tag{C.5}$$

This gives the matrix elements for the operators $a^\dagger$ and a. *It is important to note that as seen from Eqn C.5 the matrix element for the process in which a quanta is created or emitted (i.e., one involving the creation operator) yields a term $(n(initial state) + 1$, while the one where a quanta is destroyed or absorbed has a term $n(initial state)$. Thus the creation or emission rate (proportional to the square of the matrix element) has a term proportional to n(initial) (the stimulated rate) and a term independent of the initial particles (the spontaneous rate). The importance of these two rates for light emission devices is brought in Chapters 9, 10 and 11 where we discuss LEDs and LDs.

C.2 TIME DEPENDENT PERTURBATION THEORY AND THE FERMI GOLDEN RULE

The time dependent perturbation theory has been used extensively in this text to address the problem of scattering of electrons by photons. The expressions for the scattering rates are given by the Fermi golden rule, given in this appendix. The general Hamiltonian of interest is of the form

$$H = H_0 + H'$$ (C.6)

where

$$H_0 u_k = E_k u_k$$ (C.7)

and E_k, u_k are known. The effect of H' is to cause transitions between the states u_k. The time dependent Schrödinger equation is

$$i\hbar \frac{\partial \psi}{\partial t} = H\psi$$ (C.8)

The approximation will involve expressing ψ as an expansion of the eigenfunctions $u_n \exp(-iE_n t/\hbar)$ of the unperturbed time dependent functions

$$\psi = \sum_n a_n(t) u_n e^{-iE_n t/\hbar}$$ (C.9)

We write

$$\omega_{kn} = \frac{E_k - E_n}{\hbar}$$ (C.10)

We assume that initially the system is in a single, well-defined state given by

$$
\begin{aligned}
a_k^{(0)} &= \langle k|m \rangle \\
&= \delta_{km}
\end{aligned}
$$ (C.11)

To the first order we have at time t

$$a_k^{(1)}(t) = \frac{1}{i\hbar} \int_{-\infty}^{t} \langle k|H'(t')|m \rangle \, e^{i\omega_{km} t'} \, dt'$$ (C.12)

We choose the constant of integration to be zero since $a_k^{(1)}$ is zero at time $t \to -\infty$, when the perturbation is not present.

Consider the case where the perturbation is harmonic, except that it is turned on at $t = 0$ and turned off at $t = t_0$. Let us assume that the time dependence is given by

$$\langle k|H'(t')|m \rangle = 2\langle k|H'(0)|m \rangle \sin \omega t'$$ (C.13)

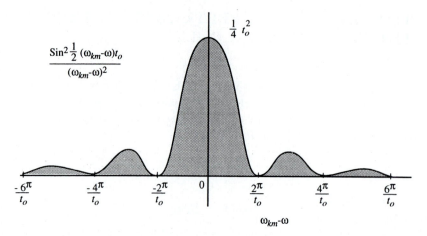

Figure C.1: The ordinate is proportional to the probability of finding the system in a state k after the perturbation has been applied to time t_0.

Carrying out the integration until time $t \geq t_0$ in Eqn. C.12, we get

$$a_k^{(1)}(t \geq t_0) = -\frac{\langle k|H'(0)|m\rangle}{i\hbar} \left(\frac{\exp[(\omega_{km}+\omega)t_0]-1}{\omega_{km}+\omega} - \frac{\exp[(\omega_{km}-\omega)t_0]-1}{\omega_{km}-\omega} \right) \quad (C.14)$$

The structure of this equation says that the amplitude is appreciable, only if the denominator of one or the other term is practically zero. The first term is important if $\omega_{km} \approx -\omega$, or $E_k \approx E_m - \hbar\omega$. The second term is important if $\omega_{km} \approx \omega$ or $E_k \approx E_m + \hbar\omega$

Thus, the first order effect of a harmonic perturbation is to transfer, or to receive from the system, the quanta of energy $\hbar\omega$. If we focus on a system where $|m\rangle$ is a discrete state, $|k\rangle$ is one of the continuous states, and $E_k > E_m$, so that only the second term is important, the first order probability to find the system in the state k after the perturbation is removed is

$$\left| a_k^{(1)}(t \geq t_0) \right|^2 = 4|\langle k|H'(0)|m\rangle|^2 \frac{\sin^2\left[\frac{1}{2}(\omega_{km}-\omega)t_0\right]}{\hbar^2(\omega_{km}-\omega)^2} \quad (C.15)$$

The probability function has an oscillating behavior as shown in Fig.C.1. The probability is maximum when $\omega_{km} = \omega$, and the peak is proportional to t_0^2. However, the uncertainty in frequency $\Delta\omega = \omega_{km} - \omega$, is non-zero for the finite time t_0. This uncertainty is in accordance with the Heisenberg uncertainty principle

$$\Delta\omega \, \Delta t = \Delta\omega \, t_0 \sim 1 \quad (C.16)$$

If there is a spread in the allowed values of $(\omega_{km} - \omega)$, which may occur either because the initial and/or final states of the electron are continuous, or because the perturbation has a spread of frequencies ω, it is possible to define a scattering rate per unit time. The total rate per unit time for scattering into any final state, is given by

$$W_m = \frac{1}{t_0} \sum_{\text{final states}} \left| a_k^{(1)}(t \geq t_0) \right|^2 \qquad \text{(C.17)}$$

If t_0 is large, the sum over the final states only includes the final states where $\omega_{km} - \hbar\omega = 0$, i.e., energy is conserved in the process.

If we assume that $|\langle k|H'|m\rangle|^2$ does not vary over the (infinitesimally) small spread in final states, we can write

$$x = \frac{1}{2}(\omega_{km} - \omega)t_0 \qquad \text{(C.18)}$$

and use the integral

$$\int_{-\infty}^{\infty} x^{-2} \sin^2 x \, dx = \pi$$

to get

$$W_m = \frac{2\pi}{\hbar} \sum_{\text{final states}} \delta\left(\hbar\omega_{km} - \hbar\omega\right) |\langle k|H'|m\rangle|^2 \qquad \text{(C.19)}$$

This is the Fermi golden rule, which is used widely in our text. It is interesting to note that an identical expression occurs for scattering rate from fixed defects when Born approximation is used for scattering.

In a given scattering problem, one has to pay particular attention to the conditions under which the golden rule has been derived. The more important condition is that the time of interaction of the pertubation be essentially infinite. If the interaction time is finite, the δ-function of the Golden Rule changes over to the broadened function of Fig.C.1. Thus, the energy conservation is not strictly satisfied and the final state of the electron is not well-defined.

C.3 EIGENVALUE METHOD TO SOLVE COUPLED EQUATIONS

A very large class of problems, in classical mechanics and quantum mechanics, can be written in the form of a matrix equation, which can be solved by eigenvalue solvers. Such eigenvalue solvers are generally available at most computer centers. These libraries provide subroutines which can be called to solve the matrix for both eigenvalues and eigenfunctions. Both the Schrödinger equation and the wave equation for photons can be addressed by the formulation given here, provided we are interested in spatially confined states. Examples are the electron problem in a quantum well and the modes of a waveguide.

The general equation to be solved is

$$H\Psi = E\Psi \tag{C.20}$$

where H is an $n \times n$ matrix, E is an eigenvalue which can have n values, and Ψ and is an n-dimensional vector. In general, the eigenfunction Ψ can be expanded in terms of an orthonormal basis set $\{\psi_n\}$

$$\Psi = \sum_n a_n \psi_n \tag{C.21}$$

A wavefunction Ψ_i is known when all the a_n's are known for that function. Substituting Eqn. C.21 in Eqn. C.20, multiplying successively by $\psi_1^*, \psi_2^*, \cdots, \psi_n^*$ and integrating, we get a set of equations

$$\begin{bmatrix} H_{11} - E & H_{12} & H_{13} & \cdots & H_{1n} \\ H_{21} & (H_{22} - E) & H_{23} & \cdots & H_{2n} \\ & & \vdots & & \\ & & \vdots & & \\ H_{n1} & N_{n2} & H_{n3} & \cdots & (H_{nn} - E) \end{bmatrix} \begin{bmatrix} a_1\psi_1 \\ a_2\psi_2 \\ \vdots \\ \vdots \\ a_n\psi_n \end{bmatrix} = 0 \tag{C.22}$$

where

$$H_{mn} = \int \psi_m^* \, H \, \psi_n \, d^3r \tag{C.23}$$

This is the standard form of the eigenvalue problem. It can be solved analytically if n is small (say less than 4), but, in general, will require numerical techniques.

The general, Schrödinger equation (the approach can be used for the wave equation also)

$$-\frac{\hbar^2}{2m}\nabla^2\Psi + V\Psi = E\Psi \tag{C.24}$$

can be expressed in terms of a matrix equation. This approach is very useful if we are looking for bound or quasi-bound states in a spatially varying potential V. Let us consider the scalar Schrödinger equation (or the single band equation often used to describe electrons in the conduction band). Let us assume that the eigenfunction we are looking for is confined in a region L as shown in Fig.C.2. We divide this region into equidistant ℓ mesh points x_i, each separated in real space by a distance Δx. The wavefunction we are looking for is now of the form

$$\Psi = \sum_n a_n \psi_n \tag{C.25}$$

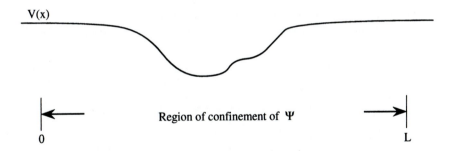

Figure C.2: A potential profile in which a wavefunction is confined.

where ψ_n are simply functions at the mesh point which are normalized within the interval centered at x_n and are zero outside that interval.

We can also write the differential equation as a general difference equation

$$-\frac{\hbar^2}{2m}\left[\frac{\Psi(x_i-1)-2\Psi(x_i)+\Psi(x_i+1)}{\Delta x^2}+V\Psi\right]=E\Psi \tag{C.26}$$

Once again, substituting for the general wavefunction Eqn. C.26 and taking an outer product with $\psi_1, \psi_2, \cdots, \psi_\ell$, we get a set of ℓ equations (remember that we are assuming $a_0 = a_{\ell+1} = 0$, i.e., the wavefunction is localized in the space L) which can be written in the matrix form as

$$\begin{bmatrix} A(x_1) & B & 0 & \cdots & 0 & 0 \\ B & A(x_2) & B & \cdots & 0 & 0 \\ 0 & B & A(x_3) & \cdots & 0 & 0 \\ & & \vdots & & & \\ & & \vdots & & & \\ 0 & 0 & 0 & \cdots & B & A(x_\ell) \end{bmatrix} \begin{bmatrix} a_1\,\psi_1 \\ a_2\,\psi_2 \\ a_3\,\psi_3 \\ \vdots \\ \vdots \\ a_\ell\,\psi_\ell \end{bmatrix} = 0 \tag{C.27}$$

with

$$A(x_i) = \frac{\hbar^2}{m\Delta x^2} + V(x_i) - E$$

$$B = -\frac{\hbar^2}{2m\Delta x^2}$$

This $\ell \times \ell$ set of equations can again be solved by calling an appropriate sub-routine from a computer library to get the eigenvalues E_n and wavefunctions ψ_n. In general we will get ℓ eigenvalues and eigenfunctions. The lowest lying state is the ground state, while the others are excited states. Note that the conditions imposed by us on the wavefunction, viz. the function goes to zero outside our confinement distance L, may not apply to some of the excited states. To check for the accuracy of the condition, one should increase L and see that there is no change in the eigenenergies and eigenfunctions of interest.

APPENDIX
D

BIPOLAR
TRANSISTOR ACTION

D.1 THE BIPOLAR TRANSISTOR

In Chapter 7 we have discussed the phototransistor which is a low noise, high gain detector of light. The working of the phototransistor depends upon the bipolar transistor which allows a small electrical signal to be amplified. In this appendix we will examine the operation of the bipolar device. Results derived in this appendix are used in Chapter 7 for the phototransistor.

The bipolar transistor is essentially made from two back to back p-n diodes. The n-p-n device shown in Fig. D.1 has an emitter, a base and a collector. The bipolar device has a complex physical orientation of current flow, as the carriers from the emitter are injected "vertically" across the base while the base charge is injected from the "side" of the device as can be seen in Fig. D.1a. However, unless the emitter area is very narrow, the device can be understood using a 1-dimensional analysis.

In general, a number of currents can be identified in the bipolar device as shown in Fig. D.1b. The symbols for the doping density are (for the n-p-n device): N_{de}—donor density in the emitter; N_{ab}—acceptor density in the base; N_{dc}—donor density in the collector.

We will consider the bipolar device under the forward active mode, i.e., when the emitter base junction (EBJ) is forward biased and the base collector junction (BCJ) is reverse biased. This produces a high current gain β.

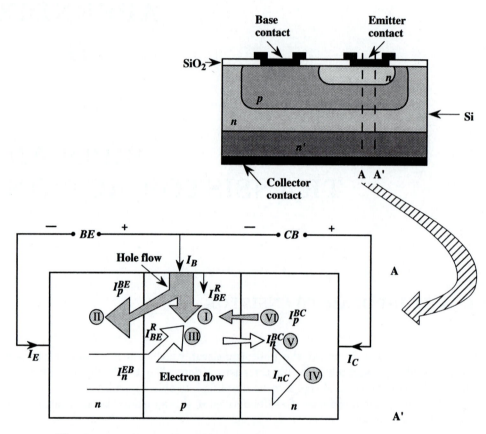

I_n^{EB} = Emitter current injected into the base $\equiv I_{En}$

I_p^{BE} = Base current injected into the emitter $\equiv I_{Ep}$

I_{BE}^R = Recombination current in the base region

I_p^{BC} = Hole current injected across reverse biased-base collector junction

I_n^{BC} = Electron current injected across reverse biased-base collector junction

I_{nC} = Electron current coming from the emitter ($\equiv I_C$)

Figure D.1: A schematic of an Si BJT showing the 3-dimensional nature of the structure and the current flow. Along the section AA', the current flow can be assumed 1-dimensional. The various current components in a BJT are also shown.

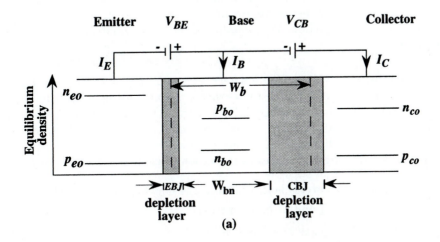

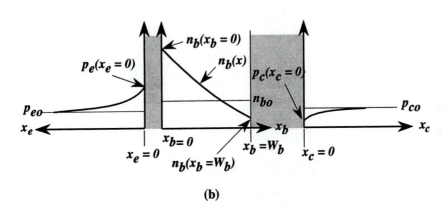

Figure D.2: (a) The equilibrium carrier concentrations of electrons and holes and positions of the junction depletion regions in the npn transistor. (b) Minority carrier distributions in the emitter, base, and collector regions. In the simple model considered initially, we assume that the actual base width W_b and the natural base width W_{bn} are equal.

D.1.1   Current Flow in a BJT

In order to derive the various current components flowing in the bipolar device we will make use of our understanding of the p-n junction developed in Chapter 5. In that chapter, we had made a number of approximations regarding the charge density and current flow across a p-n junction. We will retain those approximations for the study of the bipolar device.

For simplicity we will use the different axes and origins as shown in Fig D.2. The distances are labelled as x_e, x_b, and x_c as shown and are measured from the edges of the depletion region. The base width is W_b, but the width of the "neutral" base region is W_{bn} as shown. We assume that W_b and W_{bn} are equal. Later we will study the effect of the two widths being different. Using the p-n diode theory we have the following relations for the excess carrier densities in the various regions (see Section 5.3.1):

$$\delta p_e(x_e = 0) = \text{excess hole density at the emitter side the of EBJ}$$
$$\text{(emitter-base junction)}$$
$$= p_{eo}\left[exp\left(eV_{BE}/k_BT\right) - 1\right] \tag{D.1}$$
$$\delta n_b(x_b = 0) = \text{excess electron density on the base side of EBJ}$$
$$= n_{bo}\left[exp\left(eV_{BE}/k_BT\right) - 1\right] \tag{D.2}$$
$$\delta n_b(x_b = W_b) = \text{excess electron density at the base side of the CBJ}$$
$$\text{(collector-base junction)}$$
$$= n_{bo}\left[exp\left(-eV_{CB}/k_BT\right) - 1\right] \tag{D.3}$$
$$\delta p_c(x_c = 0) = \text{excess hole density at the collector side of the CBJ}$$
$$= p_{co}\left[exp\left(-eV_{CB}/k_BT\right) - 1\right] \tag{D.4}$$

In these expressions the subscripts p_{eo}, n_{bo}, and p_{co} represent the minority carrier equilibrium densities in the emitter, base, and the collector, respectively. The total minority carrier concentrations p_e in the emitter, n_b in the base and p_c in the collector are shown schematically in Fig. D.2b. The majority carrier densities are n_{eo} ($=N_{de}$), p_{bo} ($=N_{ab}$), and N_{co} ($=N_{dc}$) for the emitter, base, and collector. We will assume that the emitter and collector regions are longer than the hole diffusion lengths L_p, so that the hole densities decrease exponentially away from base regions.

In the base region, the excess electron density is given at the edges of the neutral base region by the Eqns. D.11 and D.12. To obtain the electron density in the base we must solve the continuity equation using these two boundary conditions. The continuity equation is

$$\frac{d^2\delta n_b(x_b)}{dx^2} - \frac{\delta n_b(x_b)}{L_b^2} = 0 \tag{D.5}$$

Upon solving for this equation we get for the excess carriers

$$\delta n_b(x_b) = \frac{n_{bo}}{sinh\left(\frac{W_{bn}}{L_b}\right)}\left\{sinh\left(\frac{W_{bn}-x_b}{L_b}\right)\left[exp\left(\frac{eV_{BE}}{k_BT}\right)-1\right]\right.$$
$$\left.+sinh\left(\frac{x_b}{L_b}\right)\left[exp\left(-\frac{eV_{CB}}{k_BT}\right)-1\right]\right\} \tag{D.6}$$

The form of the total minority carrier densities (i.e., background and excess) is shown in Fig. D.2b. The electron distribution in the base is almost linear as can be seen and is assumed to be so for devices where the base width is much smaller than the minority carrier diffusion length. Once the excess carrier spatial distributions are known, we can calculate the currents as we did for the *p-n* diode. We assume that the emitter-base currents are due to carrier diffusion once the device is biased. We have, for a device of area A and diffusion coefficients D_b and D_e in the base and emitter, respectively

$$I_{En} = I_n^{EB} = eAD_b \left.\frac{d\delta n_b(x)}{dx_b}\right|_{x_b=0} \tag{D.7}$$

$$I_{Ep} = I_p^{BE} = -eAD_e \left.\frac{d\delta p(x)}{dx_e}\right|_{x_e=0} \tag{D.8}$$

These are the current components shown in Fig. D.1b and represent the emitter current components III and IV. Assuming an exponentially decaying hole density into the emitter, we have, as in the case of a *p-n* diode

$$I_{Ep} = -A\left(\frac{eD_e p_{eo}}{L_e}\right)\left[exp\left(\frac{eV_{BE}}{k_BT}\right)-1\right] \tag{D.9}$$

Using the electron distribution derived earlier (Eqn. D.6), we have for the electron part of the emitter current

$$I_{En} = -\frac{eAD_b n_{bo}}{L_b sinh\left(\frac{W_{bn}}{L_b}\right)}\left\{cosh\left(\frac{W_{bn}-x_b}{L_b}\right)\left[exp\left(\frac{eV_{BE}}{k_BT}\right)-1\right]\right.$$
$$\left.-cosh\left(\frac{x_b}{L_b}\right)\left[exp\left(-\frac{eV_{CB}}{k_BT}\right)-1\right]\right\}\Bigg|_{at\ x_b=o}$$
$$= -\frac{eAD_b n_{bo}}{L_b sinh\left(\frac{W_{bn}}{L_b}\right)}\left\{cosh\left(\frac{W_{bn}}{L_b}\right)\left[exp\left(\frac{eV_{BE}}{k_BT}\right)-1\right]\right.$$
$$\left.-\left[exp\left(-\frac{eV_{CB}}{k_BT}\right)-1\right]\right\} \tag{D.10}$$

Note that for high emitter efficiency and high current gain the hole current into the emitter I_{Ep} *can be reduced if the emitter is very heavily doped so that* p_{eo} *is very small.* We will return to this issue later. The total emitter current is now

$$I_E = I_{En} + I_{Ep} = -\left\{ \frac{eAD_b n_{bo}}{L_b} coth\left(\frac{W_{bn}}{L_b}\right) + \frac{eAD_e p_{eo}}{L_e} \right\}$$

$$\left[exp\left(\frac{eV_{BE}}{k_B T}\right) - 1 \right] + \frac{eAD_b n_{bo}}{L_b sinh\left(\frac{W_{bn}}{L_{bn}}\right)} \left[exp\left(-\frac{eV_{CB}}{k_B T}\right) - 1 \right] \quad \text{(D.11)}$$

The collector current components can be obtained by using the same approach. Thus we have

$$I_n^{BC} = eAD_b \left. \frac{d\delta n_b(x_b)}{dx_b} \right|_{x_b = W_{bn}} \quad \text{(D.12)}$$

$$I_p^{BC} = eAD_p \left. \frac{d\delta p(x_c)}{dx_c} \right|_{x_c = 0} \quad \text{(D.13)}$$

Using Eqn. D.12 at $x_b = W_{bn} \; (\cong W_b)$, we have

$$I_n^{BC} = - \frac{eAD_b n_{bo}}{L_b sinh(W_{bn}/L_b)} \left[exp\left(\frac{eV_{BE}}{k_B T}\right) - 1 \right]$$

$$+ \frac{eAD_b n_{bo}}{L_b} coth\left(\frac{W_{bn}}{L_b}\right) \left[exp\left(-\frac{eV_{CB}}{k_B T}\right) - 1 \right] \quad \text{(D.14)}$$

The hole current from the collector side is the same as for a reverse biased *p-n* junction:

$$I_p^{BC} = -\frac{eAD_c p_{co}}{L_c} \left[exp\left(-\frac{eV_{CB}}{k_B T}\right) - 1 \right] \quad \text{(D.15)}$$

The way we have defined the currents, the two current components flow along $+x$ direction. If we define I_C as the total current flowing from the collector into the base, we have

$$-I_C = \left[\frac{eAD_c p_{co}}{L_c} + \frac{eAD_b n_{bo}}{L_b} coth\left(\frac{W_{bn}}{L_b}\right) \right] \left[exp\left(-\frac{eV_{CB}}{k_B T}\right) - 1 \right]$$

$$- \frac{eAD_b n_{bo}}{L_b sinh\left(\frac{W_{bn}}{L_b}\right)} \left[exp\left(\frac{eV_{BE}}{k_B T}\right) - 1 \right] \quad \text{(D.16)}$$

We now have all the various current components in the bipolar device ($I_B = I_E - |I_C|$). It is interesting to point out that if the base region W_{bn} is much smaller than the diffusion length, the electron gradient in the base region is simply given by the linear value

$$\frac{d\delta n_b(x_b)}{dx_b} \longrightarrow \frac{\delta n_b(x_b) - \delta n_b(x_b = 0)}{W_{bn}} \quad \text{(D.17)}$$

If we assume that the emitter efficiency is essentially unity, reverse collector current is negligible, and W_b is much smaller than L_b, we get

$$I_E \cong -\frac{eAD_b n_{bo}}{W_{bn}}\left[exp\left(\frac{eV_{BE}}{k_B T}\right) - 1\right] \tag{D.18}$$

$$I_C \cong \frac{eAD_b n_{bo}}{W_{bn}}\left[exp\left(\frac{eV_{BE}}{k_{BT}}\right) - 1\right] \tag{D.19}$$

$$I_B = \frac{eAD_b n_{bo} W_{bn}}{2L_b^2}\left[exp\left(\frac{eV_{BE}}{k_B T}\right) - 1\right] \tag{D.20}$$

D.2 BJT DESIGN LIMITATIONS: THE NEED FOR BAND TAILORING AND HBTs

In the BJT, once a material system is chosen (say Si, Ge, GaAs, etc.), the only flexibility one has in the device design is the doping levels and the device dimensions. We will show in this section why this is a serious handicap for high performance devices. Let us examine the material parameters controlling the device performance parameters. We have (α is the ratio of the collector current to the emitter current)

$$\alpha = \left[1 - \frac{p_{eo} D_e W_{bn}}{n_{bo} D_b L_e}\right]\left[1 - \frac{W_{bn}^2}{2L_b^2}\right] \tag{D.21}$$

and the current gain β, the ratio of the collector current to the base current, is

$$\beta = \frac{\alpha}{1 - \alpha} \tag{D.22}$$

For α to be close to unity and β to be high, it is essential that: i) the emitter doping be much higher than the base doping ($p_{eo} \gg n_{bo}$); and ii) base width be as small as possible. In fact, the product $p_{bo} W_b$, called the Gummel number, should be as small as possible.

A small base with relatively low doping (usually in BJTs $n_{eo} \sim 10^2$-$10^3 p_{bo}$) introduces a large base resistance which affects the device performance. However, a far greater problem arises from the bandgap shrinking of the emitter region which is very heavily doped. If the emitter has a different bandgap (due to heavy doping) than the base, the equations for the excess hole and electrons injected across the EBJ change and affect the transistor performance.

If we assume that hole injection across the EBJ is a dominant factor, the current gain of the device becomes

$$\beta = \frac{\alpha}{1 - \alpha} \simeq \frac{n_{bo} D_b L_e}{p_{eo} D_e W_{bn}} \tag{D.23}$$

If the emitter bandgap shrinks by ΔE_g due to doping, the hole density for the same doping changes by an amount which can be evaluated using the change in the intrinsic carrier concentration,

$$n_{ie}\left(E_g - \Delta E_g\right) = n_{ie}\left(E_g\right) exp\left(\frac{\Delta E_g}{2k_B T}\right) \tag{D.24}$$

where ΔE_g is positive in our case. Thus the value of p_{eo} changes as

$$p_{eo}\left(E_g - \Delta E_g\right) \quad \alpha \quad n_{ie}^2\left(E_g - \Delta E_g\right) \tag{D.25}$$

$$= \quad p_{eo}\left(E_g\right) exp\left(\frac{\Delta E_g}{k_B T}\right) \tag{D.26}$$

As a result of the bandgap decrease, the gain of the device decreases as (for the case where $L_b \gg W_{bn}$)

$$\beta = \frac{D_b N_{de} L_e}{D_e N_{ab} W_{bn}} exp\left(-\frac{\Delta E_g}{k_B T}\right) \tag{D.27}$$

where we have used the fact that

$$\frac{p_{eo}}{n_{bo}} = \frac{N_{ab}}{N_{de}} \tag{D.28}$$

where N_{ab} and N_{de} are the acceptor and donor levels in the base and emitter, respectively. As a result of this for a fixed base doping, as the emitter doping is increased, initially the current gain increases, but then as bandgap shrinkage increases, the current gain starts to decrease .

From the discussion above, it is clear that the conflicting requirements of heavy emitter doping, low base doping, small base width, etc., as shown in Fig. D.3, cannot be properly met by a single bandgap structure. This had led Shockley and Kroemer in the 50's to conceive of bipolar devices where the bandgap could change from one region to another. Typically, in these heterojunction bipolar transistors (HBJT or HBT), the *emitter could be made from a wide gap material.* This would dramatically suppress the injection of holes from the base to the emitter. If ΔE_v is the difference between the emitter and base valence band lineups, then we have once again for emitter injection limited gain (for $L_b \gg W_{bn}$)

$$\beta_{max} = \frac{D_b N_{de} L_e}{D_e N_{ab} W_{bn}} \ exp\left(\frac{\Delta E_v}{k_B T}\right) \tag{D.29}$$

The structure now has a band profile as shown in Fig. D.4. Typically $\Delta E_v / k_B T$ is chosen to be ~ 5, so that an improvement of 10^2 occurs. Now one can dope the base very heavily without paying the penalty of low gain.

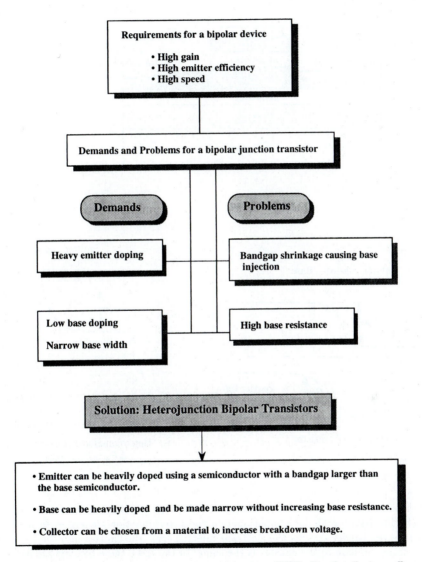

Figure D.3: Conflicting requirements for high performance BJTs. Band tailoring offers reconciliation of all these requirements.

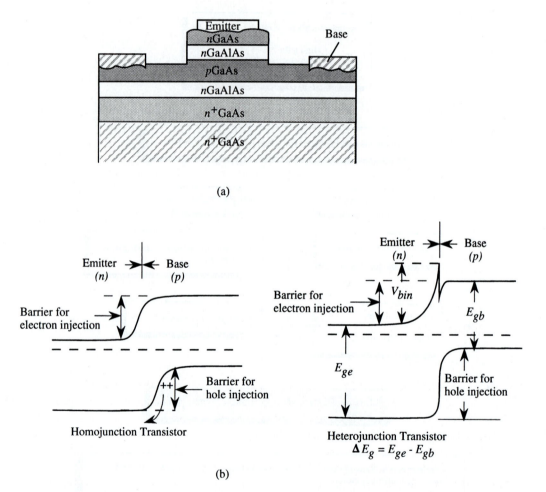

(a)

(b)

Figure D.4: (a) A schematic of a heterojunction bipolar transistor made from GaAs and Al-GaAs. (b) The band profile for a homojunction and heterojunction transistor. In the homojunction device, the barrier seen by the electrons injected from the emitter into the base and holes injected from the base to the emitter is the same. In the HBT these barriers are different. The spike shown in the HBT conduction band is smoothed out by using a grading in the bandgap between the emitter and the base.

APPENDIX
E

OPTICAL WAVES IN WAVEGUIDES AND CRYSTALS

E.1 GUIDED OPTICAL MODES IN PLANER WAVEGUIDES

In this text we have seen the use of heterostructures to alter the properties of both electronic and optical states. The heterostructures provide either a spatial variation in the potential seen by the electrons, or a spatial variation in the refractive index seen by light waves. The resulting structures are often called quantum wells and optical waveguides, respectively. We have studied the properties of quantum well structures in some detail in the text. Mathematically, the waveguide problem is similar and in this appendix, we will summarize some properties of the optical modes in planer waveguides.

Let us consider a planer waveguide structure as shown in Fig. E.1. The waveguide has a three layer structure, as shown, where a high refractive index material is surrounded by a low refractive index material. The thickness of the guide is d, as shown. Although in general, the refractive indices (dielectric constants) of regions surrounding the guide can be different, we will discuss the case where $n_{r2} = n_{r3}$, as shown in Fig. E.1.

Let us consider the case of a wave propagating along the z-axis. Assuming that there is no spatial variation of the wave in the y-direction, the wave equation for the

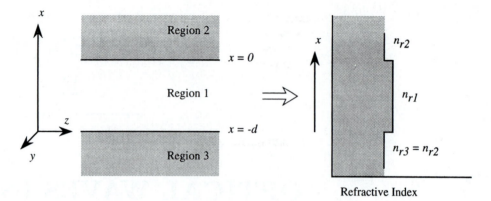

Figure E.1: A planer waveguide in which the refractive index (dielectric constant) varies in the x-direction.

electric and magnetic fields are, in general,

$$
\begin{aligned}
F &= F(x,y) \; exp \, [i(\omega t - \beta z)] \\
H &= H(x,y) \; exp \, [i(\omega t - \beta z)]
\end{aligned}
$$

(E.1)

(E.2)

The x-dependence of the fields is given from Maxwell's equations by the usual wave equation

$$
\frac{\partial^2 F_y}{\partial x^2} \; + \; (\omega^2 \epsilon \mu - \beta^2) F_y = 0
$$

(E.3)

$$
\frac{\partial^2 H_y}{\partial x^2} \; + \; (\omega^2 \epsilon \mu - \beta^2) H_y = 0
$$

(E.4)

These two equations represent waves that are called TE (electric field along y-axis) and TM (magnetic field along y-axis) polarized. Note that in our discussion here, the confinement direction is taken to be along the x-axis. Often in the text, the confinement direction has been taken to be the z-axis. This choice is not of significance and is arbitrary. Let us consider the TE polarized light case. The equation can be rewritten as

$$
\frac{\partial^2 F_y}{\partial x^2} + \left(n_r^2 k_o^2 - \beta^2 \right) F_y = 0
$$

(E.5)

where we have used the fact that

$$
\omega \sqrt{\epsilon \mu} = n_r \, k_o
$$

(E.6)

where

$$
k_o = \frac{2\pi}{\lambda_o}
$$

(E.7)

is the free space wave vector.

From Eqn. E.5 we can see that, in general, the solutions will either have the form

$$F_y \sim exp\,(i\,k_x x) \quad \text{if } n_r^2 k_o^2 > \beta^2 \tag{E.8}$$

where

$$k_x = \sqrt{n_r^2 k_o^2 - \beta^2} \tag{E.9}$$

or the form

$$F_y \sim exp(-\gamma x) \quad \text{if } n_r^2 k_o^2 < \beta^2 \tag{E.10}$$

where

$$\gamma = \sqrt{\beta^2 - n_r^2 k_o^2} \tag{E.11}$$

Equations E.8 and E.10 represent oscillatory and decaying waves. The waveguide is designed so that for some modes, the field has a general form given by Eqn. E.8 in the guiding region and the form given by Eqn. E.11 in the cladding region. Referring to Fig. E.1, we get the following solutions for the electric field:

$$F_y = \begin{cases} A\,exp - \gamma x & 0 \le x \le \infty \\ B\,cos(k_x x) + C\,sin(k_x x) & -d \le x \le 0 \\ D\,exp\,\gamma(x+d) & -\infty \le x \le -d \end{cases} \tag{E.12}$$

where γ and k_x are given by Eqns. E.9 and E.11 applied with the appropriate refractive indices.

Using the continuity of F_y and $\frac{\partial F_y}{\partial x}$ across the boundaries at $x = 0$ and $x = -d$, we get the general solution

$$F_y = \begin{cases} A'\,exp\,(-\gamma x) & 0 \le x \le \infty \\ A'[cos(k_x x) - (\gamma/k_x)\,sin(k_x x)] & -d \le x \le 0 \\ A'[cos(k_x d) + (\gamma/k_x)sin(k_x d)]\,exp\,[\gamma(x+d)] & -\infty \le x \le -d \end{cases} \tag{E.13}$$

As noted earlier

$$\gamma = \sqrt{\beta^2 - n_{r2}^2 k_o^2} \tag{E.14}$$

$$k_x = \sqrt{n_{r1}^2 k_o^2 - \beta^2} \tag{E.15}$$

The solutions of this problem (i.e., the values of γ or k_x) are given by using these solutions in the wave equation (Eqn. E.5). This gives two transcendental equations, either of which has to be satisfied by the solutions

$$\frac{k_x d}{2}\,tan\,\frac{k_x d}{2} = \frac{\gamma d}{2} \tag{E.16}$$

or

$$\frac{k_x d}{2} \cot \frac{k_x d}{2} = -\frac{\gamma d}{2} \qquad (E.17)$$

The first of these equations results in modes that have even parity, while the second equation gives modes with odd parity. The modes are denoted by $m = 0, 1, 2 \ldots$ The transcendental Eqns. E.16 and E.17 can be solved by numerical techniques. One useful approach is a graphical technique shown in Fig. E.2a. One starts out with plotting curves in the $\gamma d/2, k_x d/2$ plane which satisfy Eqns. E.16 or E.17. Note that a large number of k_x values satisfy the equations for a given value of γ. Next, we note that we have the equality

$$\left(\frac{k_x d}{2}\right)^2 + \left(\frac{\gamma d}{2}\right)^2 = (n_{r1}^2 - n_{r2}^2)\left(\frac{k_o d}{2}\right)^2$$
$$\equiv R(d)^2 \qquad (E.18)$$

We therefore draw a circle with radius $R(d)$. For a given value of n_{r1}, n_{r2} and d, there is one such circle. The intersection of this circle with the first set of curves gives the desired solutions. There may be a number of allowed solutions for a given waveguide thickness. As the waveguide thickness increases, the number of allowed modes also increases, as shown in Fig. E.2a. To find the cutoff mode for a given thickness, we note that $(k_x d/2) \tan (k_x d/2)$ intersects the $(k_x d/2)$ axis at values of $m\pi/2$. Thus the thickness at which a mode m is just allowed is given by

$$R(d_c) = \frac{m\pi}{2}$$

or

$$d_c = \frac{1}{2} \frac{m\lambda}{(n_{r1}^2 - n_{r2}^2)^{1/2}} \qquad (E.19)$$

Once k_x (or γ) is known, the propagation constant β is known for a given free space k_o from the relation

$$k_x^2 = n_{r1}^2 k_o^2 - \beta^2$$

Typical modal profile is shown in Fig. E.2b.

E.1.1   Optical Confinement Factor

An important parameter in optical waveguides is the fraction of the optical energy in the guide region (region 1 of Fig. E.1). This is the optical confinement factor Γ that

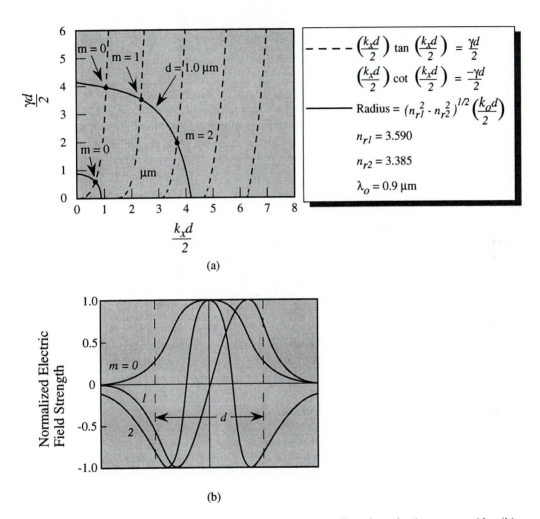

Figure E.2: (a) The graphical approach to solving for the allowed modes in a waveguide. (b) Typical solutions for the waveguide modes.

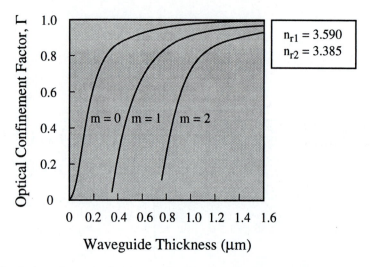

Figure E.3: Optical confinement factor as a function of waveguide thickness.

appears in the semiconductor laser performance. This factor is known from Eqn. E.13 once γ and k_x are known. The optical confinement factor is

$$\Gamma = \frac{\int_o^d E_y^2 \, dx}{\int_{-\infty}^{\infty} E_y^2 \, dx} \tag{E.20}$$

This gives the value (from Eqn. E.13)

$$\Gamma = \left\{ 1 + \frac{cos^2(k_x d/2)}{\gamma \left[d/2 + \left(\frac{1}{k_x} \right) sin\left(k_x d/2\right) cos\left(k_x d/2\right) \right]} \right\}^{-1} \tag{E.21}$$

The value of Γ is a strong function of the mode number, the thickness of the guide, and the difference between n_{r1} and n_{r2}. Some values for γ are shown in Fig. E.3 for GaAs/$Al_{03}Ga_{0.7}$As waveguides.

E.2   WAVE PROPAGATION IN CRYSTALS—A SUMMARY

In an isotropic medium, the propagation of light waves is described by a direction independent dielectric constant (or refractive index). However, in crystalline materials, the medium is not isotropic. In fact, the anisotropy of the crystalline materials and the ability to alter this anisotropy is widely exploited to fabricate polarizers and optical modulators. We have seen in Chapters 5 and 12 how optoelectronic modulators can be designed on the basis of altering the refractive index of a crystal along a particular direction. In this section we will briefly describe the physics of wave propagation in crystals.

It is useful to describe the properties of a crystal by choosing principal axes determined by the crystal symmetry. The displacement D and the electric field F of the light waves, in general, have a relation given by the dielectric tensor

$$
\begin{aligned}
D_1 &= \epsilon_{11}F_1 + \epsilon_{12}F_2 + \epsilon_{13}F_3 \\
D_2 &= \epsilon_{21}F_1 + \epsilon_{22}F_2 + \epsilon_{23}F_3 \\
D_3 &= \epsilon_{31}F_1 + \epsilon_{32}F_2 + \epsilon_{33}F_3
\end{aligned}
\tag{E.22}
$$

with $\epsilon_{ij} = \epsilon_{j1}$

A most useful concept to describe wave propagation is the concept of Fresnel ellipsoid or the equivalent concept of the index ellipsoid. The Fresnel ellipsoid is given by the relation

$$\sum \epsilon_{ij} x_i x_j = \text{constant}$$

A more useful constraint placed on wave propagation is an equivalent index ellipsoid (or indicatrix)

$$\frac{x_i^2}{n_{r1}^2} + \frac{x_2^2}{n_{r2}^2} + \frac{x_3^2}{n_{r3}^2} = \text{constant} \tag{E.23}$$

We will briefly describe how the index ellipsoid is used to describe the polarization of light propagating in a crystalline material. As shown in Fig. E.4, we have the index ellipsoid of a crystal and a light wave is propagating along a direction k. In an isotropic medium the wave can have an arbitrary polarization in the plane perpendicular to k. However, in an anisotropic medium the wave has either of two linear polarizations and the velocity of the light with each polarization is, in general, different.

To calculate the polarization, we use the construction outlined in Fig. E.4. A plane is drawn perpendicular to the k-vector and the intersection of this plane with the ellipsoid produces an ellipse. The ellipse produced has principal axes a and b as shown in Fig. E.4. The direction of polarization allowed for the wave are now given by D_a and D_b, i.e., parallel to the principal axes. The velocities of the light with the two polarizations are inversely proportional to the length of the principal axes. In particular, if the light is propagating along the axis $i = 3$, the light is polarized along $i = 1$ and $i = 2$ with velocities c/n_{r1} and c/n_{r2}, respectively. It may be noted that there is a special direction called the optic axis where the two velocities are equal. In general, the optic axis may not coincide with any of the principal axis.

As discussed in Chapter 5, one can alter the refractive index along one of the principal axis and as a result add an optical delay for one of the polarizations. This can lead to switching devices as discussed in Chapter 12.

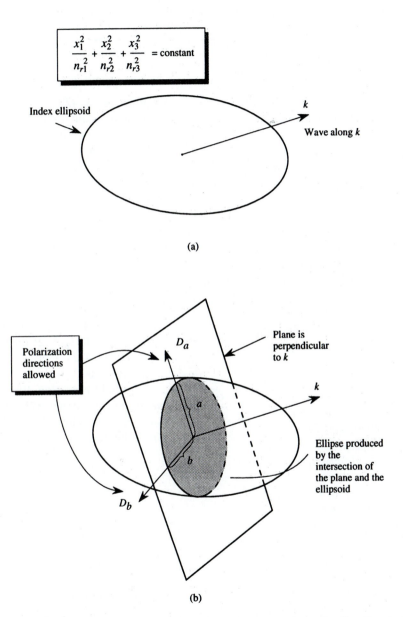

Figure E.4: (a) An index ellipsoid for a crystal. Shown is a wave along the direction k. (b) The construction used to obtain the polarization of the wave.

Index

FREQUENTLY USED QUANTITIES

QUANTITY	SYMBOL	VALUE
Planck's constant	h	6.626×10^{-34} J-s
	$\hbar = h/2\pi$	1.055×10^{-34} J-s
Velocity of light	c	2.998×10^8 m/s
Electron charge	e	1.602×10^{-19} C
Electron volt	eV	1.602×10^{-19} J
Mass of an electron	m_0	9.109×10^{-31} kg
Permittivity of vacuum	$\varepsilon_0 = \dfrac{10^7}{4\pi c^2}$	8.85×10^{-14} F cm^{-1} $= 8.85 \times 10^{-12}$ F m^{-1}
Boltzmann constant	k_B	8.617×10^{-5} eVK^{-1}
Thermal voltage at 300 K	$k_B T/e$	0.026 V

Wavelength – energy relation: $\lambda(\mu m) = \dfrac{1.24}{E(eV)}$

Linewidth $\delta E = 1$ meV $\Longrightarrow$ $\delta\nu = 0.243$ THz

Linewidth $\delta E = 1$ meV $\Longrightarrow$ $\begin{cases} \delta\lambda = 19.4 \text{ Å @ } \lambda = 1.55 \ \mu m \\ \delta\lambda = 6.2 \text{ Å @ } \lambda = 0.88 \ \mu m \end{cases}$

MATERIAL	CONDUCTION BAND EFFECTIVE DENSITY (N_c)	VALENCE BAND EFFECTIVE DENSITY (N_v)	INTRINSIC CARRIER CONCENTRATION $(n_i = p_i)$
Si (300 K)	2.78×10^{19} cm^{-3}	9.84×10^{18} cm^{-3}	1.5×10^{10} cm^{-3}
Ge (300 K)	1.04×10^{19} cm^{-3}	6.0×10^{18} cm^{-3}	2.33×10^{13} cm^{-3}
GaAs (300 K)	4.45×10^{17} cm^{-3}	7.72×10^{18} cm^{-3}	1.84×10^6 cm^{-3}

Material	Crystal Structure	Bandgap (eV)	Static Dielectric Constant	Lattice Constant (Å)	Density (gm-cm⁻³)
C	DI	5.50, I	5.570	3.5668	3.5153
Si	DI	1.1242, I	11.9	5.431	2.3290
SiC	ZB	2.416, I	9.72	4.3596	3.166
Ge	DI	0.664, I	16.2	5.658	5.323
AlN	W	6.2, D	$\bar{\varepsilon} = 9.14$	$a = 3.111$ $c = 4.981$	3.255
AlP	ZB	2.45, I	9.8	5.4635	2.401
AlAs	ZB	2.153, I	10.06	5.660	3.760
GaN	W	3.44, D	$\varepsilon_{\parallel} = 10.4$ $\varepsilon_{\perp} = 9.5$	$a = 3.175$ $c = 5.158$	6.095
GaP	ZB	2.272, I	11.11	5.4505	4.138
GaAs	ZB	1.424, D	13.18	5.653	5.318
GaSb	ZB	0.75, D	15.69	6.0959	5.6137
InN	W	1.89, D	—	$a = 3.5446$ $c = 8.7034$	6.81
InP	ZB	1.344, D	12.56	5.8687	4.81
InAs	ZB	0.354, D	15.15	6.058	5.667
InSb	ZB	0.230, D	16.8	6.479	5.775
ZnS	ZB	3.68, D	8.9	5.4102	4.079
ZnS	W	3.9107, D	$\bar{\varepsilon} = 9.6$	$a = 3.8226$ $c = 6.6205$	4.084
ZnSe	ZB	2.822, D	9.1	5.668	5.266
ZnTe	ZB	2.394, D	8.7	6.104	5.636
CdS	W	2.501, D	$\bar{\varepsilon} = 9.38$	$a = 4.1362$ $c = 6.714$	4.82
CdS	ZB	2.50, D	—	5.818	—
CdSe	W	1.751, D	$\varepsilon_{\parallel} = 10.16$ $\varepsilon_{\perp} = 9.29$	$a = 4.2999$ $c = 7.0109$	5.81
CdTe	ZB	1.475, D	10.2	6.482	5.87
PbS	R	0.41, D*	169.	5.936	7.597
PbSe	R	0.278, D*	210.	6.117	8.26
PbTe	R	0.310, D*	414.	6.462	8.219

Data given are room temperature values (300 K).
Key: DI: diamond; R: rocksalt; W: wurtzite; ZB: zinc-blende
*: gap at L point; D: direct; I: indirect; $\varepsilon_{\parallel}$: parallel to c-axis; $\varepsilon_{\perp}$: perpendicular to c-axis